Evolutionary Biology of the New World Monkeys and Continental Drift

ADVANCES IN PRIMATOLOGY

THE PRIMATE BRAIN
Edited by Charles R. Noback and William Montagna

MOLECULAR ANTHROPOLOGY: Genes and Proteins in the Evolutionary Ascent of the Primates
Edited by Morris Goodman and Richard E. Tashian

SENSORY SYSTEMS OF PRIMATES
Edited by Charles R. Noback

NURSERY CARE OF NONHUMAN PRIMATES
Edited by Gerald C. Ruppenthal

COMPARATIVE BIOLOGY AND EVOLUTIONARY RELATIONSHIPS OF TREE SHREWS
Edited by W. Patrick Luckett

EVOLUTIONARY BIOLOGY OF THE NEW WORLD MONKEYS AND CONTINENTAL DRIFT
Edited by Russell L. Ciochon and A. Brunetto Chiarelli

Evolutionary Biology of the New World Monkeys and Continental Drift

Edited by

RUSSELL L. CIOCHON

University of North Carolina
Charlotte, North Carolina

and

A. BRUNETTO CHIARELLI

Institute of Anthropology
Florence, Italy

PLENUM PRESS · NEW YORK AND LONDON

Library of Congress Cataloging in Publication Data

Main entry under title:

Evolutionary biology of the New World monkeys and continental drift.
(Advances in primatology)
"Expanded proceedings of a symposium at the seventh congress of the International Primatological Society, held at Bangalore, India, January 1979."
Includes index.
1. Monkeys—Evolution. 2. Monkeys—Anatomy. 3. Monkeys—South America—Evolution—Congresses. 4. Monkeys, Fossil—Congresses. 5. Continental drift—Congresses. 6. Paleobiogeography—Congresses. 7. Mammals—Evolution—Congresses. 8. Mammals—South America—Evolution—Congresses. I. Ciochon, Russell L. II. Chiarelli, A. B. III. International Primatological Society.
QL737.P925E86 599.8′2 80-16063
ISBN 0-306-40487-7

Expanded proceedings of a symposium at the Seventh Congress of the International Primatological Society, held at Bangalore, India, January, 1979.

A Division of Plenum Publishing Corporation
227 West 17th Street, New York, N.Y. 10011

Printed in the United States of America

Contributors

M. Baba
Department of Science and Technology
University Studies/Weekend College Program
College of Lifelong Learning
Wayne State University
Detroit, Michigan 48201

J. Bugge
Department of Anatomy
Royal Dental College
DK-8000 Aarhus C Denmark

M. Cartmill
Department of Anatomy
Duke University Medical Center
Durham, North Carolina 27710

A. B. Chiarelli
Istituto di Antropologia
Università di Firenze
Via del Proconsolo, 12
50122 Firenze, Italy

R. L. Ciochon
Department of Anthropology and Sociology
University of North Carolina at Charlotte
Charlotte, North Carolina 28223

J. E. Cronin
Departments of Anthropology and Organismal and Evolutionary Biology
Peabody Museum
Harvard University
Cambridge, Massachusetts 02138

L. Darga
Department of Anthropology
Oakland University
Rochester, Michigan 48063

E. Delson
Department of Vertebrate Paleontology
American Museum of Natural History
New York, New York 10024
and Department of Anthropology
Lehman College, CUNY
Bronx, New York 10468

D. Falk
Department of Anatomy
University of Puerto Rico Medical Sciences Campus
San Juan, Puerto Rico 00936

S. M. Ford
Department of Anthropology
Southern Illinois University
Carbondale, Illinois 62901

D. G. Gantt
Department of Anthropology
Florida State University
Tallahassee, Florida 32306

P. D. Gingerich
Museum of Paleontology
University of Michigan
Ann Arbor, Michigan 48109

M. Goodman
Department of Anatomy
School of Medicine
Wayne State University
Detroit, Michigan 48201

K. G. Gould
Yerkes Regional Primate Center
Emory University
Atlanta, Georgia 30322

R. Hoffstetter
Institut de Paléontologie
Muséum National d'Histoire Naturelle
75005 Paris, France

R. F. Kay
Department of Anatomy
Duke University Medical Center
Durham, North Carolina 27710

R. Lavocat
Ecole Pratique des Hautes Etudes
Laboratoire de Paléontologie des Vertébrès
Faculte des Sciences
Place E. Bataillon
34000 - Montpellier, France

W. P. Luckett
Department of Anatomy
School of Medicine
Creighton University
Omaha, Nebraska 68178

W. Maier
Klinikum der Johann Wolfgang Goethe-Universität
Zentrum der Morphologie
Dr. Senckenbergische Anatomie
Theodor-Stern-Kai 7
D-6000 Frankfurt 70, West Germany

D. E. Martin
Yerkes Regional Primate Center
Emory University
Atlanta, Georgia 30322
and Georgia State University
Atlanta, Georgia 30303

M. C. McKenna
Department of Vertebrate Paleontology
American Museum of Natural History
New York, New York 10024

W. C. Meyer
Department of Earth Sciences
Pierce College
Woodland Hills, California 91371

F. J. Orlosky
Department of Anthropology
Northern Illinois University
DeKalb, Illinois 60115

E. M. Perkins
Department of Biological Sciences
University of Southern California
Los Angeles, California 90007

A. L. Rosenberger
Department of Anthropology
University of Illinois at Chicago Circle
Box 4348
Chicago, Illinois 60680

V. M. Sarich
Departments of Anthropology and Biochemistry
University of California
Berkeley, California 94720

F. S. Szalay
Department of Anthropology
Hunter College
695 Park Avenue
New York, New York 10021

D. H. Tarling
Department of Geophysics and Planetary Physics
The University
Newcastle upon Tyne
NE1 7RU, England

A. E. Wood
Department of Biology
Amherst College
Amherst, Massachusetts 01002

Preface

It is now well known that the concept of drifting continents became an established theory during the 1960s. Not long after this "revolution in the earth sciences," researchers began applying the continental drift model to problems in historical biogeography. One such problem was the origin and dispersal of the New World monkeys, the Platyrrhini.

Our interests in this subject began in the late 1960s on different continents quite independent of one another in the cities of Florence, Italy, and Berkeley, California. In Florence in 1968, A. B. Chiarelli, through stimulating discussions with R. von Koenigswald and B. de Boer, became intrigued with the possibility that a repositioning of the continents of Africa and South America in the early Cenozoic might alter previous traditional conceptions of a North American origin of the Platyrrhini. During the early 1970s this concept was expanded and pursued by him through discussions with students while serving as visiting professor at the University of Toronto. By this time, publication of the *Journal of Human Evolution* was well underway, and Dr. Chiarelli as editor encouraged a dialogue emphasizing continental drift models of primate origins which culminated in a series of articles published in that journal during 1974–75.

In early 1970, while attending the University of California at Berkeley, R. L. Ciochon was introduced to the concept of continental drift and plate tectonics and their concomitant applications to vertebrate evolution through talks with paleontologist W. A. Clemens and anthropologist S. L. Washburn. These discussions resulted in a graduate research project on the early Cenozoic paleogeography and paleoclimatology of the Earth and its probable effects on primate evolution. This research, which produced a biogeographic model supporting an independent derivation of the Platyrrhini from North American Eocene lower primates, was completed in 1972 and shelved but not forgotten.

In Fall, 1975, after having read the continental drift/primate dispersal collection of articles published in the *Journal of Human Evolution*, RLC wrote to ABC suggesting that a special issue of *Journal of Human Evolution* be published that combined the previous collection of articles together with several important additional papers on comparative platyrrhine morphology. ABC's reply was an offer to coedit a detailed volume on the origin of the New World monkeys and continental drift and plan a symposium to promote further discussion on the subject.

Thus, from quite separate backgrounds and levels of experience, our collaboration in this edited volume had begun. Over the next year and one-half a list of potential contributors was drawn up, revised, and letters of invitation sent out. Initial replies were 90% affirmative. In Spring, 1977, RLC visited ABC in Turin to discuss publication of the volume and plans for the symposium. We agreed the symposium might be held at either the annual meeting of the American Association of Physical Anthropologists or at the Seventh Congress of the International Primatological Society scheduled to meet in Bangalore, India.

The pace toward volume completion began to quicken. Many of the completed manuscripts were now in our hands, and inquiries regarding the proposed symposium became more prevalent. We decided to schedule the symposium to run in conjunction with the Seventh IPS Congress in Bangalore, January 8–12, 1979. Unfortunately, we now had so many contributors that one meeting would not be adequate. ABC then decided to schedule a pre-Bangalore symposium in Turin at the Academy of Sciences for December 6–7, 1978. A grant from the Consiglio Nazionale delle Ricerche in Italy was received to support the meeting. In the meantime, RLC had received a grant from the Smithsonian Foreign Currency Program to cover travel and living expenses for participants at the Bangalore symposium. Thus we now had funds for all the volume contributors to attend at least one of the two symposia.

The first symposium in Turin was held in the 17th century elegance of the grand conference room of the Academy of Sciences. Twelve contributors attended the two-day meeting. From the very beginning it became clear that a gathering of this kind could promote useful discussion regarding possible geographic sources of the New World monkeys. D. Tarling's presentation at Turin provided an important example. Illustrating the positioning of South America *vis-à-vis* Africa and North America throughout the last 150 million years, he concluded that dispersal routes across the South Atlantic Ocean in the early Cenozoic were a distinct possibility. This proposal elicited much debate and set the stage for future discussion at the second symposium in Bangalore.

In early January, 1979, the Seventh Congress of the International Primatological Society was convened in the southern Indian city of Bangalore. The organizer, M. Mougdal of the Bangalore Institute of Science, did an excellent job of planning and integrating our symposium into the Congress proceedings. For this two-day symposium, followed by a round-table discussion, 18 participants were present. D. Tarling once again began the symposium, emphasizing the possibility of an island-hopping dispersal route across the South Atlantic, as well as the more traditional Caribbean or Middle American dispersal route. The remainder of the papers focused on comparative studies of the morphology, paleontology, and genetics of platyrrhines, catarrhines, and lower primates, with each speaker addressing the new synthesis of paleogeographic data in light of various phylogenetic hypotheses of New World monkey origins. On the second day at the concluding round-table discussion, a consensus was reached that there were really two separate but interrelated problems concerning the origin of the Platyrrhini. The first was their phylogenetic source: Are the Platyrrhini descended from tarsiiform ancestors (i.e., Omomyidae), lemuriform ancestors (i.e., Adapidae), or directly from an Old World anthropoid primate stock? A corollary of this phylogenetic question should also be raised: Is the reported great similarity of the living Platyrrhini to the Catarrhini of the Old World the result of parallel evolution *or* of a relatively recent common ancestry? Intimately tied to these phylogenetic questions but still separate from them was the problem of the geographic source of the Platyrrhini: Did the ancestors of the New World monkeys reach South America directly from North America, or from Africa, or from Asia via either North America or Africa? A corollary of this biogeographic question might also be posed: Could there have been more than one biogeographic source of the Platyrrhini of possibly multiple dispersals from one source?

The following series of chapters in this volume address these questions from a variety of perspectives. Though a clear consensus on the phylogenetic and geographic sources of the New World monkeys would have been desirable, even in its absence it seems certain that this collection of papers will result in further clarification of this controversy and the delineation of future avenues of research which may one day solve the issue.

R. L. Ciochon
A. B. Chiarelli
Florence, Italy

Acknowledgments

There are several individuals who provided invaluable scientific stimulus without which this volume would never have become a reality. Already mentioned in this regard are W. A. Clemens, B. de Boer, G. H. R. von Koenigswald and S. L. Washburn. Additionally, E. Delson, F. Clark Howell, W. P. Luckett, W. E. Meikle, P. H. Napier, and D. E. Savage should be added to this list. The following students provided invaluable technical and editorial assistance: Giuseppe Ardito, Amelia Dennis, Richard Dreiman, Joy Myers, Paula Neal, and Donna Ryan. Joy Myers deserves a special note of thanks for having taken on the enormous responsibility of compiling the author and subject indexes for the volume. A wide-ranging scientific endeavor of this sort cannot possibly be undertaken successfully without the extraordinary organizational talents of one individual responsible for all the details, and in this regard, we whole-heartedly acknowledge Chiara Bullo. Along these same lines we also thank Doris Carter, Boyd Davis, and other staff members at The University of North Carolina at Charlotte and Francine Berkowitz of the Smithsonian Institution. We wish to thank the Smithsonian Foreign Currency Program in Washington, D.C. for funding the Bangalore symposium and some publication costs, the Consiglio Nazionale delle Ricerche in Rome for funding the Turin symposium, and the L.S.B. Leakey Foundation in Pasadena for providing research funds utilized during the final phase of the volume's preparation. We also acknowledge President Norberto Bobbio for graciously providing the facilities at the Academy of Sciences in Turin and Professor M. Moudgal for the great job he did as Director of the Seventh IPS Congress in Bangalore. Lastly, we wish to thank all the symposium participants and contributors to the volume who unflaggingly put up with numerous letters and several long delays. Naturally, without their participation the present volume would have never become a reality.

RLC
ABC

Contents

VI. Evidence from Other Comparative Anatomy Studies

VII. Evidence from Karyological and Biochemical Studies

VIII. Synthesis, Perspectives, and Conclusions

Geological and Paleontological Background I

The Geologic Evolution of South America with Special Reference to the Last 200 Million Years

1

D. H. TARLING

Introduction

Ideas concerning the existence and nature of intercontinental links at different times of the Earth's history have undergone dramatic changes during the last 100 years or so. While Darwin's observations (1859) led him to an understanding of the importance of geographical isolation to the geological record of evolution, he considered that vertical motions of land masses were much more critical in creating this condition than horizontal translations. This view, of course, not only reflected his own observations relating to sea-level changes but also the extant views of the nature of the Earth in which the ocean basins were still considered to be submerged land masses. Under such circumstances, the elevation of parts of the ocean floor was conceivable and would have lead to the formation of extensive land bridges linking the main continents and, conversely, the sinking of land bridges would lead to the isolation of previously united biotic communities. During the 19th century paleontologists and biogeographers recognized that changing connections of this type could account for similarities and dissimilarities in the geological record for life on all of the known continents. In particular, these studies distin-

D. H. TARLING • Department of Geophysics and Planetary Physics, The University, Newcastle upon Tyne NE1 7RU, England.

guished a "northern" continent, Laurasia, and a "southern" continent, Gondwanaland, in which constituent continents were thought to be linked by a series of land bridges that were destroyed sometime since the Jurassic and Cretaceous periods (the last 200 million years of the Earth's 4550-million-year history). These paleontological observations were augmented by the geological climatic record although this record also showed some strange features, such as the prevalence of glacial conditions in Gondwanaland some 300 million years ago but of hot-desert and tropical conditions in Laurasia at the same time. Although attempts were made to explain this situation, which involved ice sheets somehow existing from the present South Pole to north of the equator while tropical conditions extended from the North Pole to mid-northern latitudes, most explanations only involved the climatic belts of the Earth having changed their relative positions because the Earth was thought to have then been in a different orientation relative to the sun. Although such models were, in fact, extremely difficult to reconcile with the data, such problems appeared small when compared with those produced by ideas of continental drift which were presented early in the 20th Century when Baker (1912), Taylor (1910), Wegener (1914), and Argand (1924) suggested that horizontal motions between the continents could account for all of the geological and paleontological evidence (Tarling and Tarling, 1975). This explanation was particularly unacceptable at a time when geophysicists had just shown that the interior of the Earth was "solid" to the passage of earthquake waves and only the Earth's central core was "fluid." It is not surprising, therefore, that the idea of continental drift was largely discarded in the 1920s–1930s, with some notable exceptions among geologists living in the southern hemisphere and even fewer in the northern hemisphere. A change in attitude to continental drift came very gradually during the 1940s and 1950s as more and more geological, geophysical, and paleontological evidence came forward to actually require the extremely close proximity of certain continents at particular times. It was also becoming clearer that while the Earth's interior was solid to the transmission of earthquake waves, which last for only a few minutes, it could plastically deform and virtually behave as a liquid to stresses persisting over geological time scales. However, it was essentially the study of the magnetization of rocks which eventually led to a world-wide acceptance of the continental drift theory, under the name of "plate tectonics," and this technique still provides the major information for the past location of continental blocks (Tarling, 1971*a*). However, it must be emphasized that an assessment of the paleontological evidence is still essential to test possible models of past continental relationships and particularly the nature of any links between them.

In this paper, a brief review is presented of the nature of the evidence for the past relationships between South America, Africa, Antarctica, and North America. This is partially intended as an introduction to some of the terms of plate tectonics that are a useful shorthand for describing particular geological situations in different areas at different times, and partially as an attempt to give some idea of the assumptions and sources of error which will not usually be discussed in full for each particular period. This introduction to the nature of the evidence will be followed by an appraisal of the earlier geological evolution of South America prior to more detailed discussion of the changing intercontinental relationships of South America with Africa, Antarctica, and, finally, North America. The possible biogeographical character of the different links to South America will then be considered separately.

The Nature of the Evidence

Perhaps the most convincing evidence for the contiguity of South America and Africa is the perfection of the geometric fit of the edges of their continental shelves (Fig. 1). However, the geometric fits for most of the other continents, including North America to northwestern Africa, are much less clear. In particular, the actual edges to be fitted may have been modified either by subsequent sedimentary deposition and erosion or by late volcanic activity. Of particular importance in this context is the fit of southeastern North America to West Africa. The North American shelf in this area contains continental fragments that broke off the North American continent as the Atlantic began to form (Mullins and Lynts, 1977) but have since been linked by later sedimentary depositions, for example, in the Blake Plateau region (Sheridan, 1974). However, the reliability of the geometric fit can usually be assessed by comparing the similarity of geological evolution on the adjacent continents. While joined, they would be expected to have an almost identical geological evolution, but this is likely to have been different follow-

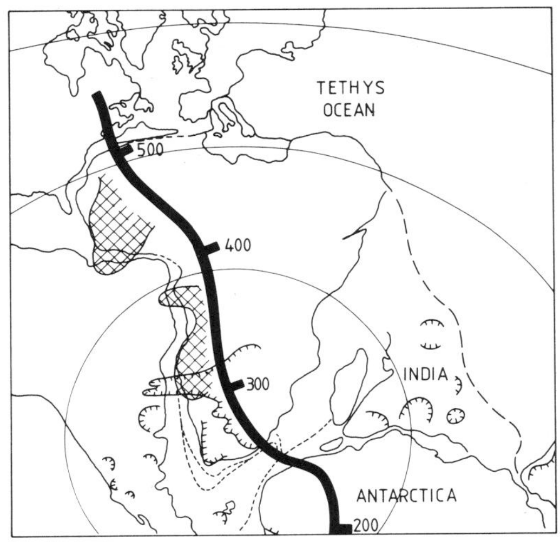

Fig. 1. The fit of the continents and their geological and geophysical properties. The fit of the edges of the continents is also mirrored by the fit of the pre-Cretaceous (earlier than 130 million years) geology and geophysics. Depicted on this fit are the boundaries between rocks older than 2000 million years and those of 700–500 million years (cross hatched) and the edges for the Gondwanan ice sheets of the Lower Permiam (ca. 265 million years). Superimposed on this are the positions for the South American (hollow) and African (solid) paleomagnetic poles between 500 and 200 million years ago.

ing their separation. In general, most fits and assessments are rather subjective and only give an order of magnitude evaluation of the relationship and age of separation of the previously contiguous parts. In most cases, the most reliable assessment of the nature of the connection comes from a detailed paleontological study to evaluate the degree of similarity between the flora and fauna and thus whether the two areas must have had good or poor contact. One problem with this approach is the paucity of data so far available for many areas, the selectivity of preservation conditions, and the subjectivity necessarily involved in determining the degree of similarity between the flora and fauna of now disparate continents.

The development of magnetometers in the early 20th century enabled measurements of the magnetic properties of certain rock types to be studied. These studies showed that, when averaged over a few thousand years, the direction of the Earth's magnetic field averages to that which would be expected if the Earth contained a bar magnet at its center which was aligned along the Earth's axis of rotation (an axial geocentric dipole field). Studies of older and older rocks in the same continent have demonstrated that this averaging is still necessary but that the position of the average magnetic pole, instead of corresponding with the Earth's present axis of rotation (the geographic pole), becomes displaced further and further away from it. Each continent thus has its own unique polar wandering path showing the position of the average geomagnetic pole relative to the continent. (The term "polar wandering curve" is, in fact, a misnomer as the continent wanders relative to the pole and not vice versa). Each of these polar wandering curves show similarities for at least the last 200–300 million years which enable them to be matched when the continents, with their corresponding polar wandering paths, are moved into different positions relative to each other. In the case of South America and Africa, for example, the movement of the two continents together to satisfy their apparent geometric similarities also results in an identity of their polar wandering paths from at least 600 million years ago until some 150 million years ago (Fig. 1). As the Earth's magnetic field is largely caused by motions within the Earth's liquid core, and these motions will be influenced by the Earth's rotation, it seems realistic to assume that the position of the average geomagnetic pole for any particular time can be used not merely to determine the relative positions of the continents but also to define the distance of the continent from the geographic (rotational) pole, i.e., the paleolatitude of the continent. That this is so is confirmed by the excellent agreement between the paleomagnetic evidence and the paleolatitudes indicated from the study of distribution of geological indicators of past climatic belts, such as evaporites and desert sands in the tropical latitudes and glacial debris in polar regions (Tarling, 1971*a*). This agreement shows that the geomagnetic pole positions for any one continent and at any one time can be used to determine the paleolatitude and orientation of that continent relative to the rotational pole. In the following reconstructions, the paleomagnetically determined latitudes are largely derived from North America data and, with the notable exception of the Jurassic, are considered to be accurate within ±5°. For the Jurassic the standard deviation for the North America paleolatitudes may be as much as ±10° and so, for Late Triassic–Early Jurassic, the paleomagnetic data used for latitude determinations were mainly derived from the more precise African studies.

While the average position of the geomagnetic pole changes very gradually with time, averaging some 0.3° per million years for the North American and European continents during the last 400 million years, dramatic changes in the polarity of this pole occur on a much shorter time scale. Some 690,000 years ago, the polarity of the geomagnetic field changed so that the North Pole became the South Pole and vice versa. Somewhat earlier, the converse polarity change occurred and it is now known that, on the average, some three polarity changes occurred every million years during the last 60 million years. For earlier times, the number and frequency of reversals of polarity have been different, with possibly 30 million years of constant polarity during the middle Cretaceous (Helsley and Steiner, 1969). The biological effects of such polarity changes may be of major importance as there is some indication for increased rates of evolution during times of polarity change (Harrison and Funnell, 1964; Hays *et al.*, 1969) possibly associated with the reduction in the strength of the geomagnetic field at such times as this could result in changes in the amount of short-wavelength radiation reaching the Earth's surface or short-term climatic changes. However, the main importance of such polarity changes, in the present context, is that they enable high-precision in location of continental blocks relative to each other. New oceanic rocks are forming continuously at the mid-oceanic ridges of the world which extend as a continuous line some 80,000 km long. At these ridges, volcanic activity is continually generating new crustal rocks which, as they cool, become magnetized in the direction of the ambient geomagnetic field. These new rocks are carried to each side of the ridge by slow convective currents operating in the interior of the Earth and the displaced rocks are gradually replaced by the continuing volcanic activity at the ridge (Fig. 2). This process produces a magnetic "tape recording" of the changes in the polarity of the Earth's magnetic field on both sides of the oceanic ridges. The determination of the magnetic polarity of the rocks of the ocean floors can be readily obtained from measurements of the strength of the geomagnetic field using magnetometers towed behind ships at the ocean surface. Where the magnetometer lies over rocks which were magnetized in the same sense as the present geomagnetic field (normal), then the measured intensity of the field is higher than normal, while the readings are reduced below average when the rocks beneath are magnetized in the opposite direction to the present field (reversed). As the age of the polarity changes can be measured by paleontologically dating the sediments overlying them, or by comparison with polarity scales determined for radiometrically dated rocks on land, then the time at which that portion of the ocean floor lay at the crest of the then mid-oceanic ridge can be determined. Moving the adjacent magnetic polarity

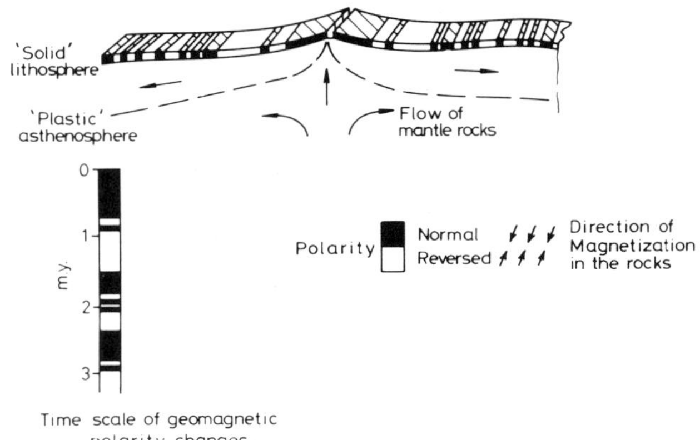

Fig. 2. The creation of oceanic magnetic anomalies. Hot mantle rocks are intruded along the center of the world's oceanic ridge systems. As these rocks cool to form new oceanic crustal rocks, they become magnetized in the direction of the ambient geomagnetic field. These solid rocks are then carried away from the center of the ridge by the motions of the mobile rocks of the mantle and are replaced by new hot rocks that again cool and become magnetized. If the geomagnetic field changes polarity (reverses) then the oceanic crustal rocks retain a recording of these polarity changes that can be measured at the ocean's surface. As these polarity changes can be dated from either the age of the overlying sediments or by comparison with dated continental rock sequences, the identification of reversal sequences in the magnetic anomalies of the ocean floors can thus be used to date the age when those rocks were cooling at the previous oceanic ridge.

strips, together with older oceanic and continental areas, until they are superimposed then provides a precise reconstruction of the relative positions of the older portions at the time that these rocks at the reconstructed ridge were becoming magnetized (Phillips and Forsyth, 1972; Pitman and Talwani, 1972; and Sclater and McKenzie, 1973). Naturally, there are uncertainties in identifying some of these magnetic anomalies as the shape of each anomaly is not always clearly diagnostic and can therefore be misidentified. However, such misidentifications are becoming rare as processing techniques improve and when possibly erroneously identified areas are becoming more evident as the total pattern becomes clearer. Nonetheless, there are still some difficulties in the dating of many anomalies although this is now considered to be accurate within some ±5% for the last 60 million years and within some ±10–20% for earlier times (Tarling and Mitchell, 1976). Ultimately, therefore, the analysis of magnetic anomaly patterns in the ocean floors offers extremely high precision for continental block reconstructions, but caution is still necessary in many areas, particularly for the areas of the Caribbean and Scotia Sea discussed below.

As the Earth is not increasing its surface area, the newly created ocean floor must be compensated by the removal of surface area elsewhere. This compensation takes place by the subduction of oceanic rocks and by the compression of continental rocks in specific areas. The descent of oceanic rocks takes place along the deep oceanic trenches where their subduction is marked by strong, deep earthquake activity as the rocks revert to minerals stable at mantle pressures and temperatures (Fig. 3). Continental rocks are generally too light to be carried down at such subduction zones and the entry of continental rocks into such a zone is generally considered to stop further subduction. As the oceanic rocks revert to mantle types, they release volatiles into the overlying rocks, causing their melting temperature to be lowered, resulting eventually in explosive volcanic activity along a broad arc bordering the subduction zone. In this arc, new continental crustal rocks gradually form. The strong geological activity along subduction zones clearly results in major mountain building occurring, such as along the Andes today, and subduction of some 200 km or more of oceanic rocks would be expected to be marked in the geological record by the associated volcanic and tectonic activity. Even more dramatic geological activity occurs when the continental rocks on either side of a subduction zone collide as the intervening oceanic rocks are subducted. Examples of such mountain building are evidenced in the mountain belt extending from the Alps to the Himalayas as Africa, Arabia, and India collided with Europe and Asia after the intervening Tethys Ocean rocks had been subducted.

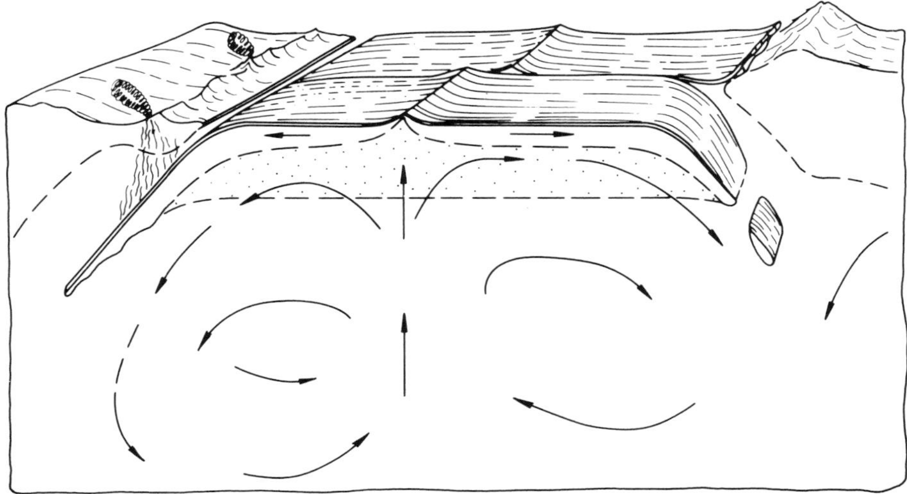

Fig. 3. Mantle and crustal convective motions. The generation of heat from the breakdown of radioactive elements within the Earth's interior causes very slow, gradual motions (centimeters per year) of the deep internal mantle rocks of the Earth. These motions carry heat to the Earth's surface and the uprising motions are associated with the formation of the oceanic ridges and the creation of new oceanic crustal rocks that are then carried away from the ridge by the convective motions of the underlying rocks. Over the descending convective currents, the oceanic rocks are carried down (subducted) into the Earth's mantle and gradually revert back to their original mantle composition—this reversion being accompanied by volcanic and earthquake activity. Continental rocks, however, are too light to be carried down in these subduction zones and eventually cause the subduction to cease. Where the subduction borders a continent, Andean-type mountains are formed; but when the subduction is within an ocean, volcanic island arcs form over the subducting rocks. (The seismic low velocity zone is shown as the dotted area.)

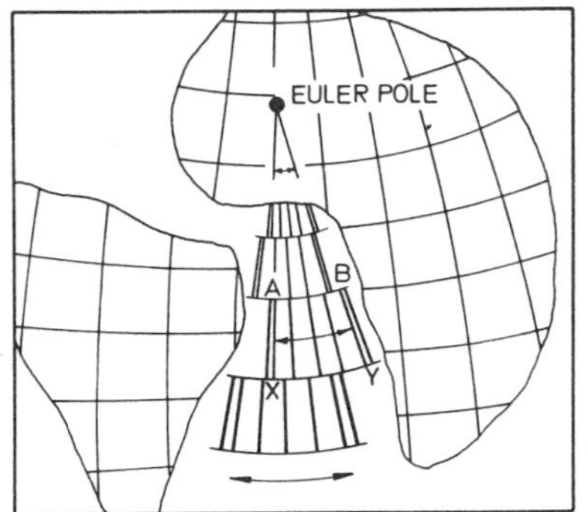

Fig. 4. The Euler pole of rotation. As two blocks of the earth's surface separate by the action of convective currents, their motion relative to each other can be completely defined in terms of a point of no relative motion, the Euler pole, and the angular rate of opening about that pole. In this cartoon the location of this pivot is defined by the radiation of oceanic magnetic anomalies away from it and by the fracture zones that form curves concentric to it. The rate of opening can then be determined by the angular separation of magnetic anomalies of the same age on opposite sides of the ocean. The analogy is essentially with the Earth's axis of rotation in which the magnetic anomalies are aligned as lines of longitude and the fracture zones form lines of latitude, but the location of the actual pole differs for each pair of blocks (plates) of the Earth's surface.

In general, the separation between two continental blocks will be accompanied by the formation of new oceanic rocks, while their approach must be compensated by the subduction of intervening oceanic rocks, with associated volcanic and tectonic activity, and eventually a collision between the two blocks with major deformation along their leading edges as they collide. The actual motions of such continental blocks can be simply described as a rotation of each of the blocks about a point specific for that block (the Euler pole) at a specified angular rate (Fig. 4). (This is similar in concept to the rotation pole of the Earth, with Africa being moved eastwards by so many degrees of longitude—the distance being a fixed number of degrees but differing in actual distance in kilometers according to how near that part of the continent is to the pole itself. Conventionally, the angular rotation from above the Euler pole is positive for an anticlockwise rotation.)

The Early Evolution of the South American Continent (Precambrian to Mid-Mesozoic)

South America began to form as a geological entity at the same time as the other continents, some 4000 million years ago. Although the conditions at that time are unclear, several major blocks had formed by some 2500 million years ago, each comprising a thickness of some 30–40 km of continental crustal rocks closely coupled with the upper 200–300 km of the Earth's mantle (Tarling, 1978*a*). These blocks were already in approximately the same relative positions to each other within both Africa and South America, but the two continents were then contiguous. The rocks lying between these different blocks finally underwent major thermal changes to weld all of them together some 500–700 million years ago. By at least this time the Afro-South American continent was part of the super continent of Gondwanaland which comprised South America, Africa, Madagascar, India, Ceylon, Australia, New Zealand, and Antarctica (Fig. 1). It seems probable, therefore, that algal life forms, recorded in South Africa (Muir and Grant, 1976), were also present in South America some 3500 million years ago as such aquatic organisms, some of which secreted calcareous material, would be able to migrate freely through Precambrian seas traversing the Gondwanan continent. However, much of the western and northern parts of South America were still not in existence, even at the time of the appearance of skeletal organisms in Cambrian times (550 million years ago).

At the start of Cambrian times, South America was inverted (relative to its present position) so that the South Pole, with polar climatic conditions, lay to the northeast of Brazil while Argentina lay in tropical latitudes (Fig. 1). As Gondwanaland drifted over the surface of the Earth, its paleolatitudes changed until the South Pole lay in the Sahara during the Llanvirn stage of the Lower Ordovician period (some 470 million years ago). Much of northeastern South America must then have lain beneath or close to the polar ice cap and any high ground on the continent was likely to have been glaciated. As time proceeded, the drift of Gondwanaland carried it so that the pole traversed Africa, although glaciation appears to have been only localized during the Devonian, and it was not until the pole lay southwest of Africa (Fig. 1) that the major Gondwana ice sheet glaciation occurred during the Sakmarian stage of the Lower Permian (some 275 million years ago). At this time vast ice sheets affected much of South America, Africa, India, Australia, and Antarctica (Tarling, 1978*b*). This ice sheet glaciation terminated some 250 million years ago but the traces of this major climatic event still provide striking evidence for the contiguity of all of these Gondwanan continents at this time. This proximity meant that there were few geographical barriers to migration between all of these continents, although obviously climatic zonations and local marine incursions provided some constraints. Nonetheless, there was essentially free interchange between these continents until after the end of the Paleozoic (ca. 230 million years ago). The possibility of migrations and exchanges between Gondwana and the other continents, which became united as a single continent (Laurasia) in mid-Carboniferous times (ca. 230 million years ago), gradually increased during the Upper Paleozoic so that floral and faunal exchanges became increasingly possible at the end of the Paleozoic, although the paleontological evidence suggests that climatic or marine barriers may have restricted the extent of exchange between the two supercontinents even though they were then essentially united as a single continent, Pangaea, from approximately mid-Carboniferous times until Jurassic times, that is, between 300–200 million years ago.

During Triassic times, the fairly flat monotonous landscape of Gondwanaland appears to have developed into a series of basins and swells on a large scale, 1000–2000 km in extent (King, 1958). The topographic features were fairly gentle on a local scale but would have a slight inhibiting effect on migration across the continent. During mid-Triassic times (220 million years ago), southern Africa was a scene of mountain building, but there was clearly no separation of Africa and South America at this time as this orogenic episode is paralleled in the Sierra de la Ventana of Argentina. More significantly, the Triassic appears to have been a time of extremely extensive fracturing of the continental blocks, possibly on a world-wide scale (Kent, 1976). This period certainly seems to mark the time of the initial development of motions in the Earth's mantle which were to finally result in the separation of the world's present continents from the Late Paleozoic supercontinent of Pangaea. The subsequent separation of South America can then be examined in terms of changing connections with (1) Africa, (2) Antarctica, and (3) North America.

The African Connection

The initial fracturing along much of the South Atlantic (Fig. 5) was contemporaneous with faulting in the Central and North Atlantic as well as in the Indian Ocean, but the earliest separation between the continents was the

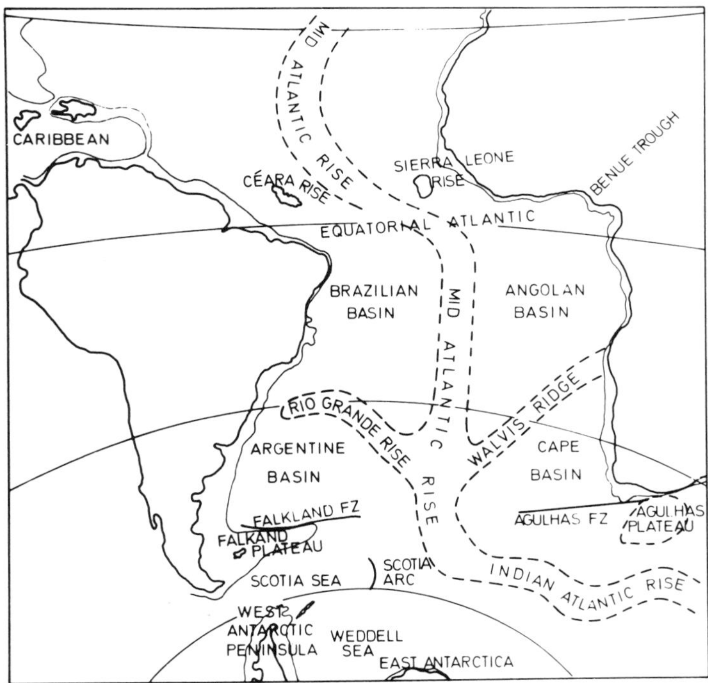

Fig. 5. The South Atlantic. This map depicts most of the geographic features discussed in the text for the opening of the South Atlantic and Scotia Sea.

motion of North America, with Greenland and Europe attached to it, away from northwestern Africa. This movement has been dated on paleomagnetic grounds as being some 180 million years ago (basal Jurassic) and this has been confirmed by the identification of Jurassic anomalies offshore from both North America and Africa (Hailwood and Mitchell, 1971; Pitman *et al.*, 1971). This indicates that sea-floor spreading was in operation in the Central Atlantic by the Pliensbachian stage of the Lower Jurassic. Such splitting slightly preceded the separation of India and Madagascar away from Africa, during the Middle Jurassic (approximately 160 million years ago), with the separation of India from Antarctica following shortly afterwards (150 million years ago). The oldest dated ocean-floor in the South Atlantic lies immediately offshore from South Africa and has been dated from magnetic anomalies (Larson and Ladd, 1973) as forming during the Valangian stage of the Jurassic, approximately 130 million years ago (Larson and Hilde, 1975).

The initial movement was by the pivoting of South America away from Africa about a point in the western Sahara (Figs. 7–11). Such a motion is clearly defined by the matching of the Falkland and Agulhas Fracture Zones around South Africa at the time of opening (Francheteau and Le Pichon, 1972; Rabinowitz *et al.*, 1976; du Plessis, 1977). Similar, but less clear, indications of this motion are also provided by the series of matching broad fracture zones in the Equatorial Atlantic (Kumar *et al.*, 1976; Delteil *et al.*, 1976; Gorini and Bryan, 1976; Mascle and Renard, 1976). The position of this pivot point (the Euler pole) meant that there was very little opening in the northern part of the South Atlantic as the Falkland Plateau and southern South America were gradually swinging away from southern Africa.

The nature of the opening was drastically different in different areas. In the south, the Falkland Plateau gradually moved past South Africa, very much in the way in which Baja California and southwestern California are now passing each other along the San Andreas fault system. In contrast, the Cape and Argentine Basins and the Angolan and Brazilian Basins were all forming by the separation of the faulted edges of the continents. The continental areas that now border the Walvis Ridge and Rio Grande Rise were the sites of uplift and volcanic activity particularly in Namibia (southwest Africa) and Uruguay-southern Brazil. These two areas were exactly contiguous prior to separation and were the locus for a major updoming of the continental crust. This rise appears to have commenced in Triassic times and was accompanied by some restricted volcanicity around 147 million years B.P. (Herz, 1977). The rise was also accompanied by some faulting and fracturing and was associated with the formation of smaller rises and basins to both the north and south, some of these swells being quite substantial [in excess of 7.5 km elevation in the Ponta Grossa arch near the Santos basin of the Brazilian Coast (Bowen, 1966; Torquato, 1976; de Almeida, 1976)]. The onset of separation in these areas was contemporaneous with the spread of vast plains of flood basalts in Namibia (the Kaoko Volcanics) and particularly in southern Brazil, Paraguay, and Uraguay (the Serra Geral Volcanics). These flood basalts were probably erupted over a geologically short period, probably less than 10 million years if allowance is made for inaccuracies in radiometric determinations which give an age for the Serra Geral volcanics of 122–133 million years (Amaral *et al.,* 1966) and 110–136 million years for the Kaoko Volcanics (Siedner and Miller, 1968). The basalts, which covered over 1,000,000 km^2 in South America and in Africa, were clearly erupted in response to the conditions existing during the initial separation of the two continents in the same way as the Karroo Basalts in Africa, the Rajmahal Traps in India, the Tasmanian Dolerite in Australia and the Ferrar Dolerites in Australia are all flood basalts that were erupted in association with the initial formation of the Indian Ocean or movements between eastern and western Antarctica. However, while the causative relationship is clear, there is still no satisfactory explanation as to why such "hot spots" should occur or why they should persist—in this case forming the Walvis Ridge and Rio Grande Rise as the South Atlantic continued to open. Furthermore, the original doming appears to have persisted as a major topographic feature while the South Atlantic formed as these domed areas remained the major sources for the sediments which were being carried into the South Atlantic until about mid-Tertiary times.

To the north and south of the Walvis–Rio Grande ridges the split of the South Atlantic was less spectacular and was largely free of volcanic activity. The initial breaking of the Afro-South American continent took place during the Lower Triassic as a series of disconnected rift valleys, for example in Brazil (Urien *et al.,* 1976), analogous to the Rift Valleys of East Africa today. As time progressed, these rift valleys very slowly extended and linked together, although it is unlikely that a continuous rift valley formed until ocean rocks began to be generated, probably during the Valangian stage of the Jurassic (ca. 120 million years ago). The entire rift, however, was probably still above sea level and formed local basins (Francisconi and Kowsmann, 1976) in which fresh-water evaporites and continental sediments accumulated. As spreading continued, the rift expanded and also deepened, probably falling below sea level before marine waters were able to pass over the Falkland Plateau which, although now some 2–3 km deep, was then close to sea level and acted as a barrier to the Indian Ocean waters (Seisser *et al.,* 1974) which had then reached as far as South Africa. The nature of the Agulhas Plateau is uncertain (Tucholke and Carpenter, 1977) and may be either very thin continental or thickened oceanic, possibly an extinct spreading ridge (Scrutton, 1973). An alternative explanation is that it was part of the Falkland Plateau, but broke off and remained in its present position from about Albian times. Such a suggestion would imply that its structure is identical to the Mozambique Ridge and the eastern Falkland Plateau, and that it was originally filling the Transkei Basin off eastern South Africa. However, irrespective of the precise origin and date of this plateau, the Falkland Plateau alone would have provided a significant barrier to the migrating Indian Ocean waters. By at least Aptian times, however, changes in relative sea level allowed Indian Ocean waters to pass over the Falkland Plateau into the Argentine and

Cape Basins as far as the Walvis and Rio Grande barrier, but such waters were frequently cut off and stagnated. This volcanic barrier was largely subaerial, but the irregular volcanic topography allowed some water to pass into the Angolan–Brazilian Basin which was, at this stage, a wide, but very shallow rift valley. The very restricted flow of water over the Walvis-Rio Grande barrier, combined with a shallow, low-latitude environment to the north, resulted in rapid evaporation of the marine waters in the Angolan–Brazilian Basins (Campos *et al.*, 1974; Driver and Pardo, 1974), forming major salt deposits along this area as far as Nigeria (Fig. 6), although the seas to the south only occasionally stagnated. The gradual increase in the width and depth of the South Atlantic, combined with a eustatic increase in sea level, led to full marine conditions developing in both the southern and northern basins by Aptian times (ca. 100 million years ago), with somewhat deeper waters along the continental edges in the Cape and Argentine Basins (Fig. 7).

The opening for the next 15–20 million years was essentially a continuation of the initial pattern (Fig. 8). Both the northern and southern basins continued to widen and deepen, with the northern basins both shallower and strongly controlled by water passage over the Walvis–Rio Grande barrier. By

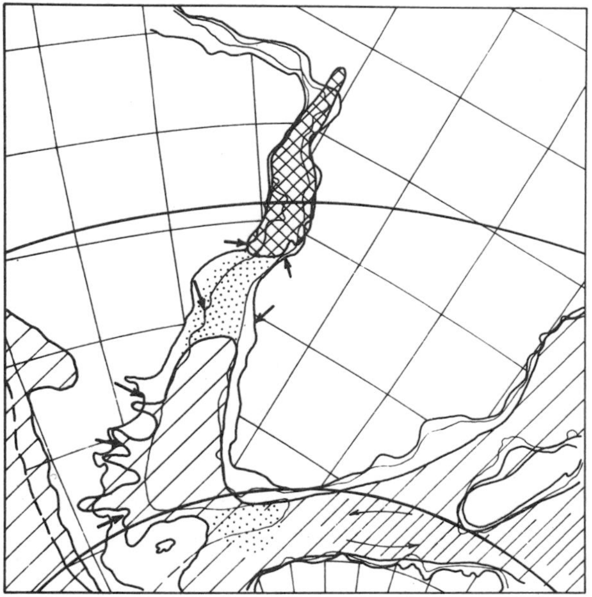

Fig. 6. The South Atlantic during the Basal Aptian (ca. 110 million years ago). On this, and the following three figures, the relative positions of the two continents are shown with their ancient latitudes. The coarse hatching is thought to be shallow seas and the fine hatching is deep oceanic water, with fine arrows indicating postulated surface water circulation. Areas of intermittent or extremely shallow water are shown as dotted, with thick arrows indicating the main sources of sediment. The cross hatching (on this figure only) indicates the area of extensive evaporite accumulation.

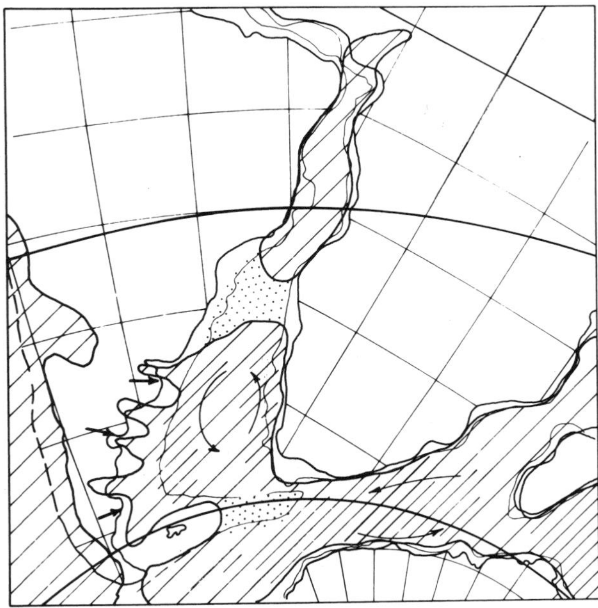

Fig. 7. The South Atlantic during the Albian (ca. 100 million years ago). See legend to Fig. 6.

Turonian times (ca. 90 million years ago) the marine seas extended along the entire coastlines of the South Atlantic, and restricted oceanic circulation developed, although bottom water activity was still restricted by the Falkland Plateau, the Walvis–Rio Grande barrier and by the shallow nature of the basins. During Turonian times, a world-wide rise in sea level allowed very shallow waters to flood along the Benue Trough in Nigeria, linking with shallow epicontinental seas spreading southwards from the Mediterranean, and also with very shallow waters passing from the Central Atlantic towards the southwest. Such marine connections would initially be very shallow and discontinuous, but by lower Turonian times (Reyment, 1969; Reyment *et al.*, 1976) the fracturing of the continents bordering the Equatorial Atlantic allowed slightly deeper but still very shallow waters to pass between the Central and South Atlantic Oceans allowing a mixing of "Mediterranean" and "Indian–South Atlantic" ammonite faunas. The mixture of these fauna demonstrates the continuity of marine waters between Africa and South America at this time. At the end of Turonian times, a world-wide fall of sea level broke the trans-Saharan connection, but by then the gradual separation of South America from Africa had allowed the Equatorial Atlantic to maintain a marine link that was to be retained and expanded to the present day.

The subsequent history of the South Atlantic (Berggren and Hollister, 1977; Van Andel *et al.*, 1977) between Africa and South America, has been one of gradual opening and deepening (Fig. 9). By Campanian–Santonian times (80–85 million years ago), the separation, even in the north, amounted

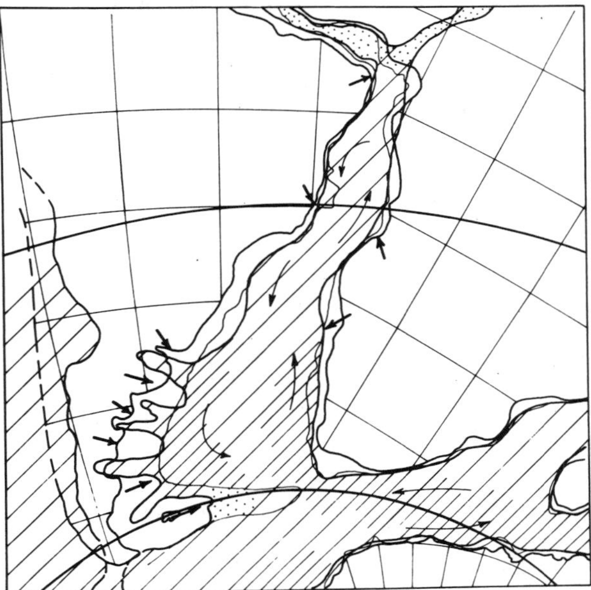

Fig. 8. The South Atlantic during the Turonian (85–90 million years ago). See legend to Fig. 6.

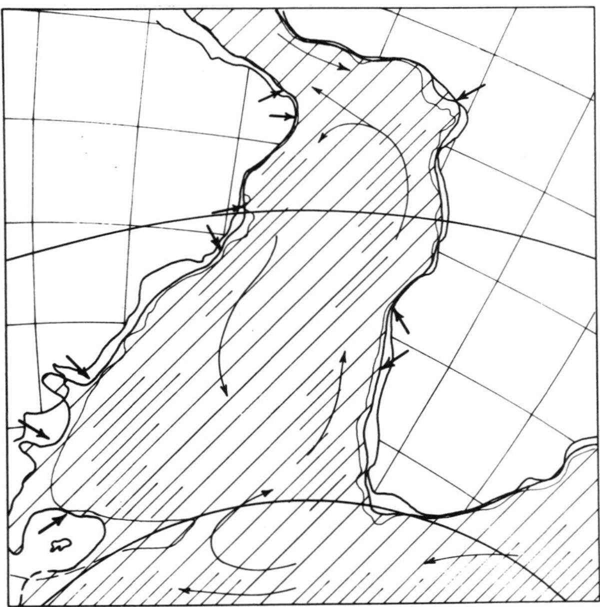

Fig. 9. The South Atlantic during the Campanian (ca. 74 million years ago). See legend to Fig. 6.

to some 500 km and a slight change in the direction of relative movement had taken place. This change was only slight in its effects on the opening between southern South America and southern Africa, but resulted in a more east–west motion of the two continents so that the entire South Atlantic was opening at almost the same rate, rather than the previously small rate of opening in the north compared with the south. This change is likely to have resulted from the linkage of the mantle convection motions in the Central Atlantic with those of the South Atlantic and had major impact on the Caribbean region (see below). In the South Atlantic, however, it seems to have been reflected by a slightly accelerated rate of subsidence as the post-Campanian sediments in the South Atlantic became more generally of a deep-water nature with the first mixing of ostracod fauna, and increasingly clear evidence of both surface and bottom water circulation (Berggren and Hollister, 1977). The Walvis Ridge and Rio Grande Rise, however, continued to form major barriers to bottom water circulation. In the Equatorial Atlantic, the uplift of the oceanic Sierra Leone and Céara Rises of 80 million years ago appears to have commenced in the late Cretaceous and even Paleo–Eocene times, possibly in response to changes in motion between the North Atlantic, South Atlantic, and Africa plate motions at about this time. The creation of these higher oceanic areas may have formed a significant restriction to deep oceanic circulation between the Central and South Atlantic from this period into Oligocene times. The final event in the South Atlantic was the initiation of major bottom water circulation resulting from the formation and melting of the vast ice sheets on Antarctica during the last 10 million years. Such circulation is likely to have had a major effect on the climate of continents bordering the South Atlantic, but it is unlikely that the changes in relative sea level during the Plio–Pleistocene had any significant effect on possible dispersion routes between Africa and South America.

The Antarctic Connection

The geological similarity between the Antarctic Peninsula and southernmost South America (Fig. 10) has been recognized for over a century (Barrow, 1831; Arctowski, 1895; Suess, 1885) and most subsequent workers have recognized a fundamental similarity in the evolution of the two areas that has required a previous continuity of their structures (Hawkes, 1962; Frakes, 1966; Dalziel and Elliott, 1971, 1973; Adie, 1971; Dott, 1972; Barker and Griffiths, 1977). Both areas have a Paleozoic basement on which has been imposed a similar series of metamorphic and volcanic episodes that are generally interpreted in terms of an active subduction zone bordering both areas during at least the Mesozoic (Suarez, 1976), but some difference appears to have developed by the end of the Mesozoic when most of the area was subjected to a strong tectonic activity and volcanism, except for the Antarctic Peninsula, the South Orkney, and the South Shetland Islands, in which block

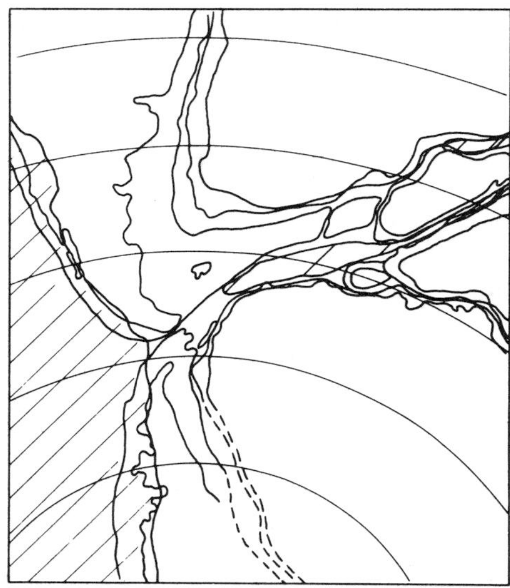

Fig. 10. South America–Antarctica in Jurassic-Cretaceous times (ca. 130 million years ago). The continental shelf edges and present-day coastline are shown in their previous relationship and in their paleolatitudinal positions. Oceanic depths of water are shown by fine hatches and shallow, epicontinental seas by coarse hatching. The relative position of West Antarctica in this and the following reconstruction is distinctly uncertain.

faulting appears to have been predominant (Katz, 1972; Dalziel and Elliott, 1973). The precise evolution of these different areas is, however, still very unclear with many viable hypotheses in existence. It is quite clear, however, that all of the Gondwanan continents were still linked in early Jurassic times, and the distribution of marsupial and placental mammals clearly indicates the continuity of terrestrial migration routes between South America, Antarctica, and Australia until late Cretaceous times (Keast, 1971; Tedford, 1974; Hoffstetter, 1976), but it is unclear whether the connections between South America and Antarctica necessitated the presence of relatively flat, stable subaerial continental blocks or whether the existence of land routes along an irregular mountain chain, analogous to the Malayan Peninsula, could have been adequate to account for such a dispersal pattern. The proposed evolution of this area is, therefore, distinctly uncertain and liable to significant modifications as new data become available.

While the relative position of South America and Africa in Lower Jurassic times is well established, the relationship between these West Gondwanan continents and those of the East Gondwanan (Madagascar, India, Ceylon, Australia, and Antarctica) is uncertain, although the reconstruction for the East Gondwanan continents themselves appears to be generally accepted. The way in which these two halves of Gondwanaland fit is critically determined by the position of Madagascar (Tarling, 1971*b*). There appears to be strong geological evidence against a location, in Jurassic times, of Madagascar against East Africa as undisturbed Carboniferous and later sediments occur in this location. On this basis, Madagascar was either in its present position relative to Africa (Tarling, 1971*b*) or in a somewhat more southerly position (Flores, 1970; Tarling, 1972). Either position places eastern Gondwanaland further south than on conventional reconstructions (du Toit, 1937; Smith and Hallam, 1970), but the retention of Madagascar in its present position relative to Africa is considered to be more geologically and geophysically realistic. Such a reconstruction necessitates movement between eastern and western Antarctica, as is required on most Gondwanan reconstructions, although the evidence for or against such a previous relationship is poor (Elliott, 1975). Nonetheless, the proposed reconstruction (Fig. 1) appears to be a reasonable approximation for the Lower Jurassic, and earlier, geography of Gondwanaland.

As East Gondwanaland began to break away from Africa–South America to initiate the Indian Ocean, there appears to have been a simultaneous movement between East and West Antarctica (Fig. 10). Both movements were accompanied by flood basalts in the same way as the Serra Geral and Kaoko Volcanics accompanied the initiation of the South Atlantic. This suggests that the separation of East Antarctica with Australia, from Africa, South America and West Antarctica, commenced some 160 million years ago (the age of the Ferrar Dolerites in eastern Antarctica) and that this dates the initial formation of the Weddell Sea. The magnetic anomalies in the Weddell Sea are thought to be of Jurassic age (Jahn, 1978) but the direction of motion is unclear. The Weddell Sea could have formed as "normal" ocean or have been associated with marginal basins upwelling behind the Pacific margin island arc system which paralleled the subduction zone lying along the southern Andes and Antarctic Peninsula. Possibly the Weddell Sea originated as an earlier, less organized system of sea-floor spreading within a marginal basin, shortly after

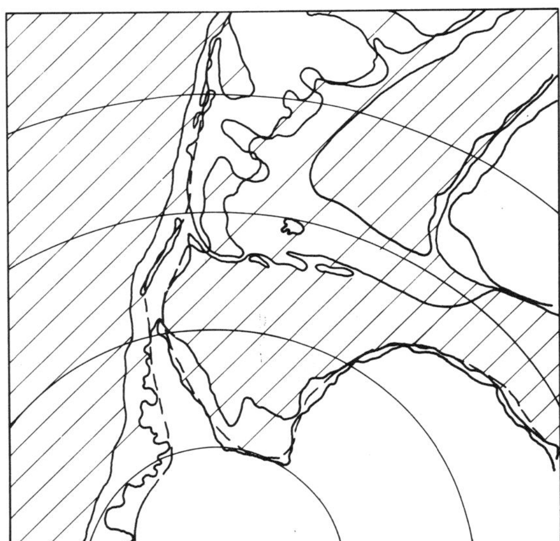

Fig. 11. South America–Antarctica in Albian times (ca. 100 million years ago). See legend to Fig. 10.

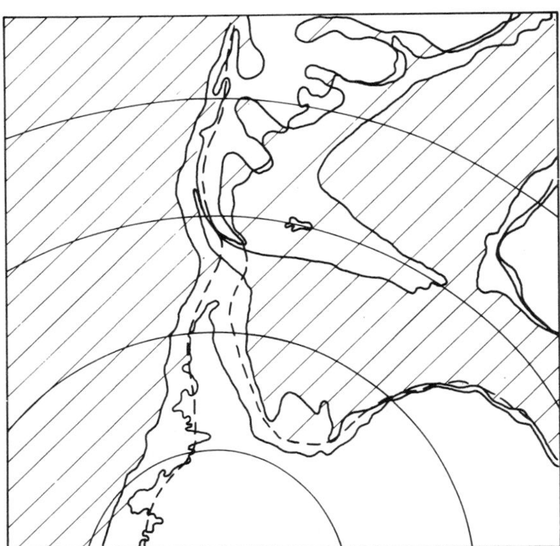

Fig. 12. South America–Antarctica in the Turonian (85–90 million years ago). See legend to Fig. 10.

the Phoenix–Farallon Rise was subducted beneath the Antarctic Peninsula (Herron and Tucholke, 1976), and subsequently developed normal ocean-floor spreading. Such a complex history may eventually be decipherable when the magnetic anomaly patterns in the Weddell Sea are established. The motion between East and West Antarctica is envisaged as a largely transcurrent motion in the Australian–New Zealand sector, but the Ellsworth Mountains (Clarkson and Brook, 1977) probably acted as a "roller" bearing between the two and rotated clockwise during their relative movement which is assumed to have largely ceased in Cretaceous times.

The Cretaceous evolution of the area is possibly even more obscure than for the Jurassic situation (Figs. 11 and 12). The main feature appears to have been the formation of a small marginal basin between the island arc system, off southernmost South America, and the mainland. This marginal basin closed in Late Cretaceous times (Fig. 13), giving rise to the Rocas Verdes Complex in Patagonia (Dalziel *et al.*, 1974; Tarney *et al.*, 1976; de Wit, 1977); this closure seems to have been part of the Laramide orogeny which affected large areas of western South America. It seems probable that a southerly extension of this marginal basin also formed which may now form part of the eastern half of the Scotia Sea, lying between the Oligocene magnetic anomalies to the west and the Miocene anomalies to the east. [Such a hypothesis has been suggested by de Wit (1977) but with a mainly Jurassic age for the entire central part of the Scotia Sea. More recent analyses of the magnetic anomalies in the western part of this area (Barker and Burrell, 1977) show that any preexisting oceanic floor must have been much smaller in

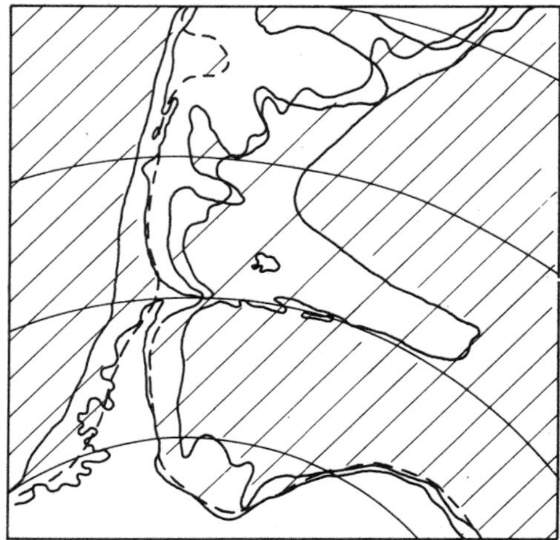

Fig. 13. South America–Antarctica in the Upper Campanian (ca. 74 million years ago). See legend to Fig. 10.

extent than proposed by de Wit.] By the end of Cretaceous times, therefore, the Weddell Sea existed in full, and possibly some 20% of the Scotia Sea, but the Andean–Antarctic connection was still present and was still probably a continuous land bridge (Fig. 14).

The final break of the South American–Antarctic connection has now been established by an analysis of magnetic anomalies in the Drake Passage area, combined with inferences from the sedimentation history in this and bordering areas (Barker and Burrell, 1977). Sea-floor spreading was initiated

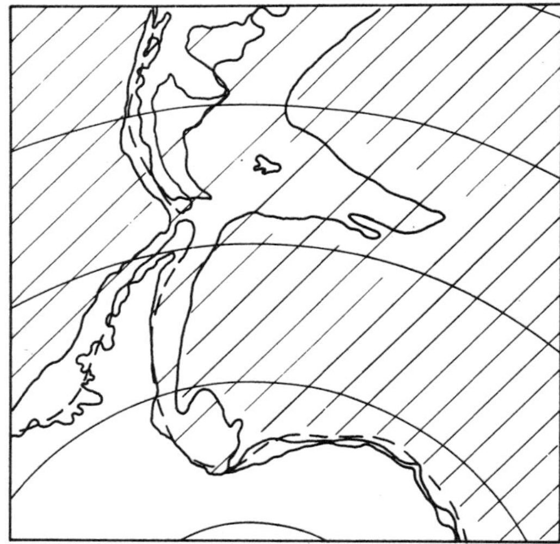

Fig. 14. South America–Antarctica in Paleocene times (ca. 60–62 million years ago). See legend to Fig. 10.

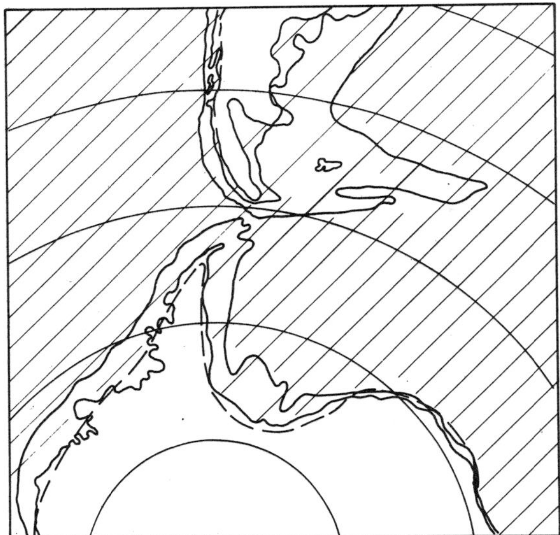

Fig. 15. South America-Antarctica in Oligocene times (ca. 34-35 million years ago). See legend to Fig. 10.

in the area in upper Oligocene times, about 27 million years ago or possibly very slightly earlier (Fig. 15). This may have been a response to the subduction of the Chile Rise and then the Aluk Rise in this area, creating a very weak zone in which an initial marginal basin developed normal ocean floor generation

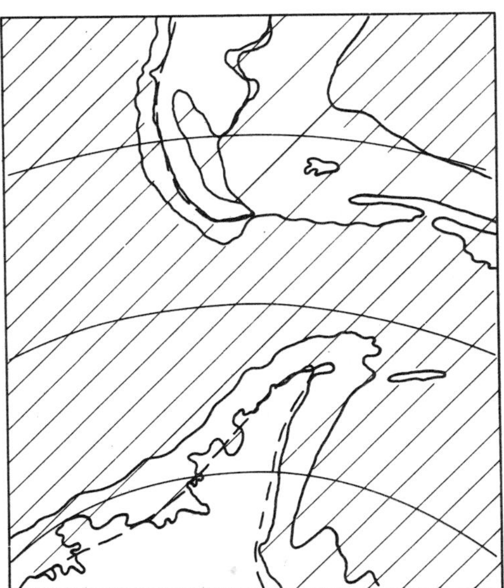

Fig. 16. South America–Antarctica in Miocene times (ca. 10 million years ago). See legend to Fig. 10.

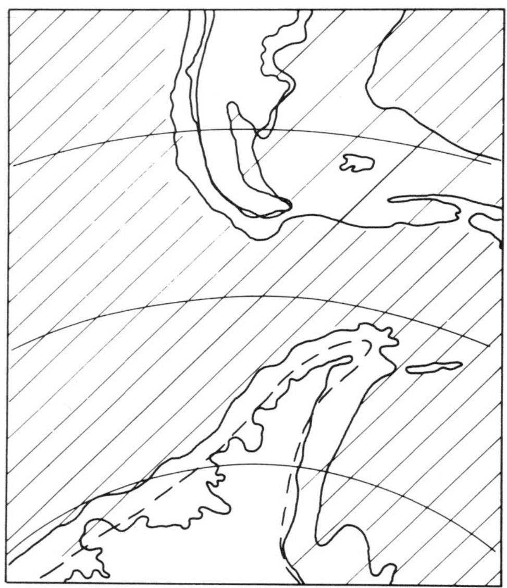

Fig. 17. South America–Antarctica in Pliocene times (ca. 3 million years ago). See legend to Fig. 10.

for some 10–20 million years, with subsequent increasing separation between Patagonia and the Antarctic Peninsula until about mid to late Miocene times, possibly around 10 million years ago (Figs. 15–17).

The gradual separation of Patagonia and the Antarctic Peninsula allowed the passage of deeper and deeper water, leading to the formation of the circum-Antarctic Current at the Oligocene–Miocene boundary, with cool waters entering the Australian region slightly before that (Foster, 1974) indicating continuous circum-Antarctic waters. Spreading then appears to have ceased in the western part of the Scotia Sea in mid to late Miocene times, approximately 10 million years ago. This cessation is almost simultaneous with the initiation of the main alpine orogeny along the Andes and the commencement of sea-floor spreading on the extreme eastern edge of the Scotia Sea. This spreading subsequently gave rise to the present Scotia Arc which has formed a very discontinuous link since Miocene times, the last 7.5–8 million years, between the remanents of the Andean–Antarctic connection (Barker, 1970, 1972).

The Caribbean Connection

The origins of both the Gulf of Mexico and the Caribbean (Figs. 18–21) are extremely unclear at the moment. The Gulf of Mexico is surrounded by ancient rocks (Precambrian and Paleozoic) with major areas of Jurassic and, especially, Cretaceous limestones (Dengo, 1975; Banks, 1975; Helwig, 1975; King, 1975). Within the Gulf itself, the basement rocks are thought to be oceanic, although the presence of thin continental crust cannot be completely discounted (Uchupi, 1975; Martin and Case, 1975). The oceanic rocks appear to be overlain directly by salt deposits but were possibly laid down on thin terrestrial sediments (Wilhelm and Ewing, 1972). The evaporites appear to have formed at a time when the Gulf was very low lying and they are overlain by later Jurassic and Cretaceous sediments, with thick Tertiary sediments largely confined to the northern and western edges of the Basin. The actual subsidence of the Basin appears to have taken place fairly quickly in Late Jurassic times. It is clear that the Gulf of Mexico, adjacent Mexico, and the Yucatan Peninsula (Ramos, 1975) must be considered as part of the North American plate today and to have behaved as part of this plate since at least Late Triassic times. However, the tectonics of western Mexico, in particular, have been drastically modified by successive subduction zones on its Pacific margin. The basic problem of the later evolution of the Central American region therefore largely hinges on an understanding of the origin of the Caribbean Basin (Case, 1975, 1977; Fox and Heezen, 1975; Donnelly, 1975; Malfait and Dinkelman, 1972; Ladd, 1976). It seems probable that the extreme northern and northwest edges of South America were not in existence in the Triassic and there is increasing, although still restricted, evidence that parts of Central America lay away from their present location. Consideration of all reasonable reconstructions for the continents in Early Jurassic times

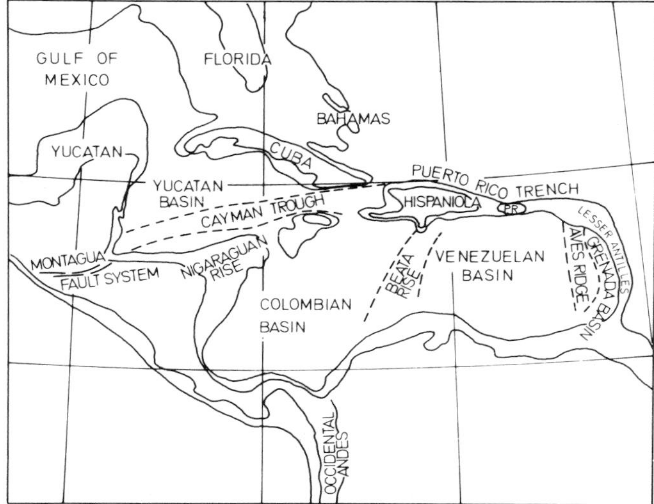

Fig. 18. The Caribbean Region. The location of most areas mentioned in the text are shown on this map.

(Figs. 19–20) indicate that most of the Caribbean must have originated since the Triassic and formed predominantly in the Late Jurassic and Cretaceous.

Reconstructions for the Early Jurassic period for the North Atlantic involve some overlap between southern Mexico and the Guyanan shield of South America. This overlap is some 600 km in the mid-Jurassic (Fig. 19) and cannot, at the moment, be explained in terms of motions within southern Mexico. Possibly the boundary between them in Triassic–Jurassic times lay along the zone of weakness now marked by the Volcanic Belt of Mexico, but such a suggestion is purely speculative. (It is conceivable that a pre-Jurassic closure could include some anticlockwise motion of Gondwanaland relative to North America, possibly along the line of the Brevard Fault, but such an explanation seems unlikely for Late Triassic and Early Jurassic reconstructions.) However, in terms of Jurassic migration routes between North and South America and, of course, between Europe and Africa, there could have been no deep oceanic barrier between these continents, although access would obviously be restricted by any shallow, epicontinental seas and, for example, the occasionally extremely high salinity of waters in the area destined to sink to form the Gulf of Mexico. It seems likely, however, that it was not until latest Jurassic times that any major oceanic barrier existed between North and

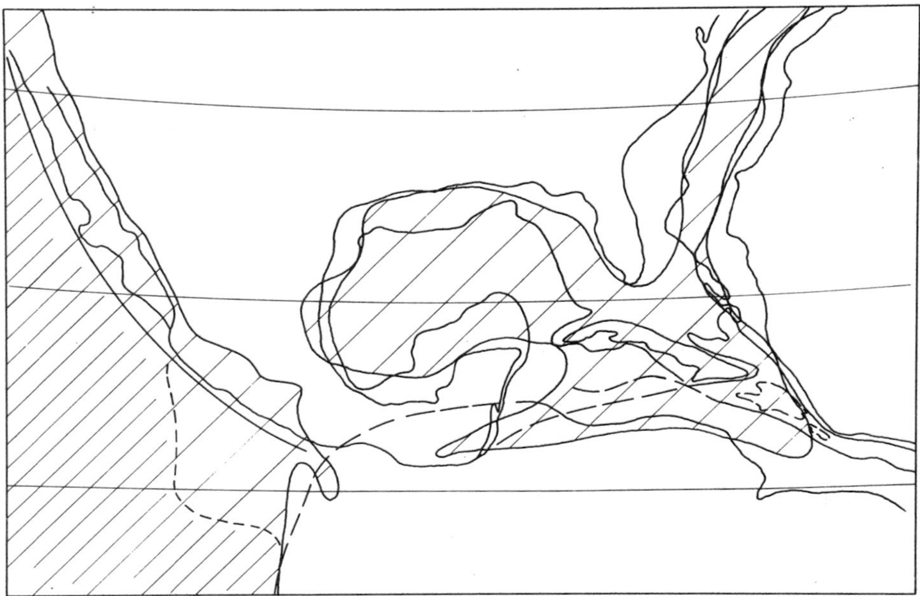

Fig. 19. The Caribbean Region in Triassic-Jurassic times (ca. 195 million years ago). The present continental shelf edge and coastlines are shown with coarse hatches indicating the presence of shallow, epicontinental seas, and postulated oceanic depths by fine hatchings. Although there are many uncertainties about specific areas, those in most doubt are shown by dashed lines, but it is emphasized that there are major uncertainties even concerning areas shown in full.

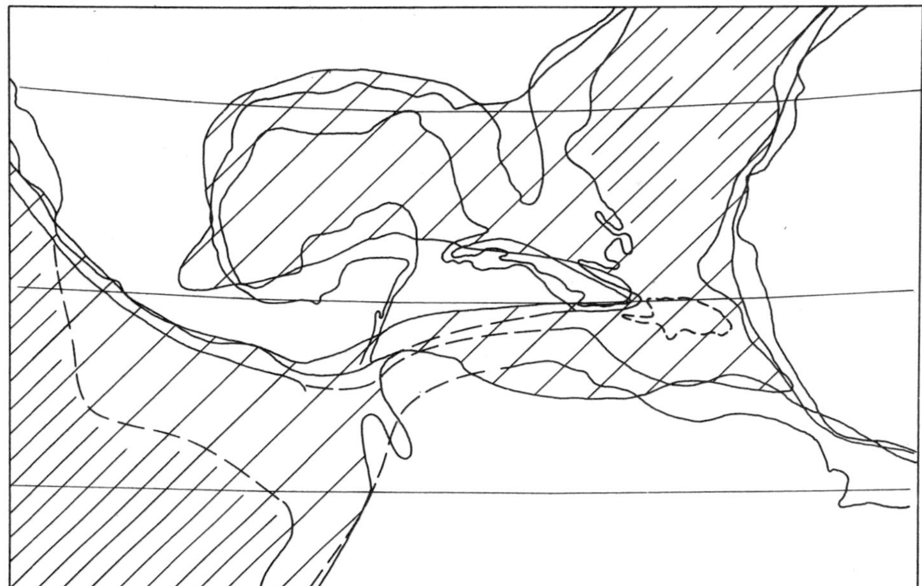

Fig. 20. The Caribbean Region in Jurassic-Cretaceous times (ca. 140 million years ago). See legend to Fig. 19.

South America and that such an early oceanic basin was initially shallow and narrow, similar to that for the early South Atlantic ocean. This oceanic trough extended rapidly during the Late Jurassic, with South America moving towards the east-southeast relative to North America at this time.

At the beginning of the Cretaceous (Figs. 20–21), South America was still linked to Africa, although the fractures leading to separation were mostly present (Nagy *et al.*, 1976; Mullins and Lynts, 1977), and North America was separating from Africa-South America in a northwestern direction (present coordinates). This separation must have been associated with the formation of a new oceanic floor between the Yucatan-Cuba block and South America and it seems logical to assume that there was normal sea-floor spreading as an extension of the spreading in the Central Atlantic through to the Pacific. It is suggested here that the Venezuelan Basin and part of the Colombian Basin largely formed during this Late Jurassic interval, although it must be emphasized that there is still little data to confirm or reject this postulate. The opening of South America from Africa, however, commencing in earliest Cretaceous times (ca. 130 million years ago), changed the direction of relative motion between North and South America into an essentially east–west motion (Fig. 21), fixing the latitudinal separation of the main continental blocks at some 1500 km until mid-Cretaceous times.

The changing location of Central America, south of Mexico, is critical to any assessment of migration routes (Fig. 18). Paleozoic rocks form the base-

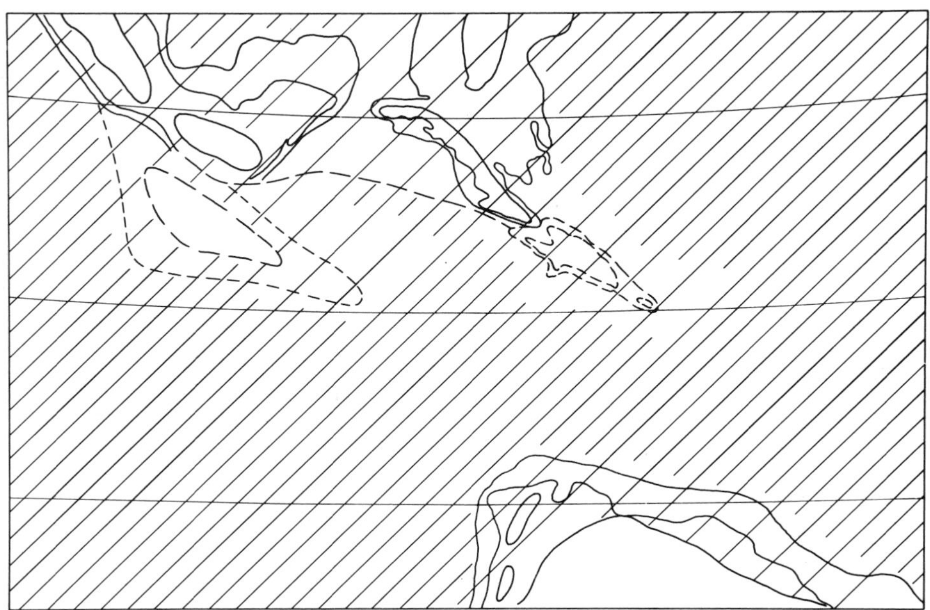

Fig. 21. The Caribbean Region in the Lower Aptian (ca. 110 million years ago). See legend to Fig. 19.

ment of much of southern Mexico, Nicaragua, and Honduras (Banks, 1975) and, although not exposed or identified, it seems reasonable to assume that rocks of this age underlie much of the Nicaraguan Rise as far as Jamaica (Arden, 1975; Perfit and Heezen, 1978). South of the Honduras-Nicaragua region, the oldest rocks appear to be entirely of Cretaceous or younger ages, with the basement apparently having essentially oceanic features (Fox and Heezen, 1975). To a first approximation, therefore, the model proposed here will presume that the Paleozoic-Early Mesozoic block of the southern Mexico-Nicaragua-Honduras-Nicaraguan Rise has behaved as a single tectonic unit and that the areas to the south of this unit have been created during the Cretaceous and Cenozoic. The junction between the North American plate and the Central American plate now lies along the line of the Montagua Fault system and Cayman Trough. As this plate cannot be fitted into the North-South American reconstructions for Jurassic and older times, and there is evidence for it having entered the area from some 1000 km to the west (Pinet, 1972; Gose and Swartz, 1977), it is suggested that this block was positioned just off western Mexico in Triassic-Jurassic times. (Such a location does not appear inconsistent with the Paleozoic trends in Mexico, but the precise position of this block is uncertain.) During the Cretaceous, it is suggested that this block rotated along the Mexican coastline, reaching its position a little further north than its present location by at least Campanian times (ca. 80 million years ago). After reaching this position (Figs. 22–25), the motions in

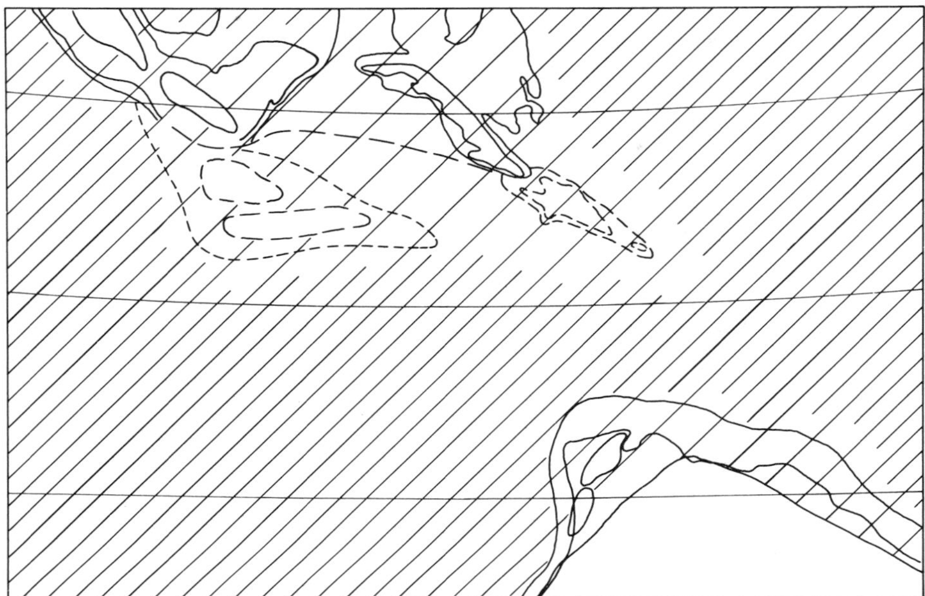

Fig. 22. The Caribbean Region in the Albian (ca. 100 million years ago). See legend to Fig. 19.

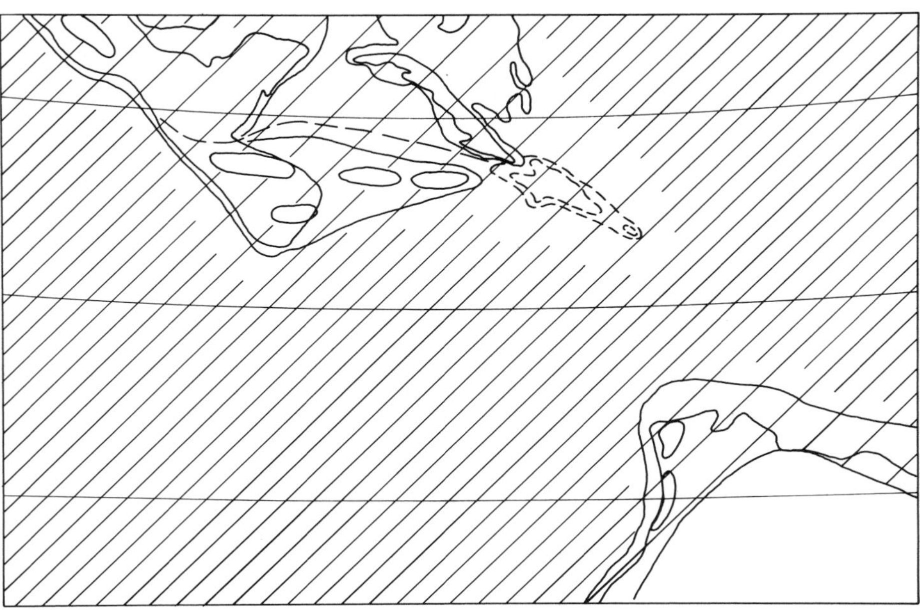

Fig. 23. The Caribbean Region in the Turonian (ca. 85–90 million years ago). See legend to Fig. 19.

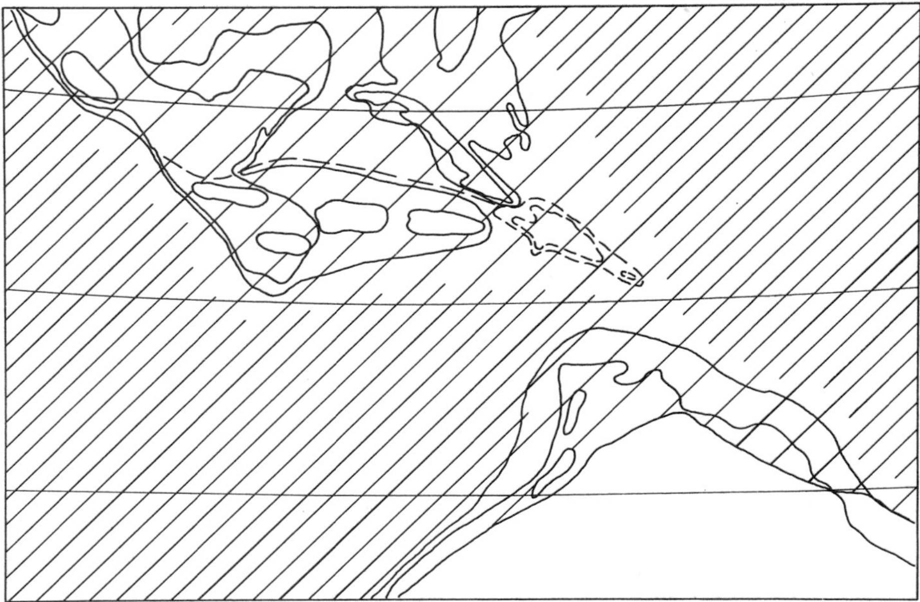

Fig. 24. The Caribbean Region in the Upper Campanian (ca. 74 million years ago). See legend to Fig. 19.

the Central East Pacific, which had been moving it, changed simultaneously with a change in the direction of motion between North and South America.

These mid-Cretaceous (80–90 million years) changes in motion probably reflect the linkage of the spreading system in the Central Atlantic with that of the South Atlantic and in the consequent cessation of spreading in the Venezuelan Basin. This cessation of spreading, combined with a drastic change in the directions of stress, probably accounts for extensive basaltic eruptions on the Caribbean ocean floor and initiation of the older Lesser Antilles arc at about this time. The new motion caused strong compression within the Caribbean, reducing the separation between the North and South American blocks to some 500–600 km by Santonian times. This required the destruction of some 800–1000 km of Venezuelan and Colombian Basin oceanic rocks along subduction zones on both the northern and southern margins. The new motion rotated the Central American block against the Yucatan Basin and gradually closed the subduction zone along this line, commencing in the west and gradually working eastward. This meant that more subduction and greater tectonic activity took place near Cuba than southern Mexico and was also later further eastwards (Pardo, 1975; Bowin, 1975). The Venezuelan Basin was essentially passive at this time, but some spreading continued in the southern Colombian Basin until latest Cretaceous times (Christoffersen, 1976)—the differential motion between the two oceanic places giving rise to the intervening Beata Rise.

All this activity ceased in Late Eocene times (Fig. 26) when the motion of North America again changed, possibly in response to the extension of spreading from the Central Atlantic into the Norwegian Sea and its cessation in the Labrador Sea. The motion within the Caribbean reverted to a simple response to the east–west motion between North and South America, although this motion was in the opposite sense to that of Early-Middle Cretaceous times. Both the eastern and western sides of the Caribbean were active, reflecting subduction of the Atlantic oceanic plate to the east and the Pacific plate to the west (Atwater, 1970; Herron and Hayes, 1969; Herron, 1972; Hey *et al.,* 1977; Noble and McKee, 1977). To the east, the original line of the Lesser Antilles, that had already begun to form 80 million years ago, continued to evolve while, to the west, the Central American plate was being added to by subductive activity on its western and southwestern flanks. During the Oligocene (Fig. 27) there was continuing east–west motion between the North and South American plates and changes in the spreading in the Pacific (Hey, 1977; Hey *et al.,* 1977; Noble and McKee, 1977) which resulted in the southwestern edge of the Nicaraguan Rise being extended further and further southwards so that the Colombian Basin became mechanically isolated from the Pacific spreading systems from Early Tertiary times onwards. Within the Caribbean itself, renewed activity along the Lesser Antilles (Tomblin, 1975; Arculus, 1976) and the continued growth of the Central American

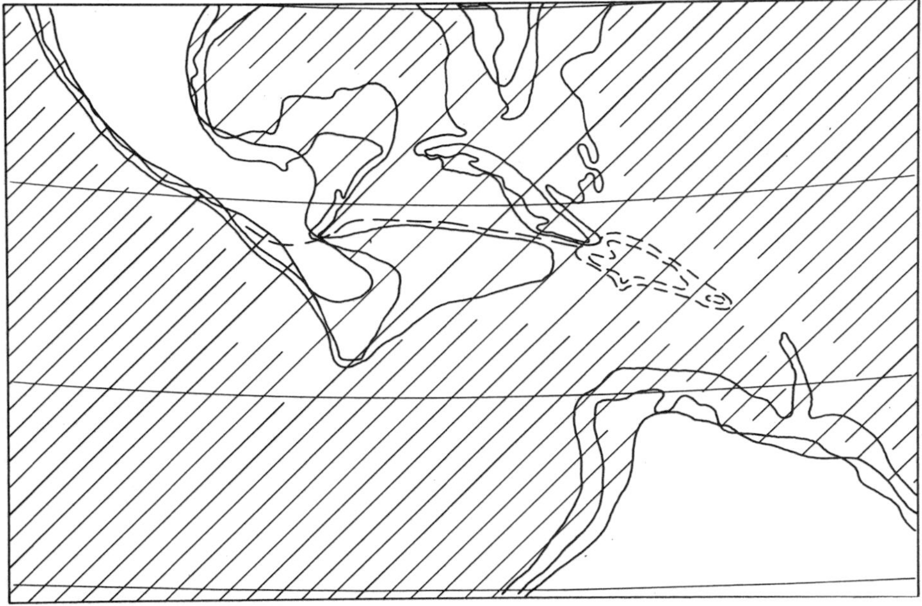

Fig. 25. The Caribbean Region in the Paleocene (ca. 60–62 million years ago). See legend to Fig. 19.

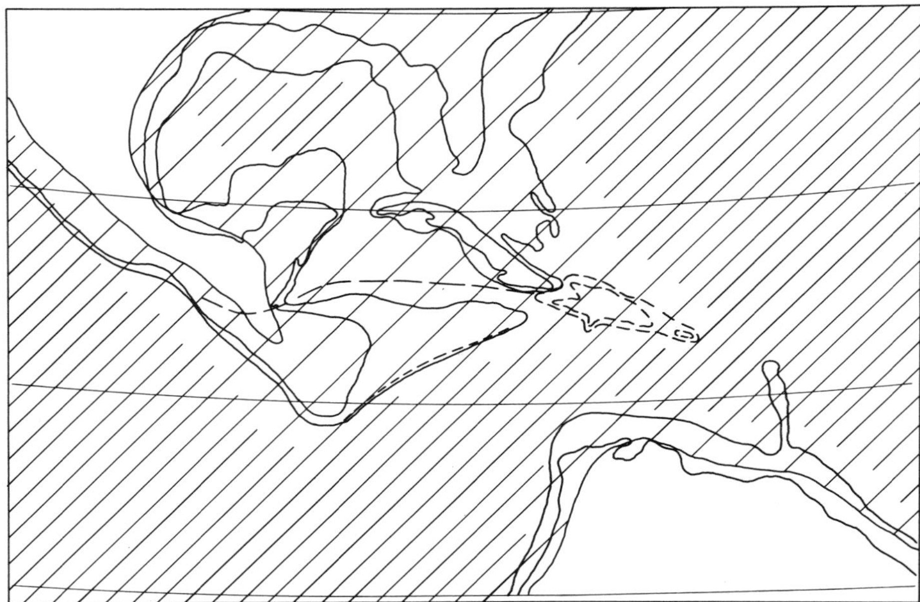

Fig. 26. The Caribbean Region in the Eocene (ca. 50 million years ago). See legend to Fig. 19.

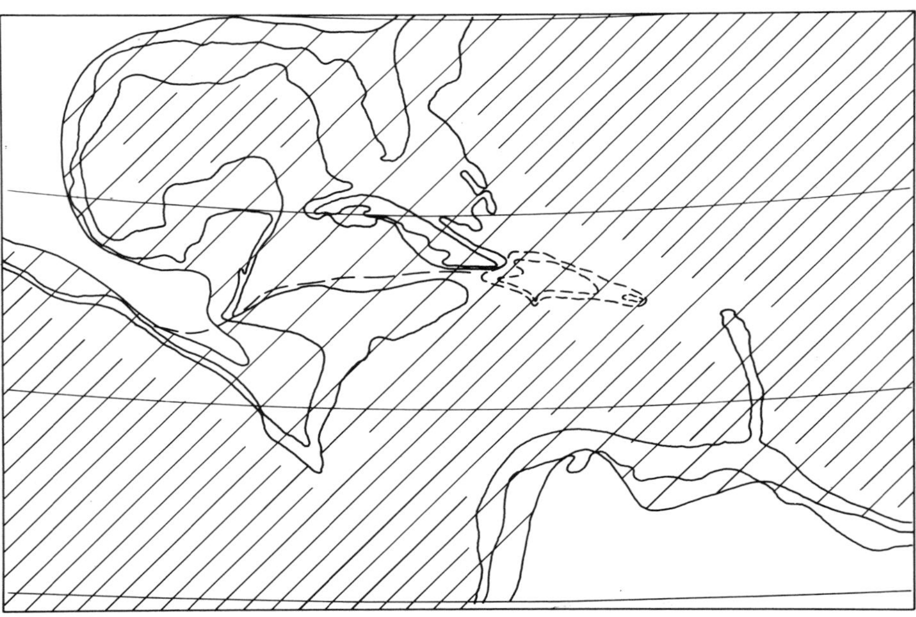

Fig. 27. The Caribbean Region in Oligocene times (ca. 34–35 million years ago). See legend to Fig. 19.

block gradually increased the proximity of the two land masses (Figs. 28 and 29). The reorganization of spreading in the East Pacific, with the creation of the present East Pacific Rise, meant that the final link between Central America and South America resulted from the final formation of the Occidental Andes (Julivert, 1970; Shagam, 1975) in Colombia, with simultaneous volcanic eruptions in Central America, along the line of the Middle American Trench, increasing the size of the Panamanian Isthmus. Finally, the increasing elevation of northwestern Colombia gradually improved the terrestrial connection between North and South America. However, coincidental with the creation of this terrestrial link (Whitmore and Stewart, 1965; Porter, 1972), the actual separation between the North and South American blocks increased slightly (Fig. 29) and it seems probable that this extension was essentially taken up by the separation between the Nicaraguan Rise and the Yucatan Basin. It is difficult to reconcile the undoubted evidence for east–west spreading within the Cayman Trough with this evidence for north–south separation (McDonald and Holcome, 1978; Perfit and Heezen, 1978), but the creation of this new oceanic floor must necessarily require separation between the Nicaraguan Plateau and the Colombian-Venezuelan Basins which, if continued into the future would eventually give rise to a separation between North and South America along the line of the Montagua Faults and Cayman Trough.

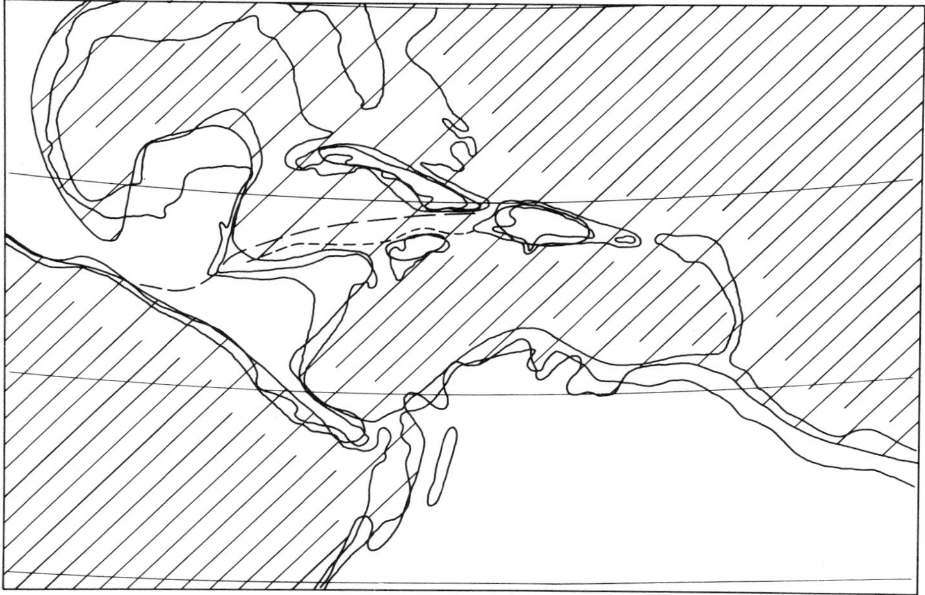

Fig. 28. The Caribbean Region in Miocene times (ca. 10 million years ago). See legend to Fig. 19.

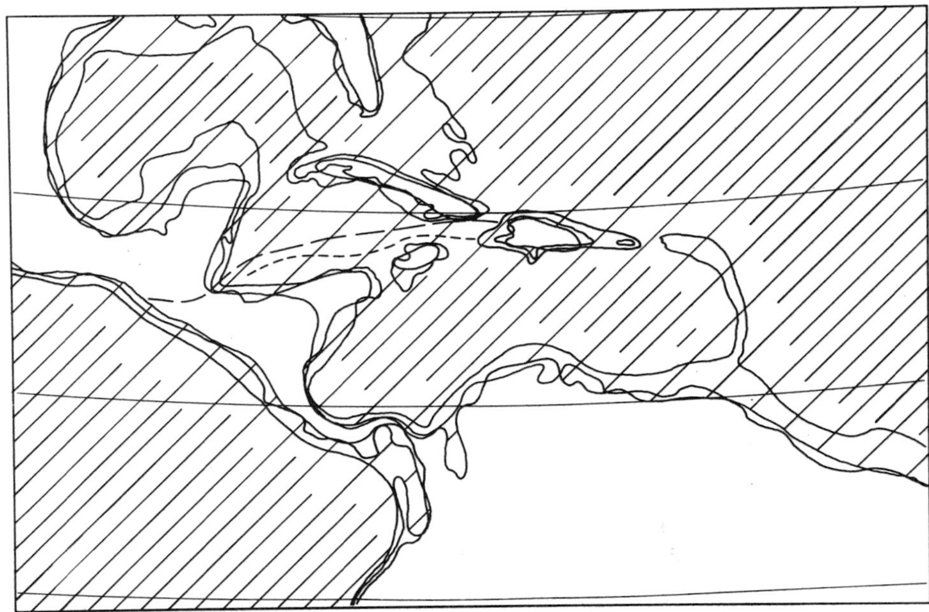

Fig. 29. The Caribbean Region in Pliocene times (ca. 3 million years ago). See legend to Fig. 19.

The Nature of the Possible Terrestrial Migration Routes to South America

Before the initiation of the Atlantic Ocean, in Late Carboniferous–Permian times, access between the European, North and South American, African, and Antarctic continents was completely free other than the occasional shallow sea and the vestiges of mountain chains along, for example, the Appalachians. Obviously such barriers could be extremely effective for some organisms but the absence of different types of ecological niche would enhance the lack of geographical isolation and few differences would be expected between the fauna and flora of the different continents. The initiation of rifting and doming, mostly in Late Triassic and Early Jurassic times would increase the number of niches and would eventually develop into a barrier directly analogous with the East African Rift Valleys in which the topography may locally be extremely abrupt and persist for long distances, with very low-lying ground between, often with internal drainage and saline lakes. In the case of the South Atlantic the area lay in low latitudes and is likely to have been highly susceptible to strong evaporation conditions, initially evaporating fresh water drainage systems, but, as the area deepened, marine waters provided the main source of water. It is probable that, at this particular time, the very high salinity in the central rift zones was a more effective barrier to intercontinental migration than in the times when the rift valleys were effectively flooded as they gradually sank below sea level. At such a time, the presence of fairly continuous water is likely to have had a local ameliorating effect on the climate on both sides of the incipient ocean. Naturally, such climatic change would be inimical to some forms of life but would generally provide similar conditions at fairly short distances apart that would, at this stage, be fairly accessible across the occasional higher pieces of ground. As the continental sides gradually separated and the waters between them deepened, migration routes would become increasingly restricted to particularly anomalous areas, such as the Walvis–Rio Grande Rises in the South Atlantic. The early ocean would be shallow, as the oceanic ridge, from which the ocean floor was being generated, would stand close to sea level with rare, but possibly significant, volcanic islands along its length. In general, however, the separation of the two sides would lead to the inexorable inhibition of transoceanic migration (Tarling, 1980), although such inhibition obviously depends on the characteristics of each particular organism.

If it is assumed that migration between facing rifted edges of the continents is essentially inhibited when the oceanic separation exceeds some 200–300 kms, it is clearly the other links that are of major importance. In the case of the African connections, the two continents were close together between northwestern Brazil and West Africa until some 80 million years ago, but the main part of the South Atlantic had formed from the opening of rift edges. The Late Cretaceous–Early Tertiary migration routes therefore must have been concentrated along (1) southern Africa onto the Falkland Plateau, (2) the Walvis–Rio Grande Rise, and (3) the northern proximity. (1) The southern route was initially wide, but clearly became increasingly constricted as the Falkland Plateau slid past the Cape, and eventually separated from it. During the motion it seems likely that epicontinental seas provided the main barrier to migration between the continents, and the key problem, in this context, is the depth of the eastern part of the Falkland Plateau at this time. It seems likely that this was then much higher than now and probably provided a good migration route until the actual separation between the two continental masses was completed in Late Albian times. (2) The Walvis–Rio Grande Rise, however, is likely to have been a much more rugged, intermittent migration route with the gradual decrease in its elevation leading to greater difficulty in "island-hopping," but the occasional volcanic eruption creating temporary new land and providing improved connections for short periods. This route must, therefore, have acted as a major filter on trans-Atlantic migration routes until its general submergence in mid-Oligocene times. Subsequently, the occasional volcanic island probably formed along the axial ridge in the central southern Atlantic, but this was far distant from the continental edges. (3) The separation of the Brazilian and West African coasts gradually increased especially following the change in the direction of spreading some 80–85 million years ago. The Equatorial Atlantic in this area, however, had been subjected to major stresses both during the initial pivoting of South America from Africa and during the change of spreading direction. The oceanic crust is likely to have been somewhat higher than average, reflecting the numerous offsets of the spreading ridge caused by this degree of shattering. The Céara Rise, off Brazil, and the Sierra Leone Rise, off West Africa, were also strong positive features in the Early Tertiary. It seems probable, therefore, that oceanic islands, some hundreds of square kilometers in size, existed offshore from both Brazil and West Africa, and that the mid-oceanic rise was also subaerial for much of this period. The availability of these "stepping stones" at specific times is not clear as the precise sequence of sea-level changes is not yet known in detail, but it seems probable that these oceanic islands were present spasmodically up to and including the Eocene–Oligocene boundary (Fig. 30), but they became submerged following a eustatic rise of sea level and subsequently remained largely or entirely submarine. Such islands would not provide a complete terrestrial migration route as it is clear that shallow marine waters continued to circulate between the Central and South Atlantic from basal Turonian times onwards. The areas between the continental blocks, the Céara and Sierra Leone Rises, and the mid-oceanic ridge system were certainly marine, and provided barriers some 200–300 km wide, thereby acting as major filters on any migrating organisms. Under such circumstances, surface oceanic currents are likely to play a major part in any migration pattern and such currents would strongly inhibit west to east migration, but would amplify any migration from Africa towards South America even after extensive water exchange between the Central and South Atlantic became important as this region still straddled the Equator and would thus have surface currents flowing westwards.

To the south, interchange between South America and the Antarctic Peninsula is likely to have been free and easy although mountain-building episodes, particularly on the western edge, may have created rain shadow effects at some times. In general, however, the route appears to have been fairly low lying and the topography paralleled the direction of faunal and floral exchanges until the final, geologically sudden cessation of all land connection from some 35 million years ago.

The Caribbean connection differs from the other connections in that a rifted separation must have occurred between North and South America in Jurassic-Early Cretaceous times, but the reestablishment of links came about by the actual growth of continental material, rather more than by changes in the total separation of the two major tectonic blocks. In the Caribbean, the separation in Late Eocene-Early Oligocene times may have been as little as 500 km between the continental edge of Colombia and Central America, but much of northwestern Colombia and Central America were submarine at this period and the separation between exposed land areas appears to have exceeded 1200 km. Unfortunately, the exact nature of the volcanism in Central America, as far as Panama, is not clear and some of this may have been subaerial, giving rise to volcanic islands leading to nearly 1000 km of the exposed land of central Colombia. The Antilles were also continuing their volcanic activity, giving rise to volcanic islands leading northwards, and could have been within 500 km of Puerto Rico, but this and most of the Caribbean were submarine at this period, although occasional islands cannot be excluded from this picture. Oceanic currents would be moving westwards along northern South America, possibly assisting drifting that could encounter volcanic islands in the Panama region, and hence to North America, but the currents from the Panamanian area are likely to have carried materials into the Gulf of Mexico and the Gulf Stream. Migration from South America would therefore be possible by means of drifting and then island hopping, but most factors seem to inhibit possible migration from North to South America at this

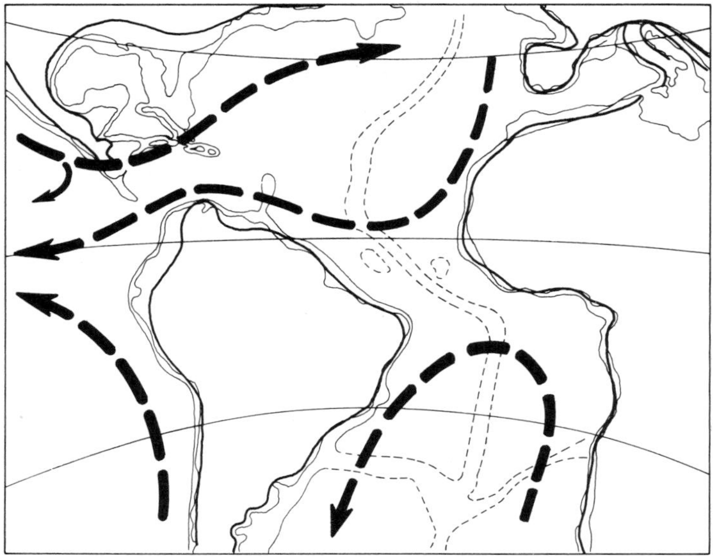

Fig. 30. Early Oligocene migration routes. At the start of the Oligocene, sea level fell and it was probable that discontinuous parts of the Mid-Atlantic Rise lay close to, or above, sea level. The Walvis and Rio Grande Rises were also exposed in parts, although probably not immediately adjacent to the continents, and would provide a series of "stepping stones," albeit of a very discontinuous nature. Further north, the fall in sea level is likely to have made parts of the oceanic Céara and Sierra Leone Rises close to sea level, so that deep oceanic waters rarely exceeded 200 km in width, although it is not yet clear how much of these Rises, including the Mid-Atlantic Rise, actually lay above sea level at any particular time. The lower sea level probably exposed appreciable parts of the Antilles arc and also the volcanic arc then developing from southern Mexico to Panama, but most of the Caribbean and Gulf of Mexico regions still lay below sea level. The postulated oceanic currents (heavy arrows) would clearly assist the migration of floating organisms from Africa towards South America in equatorial regions and the same currents are likely to have inhibited migration from North to South America. A rise in sea level, still in Early Oligocene times, flooded most of the trans-Atlantic connections and they have remained below sea level, except for very isolated islands (such as Tristan de Cunha today), for the rest of the Cenozoic. In the Caribbean, it seems probable that the oceanic currents continued to inhibit trans-Caribbean migration until immediately prior to the formation of a continuous land bridge in the late Miocene.

period. It is also pertinent that the subsequent motions of North and South America would continually decrease the separation and thus any earliest Oligocene migration routes between these continents would be enhanced. As there appears to be evidence against such later migration routes until Miocene and Pliocene times, it seems most unlikely that significant migration could have occurred from North to South America at this earlier time, but converse migration seems possible throughout most of Tertiary time. It was, however, local uplift of parts of Colombia in Mio-Pliocene times that was mainly responsible for completing the terrestrial connection to the Panamanian Isthmus, even though much of the isthmus was itself built from volcanic activity immediately prior to the land rising in Colombia.

In general, therefore, the geological evolution of South America has lead to the formation and destruction of a range of different migration routes which must have acted as varying degrees of filter at different times. However, while such an evolution is only just becoming clearer and is by no means fully established, there is now a general geological background against which the interchange of biota with the adjacent continents can be examined. Obviously such studies will cause revisions of the geological story and a study of the nature of the exchange may yield valuable information on the precise nature of the connection. It is also interesting to note how the model proposed here necessarily involves a reconsideration of the existence and destruction of land bridges, but the nature of these bridges differs drastically from those proposed in the 19th century.

ACKNOWLEDGMENTS

This article was prepared at the request of Professors R. L. Ciochon and A. B. Chiarelli, organizers of the symposium on "Origin of the New World Monkeys and Continental Drift" at the Seventh Congress of the International Primatological Society. I am grateful, therefore, for being drawn into this intriguing area. I would also like to thank all those attending the Symposium for their comments, Dr. P. F. Barker and his colleagues in the Department of Geology, Birmingham University, for comments on the Antarctic route, and Dr. G. K. Westbrook of the Department of Geological Sciences, Durham University, for his comments on the Caribbean. I have benefited from all of these criticisms but have not accepted all of them, so the undoubted errors of judgment are mine, not theirs.

References

Adie, R. J., 1971, Review of Antarctic geology, in: *Second I.U.G.S. Gondwanan Symposium, Capetown,* pp. 15–22.

Amaral, G., Cordani, U. G., Kawashita, K., and Reynolds, J. H., 1966, Potassium argon data of basaltic rocks from southern Brazil, *Geochim. Cosmochim. Acta* **31:**117-142.

Arctowski, H., 1895, Observations sur l'intérêt que présente l'exploration géologique des terres australes, *Bull. Soc. Géol. Fr., Sér.* 3 **23:**589-591.

Arculus, R. J., 1976, Geology and geochemistry of the alkali basal-andesite association of Grenada, Lesser Antilles island arc, *Bull. Geol. Soc. Am.* **87**:612-624.

Arden, D. D., Jr., 1975, Geology of Jamaica and the Nicaragua Rise, in: *The Ocean Basins and Margins,* Vol. 3, *The Gulf of Mexico and the Caribbean* (A. E. M. Nairn and F. G. Stehli, eds.), pp. 617-661, Plenum Press, New York.

Argand, E., 1924, la Tectonique de l'Asie, in: *Proc. Intl. Geol. Cong.* **13:**1171.

Atwater, T., 1970, Implications of plate tectonics for the Cenozoic tectonic evolution of western North America, *Bull. Geol. Soc. Am.* **81**:3513-3536.

Baker, H. B., 1912, The origin of continental forms, *Mich. Acad. Sci. Ann. Rep.* 116-141.

Banks, P. O., 1975, Basement rocks bordering the Gulf of Mexico and the Caribbean Sea, in: *The Ocean Basins and Margins,* Vol. 3, *The Gulf of Mexico and the Caribbean* (A. E. M. Nairn and F. G. Stehli, eds.), pp. 181-199, Plenum Press, New York.

Barker, P. F., 1970, Plate tectonics of the Scotia Arc region, *Nature* **228**:1293-1296.

Barker, P. F., 1972, Magnetic lineations in the Scotia Sea, in: *Antarctic Geology and Geophysics* (R. J. Adie, ed.), pp. 17-26, Universitetsforlaget, Oslo.

Barker, P. F., and Burrell, J., 1977, The opening of Drake Passage, *Marine Geol.* **25**:15-34.

Barker, P. F., and Griffiths, D. H., 1977, Towards a more certain reconstruction of Gondwanaland, *Phil. Trans. Roy. Soc. Lond., Ser. B* **279:**143-159.

Barrow, J., 1831, Introductory note, *Roy. Geogr. Soc.* **1**:62.

Berggren, W. A., and Hollister, C. D., 1977, Plate tectonics and palaeocirculation—Commotion in the ocean, *Tectonophysics* **38**:11-48.

Bowen, R. L., 1966, Tectogenetic relations of major Late Jurrasic-Early Cretaceous arching and lava extrusion in Southern Brazil, *Abs. Trans. Am. Geophys. Union* **47**:177-178.

Bowin, C., 1975, The geology of Hispaniola. in: *The Ocean Basins and Margins,* Vol. 3, *The Gulf of Mexico and the Caribbean* (A. E. M. Nairn and F. G. Stehli, eds.), pp. 501-552, Plenum Press, New York.

Campos, C. W. M., Ponte, F. C., and Miura, K., 1974, Geology of the Brazilian continental margin, in: *The Geology of Continental Margins* (C. A. Burke and C. L. Drake, eds.), pp. 447-461, Springer-Verlag, Berlin.

Case, J. E., 1975, Geophysical Studies in the Caribbean Sea, in: *The Ocean Basins and Margins,* Vol. 3, *The Gulf of Mexico and the Caribbean* (A. E. M. Nairn and F. G. Stehli, eds.), pp. 107-180, Plenum Press, New York.

Case, J. E., 1977, Geologic framework of the Caribbean region, in: *Geology, Geophysics and Resources of the Caribbean* (J. D. Weaver, ed.), pp. 3-26, Intergovernmental Oceanographic Commission, Unesco.

Christofferson, E., 1976, Colombian Basin magnetism and Caribbean plate tectonics, *Bull. Geol. Soc. Am.* **87**:1255-1258.

Clarkson, P. D., and Brook, M., 1977, Age and position of the Ellsworth Mountains crustal fragment, Antarctica, *Nature* **265:**515-516.

Dalziel, I. W. D., and Elliot, D. H., 1971, Evolution of the Scotia Arc, *Nature* **233**:246-252.

Dalziel, I. W. D., and Elliot, D. H., 1973, The Scotia Arc and Antarctic Margin, in: *The Ocean Basins and Margins,* Vol. 1, *The South Atlantic* (A. E. M. Nairn and F. G. Stehli, eds.), pp. 171-246, Plenum Press, New York.

Dalziel, I. W. D., de Wit, M., and Palmer, K. F., 1974, Fossil marginal basin in the southern Andes, *Nature* **250**:291-294.

Darwin, C., 1859, *The Origin of Species by Means of Natural Selection,* John Murray, London.

De Almeida, F. F. M., 1976, The system of continental rifts bordering the Santos Basin, Brazil, *An. Acad. Brasil. Ciências* **48**:15-26.

Delteil, J. R., Rivier, F., Montadert, L., Apostolesw, V., Didier, J., Goslin, M., and Patriat, P. H., 1976, Structure and sedimentation of the continental margin of the Gulf of Benin, *An. Acad. Brasil. Ciências* **48**:51-66.

Dengo, G., 1975, Paleozoic and Mesozoic tectonic belts in Mexico and Central America, in: *The Ocean Basins and Margins,* Vol. 3, *The Gulf of Mexico and the Caribbean* (A. E. M. Nairn and F. G. Stehli, eds.) pp. 283–323, Plenum Press, New York.

De Wit, M. J., 1977, The evolution of the Scotia Arc as a key to the reconstruction of southwestern Gondwanaland, *Tectonophysics* **37**:53–81.

Donnelly, T. W., 1975, The geological evolution of the Caribbean and Gulf of Mexico—some critical problems and areas, in: *The Ocean Basins and Margins,* Vol. 3, *The Gulf of Mexico and the Caribbean* (A. E. M. Nairn and F. G. Stehli, eds.), pp. 662–689, Plenum Press, New York.

Dott, R. H., Jr., 1972, The antiquity of the Scotia Arc, *Trans. Am. Geophys. Union* **52**(2):178–179.

Driver, E. S., and Pardo, G., 1974, Seismic traverse across the Gabon continental margin, in: *The Geology of Continental Margins* (C. A. Burke and C. L. Drake, eds.), pp. 293–295, Springer-Verlag, Berlin.

Du Plessis, A., 1977, Seafloor spreading south of the Agulhas Fracture zone, *Nature* **270**:719–721.

DuToit, A. L., 1937, *Our Wandering Continents,* Oliver and Boyd, London.

Elliot, D. H., 1975, Gondwana basins of Antarctic, in: *Gondwana Geology* (K. S. W. Campbell, ed.), pp. 493–536, Australian National University Press, Canberra.

Flores, G., 1970, Suggested origin of the Mozambique Channel, *Trans. Geol. Soc. S. Afr.* **73**:1–16.

Foster, R. J., 1974, Eocene echinoids and the Drake Passage, *Nature* **249**:751.

Fox, P. J., and Heezen, B. C., 1975, Geology of the Caribbean Crust, in: *The Ocean Basins and Margins,* Vol. 3, *The Gulf of Mexico and the Caribbean* (A. E. M. Nairn and F. G. Stehli, eds.), pp. 421–466, Plenum Press, New York.

Frakes, L. A., 1966, Geologic setting of South Georgia Island, *Bull. Geol. Soc. Am.* **77**:1463–1468.

Francheteau, J., 1973, Plate tectonic model of the opening of the Atlantic Ocean south of the Azores, in: *Implications of Continental Drift to the Earth Sciences* (D. H. Tarling and S. K. Runcorn, eds.), pp. 197–202, Academic Press, London.

Francheteau, J., and Le Pichon, X., 1972, Marginal fracture zones as structural framework of continental margins in the South Atlantic Ocean, *Bull. Am. Assoc. Petrol. Geol.* **56**:991–1007.

Francisconi, O., and Kowsmann, R. O., 1976, Preliminary structural study of the South Brazilian continental margin, *An. Acad. Brasil. Ciêncas* **48**:89–100.

Gorini, M. A., and Bryan, G. M., 1976, The tectonic fabric of the equatorial Atlantic and adjoining continental margins: Gulf of Guinea to north-eastern Brazil, *An. Acad. Brasil. Ciêncas* **48**:101–120.

Gose, W. A., and Swartz, D. K., 1977, Palaeomagnetic results from Cretaceous sediments in Honduras: Tectonic implications, *Geology* **5**(8):505–508.

Hailwood, E. A., and Mitchell, J. G., 1971, Palaeomagnetic and radiometric dating results from Jurassic intrusions in South Morocco, *Geophys. J.* **24**:351–364.

Harrison, C. G. A., and Funnell, B. M., 1964, Relationship of palaeomagnetic reversals and micropalaeontology in two Late Cenozoic cores from the Pacific Ocean, *Nature* **204**:566.

Hawkes, D. D., 1962, The structure of the Scotia Arc, *Geol. Mag.* **99**:85–91.

Hays, J. D., Sato, T., Opdyke, N. D., and Burckle, L. H., 1969, Plio-Pleistocene sediments of the equatorial Pacific: Their paleomagnetic, biostratigraphic and climatic record, *Bull. Geol. Soc. Am.* **80**:1481–1514.

Helsley, C. E., and Steiner, M. D., 1969, Evidence for long intervals of normal polarity during the Cretaceous period, *Earth Planet Sci. Lett.* **5**:325–332.

Helwig, J., 1975, Tectonic evolution of the southern continental margin of North America from a Paleozoic perspective, in: *The Ocean Basins and Margins,* Vol. 3, *The Gulf of Mexico and the Caribbean* (A. E. M. Nairn and F. G. Stehli, eds.), pp. 243–255, Plenum Press, New York.

Herron, E. M., 1972, Sea-floor spreading and the Cenozoic history of the East-Central Pacific, *Bull. Geol. Soc. Am.* **83**:1671–1691.

Herron, E. M., and Hayes, D. E., 1969, A geophysical study of the Chile Ridge, *Earth Planet Sci. Lett.* **6**:77–83.

Herron, E. M., and Tucholke, B. E., 1976, Sea-floor magnetic patterns and basement structure in the southeastern Pacific, *Init. Rep. Deep Sea Drilling Project* **35**:263–278.

Herz, N., 1977, Timing of spreading of the South Atlantic: Information from Brazilian alkali rocks, *Bull. Geol. Soc. Am.* **88**:101–112.

Hey, R., 1977, Tectonic evolution of the Cocos-Nazca spreading centre, *Bull. Geol. Soc. Am.* **88**(10):1404–1420.

Hey, R., Johnson, G. L., and Lowrie, A., 1977, Recent plate motions in the Galapagos area, *Bull. Geol. Soc. Am.* **88**:1385–1403.

Hoffstetter, R., 1976, Histoire des mammifères et dérive des continents, *La Recherche* **7**(64):124–138.

Jahn, R. A., 1978, A preliminary interpretation of Weddell Sea magnetic anomalies, *Geophys. J. Roy. Astr. Soc.* **53**:164.

Julivert, M., 1970, Cover and basement tectonics in the Cordillera Oriental of Colombia, South America, and a comparison with some other folded chains, *Bull. Geol. Soc. Am.* **81**:3623–3646.

Katz, H. R., 1972, Plate tectonics and orogenic belts in the south-east Pacific, *Nature* **237**:331–332.

Keast, A., 1971, Continental drift and the evolution of the biota of the southern continents. Evolution, mammals, and southern continents, *Q. Rev. Biol.* **46**:335–378.

Kent, P. E., 1976, Major synchronous events in continental shelves, *Tectonophysics* **36**:87–91.

King, L. C., 1958, Basic palaeogeography of Gondwanaland during the Late Palaeozoic and Mesozoic Eras, *Quart. J. Geol. Soc. London* **14**:47–78.

King, P. B., 1975, The Ouachita and Appalachian orogenic belts, in: *The Ocean Basins the Margins,* Vol. 3, *The Gulf of Mexico and the Caribbean* (A. E. M. Nairn and F. G. Stehli, eds.), pp. 201–241, Plenum Press, New York.

Kumar, N., Bryan, G., Gorini, M., and Carvolho, J., 1976, Evolution of the continental margin off northern Brazil: Sedimentation and carbon potential, *An. Acad. Brasil. Ciências* **48**:131–145.

Ladd, J. W., 1976, Relative motion of South America with respect to North America and Caribbean tectonics, *Bull. Geol. Soc. Am.* **87**:969–976.

Larson, R. L., and Hilde, T. W. C., 1975, A revised time scale of magnetic reversals for the Early Cretaceous and Late Jurassic, *J. Geophys. Res.* **80**(17):2586–2594.

Larson, R. L., and Ladd, J. W., 1973, Evidence for the opening of the South Atlantic in the Early Cretaceous, *Nature* **246**:209–212.

Le Pichon, X., Sibuet, J.-C., and Francheteau, J., 1977, The fit of the oceans around the North Atlantic Ocean, *Tectonophysics* **38**:169–209.

Malfait, B. T., and Dinkelman, M. G., 1972, Circum-Caribbean tectonics and igneous activity and the evolution of the Caribbean plate, *Bull. Geol. Soc. Am.* **83**:251–272.

Martin, R. G., and Case, J. E., 1975, Geophysical studies in the Gulf of Mexico, in: *The Ocean Basins and Margins,* Vol. 3, *The Gulf of Mexico and the Caribbean* (A. E. M. Nairn and F. G. Stehli, eds.), pp. 65–106, Plenum Press, New York.

Mascle, J., and Renard, V., 1976, The Marginal São Paulo Plateau: Comparison with the southern Angolan margin, *An. Acad. Brasil. Ciências* **48**:179–190.

McCoy, F. W., and Zimmerman, H. B., 1977, A History of sediment lithofacies in the South Atlantic Ocean, *Init. Rep. Deep Sea Drilling Project* **39**:1047–1079.

McDonald, K. C., and Holcombe, T. L., 1978, Inversion of magnetic anomalies and sea-floor spreading in the Cayman Trough, *Earth Planet Sci. Lett.* **40**:407–414.

McKenzie, D. P., and Sclater, J. G., 1971, The evolution of the Indian Ocean since Late Cretaceous, *Geophys. J. Roy. Astr. Soc.* **24**:437–528.

Muir, M. D., and Grant, P. R., 1976, Micropalaeontological evidence from the Onverwacht Group, South Africa, in: *The Early History of the Earth* (B. F. Windley, ed.), pp. 595–604, Wiley, London.

Mullins, H. T., and Lynts, G. W., 1977, Origin of the northwestern Bahama Platform: Review and reinterpretation, *Bull. Geol. Soc. Am.* **88**:1447–1461.

Nagy, R. M., Ghuma, M. A., and Roger's, J. J. W., 1976, A crustal structure and lineament in North Africa, *Tectonophysics* **31**:T67–72.

Noble, D. C., and McKee, E. H., 1977, Spatial distribution of earthquakes and subduction of the Nazca plate beneath South America: Comment, *Geology* **5**:576–577.

Pardo, G., 1975, Geology of Cuba, in: *The Ocean Basins and Margins,* Vol. 3, *The Gulf of Mexico and the Caribbean* (A. E. M. Nairn and F. G. Stehli, eds.), pp. 553–615, Plenum Press, New York.

Perfit, M. R., and Heezen, B. C., 1978, The geology and evolution of the Cayman Trench, *Bull. Geol. Soc. Am.* **89**:1155–1174.

Phillips, J. D., and Forsyth, D., 1972, Plate tectonics, paleomagnetism, and the opening of the Atlantic, *Bull. Geol. Soc. Am.* **83**:1579–1600.

Pinet, P. R., 1972, Diapirlike features offshore Honduras: Implications regarding tectonic evolution of Cayman Trough and Central America, *Bull. Geol. Soc. Am.* **83**:1911–1922.

Pitman, W. C., and Talwani, M., 1972, Sea-floor spreading in the North Atlantic, *Bull. Geol. Soc. Am.* **82**:619–649.

Pitman, W. C., Talwani, M., and Heirtzler, J. R., 1971, Age of the North Atlantic from magnetic anomalies, *Earth Planet Sci. Lett.* **11**:195–200.

Porter, J. W., 1972, Ecology and species diversity of coral reefs on opposite sides of the Isthmus of Panama, *Bull. Biol. Soc. Wash.* **2**:89–116.

Rabinowitz, P. D., Cande, S. C., and La Brecque, J. L., 1976, The Falkland Escarpment and Agulhas Fracture Zone: The boundary between oceanic and continental basement at conjugate continental margins, *An. Acad. Brasil. Ciências* **48**:241–52.

Ramos, E. L., 1975, Geological summary of the Yucatan Peninsula, in: *The Ocean Basins and Margins,* Vol. 3, *The Gulf of Mexico and the Caribbean* (A. E. M. Nairn and F. G. Stehli, eds.), pp. 257–282, Plenum Press, New York.

Reyment, R. A., 1969, Ammonite biostratigraphy, continental drift and oscillatory transgressions, *Nature* **224**:137–140.

Reyment, R. A., Bengtson, P., and Tait, E. A., 1976, Cretaceous transgressions in Nigeria and Seripe-Alagoas (Brazil). *An. Acad. Brasil. Ciências* **48**:253–264.

Sclater, J. G., and McKenzie, D. P., 1973, Paleobathymetry of the South Atlantic, *Bull. Geol. Soc. Am.* **84**:3203–3216.

Scrutton, R. A., 1973, Structure and evolution of the sea floor south of South Africa, *Earth Planet Sci. Lett.* **19**:250–256.

Seisser, W. G., Scrutton, R. A., and Simpson, E. S. W., 1974, Atlantic and Indian Ocean margins of Southern Africa, in: *The Geology of Continental Margins* (C. A. Burke and C. L. Drake, eds.), pp. 641–654, Springer-Verlag, Berlin.

Shagam, R., 1975, The northern termination of the Andes, in: *The Ocean Basins and Margins,* Vol. 3, *The Gulf of Mexico and the Caribbean* (A. E. M. Nairn and F. G. Stehli, eds.), pp. 325–420, Plenum Press, New York.

Sheridan, R. E., 1974, Atlantic continental margin of North America, in: *The Geology of Continental Margins* (C. A. Burke and C. L. Drake, eds.), pp. 291–407, Springer-Verlag, Berlin.

Siedner, G., and Miller, J. A., 1968, K-Ar age determinations on basaltic rocks from South West Africa and their bearing on continental drift, *Earth Planet Sci. Lett.* **3**:451–458.

Smith, A. G., and Hallam, A., 1970, The fit of the southern continents, *Nature* **225**:139–144.

Suarez, M., 1976, Plate tectonic model for southern Antarctic Peninsula and its relation to southern Andes, *Geology* **4**:211–214.

Suess, E., 1885, *Das Antilitz der Ede,* G. Frenlag, Leipizig.

Tarling, D. H., 1971*a*, *Principles and Applications of Palaeomagnetism,* London, Chapman and Hall.

Tarling, D. H., 1971*b*, Gondwanaland, palaeomagnetism and continental drift, *Nature* **229**:17–21.

Tarling, D. H., 1972, Another Gondwanaland, *Nature* **283**:92–93.

Tarling, D. H., 1978*a*, Plate tectonics: Present and past, in: *Evolution of the Earth's Crust* (D. H. Tarling, ed.), pp. 361–408, Academic Press, London.

Tarling, D. H., 1978*b*, The geological framework of ice ages, in: *Climatic Change* (J. Gribben, ed.), pp. 3–24, Cambridge University Press, Cambridge.

Tarling, D. H., 1980, *Continental Drift and Biological Evolution,* p. 32, Oxford University Press, Oxford.

Tarling, D. H., and Mitchell, J. G., 1976, A revised Cenozoic polarity scale, *Geology* **4**:133–336.

Tarling, D. H., and Tarling, M. P., 1975, *Continental Drift,* 2nd Edn., Doubleday, New York.

Tarney, J., Dalziel, I. W. D., and de Wit, M. J., 1976, Marginal basin "Rocas Verdes" Complex from S. Chile: A model for Archaean Greenstone Belt Formation, in: *The Early History of the Earth* (B. F. Windley, ed.), pp. 131–146, Wiley, London.

Taylor, F. B., 1910, Bearing of Tertiary mountain belts on the Earth's plan, *Bull. Geol. Soc. Am.* **21**:179–226.

Tedford, R. H., 1974, Marsupials and the new paleogeography, in: *Palaeogeographic Provinces and Provinciality* (C. A. Ross, ed.), *Soc. Econ. Paleontol. Mineral. Spec. Publ.* **No. 21**:109–126.

Tomblin, J. F., 1975, The Lesser Antilles and Aves Ridge, in: *The Ocean Basins and Margins,* Vol. 3, *The Gulf of Mexico and the Caribbean* (A. E. M. Nairn and F. G. Stehli, eds.), pp. 467–500, Plenum Press, New York.

Torquato, J. R., 1976, Geotectonic correlations between S. E. Brazil and S. W. Africa, *An. Acad. Brasil. Ciêncas* **48**:353–364.

Tucholke, B. E., and Carpenter, G. B., 1977, Sediment distribution and Cenozoic sedimentation patterns on the Agulhas Plateau, *Bull. Geol. Soc. Am.* **88**:1337–1346.

Uchupi, E., 1975, Physiography of the Gulf of Mexico and Caribbean Sea, in: *The Ocean Basins and Margins,* Vol. 3, *The Gulf of Mexico and the Caribbean* (A. E. M. Nairn and F. G. Stehli, eds.), pp. 1–64, Plenum Press, New York.

Urien, C. M., Martins, L. R., and Zambrano, J., 1976, The geology and tectonic framework of southern Brazil, Uruguay and northern Argentina continental margin: Their behaviour during the southern Atlantic opening, *An. Acad. Brasil. Ciêncas* **48**:365–376.

Van Andel, T., Thiede, J., Sclater, J. G., and Hay, W. W., 1977, Depositional history of the South Atlantic Ocean during the last 125 million years, *J. Geol.* **85**:651–598.

Wegener, A., 1914, *Die Entstehung der Kontinente und Ozeane,* 94 pp., Braunschweig, Berlin.

Whitmore, F. C., Jr., and Stewart, R. H., 1965, Miocene mammals and Central American seaways, *Science* **148**:180–185.

Wilhelm, O., and Ewing, M., 1972, Geology and history of the Gulf of Mexico, *Bull. Geol. Soc. Am.* **83**:575–600.

Early History and Biogeography of South America's Extinct Land Mammals

2

M. C. McKENNA

Introduction

South America's peculiar extinct mammalian fauna has been a source of fascination since the late 1700's when a Pleistocene skeleton of the giant ground sloth *Megatherium,* later described by Cuvier (1796, 1812), was unearthed and sent to Spain by the Dominican Manuel Torres. Strange new Pleistocene discoveries continued to be made throughout the 19th Century (see, for instance, Darwin, 1839; Lund, 1841; Owen, 1842). Toward the end of the 1800's the Tertiary faunal history of South America began to be documented, notably by the famous Ameghino brothers. The history of 19th- and 20th-Century vertebrate paleontology in South America, however, has been the subject of excellent summaries elsewhere (e.g., Simpson, 1940, 1948, 1967, 1978; Marshall *et al.,* in press *a,b*) and will only be mentioned briefly where appropriate here. The purpose of this study is to provide a critique of theories of the origin, rather than the full history, of the early mammalian fauna of a southern land mass that has changed its geological affinities profoundly since the Jurassic. I shall not be much concerned with the effects of

M. C. McKENNA • Department of Vertebrate Paleontology, American Museum of Natural History, New York, New York 10024.

reconnection of South with Central America during the Pliocene. For discussions of the resulting "Great American Interchange" at the end of the Cenozoic, one is referred to Patterson and Pascual (1968*a*), Webb (1976, 1978*a,b*), Marshall and Hecht (1978), Marshall (1979), and Marshall *et al.* (1979).

Many hypotheses about the origin of South America's extinct mammalian faunal strata have been published. Generally, as time has passed, the history of South American mammalian faunas has been admitted to be more and more complex. Almost all students who have dealt with the subject have prefaced their works with remarks about how little is known at the time of writing and how much remains to be found out concerning the extinct South American fauna. As always, new discoveries and new analysis of previous hypotheses are needed to elucidate this problem.

In the first decade of the 20th Century two different and competing hypotheses appeared, authored, respectively, by Florentino Ameghino (1906) and Albert Gaudry (1902, 1904, 1906*a,b,* 1908, 1909). Both hypotheses drew upon newly collected Cenozoic faunas from Argentina. Ameghino's was based on the assumption that Argentina was a sort of Garden of Eden, whose early Cenozoic deposits he supposed to be Cretaceous, the fossils from these deposits representing ancestors of groups that later spread elsewhere. Names coined by Ameghino for some of the animals occurring in these deposits, such as *Archaeohyrax* and *Notopithecus*, signal his unwavering dedication to this view, which even included the South American origin of our own species. Gaudry, on the other hand, originated the island continent hypothesis, which holds that South America's fauna was isolated for most of the Cenozoic and had nothing to do during that time with the origin of other groups elsewhere.

Ever since Matthew's (1915) influential book, *Climate and Evolution,* various northern authors have supported, with nearly Ameghinian fervor, a northern Garden of Eden, generally believed to be located somewhere in Holarctica or, more specifically, in North America. The paleontological record of the Paleocene and late Cretaceous is well known there relative to the other continents, and a present-day bridge exists to South America which has been presumed to be a reactivation of a former, much earlier connection at the same site. From the north various kinds of organisms are hypothesized to have spread southward to populate an essentially empty southern group of discrete land masses in a number of distinct invasions. Hershkovitz (1969, 1972) has aptly termed this the "Sherwin-Williams effect." The idea has rested, in part, on a fixed world in which continental drift and other features of plate tectonics played no role.

The ideas originally distilled by Matthew have been developed eloquently in a series of studies by George Gaylord Simpson. Simpson's early ideas on the fixity of continents have now changed, as have those of most scientists formerly in opposition to mobilist notions, and he has thus made various adjustments in Matthew's scheme. Simpson's latest summary of the subject (1978) contains much valuable historical information relating to the development of the present widely (but not universally) accepted system of hypotheses which he and others have constructed and elaborated over the years to explain the origin and development of the peculiar South American mammalian fauna, especially during the earlier parts of the Cenozoic.

Why write another brief summary? Aside from the need for a "backdrop" chapter in this book, I think it important to examine, even if only initially, some of the concepts put forth by authors such as Hoffstetter, Lavocat, and Hershkovitz regarding early South American mammalian faunas, and to re-examine some of the assumptions of the Matthew hypothesis insofar as it deals with South America. It seems to me that perhaps the full impact of new ideas in both geology (plate tectonics) and biology (cladistic analysis, vicariance biogeography) on Matthew's ideas has not yet occurred. For detailed biostratigraphic information and radioisotopic dating, interested readers should consult Marshall *et al.* (1977) and Marshall *et al.* (in press *b*).

Thumbnail Sketch of South America's Tertiary Mammalian Faunas

Southern South America

From the very beginning, Argentina has played the major role in supplying fossils and, through its paleontologists, the interpretation of much of what is known of the South American Cenozoic mammalian fauna. At first, investigations centered on the rich Pleistocene deposits in the vicinity of Buenos Aires and various Tertiary deposits around the country also became known. In the summer of 1894–1895 Carlos Ameghino discovered the Casamayor (*Notostylops*) fauna of Patagonia, presently at about 46° S. His brother Florentino first published on the fauna in 1897 (Ameghino, 1897). By 1899, Carlos had determined that the Casamayor fauna was the oldest then known in the country and had found most of the other early Cenozoic faunas of the area as well. After 1903, Carlos ceased collecting Argentina's early mammal faunas, leaving the field to others while he concentrated on younger assemblages. Since the beginning of the 20th Century, expeditions led by some of the most famous names in paleontology have filled in the Cenozoic history of terrestrial vertebrates in Argentina, much of the information coming from Patagonia at South America's narrow southern end. Patagonia, however, is now, and was probably also in the past, not fully provided with the entire diversity of South America's fauna.

The three earliest fossil Mammal Ages of Argentina—Riochican, Casamayoran, and Mustersan—occupy parts of the interval from about late Paleocene through middle Eocene (*sensu lato*), but thus far no reliable radioisotope dates have allowed them to be calibrated accurately in relation to the rest of the world. These early chapters at the base of the Argentine Cenozoic land-mammal-bearing sequence are poorly known (the Casamayoran better than the others), but they are very important in that they occur in sequence and are preceded by the late Danian marine Salamanca Formation (Bertels, 1975). Biological and presumably temporal gaps separate them, both among themselves and from the faunas of the overlying Deseadan Mammal Age. On a slim biostratigraphic basis the Riochican is generally believed to be close in age to the Paleocene/Eocene boundary, approximately correlated with the late Tiffanian and Clarkforkian Mammal Ages of North America [see Berggren *et al.* (1978) for radioisotopic calibration of the Eocene]. Exact correlation is not possible at present and will not be possible until the Riochican produces a radioisotopic date, a paleomagnetic correlation, or better biostratigraphic evidence. Attempts by L. G. Marshall and his colleagues to accomplish all three aims are in progress, but so far only the last 35 million years of the South American Cenozoic have produced reliable radioisotopic dates (Marshall *et al.*, 1977; Marshall *et al.*, in press *b*).

Above the Mustersan and below the Deseadan is a paleontological hiatus, which, farther north in Mendoza, may be filled, in part, by the Divisadero Largo fauna of probable latest Eocene age. The Divisadero Largo, described by Simpson, Minoprio, and Patterson (1962), and amplified by several short papers since then, is notable for some primitive retentions possibly having something to do with its more northerly position compared to that of Patagonia.

The nonmarine Oligocene of Argentina is represented by two recognized faunal levels, the Deseadan Mammal Age below and the Colhuehuapian Mammal Age above. Above the last marine deposits of the Oligocene Monte Léon Formation (Bertels, 1970, 1975; Camacho, 1974), left by the incursion of epicontinental seas in the Austral marine basin, the earlier Miocene is well represented in southern Patagonia by the Santa Cruz Formation and its fauna, occurring as far south as about 52° S. Later continental Miocene also occurs in the eastern Patagonian Andes, but that part of the Miocene is much better known from Colombia at the tropical end of the continent. Finally, Pliocene and Pleistocene faunas are well documented from many sites in central and northern Argentina.

Very recently, Casamayoran and possibly significantly earlier marsupials and notoungulates have been found in northwestern Argentina near the Bolivian border, but these have not yet been fully described. Their importance is great in that they occur in rocks whose current position is at about 25° S.; they are thus potentially a better source of information about the total diversity of the South American fauna than are the more peripheral Patagonian localities farther south.

Argentina, therefore, produces important information about Cenozoic mammals from approximately late Paleocene through Recent. The earliest levels within the Riochican at the beginning of the record are sparingly fossiliferous, producing primarily the xenungulate genus *Carodnia* at a single locality. Below the Riochican in Argentina, no Cretaceous or Jurassic mammals are known except for some tantalizing Jurassic footprints found some years ago in Patagonia by Casamiquela (1961).

Brief reference to the very poorly known Cenozoic mammals of Chile and the somewhat better known faunas of Uruguay is appropriate here. Two specimens, referable to *Astrapotherium* and *Nematherium*, have been found at about 52° S. in an extension of the Santa Cruz Formation close to the Argentine border (Simpson, 1943). The type Friasian (middle Miocene) also occurs in Chile (Marshall *et al.*, in press *b*). The Cenozoic faunas of Uruguay have been reviewed recently by Francis (1975). The middle to late Cenozoic, possibly ranging from Deseadan to Recent, is exposed in Uruguay.

Brazil

Brazil played an important early role in revealing the peculiar fauna of the South American Pleistocene, especially at the caves of Lagoa Santa (see Lund, 1841; Paula Couto, 1956*b*). But surely the most important single fauna known from the Cenozoic of Brazil is that from infillings in the limestone quarry at São José de Itaboraí in the State of Rio de Janeiro at a latitude of approximately 24° S. The geology of the site has been discussed extensively by Francisco and de Souza Cunha (1978), who presented a useful bibliography. The Itaboraí fauna produces Riochican vertebrates of many kinds (Paula Couto, 1952*a,b,* 1958), including a pipid and a caecilian with African affinities (Estes and Wake, 1972; Estes, 1975). Numerous specimens of fairly well-preserved mammal bones and teeth occur there as well, deposited in a marl filling former caverns in an earlier Paleocene fresh-water limestone. Hyopsodont-like ungulates, macraucheniid and proterotheriid litopterns, a uintathere-like xenungulate and a trigonostylopoid, three kinds of primitive notoungulates (*Colbertia, Camargomendesia, Itaboraitherium*), and 14 described species of marsupials are present (Paula Couto, 1970*a,b,* 1978*a,b*). The marsupials are primarily didelphoids, but there are also caenolestoid-like forms (*Derorhynchus*), polydolopoids (*Epidolops*), and early borhyaenoids (*Patene*). The ecological situation at Itaboraí is different from that of the known Patagonian Riochican, 22° farther south.

Cenozoic sediments in western Brazil near the Peruvian border have yielded the astrapothere *Synastropotherium,* thought to be Oligocene in age by its describer (Paula Couto, 1977).

Bolivia

The Pleistocene at Tarija in southern Bolivia has long been known (Boule and Thevenin, 1920), but in recent years the Tertiary has produced a number of new localities (Hoffstetter, 1970*g*, 1977; Hoffstetter *et al.*, 1971, 1972; Marshall *et al.*, in press *b*).

The most significant Bolivian fauna, and certainly one of the most intriguing in South America, is Deseadan in age. Discovered by G. Bejarano in 1962, it occurs in deposits of the Salla-Luribay Basin near Sicasica (Hoffstetter, 1968). Abundant and well-preserved specimens of *Pyrotherium, Parastrapotherium,* at least four genera of marsupials, edentates including orophodonts, megalonychids, glyptodonts, and armadillos, a series of rodents similar to phiomyoids, various kinds of notoungulates and litopterns, and birds, turtles, and frogs have been mentioned in the literature. One of the most interesting members of this fauna, however, is the primate *Branisella* (Hoffstetter, 1969).

Although no mammals have been found there yet, the extensive Puca Group of the Bolivian Andes should be mentioned here as an attractive target for future exploration for Cretaceous mammals. The El Molino Formation of the Puca Group would be a good place to begin (Sigé, 1972).

Peru

Although mammalian footprints have been found in the Jurassic of Patagonia (Casamiquela, 1961), they provide almost no information of value. The earliest actual mammal remains from South America are found in Peru at Laguna Umayo, near Lake Titicaca (Grambast *et al.,* 1967; Sigé, 1972; Portugal, 1974). This mammalian fauna has been dated as Late Cretaceous in age on the basis of dinosaur egg shell fragments and charophytes. The mammals are represented mostly by teeth obtained by screen-washing two tons of sediments of a unit correlated with Newell's Vilquechico Formation (1949). Two major taxonomic divisions of the Mammalia are represented: marsupials and a primitive ungulate. The marsupials are similar to the otherwise North American genera *Alphadon* and *Pediomys*, whereas the ungulate is too poorly known to discuss in much detail.

An Eocene pyrothere, *Griphodon,* was described by Anthony in 1924 from eastern Peru. Anthony thought he was describing a new perissodactyl, but the true affinities and precise locality of the fossil were later made known by Patterson (1942, 1977). No additional remains have been found at the site other than a few bone scraps.

A few late Cenozoic localities have turned up in recent years, including some Miocene marine beds with pinnipeds, whales, and a sloth from the west coast and a scattering of late Tertiary and Pleistocene sites east of the Andes (Hoffstetter, 1970*c*).

Ecuador

Hoffstetter (1970*b*) has summarized the Cenozoic vertebrate faunas of Ecuador. Almost the entire record is restricted to the Pleistocene, with only two known sites from the late Tertiary showing much promise. These are near Nabón and Azogues (Repetto, 1977) in the southern part of the country, where the Tertiary is apparently less covered by Quaternary volcanics. More than fifty years ago, Anthony (1922) named a supposedly new genus of rodents, *Drytomomys*. The specimen on which the type species was based was reported to have come from the Pleistocene out of a cave in Tertiary volcaniclastics near Nabón, but *Drytomomys* has since proven to be merely a synonym of *Olenopsis*, known from the Miocene elsewhere in South America (Fields, 1957). It is probable that Anthony's specimen came from the volcaniclastics themselves rather than the cave fill.

Colombia

Colombia has provided a record of fossil mammals ranging from Eocene to Recent, although the Pleistocene and middle Miocene are by far the best-represented parts of the Cenozoic. Most of the sites occur in the Magdalena Valley, (Wellman, 1970; Van Houten, 1976). Unexplored early Cenozoic continental deposits occur in eastern Colombia.

The earliest Colombian fossil mammal is apparently from the Eocene Mugrosa Formation near Tama, Department of Santander (Stirton, 1953), but this consists of a single astrapothere tooth close to the genus *Astraponotus*. Also of Eocene age, apparently late Eocene, is a well-preserved primitive pyrothere palate (Hoffstetter, 1970*d*) from the lower part of the Guandalay Group (or Formation) near Guandalay in the Department of Tolima.

Higher in the Guandalay Group (Tuné Formation), the Chaparral fauna occurs in the Department of Tolima. Its age appears to be Deseadan on the basis of a few specimens of sloths, litopterns, notoungulates, and astrapotheres. In addition, Stirton (1947*b*) described a genus of low-crowned but somewhat lophodont ungulates, *Lophiodolodus*, that is possibly, but not necessarily, an early sirenian (McKenna, 1956). Another apparently Deseadan locality occurs east of the Andes in the drainage of the Río Caquetá just north of the border with northern Ecuador and Peru (Stirton, 1953; Hoffstetter, 1970*a*). The site has potential, but only one rodent premolar referred to *Eosteiromys* has been found so far.

Late Oligocene or possibly early Miocene mammals are known from the Coyaima fauna from the Honda Group, Department of Tolima. Notable for the presence of true gavials, the fauna also produces rodents, litopterns, astrapotheres, and three families of notoungulates (Stirton, 1953).

The best-known Tertiary mammal fauna in Colombia is the La Venta/Carmen de Apicalá fauna of the Honda Group, Departments of Huila and Tolima (Hirschfeld and Marshall, 1976). Most vertebrate classes are represented in this fauna, whose age is apparently Friasian and therefore approximately middle Miocene. The La Venta is a very rich fauna and is especially important because of its tropical diversity and nearness to the present-day site of a land connection to North America. No North American contamination by terrestrial forms is evident, nor is any South American form known from the approximately contemporary Cucaracha Formation of Panama, which was then on the other side of the "Bolívar geosyncline." Among the La Venta mammals thus far known are bats, marsupials, ceboid primates, various kinds of edentates, a mix of caviomorph rodents, a sirenian, various notoungulates, two astrapotheres, and most interestingly, three primitive holdovers from earlier times represented by a didolodont, a primitive notoungulate, and a polymorphine proterotheriid litoptern (McKenna, 1956; Hoffstetter, 1970*a*).

A few other Miocene sites, fewer Pliocene, but a good many Pleistocene ones occur as well in Colombia, mostly in the Magdalena Valley.

Unfortunately, with one faintly possible exception, no continental Tertiary faunas have been located in the Cordillera Occidental or from the Pacific coast of Colombia. Western Colombia and Ecuador were the site of the "Bolívar geosyncline" and some possible continental accretion, so it is important to scour that area for Tertiary vertebrates. The only possible site now known, alluded to above, lies at the headwaters of the southeast fork of the Río Patria about 150 km from the Pacific near the Ecuadorian border. Stirton (1947*a*) described a rodent and a peccary from there, either Pliocene or Pleistocene. This is probably too young a locality to be really interesting; a Miocene fauna from the Colombian Pacific coast, permitting assessment of some of Hershkovitz's biogeographic hypotheses on a more factual basis than is possible at present, would be fascinating.

Venezuela

The Venezuelan Pleistocene fauna is fairly well known but Venezuelan Tertiary mammals are not, even though a few late Tertiary sites have been found (see Marshall *et al.,* in press *b,* for summary). The only early Cenozoic mammal specimen so far described (Patterson, 1977) is a low-crowned and somewhat bilophodont early pyrothere dentition, approximately of early Eocene age, from the thick Trujillo Formation in the State of Lara.

Isolation of South America

Ocean-floor spreading in the central Atlantic between North America and Africa evidently began at about 180 ma (million years before the present), but the beginning of the active separation of South America from Africa did not occur until about 127 ma (Rabinowitz, 1976), at which time South America was far to the southeast of its former position relative to North America. This was due to South America remaining firmly attached to Africa for more than 50 million years while the latter separated from North America (Pitman and

Talwani, 1972; Ladd, 1974). From 127 ma onward both Americas moved essentially westward with respect to Europe and Africa, but they also possessed their own relative motions. As Africa and South America pulled apart, however, their coastlines in the equatorial zone separated more slowly because of transform fault zones involving large east–west dislocations nearly parallel with the coasts. Sibuet and Mascle (1978) have published updated reconstructions of various stages of separation between Brazil and West Africa. The newly formed crust between Africa and South America in the equatorial Atlantic south of the Ceará and Sierra Leone rises, extending from there all the way to the Walvis and Rio Grande rises in the central part of the South Atlantic, was the site of several deep basins. In the Angola–Brazil basin south of the Romanche Fracture Zone, at first, continental sediments, and then, a very thick accumulation of Aptian evaporites developed until approximately 106 ma (Wardlaw and Nicholls, 1972; Burke, 1975). The situation in that area was probably not unlike the Red Sea rift system and possibly also the Messinian Mediterranean. Cessation of evaporite deposition after Aptian time indicates open oceanic conditions either to the north or to the south but not necessarily both ways.

From 127 ma to the present the motion of South America with respect to both Africa and North America is known rather exactly, in the former case directly from South Atlantic magnetic anomaly data (Le Pichon and Hayes, 1971; Sibuet and Mascle, 1978) and in the latter case from vector summation of the NA-Af and Af-SA anomaly data to yield the NA-SA result (Ladd, 1976). Anomaly data in the Caribbean area itself would be desirable, but unfortunately the appropriate oceanic crust producing the anomalies is often deeply buried under sediments. Some of it has already been subducted, and in any case magnetic anomalies are of increasingly vanishing amplitude near the magnetic equator, hence the vector sum approach. It should be remembered that all vector sums of this sort are net and cover a certain period of time. Complex motions may have occurred within the intervals studied. A sequence of plate motions about differing instantaneous poles of tectonic rotation on the Earth's essentially spherical surface may be imbedded in a net result. These motions are Eulerian rotations, assuming that as a first approximation the Earth's radius has remained constant.

From about 127 ma to about 80 ma (Valanginian to the end of the Santonian in the Cretaceous, based on the youngest part of anomaly 34), net motion of South America with respect to North America was essentially eastward. At about 80 ma, however, their motion took a new turn, i.e., South America began to move relatively south-southeastward. This coincides with the time of the beginning of continuous marine faunal connection of the South Atlantic with the North Atlantic (Reyment and Tait, 1972; Reyment, 1974).

It would seem, then, that the development of South America's isolation from Africa and North America passed through a number of stages over about 50 million years before significant water gaps appeared and that the separations were complex in nature, involving various second-order phenomena that should be examined. These would include: (1) local compression imbedded in net dilation, (2) contact maintained along transform fault zones, (3) excessive production of mid-oceanic ridge crust to form transverse aseismic ridges reaching sea level and in the early stages of continental separation actually filling the gap, (4) temporarily elevated sections of fracture zones, and (5) off-ridge volcanic islands.

Stepping Stones?

From about 80 ma to about 38 ma, net motion of South America consisted of southeastward separation relative to North America, but this net motion was evidently complex (Ladd, 1976). Very likely, for a time within this interval, it contained a northward convergent component. If Ladd's dates are modified on the basis of more recent calibration (Berggren *et al.*, 1978), it would seem that at some time between about 80 ma and about 56 ma North and South America may actually have approached one another for a short period, only to separate again until that separation was later reversed. The pre-Andean orogenic phase affecting the Guajira Peninsula, the Sierra Nevada de Santa Marta, and part of the Cordillera Central of Colombia occurred at this time (Van der Hammen, 1958; Irving, 1975). I mention this early convergent episode (if it really existed) because it corresponds with a time at about the end of the Cretaceous when various organisms have been hypothesized to have dispersed one way or another between the two Americas. Presumably, continental convergence would have helped, not hindered, such dispersal by creating subduction zones, volcanic arcs, and transpression and elevation of collision zones, but details of these plausible events are lacking or at least currently unidentified. From about 38 ma to the present, relative northward motion of South America has again prevailed.

It must also be remembered that the two Americas were plowing westward as well over the decoupled Farallon and Phoenix plates, essentially leaving behind the Caribbean between them as they went. However, in spite of valiant attempts by numerous highly qualified geologists, complete understanding of this movement is still vague. Furthermore, the time scale of all these events must be increased by about 2.4–2.7% if the IUGS Subcommission on Geochronology's newly recommended constants are employed (Steiger and Jäger, 1977; Mankinen and Dalrymple, 1979).

As continents pull apart they may remain connected or nearly in contact for a time. The most obvious situation is a transform fault zone where continental margins are at essentially right angles to spreading centers. In the earliest stages of separation the equatorially located Romanche Fracture Zone evidently acted in this manner between Africa and South America, but later its role was evidently more complex. Similarly, far to the south, the Falkland Plateau, an eastward extension of South America at approximately a right angle to the mid-Atlantic Ridge, extended past the southern tip of Africa in its predrift position and would have taken a certain amount of time to slide by Africa before the southern South Atlantic opened completely. Nevertheless, this land continuity was probably broken very early, during or before the middle Jurassic. The eastern end of the Falkland Plateau rests on metasedimentary gneiss and granite (JOIDES site 330; Scientific Staff, 1974) above which is nothing but initially shallow and thereafter deep-water marine sediment of Jurassic, Cretaceous, and Cenozoic age. Cretaceous and Cenozoic sediments west of site 330 are likewise deep-water marine. The Falkland Plateau, therefore, is not a candidate for a land bridge or series of island stepping stones between Africa and South America, even though it supplies a missing chunk of Gondwanaland (Paula Couto, 1974).

Several other second-order phenomena may have been important as producers of possible islands: (1) excessive production of mid-Atlantic Ridge lavas to form transverse aseismic ridges reaching sea level, (2) temporarily elevated sections of fracture zones, and (3) off-ridge volcanic islands. The first of these three phenomena is exemplified by the Walvis and Rio Grande aseismic ridges. As the South Atlantic widened initially from a rift to an ocean, its northern end, the Angola–Brazil basin, was apparently either partially or wholly blocked from connection with the Cape–Argentine basin to the south by an Iceland-like lava pile, north of which extensive evaporites up to 2 km thick built up. Was any part of this lava pile above sea level later on, as a sort of southern version of Iceland? Yes, probably, because after salt deposition ceased, the Angola–Brazil basin nevertheless remained euxinic, whereas the Cape–Argentine basin developed oxygenated bottom water after late Aptian time. Euxinic conditions occurred again in the Angola–Brazil basin from Albian to Santonian time while open ocean existed to the south, indicating that the Walvis–Rio Grande ridge during that time was still a substantial barrier (Smith, 1977; Kumar and Gamboa, 1979; Le Pichon *et al.*, 1978). Similar aseismic ridges, the Ceará and Sierra Leone, existed north of the equator at about 80 ma (Kumar and Embley, 1977). At least some part of each of these aseismic ridges subsided beneath sea level prior to the end of Campanian time in the Cretaceous, although they remained close enough to sea level to affect bottom water circulation for a long time thereafter.

Temporarily elevated blocks within fracture zones also have occurred, presumably as the result of jostling as instantaneous poles of tectonic rotation shifted. An example of a Mio-Pliocene sea-level carbonate reef that temporarily may have been partly above water in the Romanche Fracture Zone at 0° N., 17° W. was discussed by Bonatti *et al.* (1977). This tectonic block, midway between Africa and South America, has now subsided once more to a depth of more than 950 m. How often this sort of thing has occurred in the past is unknown, but there are a number of other major fracture zones between Africa and South America in addition to the Romanche. For instance, the summit of the Vema Ridge, latitude 10° 41′ N., longitude 44° 18′ W., appears to have been subaerial at some time in the past (Bonatti and Honnorez, 1971).

There is no present way to know what, if anything, colonized or dispersed from such islands, but the possibility that many more examples of various ages will be discovered as exploration progresses is sobering. Why not raised blocks in the early Cenozoic, when the Atlantic was narrower?

A third second-order phenomenon, off-axis volcanoes, produced a few islands available as potential stepping stones in the South Atlantic. For instance, after the Walvis and Rio Grande rises split apart, there were nevertheless a few volcanic outbursts off the western end of the Walvis Ridge, between there and the mid-Atlantic Ridge. The tops of these volcanoes, now flat-topped guyots, were at sea level as late as the early Tertiary, as indicated by shallow-water late Paleocene and mid-Eocene foraminifera recovered there (Connary, 1972).

All of these second-order phenomena may have helped to provide stepping stones in the Atlantic for certain terrestrial organisms' dispersal, but there is no proof that any of them actually did so at any time. I do not conclude that they necessarily played a part in primate or rodent dispersal during the Cenozoic. Nevertheless, it is important to realize that the first-order tectonic phenomena shown on most paleogeographic maps do not tell the whole story.

Major South American Taxa

I shall begin this section with a brief review of what mammals *are not* in South America.

Multituberculata

Multituberculates are an order of extinct mammals sharing certain derived features with the living Australian monotremes: (1) jugal greatly reduced, (2) orbitonasal foramen present, (3) alisphenoid bone reduced and contact with parietal lost, and (4) anterior lamina of petrosal contacting orbitosphenoid if the latter is ossified. Very common from the Jurassic until the early Eocene but lasting until about 37 ma in parts of Holarctica, the multituberculates are nevertheless unknown from any southern continent. If they were ever present in South America, they would doubtless have occupied the niche held by the polydolopoid marsupials in the earliest known Cenozoic faunas.

Eometatherian Marsupials

Although South America has a number of didelphoids and a few caenolestoids living today, to which may be added the borhyaenoids and polydolopoids of its extensive fossil record and also some questionable groups such as the argyrolagoids and groeberioids (see below), nevertheless, Australia is far richer in the diversity of its major marsupial taxa. Moreover, none of the major Australian taxa, including all the syndactylous families, is present in South America at any point in the known fossil record of that continent, from the Cretaceous to the Recent. The Tasmanian wolf, *Thylacinus*, once widespread from Tasmania to New Guinea, is thought by some to be derived from early borhyaenoid or didelphoid stock, which, if true, would make either Australian or South American marsupials polyphyletic (Kirsch, 1977). Occasionally, someone suggests other special relationships of South American with Australian groups, but at present it seems logical to regard the Australian marsupials as a natural group on the basis of the development of an epitympanic sinus and loss of a lower incisor (Eometatheria—Simpson, 1970*a*, p. 38). Under this scheme South American marsupials and marsupial-like forms in which anterior teeth are enlarged are considered convergent with the Australian diprotodonts in this one feature. Both the biological and the plate tectonic evidence points to early separation of the Australian marsupials, either in the earliest Cenozoic or, more likely, at some point in the Cretaceous. Until more details of the structural history of the Scotia Sea, West Antarctica, etc., are forthcoming, it is not possible to say whether this was a simple vicariant event or involved filtered dispersal.

Many Eutherians

Most of the major kinds of eutherian mammals were absent from South America's early Cenozoic faunas. When eutherians first appear in South American faunas they are usually advanced, not primitive, members of their groups, and they almost all occur late in the Cenozoic record. Thus, advanced insectivorans (*Cryptotis*), rodents, carnivorans, artiodactyls, perissodactyls, and proboscideans appear abruptly in the late Cenozoic record. In keeping with their advanced biological condition they do not occur early in South America, before their appearance elsewhere. Primitive eutherians, with the exception of edentates, are totally absent from known South American Paleocene and Eocene faunas, whether in Patagonia or in the tropics. Early true insectivorans (Lipotyphla), dermopterans, tupaiids, leptictids, mixodectids, macroscelidoids, lagomorphs, primitive carnivores, pantolestids, didelphodonts, pantodonts, apatotheres, taeniodonts, paramyid and sciuravid rodents, the whole range of plesiadapoid primates, even many kinds of primitive ungulates—all are missing from, and probably never were present in, the South American early Cenozoic record. Reasons could include the following: (1) they never reached South America because of some barrier, the most obvious kind being an ocean; (2) if they reached the Promised Land they turned into something else by the time that the South American record begins; (3) they arrived but were outcompeted by similarly adapted earlier arrivals; or (4) they are there in the record somewhere but for unknown reasons no trace of them has yet been found. The first hypothesis seems most likely. Solution two is partly a semantic problem peculiar to non-cladistic thinking and partly an artifact of the outmoded ideas that the "Ferungulata" are a natural group and that South American ungulates are descended from early "ferungulate" immigrants (McKenna, 1975). Solutions three and four are related but are not testable or are not supported by any evidence. They become less and less likely the more South America is paleontologically explored. In view of the fact that most eutherian groups with a northern distribution are absent from South America until the introduction of advanced forms late in the history of those groups, it is most sensible to hypothesize that South America was well isolated at the time most major northern groups differentiated, in the late Cretaceous and part, at least, of the Paleocene.

Now let us discuss what mammals *are* in South America.

Marsupials

If one constructs a cladogram of therian mammals, one is immediately struck by how few shared-derived characters are known for marsupials in comparison to the eutherians ("placentals"). Clearly, the ancestors of both were more similar to primitive marsupials than to primitive eutherians. Many of these marsupial characters are known from the soft anatomy of the living animals, such as the structure of the deltoid muscle, blastocyst arrestment, and the possession of a pseudovagina as a birth canal. These are common to both the Australian and the American living marsupials and attest to the monophyly of these living animals in which such structures can be studied. Other often-cited features of the soft anatomy are not common to all marsupials, such as the pouch, which seems to have developed several times in different ways, or the famed *fasciculus aberrans* of the brain, which seems to do essentially the same job as the differently constructed eutherian *corpus callosum* but is restricted to the Australian diprotodonts. All of these soft anatomical characters, together with karyological and molecular studies, etc., are quite helpful in working out the relationships of living marsupials but are of no use whatsoever to the perplexed paleontologist with a handful of bones or teeth to identify and fit into the system of relationships. Detailed studies of dental formulas, tooth patterns, ear regions, and foot structure come most into play, simply because these features fossilize well and are, therefore, the most available. The information that these hard parts supply is indirect regarding the definition of such terms as "marsupial" and "placental" and may simply indicate that a fossil is primitive rather than that it is a member of either the Marsupialia or the Eutheria. Older diagnoses of the Marsupialia emphasized the marsupial bones, the inflected angle of the jaw, and possession of more than three incisors and more than three molars, but these particular features are just primitive and characterized various eutherians as well until they were lost or modified (McKenna, 1975). A mode of anterior tooth replacement in marsupials differing from that in eutherians has potential value for distinguishing various fossil groups, but far greater numbers of fossils than are now available must be collected before adequate knowledge of dental replacement will be of much help in working out the relationships of fossil groups. I offer these comments as a word of caution to those who would place certain fossil groups into the marsupials on the inadequate basis of shared-*primitive* characters. Instead, one should ask what are the shared-*derived* features that ally various fossil groups with the monophyletic living Marsupialia? Does adding those groups make the Marsupialia paraphyletic? The matter is in need of further analysis.

As Hoffstetter (1970*e,f,* 1972, 1973) and others have pointed out, marsupials were distributed in a garland on the ancient Earth, an Earth whose continental configuration was different from that of the present. At the north end of this garland North America in the mid-Cretaceous was split by a north–south epicontinental sea. *Holoclemensia*, from the Albian of Texas, has been identified as a marsupial and, if this were true, would therefore be the world's oldest. But its identification as a marsupial seems to me to be based on primitive rather than derived characters. I do not know whether it is a marsupial or just a primitive therian. By the Aquilan in the Late Cretaceous didelphids, pediomyids, and stagodonts are known from North America west of this sea. Eastern North America doubtlessly had its own distinctive fauna, but except for a multituberculate-like femur and two mammal teeth (Krause and Baird, 1979), no mammals are known from the Cretaceous of America's eastern half. Cretaceous mammals of Europe are either too early (Wealden) or too poorly known (France and Spain) to permit comments about possible marsupials there, but by the early Eocene a didelphid apparently reached Europe from North America. Thereafter certain rather stereotyped didelphids made up a minor element of the European fauna until the Miocene, just as in North America (Crochet, 1978). Clearly, Europe was an outpost at the very end of Hoffstetter's garland. There is no evidence of marsupials in any of the increasingly well-known Late Cretaceous or Cenozoic deposits of Asia, despite the fact that some taxa show similarities to marsupials based on primitive characters or on convergence. Marsupials are generally believed not to have occupied Africa at any time, but this uncertain conclusion is based on their absence from the poorly known early Cenozoic record. They could have reached Africa but, if so, why didn't they occur in the Egyptian Oligocene or later in the tropical East African Miocene? Both Asia and Africa are therefore excluded from Hoffstetter's garland.

The marsupial record of the north end of Hoffstetter's garland, then, is traceable with relative certainty back only to the beginning of the Aquilan, which is not very far back in the Cretaceous. North American marsupials (and European ones when they reached there) are fairly similar to each other insofar as can now be determined, the stagodonts and glasbiine didelphids departing most widely from the primitive condition displayed by *Alphadon*. It is possible to imagine these North American marsupials as themselves representing a northern extension of a primarily southern group of marsupials (Tedford, 1974) and, possibly, protomarsupials, with a certain amount of phylogenetic branching having gone on *before* the isolation of South American from other marsupials.

Peru has produced evidence that marsupials were already a major element in the South American fauna in the latest Cretaceous, when the South American record really begins. Sigé (1972) has described two Peruvian taxa, both of which resemble North American genera (*Alphadon* and *Pediomys*).

By the late Paleocene, a swarm of primitive marsupial taxa had differentiated in South America and whatever land was connected to it, exemplified by the rich marsupial diversity of the tropical Itaboraí fauna of Brazil (Paula Couto, 1952*a,b,* 1961, 1962) and even the Río Chico fauna of Patagonia. By this time South America had few, if any, direct connections with North America. Early Cenozoic didelphoid stock in South America split into several subgroups: Caluromyinae (Kirsch, 1977), Didelphinae, Microbiotheriinae, Caroloameghiniinae, and Sparassocyninae; these lasted for various lengths of time in the South American Cenozoic. Also derived from didelphoid stock and technically within it were South America's mammalian carnivores, the borhyaenids, recently monographed by Marshall (1978) and long well known because of the early work of Sinclair (1906). In addition, descriptively plagiaulacoid marsupials called the Polydolopoidea and probably also the Caenolestoidea, some of whose members are also descriptively plagiaulacoid, arose by late Paleocene time from primitive marsupial stock.

Possible Marsupials

In addition to the forms mentioned previously, three primitive but nevertheless highly distinctive autapomorphous groups—Groeberiidae (Patterson, 1952; Simpson, Minoprio, and Patterson, 1962; Simpson, 1970*d*), Argyrolagidae (Simpson, 1970*a,c*), and Necrolestidae (Patterson, 1958)—appeared briefly at one time or another in the South American Cenozoic and then vanished. *Groeberia*, from the late Eocene Divisadero Largo fauna, is rodent-like but is evidently not a member of the Rodentia. That it is a marsupial is possible, although I know of no shared-derived characters. To place it there on the basis of presently available anatomical evidence requires an act of faith based on its geography and stratigraphic position rather than on its biology. If the genus were found anywhere else, such as in North America, China, or Africa, I doubt that it would have been placed in the Marsupialia. I prefer to regard it as an interesting and odd element on the South American therian scene, but I do not know if it is a marsupial. In contrast to *Groeberia,* the Pliocene and Pleistocene Argyrolagidae are remarkedly well known thanks to Simpson's careful descriptions. *Argyrolagus* and the closely similar

Microtragulus are bipedal, ricochetal animals, superficially not unlike kangaroo rats (*Dipodomys*). Simpson has made them a superfamily equivalent to the caenolestoids and didelphoids (in which he includes the borhyaenids). They are certainly worthy of high taxonomic rank, but after careful reading of Simpson's descriptions, I wonder if the rank should not be still higher or perhaps they should be regarded as just Theria, *incertae sedis.* With the possible exception of an expanded bulla fused to other bones, identified by Simpson as being composed of an alisphenoid element and therefore a similarity to marsupials, I have not been able to single out any potentially shared-derived features allying the argyrolagids with any likelihood to any particular therian group. Argyrolagids are obviously highly autapomorphous therians, but they could easily be a *tertium quid* with respect to the eutherian–marsupial dichotomy. Finally, *Necrolestes*, an early Miocene genus from the Santa Cruz Formation (Patterson, 1958), has been placed in the Marsupialia on the basis of shared-primitive characters such as retention of more than three incisors and various primitive cranial characters. *Necrolestes* is one of the strangest mammals known, but it is better, I think, to regard it as a highly autapomorphous therian, *incertae sedis*, than to ally it to the living marsupials on the inadequate basis of symplesiomorphy.

Edentates

Edentates (Xenarthra) have been a feature of the South American record from Riochican to Recent, although only the armadillo-like forms are present in the earliest levels of the Cenozoic. Sloths and anteaters are not known in South America's earliest faunas and are rare or absent before the Deseadan. In the past the Edentata have been thought to be allied with a North American group, the Palaeanodonta, known there from the Paleocene to the Oligocene. Emry (1970) has painstakingly criticized this earlier view and has provided cogent and detailed arguments that palaeanodonts are related to pangolins rather than to edentates. His arguments have thus far remained unchallenged by either new data or analyses by entrenched proponents of earlier views. The palaeanodonts appear to be derived from more or less pantolestid-like Paleocene mammals known from North America (Rose, 1978), so that pangolins and palaeanodonts need not be thought of as having a southern origin. Rather, they seem to be early differentiates from the Ferae.

Simpson (1978, p. 324) has recently come to the conclusion that the Edentata, in contrast, do probably have a South American origin. If they do, which Hoffstetter (1970*e*) and I also think likely, then the question to be resolved is whether they originated from Paleocene or earlier North American non-edentate immigrants by over-water dispersal followed by morphological divergence or represent a more ancient differentiate, long native in South America and isolated there by the split of South America from either North America or Africa. The first scheme requires active dispersal through a preexisting barrier; the second scheme does not.

In an earlier paper (McKenna, 1975), I listed retention of such primitive features as septomaxillary bones, ossified ribs reaching the sternum, low body temperature, poor thermoregulation, and apparent lack of a well-differentiated vagina and uterus as evidence of a truly ancient origin of the Edentata in South America, perhaps as Gondwanaland fragmented. Simpson (1978, p. 326) erroneously cited me as not holding this last view. One way to challenge my 1975 hypothesis might be to study the urogenital system of edentates more closely or to engage in various molecular studies such as of lens proteins, hemoglobin, myoglobin, etc. The latter studies have, in fact, already begun. If, on the other hand, the edentates are a differentiate from palaeanodont stock, it should be possible to demonstrate unique derived characters shared by only these two groups. This would constitute falsification of the detailed arguments presented by Emry (1970).

Edentates are divisible into two great groups, the Cingulata and the Pilosa. The former occur in the record earlier, beginning just above the Salamancan late Danian of Argentina. Included in the Cingulata are the glyptodonts, armadillos, and chlamytheres. All of these were equipped with heavy dermal armor. The Pilosa include anteaters and the huge and diverse sloth group, of which the two living genera are not each other's closest relatives (Patterson and Pascual, 1968*a*).

Toward the end of the Cenozoic the edentates spread into North America as far as Alaska. They also underwent a minor evolutionary radiation on various Caribbean islands.

Bats

I would like to discuss the Chiroptera at this point, since today they make up more than a quarter of the species of Neotropical mammals and are widely distributed and taxonomically well differentiated in South America, surely having existed there a very long time. There is not much to say, however. The earliest bat in the South American record is *Tadarida faustoi,* a mollosid from the late Oligocene or early Miocene near São Paulo, Brazil, that was originally thought to be Pleistocene in age (Paula Couto, 1956*a*; Paula Couto and Mezzalira, 1971).

Phyllostomatoids, a prominent element of the South American bat-fauna, have been found as far back as the middle Miocene of Colombia (*Notonycteris*), but elsewhere in South America the fossil record of bats is restricted to a few Pleistocene range extensions of Recent genera. I have no idea when or how the Chiroptera reached South America, but I can suggest a way to analyze the possibilities: cladistic analysis of the higher taxa of bats. Sister taxa to various South American groups exist in a variety of places and at various times that might help in deciphering early chiropteran history.

Primates

The fossil record of the arboreal platyrrhine (ceboid) primates in South America is very poorly known. Evidently they have always preferred the more tropical parts of the continent. The earliest and also the most primitive primate discovered in South America so far is the Deseadan genus *Branisella* from Salla-Luribay, Bolivia (Hoffstetter, 1969, 1974). The genus is more primitive than any other known South American primate in the features so far published, which, unfortunately, are confined to those shown by the type specimen: a maxilla of a single individual bearing two molars, a premolar, roots or alveoli of a third molar and two cheek teeth anterior to the preserved premolar. Hoffstetter (1969) has inferred from the broken roots of what he identified as P^2 that this tooth was single-rooted and small. At least it appears not to have been transversely oriented. A lower jaw fragment possibly belonging to the same species has been identified in the Princeton Collection from Salla-Luribay by L. G. Marshall and will be described by A. L. Rosenberger.

Although favorable comparisons have been made with *Apidium* of the Egyptian Oligocene (Simons, 1972), I note that *Branisella* lacks any trace of the large cuspule between the protocone and paracone of the last upper premolar, seen in premolars of *Apidium.* The molar pattern is very similar to, but perhaps more primitive than, that of *Rooneyia,* a *Microchoerus*-like omomyoid from the Oligocene of Texas [that genus as well has a large (?) paraconule on its last upper premolar].

I believe *Branisella* presents evidence that ceboids (platyrrhines) are derived from some omomyoid, but it is not clear yet just which known omomyoid is the closest, nor is it clear when or where the last common ancestor between *Branisella* and that omomyoid existed. Certainly, it does appear plausible to conclude that primates were not living in the main mass of South America in the Riochican, for otherwise one of them would probably have been found by now at Itaboraí in tropical Brazil if not in Patagonia. But it is, of course, not possible to be certain that with Deseadan *Branisella* we have reached the earliest example of a South American primate. Before *Branisella*'s discovery, the same argument applied to *Tremacebus* and *Homunculus* (including *Dolichocebus*), from the Colhuehuapian and Santacrucian, but now the primate record has been pushed back a little more. Moreover, there is no compelling argument for any continent other than North America as a candidate for the homeland of that sister group (*contra* Hoffstetter, 1973, 1974, etc.). Exactly how and when the primates entered South America, however, is unknown. I would prefer a model in which primates arrived in northern South America in the early Cenozoic via the Caribbean area, when such structures as a subaerial ridge or volcanic arcs related to the westward motion of the Americas may have increased the chances of dispersal. The Caribbean has long been an area of complex motions, including various examples of plate convergence, and probably never presented the kind of open oceanic barrier created by the Atlantic, whose primary tectonic mode was separation and subsidence. Various broad-brush maps published by Hoffstetter and by Lavocat (1977) showing North and South America as widely separated by water are useful regarding first-order phenomena, but the second-order phenomena of plate tectonics must also be taken into account if analysis is to proceed beyond the superficial.

Following *Branisella*'s tantalizing appearance in the Bolivian early Oligocene (Deseadan), primates of a much more derived ceboid aspect have been encountered in the Colhuehuapian, Santacrucian, and Friasian (approximately late Oligocene to middle Miocene) of South America (Stirton and Savage, 1950; Stirton, 1951; Hershkovitz, 1970, 1974), but I know of no Pliocene records in South America and almost no Pleistocene ones. Recent platyrrhine diversity is of course rich in northern South America, with an extension in Central America, but only *Callicebus, Alouatta,* and *Cebus* have been reported to occur in the South American Pleistocene. Moreover, two ceboids only distantly related to each other, *Xenothrix* and a species referred to *Saimiri,* have been reported from the Pleistocene of Jamaica and Hispaniola, respectively, in the West Indies (Williams and Koopman, 1952; Rosenberger, 1977; Rímoli, 1977). This presents something of a paleogeographic puzzle in need of analysis.

Rodents

The record of the caviomorph (nototrogomorph) Rodentia in South America begins abruptly in the Deseadan, approximately thirty-five million years ago. Both Patagonia and Bolivia have contributed to what we know. The Patagonian forms (Wood and Patterson, 1959; Patterson and Pascual, 1968*b*) have become increasingly well known ever since Ameghino's day, but the Bolivian genera from Salla-Luribay (Hoffstetter and Lavocat, 1970) have been described in the last decade and have been instrumental in changing the minds of many paleontologists concerning the relationships of the caviomorphs.

There are now ten described genera of Deseadan rodents: *Protosteiromys* represents the Erethizontoidea, *Luribayomys* is of uncertain relationship, and the rest consist of luantine and eocardiine cavioids, a chinchilloid, an acaremyine octodontoid and four genera of dasyproctine cavioids.

This amount of Deseadan diversity suggests that a certain amount of cladogenesis had occurred prior to the Deseadan, but the site of that evolutionary fragmentation is unknown. If it was in South America proper, it left no known trace in the latest Eocene Divisadero Largo fauna of Mendoza or in the late Eocene Musters fauna farther south. Because rodents elsewhere are highly successful on whatever land masses they occur, I have difficulty in believing that they had unused access to Argentina in the Eocene; I also have difficulty in believing that their diversity in the Deseadan was achieved overnight. Therefore, I suggest that in the Eocene a single kind of rodent may have reached some intermediate staging area, possibly over water but equally possibly on some sort of small tectonic "Noah's ark" (McKenna, 1973). After a certain amount of local taxonomic differentiation, the staging area may have tectonically coalesced with South America, releasing the already somewhat differentiated inhabitants. The Atlantic part of South America is a trailing edge coast, but the northwestern and Caribbean sides are instrinsically more interesting with regard to "Noah's arks" because of the westward motion of the Americas over the old Farallon plate.

Landry (1957), Lavocat (1971, 1973, 1974), and others have produced what to many paleontologists is by now almost overwhelming evidence of the monophyly of the hystricognath rodents, most of which occur in Africa and South America. I share that view. But it is not established that long-distance transatlantic rafting was the cause of hystricognath distribution on these two southern land masses as Lavocat (1971, 1973, 1974) and Hoffstetter (1975) insist. Now that hystricognath rodents are at last beginning to turn up in the Eocene of North America (Wood, 1972, 1973, 1974; Black and Stephens, 1973; Wahlert, 1973), it appears possible that the early hystricognaths may once have reached South America passively from North America, perhaps even with a time delay, on either dry land with sequentially opened and closed water gaps or after minor differentiation among the components of an early Tertiary island chain colliding with South America. Such a scenario seems intrinsically more likely to me than thousands of kilometers of over-water transport of a single species across the Atlantic, followed by cryptic differentiation on the same land mass on which ten genera of rodents suddenly appeared in the Deseadan.

Later in the Cenozoic additional rodents reached South America from North America; others reached North America and parts of the West Indies from South America. Hershkovitz (1966, 1969, 1972) and Patterson and Pascual (1968*a*) started a lively debate on this subject although neither camp began with the assumption of a mobile earth in which large scale horizontal motions occur. Hershkovitz and also Marshall (1979) hold that cricetids must have been cryptically evolving in northwestern South America for a long time before appearing in the record; Patterson and Pascual aver that this is not possible because there are no such rodents in the various known pre-Montehermosan late Tertiary faunas of South America. Marshall (1979) has countered Patterson and Pascual's argument with a scenario involving chance dispersal to South America during the Messinian worldwide low sea-level episode in the late Miocene, followed by spread of advanced savannah-grassland-loving genera to areas south of the Amazon Basin during the Montehermosan, about three and a half million years ago when the northern and southern savannah-grassland areas of South America coalesced. This scenario implies that *Thomasomys*-like primitive South American cricetids will be found in the late Miocene of northern South America. Thus far, they have not been collected, but there are very few well-documented faunas of that age there. Another solution that seems plausible is that if Colombia is tectonically composite, with any substantial continental accretion having occurred as the result of crustal blocks repeatedly having been swept up off the Farallon plate by the westward motion of South America, then Hershkovitz's requirement of a holding pen and Patterson and Pascual's requirement of no northern contamination might both be met. Why is not part of the present Pacific Coast of northern South America a fairly recent addition to that continent, having formerly been the isolated extreme south end of the late Tertiary Panamanian volcanic arc? The recently discovered Atrato fault (Irving, 1975, pp. 32, 33, and Pl. 1) running almost due south from the Gulf of Urabá is an obvious candidate for a late Cenozoic suture between the Panamanian arc and northwestern South America. Possibly various kinds of cricetids inhabited that area before it became part of South America.

?Dinocerata (Uintatheres)

Simpson (1935) described two genera of highly distinctive, somewhat bilophodont paenungulates (*Carodnia, Ctalecarodnia*) on the basis of a fragmentary lower jaw and some teeth collected a few years earlier at Bajo Palangana from the lower part of the Río Chico Formation, on the coast immediately north of Pico Salamanca, Patagonia. Simpson remarked that these Paleocene specimens resembled a primitive uintathere more closely than any other animal known to him. Since 1935 no additional remains of *Carodnia* have been reported from Patagonia, but in 1952 Paula Couto reported much more complete material from the Riochican of São José de Itaboraí, Brazil. On the basis of these new specimens *Ctalecarodnia* was placed in synonymy with *Carodnia* and a new family, Carodniidae, and a new order, Xenungulata, were added to the literature (Paula Couto, 1952*b*). The new Brazilian material included the complete lower and most of the upper dentition, a few bits of the skull, the complete mandibular ramus, several vertebrae, ribs, scapula, the rest of the front leg and foot bones, and from the hind leg a tibia and an astragalus. Discovery of these new anatomical features tended to increase the known similarities of *Carodnia* to uintatheres, which elsewhere in the world are known only from the Eocene of western North America (Clarkforkian to Uintan) and Asia and have no obvious sister group there except possibly the Arctocyonidae. At present much more is known about the primitive early Eocene uintatheres, thanks to the descriptive but not comparative work of Flerov (1957), so that further progress might well be made by making detailed comparisons on a cladistic basis. It might even be that uintatheres had a southern origin, as Ameghino once thought. In any case, if *Carodnia* is some sort of autapomorphously bilophodont but, nevertheless, primitive member of the Dinocerata lacking the metastylid-hypoconid crest and retaining nonmolariform premolars, it implies a distribution that bridged the Caribbean gap at some point during the Paleocene.

It may also be that the Casamayoran genus *Carolozittelia* is related to the Riochican *Carodnia.* This is suggested by the striking similarity between the left lower molar figured by Simpson (1967, Pl. 46, Figs. 1, 2) and that of *Carodnia* figured by Paula Couto (1952b, Pl. 37, Fig. 1). *Carolozittelia* has a bilophodont M^3, however, more like that of the primitive Eocene pyrothere genus *Colombitherium* from northern South America than like M^3 of *Carodnia*, which has the basic uintathere pattern. *Colombitherium* and other pyrotheres had already molarized their posterior premolars when they first appeared in the record.

Pyrotheres

The Deseadan Oligocene sediments of South America were originally known as the *Pyrotherium* beds, after the spectacular large ungulate genus of that name whose remains occur in rocks of that age. *Pyrotherium* was evidently the last of its family, whose earlier members occur back to the Casamayoran in Patagonia and also occur in the Eocene of Peru (*griphodon* Anthony, 1924). Eocene pyrotheriids (Simpson, 1967) remain very poorly known compared with *Pyrotherium* itself, from which we have much of the skeleton and the skull, described by Loomis (1914) and much more adequately by Patterson (1977). Another pyrothere family, Colombitheriidae, was added to the order by Hoffstetter (1970*d*). These animals occur in the Eocene of northern South America, *Colombitherium* in Colombia and *Proticia* in nearby Venezuela (Patterson, 1977). They are, respectively, more primitive and much more primitive than previously described pyrotheriids in that their low-crowned but incipiently lophodont teeth still wore apically on the lophs, which did not arch forward on the upper teeth or backward on the lowers. Only two premolars are supposed to occur in *Colombitherium,* but this may well be because of breakage at the front of the one known specimen. Inasmuch as the family Colombitheriidae is not united by synapomorphy, it should be abandoned.

Ameghino (1906) thought that pyrotheres were near the ancestry of the Proboscidea, a view held briefly also by Loomis (1914). Various opinions have also appeared concerning possible relationships with notoungulates and "amblypods," but the easiest way out was taken by Gaudry (1909, p. 28) who stated that pyrotheres weren't related to *anything*. Gaudry's nonview has actually prevailed, but recently Patterson (1977) has come out strongly again for notoungulate affinities on the basis of certain features of the ear region of the Oligocene genus *Pyrotherium.* I suspect that some, at least, of these similarities are convergent, namely the epitympanic sinus and foramen pneumaticum. In view of the lack of similarity of any pyrothere dentition to that of any notoungulate and the lack of notoungulate ear specializations in the trigonostylopoids, which on dental evidence seem to me to be a sister group of notoungulates, I join Simpson (1978, p. 325) in being skeptical of Patterson's conclusions. The dental patterns of *Proticia* and *Colombitherium* are instead suggestive of dentitions of didolodonts and especially of the group that I (McKenna, 1975) call tethytheres: sirenians, desmostylians, and various elephant-like groups. But, like sirenians, pyrotheres have gone much farther than such forms as *Moeritherium* and *Barytherium* in molarizing the premolars. Detailed comparisons of the ear regions of those groups and also primitive uintatheres are in order before dental similarities can be written off as useless, however. Early tethytheres included aquatic forms and some of them were distributed in the Caribbean as well as along the eastern shores of Tethys. They may even have been in South America (see below). Further, a morphological link between these groups and *Carodnia* may well exist in the form of *Carolozittelia* from the Casamayoran of Argentina.

Sirenia

Although not land mammals, the Sirenia have a long history in South America that dates back with some certainty to the Santacrucian and with doubt to the Deseadan and even the Casamayoran. *Sirenotherium* Paula Couto, 1967, is the earliest of the secure identifications and is known from the early Miocene of the State of Pará, Brazil. Less certain is the identification as a possible sirenian of the Deseadan genus *Lophiodolodus* Stirton, 1947*b* (McKenna, 1956). Still less certain is the tentative identification, made here, that *Florentinoameghinia,* from the Casamayoran of Patagonia, is related to or actually belongs to the Sirenia. Simpson (1932, 1967) described the genus, mentioning but dismissing out of hand similarities to early pyrotheres. This seems reasonable because the retention of conules on the molars (or are they milk teeth?) is a more primitive condition than shown by the lophodont molars of the most primitive pyrothere in which the upper molars are known, *Colombitherium.* Simpson noted the cryptlike structures in the maxilla above the upper molars of *Florentinoameghinia* but decided against their identification as crypts. Although Simpson did not suggest what the affinities of the animal might be, the cheek teeth of *Florentinoameghinia* seem to resemble cheek teeth of one known group of mammals not considered in Simpson's papers, namely the Sirenia. The pattern is similar to that of *Trichechus* and *Sirenotherium,* for instance, although in the anterior upper cheek teeth of sirenians in which the entire sequence of cheek teeth is molarized, the protocone and hypocone are farther apart. If *Florentinoameghinia* is truly a sirenian, it is the world's oldest in that the Casamayoran is thought to predate the middle Eocene, during which time early sirenians are known from Africa, Europe, and Jamaica. Verification of the occurrence in South America of a true sirenian at such an early date would be a fitting comfort to the man for whom *Florentinoameghinia* was named.

Other Ungulates

The earliest and most primitive South American ungulate is *Perutherium,* from the latest Cretaceous at Laguna Umayo, Peru. Thus far only a fragmentary lower jaw and part of an upper cheek tooth are known, but more work at the site will surely produce additional material. *Perutherium* retains a distinct but lingually placed paraconid on at least its M_2. A peculiar metastylid-like crest is also present. Judged from the illustration, the upper molar fragment possesses a posterior cingulum but lacks a hypocone (*contra* Sigé, 1972, p. 394, text). The genus seems to be some sort of primitive and very small ungulate, but its affinities are still unsettled. Van Valen (1978, pp. 65, 67) believed that *Perutherium* is a primitive periptychid, distinctive enough to require creation of a new subfamily, Perutheriinae.

After *Perutherium,* one next encounters ungulates in the Riochican. The primitive didolodonts found there were about the size of the Holarctic hyposodonts and in fact resemble them closely enough to warrant in the near future a detailed study of their comparable features. Paula Couto (1952*b*) actually placed the genus *Asmithwoodwardia* in the hyopsodont subfamily Hyopsodontinae, which is quite possible (Rigby, 1980) in spite of doubts later held by McKenna (1956), Paula Couto (1958, p. 18, addendum; 1978*a*), Patterson and Pascual (1968*a*; in contrast to Pascual, 1965) and Van Valen (1978). It now seems to me that: (a) *Asmithwoodwardia* is closely similar to *Didolodus, Ernestokokenia,* and *Lamegoia* in many features; (b) *Hyopsodus* itself is the North American genus most similar to *Asmithwoodwardia;* and (c) *Hyopsodus* and *Asmithwoodwardia* could have had a common ancestor that somehow crossed the Caribbean gap, possibly even from south to north. For this reason I am no longer convinced by my earlier (McKenna, 1956) argument, although it may yet be correct.

Advanced didolodonts acquired hypocones, mesostyles, and centroconids, and molarized the premolars (Simpson, 1948, 1970*b*; McKenna, 1956). The last known didolodont is *Megadolodus*, from the middle Miocene (Friasian) of Colombia.

Litopterns are another endemic South American group, often regarded as a sort of continuation of the didolodonts but better viewed as a didolodont sister-group. The details of this supposed transition have never been worked out, but *Oxybunotherium* of the Casamayoran (Pascual, 1965) appears to be morphologically transitional between the polymorphine proterotheriid litopterns and some early, probably pre-Riochican ungulate in which the paraconids of the lower molars were not yet connate with the metaconids. Pascual (1965) regarded *Oxybunotherium* as questionably a didolodont; cladistically it is a very small proterotheriid litoptern. Most of the early proterotheriid litopterns are known from rather poor material, but dentally they do seem to be similar to (?) didolodonts like *Proectocion* and also to trigonostylopoids, notoungulates, and astrapotheres. They occur at least as early as any Riochican didolodont, however. The macraucheniid litopterns, in contrast to the somewhat horselike proterotheriids, are rather distinctive from their beginnings in the Casamayoran (or possibly Riochican). Both types of litopterns lasted well up into the Cenozoic, *Macrauchenia* itself being a well-known somewhat camellike genus from the Pleistocene in which a proboscis was developed.

The trigonostylopoids are an odd but important Riochican and Casamayoran group containing at least two and probably three or four genera of early Cenozoic mammals with enlarged anterior tusks. Their molars and premolars resemble those of adapid primates. They have been discussed by Simpson (1933, 1935, 1967) and Paula Couto (1952*b*, 1963). These animals are best known from *Trigonostylops* and *Tetragonostylops,* from the Casamayoran of Argentina and the Riochican of Brazil, but Simpson (1935) tentatively added the Argentine Riochican genus *Shecenia.* I would also add Ameghino's Casamayoran and Riochican genus *Othnielmarshia* (? = *Shecenia*) and subtract *Albertogaudrya, Hedralophus*, and a specimen figured by Simpson (1967, p. 235, Fig. 49), all of which appear to me to be isotemnid notoungulates instead. *Othnielmarshia* has been considered to be a notoungulate by most workers who have mentioned it at all, but it lacks the characteristic notoungulate entoconid on its lower molars, having incorporated it into the talonid rim.

The mysterious genus *Acamana* of the Divisadero Largo fauna in the latest Eocene might conceivably be a trigonostylopoid.

As is evident from these confusions, in details of the trigonostylopoid dental pattern there is a strong similarity to primitive isotemnid notoungulates and also to astrapotheres. Elaboration of the central part of the upper molar crowns had barely begun but was closely similar to that seen in notoungulates, litopterns, and astrapotheres alike. In the ear region and elsewhere in the skull, however, *Trigonostylops* possess a mixture of plesiomorphous and autapomorphous features that establish the Trigonostylopoidea as a major side-shoot from early ungulate ancestry (Simpson, 1933, 1967).

Notoungulates are a major extinct ungulate group which had a primarily South American distribution while that continent was isolated in the Cenozoic. In the Pleistocene, however, *Mixotoxodon* occurred as far northward as El Salvador in Central America. There were notoungulates in Wyoming, Mongolia, and China as well for a time in the Paleocene and early Eocene. *Mixotoxodon* is a clear case of an advanced toxodont escaping overland from its southern prison sometime after the establishment of the Panamanian isthmus three or four million years ago, but the Paleocene and Eocene notoungulates (Arctostylopidae) of western North America and eastern Asia are not so easily explained. These early northern notoungulates differ from those of South America in several autapomorphous characters, such as the greatly reduced trigonid and the anterolabial orientation of the entoconid crest of lower molars and the hypertrophy of the upper molar ectoloph. But in other features the arctostylopids show similarities to various primitive members of early Cenozoic South American families. For instance, they possess a deep notch in the palate like that of *Notopithecus*. In still other aspects, however, such as the nonmolariform premolars, arctostylopids are clearly very primitive notoungulates.

Matthew and Granger (1925) and other authors, for example Patterson and Pascual (1968*a*), have concluded that the early northern notoungulates were evidence for a northern origin. Others, such as Simpson (1942; 1971, p. 115) and Hoffstetter (1970*e*, 1971) have stated that notoungulates probably had a southern origin. Gingerich and Rose (1977) took an intermediate stand. They held that notoungulates probably arose in Central or South America and then spread northward from there in the early Eocene (Clarkforkian). Actually, notoungulates were in the Paleocene of both China (*Sinostylops, Anatolostylops*) and Wyoming (a new Tiffanian specimen of *Arctostylops:* C. Schaff, personal communication) prior to the Clarkforkian. The place of origin of notoungulates is still unknown.

Early in their evolution notoungulates in which the skull is known acquired an enlarged epitympanic sinus that invaded the posterior part of the squamosal bone. This sinus, plus several other features of the ear region and a rather distinctive dental plan that in some groups was further modified as the teeth became very high-crowned, make notoungulate skulls rather easy to distinguish from those of most mammalian groups. Nevertheless, the dental plan can be seen to share basic characters with other South American groups once notoungulate autapomorphies are stripped away. It is clear from isotemnids and a few other primitive forms, for instance, that in the lower molars of early notoungulates the entoconids were once conical and separate. Premolar molarization must not have begun until after notoungulates were distinct, nor elaboration of the crochet and other modifications of the dental pattern that led to the formation of complex enamel lakes in high-crowned taxa. Trigonostylopoids and astrapotheres, neither of which have the notoungulate ear region or morphologically advanced premolars, appear to me to be the closest sister-group to the notoungulates, beyond which the litopterns and didolodonts would seem to be the next most closely related South American mammals.

Within the notoungulates it seems possible to divide the bewildering array of morphological diversity into two main groups, with a few stubborn primitive taxa like the arctostylopids simply not yet fitting well into any scheme. Simpson (1934) has referred a number of the primitive forms to a horizontal and arbitrary taxon, Notioprogonia. It seems to me that just two major clades exist: (a) the large toxodonts with characteristic trigonids of the lower molars, broadly including isotemnids, homalodotheres, leontiniids, notohippids, and toxodontids, and (b) almost all other notoungulates, which early reduced the canine teeth and most of which in one way or another became superficially rodentlike and rabbitlike. Within this second group a number of early twigs split off to yield genera such as *Seudenius, Peripantostylops, Allalmeia, Xenostephanus, Brachystephanus, Henricosbornia, Oldfieldthomasia,* etc. Then further differentiation occurred to yield the notostylopids, hegetothere-like forms (e.g., *Kibenikhoria,* archaeohyracids, and hegetotheriids), and typotheres (mesotheres, interatheres, archaeopithecids, notopithecids). Both major notoungulate groups were already present by Riochican time. I presume their differentiation took place earlier in the Paleocene, just after notoungulates originated from ancestors with trigonostylopoid-like teeth but lacking the otic and other autapomorphies of Riochican trigonostylopoids.

Astrapotheres are a final ungulate group that was restricted to South America from the Casamayoran to the Friasian, approximately early Eocene to middle Miocene. *Parastrapotherium* from the Deseadan and later Cenozoic was large and elephantine, with slightly elephantlike tusks and a short proboscis, but the dentition of early primitive astrapotheres is evidence of relationship to both trigonostylopoids and primitive notoungulates. The astrapothere ear region lacks the characteristic posterior epitympanic recess of the notoungulate squamosal bone, and astrapothere dentitions, especially the premolars, depart from known trigonostylopoids or notoungulates in various derived features acquired in astrapotherian history. Pre-Deseadan astrapotheres are rather poorly known (Simpson, 1967).

Acknowledgments

I thank R. Pascual, D. Domning, N. A. Neff, L. G. Marshall, and R. H. Tedford for their helpful comments on an earlier draft of the manuscript.

References

Ameghino, F., 1897, Mammifères crétacés de l'Argentine. (Deuxième contribution à la connaissance de la faune mammalogique des couches à *Pyrotherium*), *Bol. Inst. Geogr. Argentina, Buenos Aires,* **18**:406–429, 431–521.

Ameghino, F., 1906, Les Formations sédimentaires du Crétacé Supérieur et du Tertiaire de Patagonie avec un parallèle entre leurs faunes mammalogiques et celles de l'Ancien continent, *An. Mus. Nac. Buenos Aires* **15**(3):1–568.

Anthony, H. E., 1922, A new fossil rodent from Ecuador, *Am. Mus. Novit.* **No. 35**:1–4.

Anthony, H. E., 1924, A new fossil perissodactyl from Peru, *Am. Mus. Novit.* **No. 111**:1–3.

Berggren, W. A., McKenna, M. C., Hardenbol, J., and Obradovich, J. D., 1978, Revised Paleogene polarity time scale, *J. Geol.* **86**:67–81.

Bertels, A., 1970, Sobre el "Piso Patagoniano" y la representación de la época del Oligoceno en Patagonia, República Argentina, *Rev. Asoc. Geol. Arg.* **25**(4):495–501.

Bertels, A., 1975, Bioestratigrafia del Paleogeno en la República Argentina, *Rev. Española Micropal.* **7**(3):429–450.

Black, C. C., and Stephens, J. J., III, 1973, Rodents from the Paleogene of Guanajuato, Mexico, *Occ. Pap., Mus., Texas Tech Univ.* **No. 14**:1–10.

Bonatti, E., and Honnorez, J., 1971, Nonspreading crustal blocks at the Mid-Atlantic Ridge, *Science* **174**:1329–1331.

Bonatti, E., Sarnthein, M., Boersma, A., Gorini, M., and Honnorez, J., 1977, Neogene crustal emersion and subsidence at the Romanche Fracture Zone, equatorial Atlantic, *Earth Planet. Sci. Lett.* **35**:369–383.

Boule, M., and Thevenin, A., 1920, *Mammifères Fossiles de Tarija,* Soudier, Paris.

Burke, K., 1975, Atlantic evaporites formed by evaporation of water spilled from Pacific, Tethyan, and Southern oceans, *Geology,* **November**:613–616.

Camacho, H. H., 1974, Bioestratigrafia de las formaciones marinas del Eoceno y Oligoceno de la Patagonia, *Anal. Acad. Cien. Ex. Fis. Nat. Buenos Aires* **26**:39–57.

Casamiquela, R. M., 1961, El hallazgo del primer elenco (icnologico) Jurasico de vertebrados terrestres de Latinoamerica (noticia), *Rev. Asoc. Geol. Arg.* **15**(1–2):5–14.

Connary, S. D., 1972, Investigations of the Walvis Ridge and environs, Ph.D. Thesis, Columbia University, New York.

Crochet, J.-Y., 1978, *Les marsupiaux du Tertiare d'Europe,* Thèse, Acad. Montpellier, Univ. Montpellier, Montpellier.

Cuvier, C. L. C. F. D., 1796, Notice sur le squelette d'une très-grande espèce de quadrupède inconnue jusqu' à présent, trouvé au Paraguay, et déposé au cabinet d'histoire naturelle de Madrid, *Mag. Encycl. (Paris)* **1**:303–310, **2**:227–228.

Cuvier, C. L. C. F. D., 1812, *Recherches sur les ossements fossiles de Quadrupèdes,* Vol. 4, Deterville, Paris.

Darwin, C., 1839, *Journal of Researches into the Geology and Natural History of the Various Countries Visited by H. M. S. Beagle under the Command of Captain Fitzroy, R. N. from 1832 to 1836,* Colburn, London.

Emry, R. J., 1970, A North American Oligocene pangolin and other additions to the Pholidota, *Bull. Am. Mus. Nat. Hist.* **142**(6):455–510.

Estes, R., 1975, Fossil *Xenopus* from the Paleocene of South America and the zoogeography of pipid frogs, *Herpetologica* **31**:263–278.

Estes, R., and Wake, M., 1972, The first fossil record of caecilian amphibians, *Nature* **239**:228–231.

Fields, R. W., 1957, Hystricomorph rodents from the late Miocene of Colombia, South America, *Univ. Calif. Publs. Geol. Sci.* **32**(5):273–404.

Flerov, K. K., 1957, Dinotseraty Mongolii, *Trudy Pal. Inst. Akad. Nauk. S.S.S.R.* **77**:1–82.

Francis, J. C., 1975, Esquema bioestratigrafico regional de la Republica Oriental del Uruguay, *Actas del Primer Cong. Arg. Paleontol. Bioest., Tucumán* **2**:539–568.

Francisco, B. H. R., and de Souza Cunha, F. L., 1978, Geologia e estratigrafia da Bacia de São José, Municipio de Itaboraí, R. J. *An. Acad. Brasil. Ciên.* **50**(3):381–416.

Gaudry, A., 1902, Recherches paléontologiques de M. André Tournouër en Patagonie, *Bull. Soc. Hist. Nat. Autun* **15**:117–123.

Gaudry, A., 1904, Fossiles de Patagonie. Dentition de quelques mammifères, *Mém. Soc. Géol. France (Paléontol.)* **No. 31**:1–27.

Gaudry, A., 1906*a*, Fossiles de Patagonie: Les attitudes de quelques animaux, *C. R. Acad. Sci. Paris* **141**:806–808.

Gaudry, A., 1906*b*, Fossiles de Patagonie. Étude sur une portion du monde antarctique, *Ann. Paléontol. (Paris)* **1**:1–42.

Gaudry, A., 1908, Fossiles de Patagonie. De l'économie dans la nature, *Ann. Paléontol. (Paris)* **3**:1–28.

Gaudry, A., 1909, Fossiles de Patagonie. Le *Pyrotherium, Ann. Paléontol. (Paris)* **4**:1–28.

Gingerich, P. D., and Rose, K. D., 1977, Preliminary report on the American Clark Fork mammal fauna, and its correlation with similar faunas in Europe and Asia, *Géobios, Mém. Spéc.* **1**:39–45.

Grambast, L., Martinez, M., Mattauer, M., and Thaler, L., 1967, *Perutherium altiplanense,* nov. gen., nov. sp., premier mammifère mésozoïque d'Amérique du Sud, *C. R. Acad. Sci. Paris, Sér. D* **264**:707–710.

Hershkovitz, P., 1966, Mice, land bridges, and Latin American faunal interchange, in: *Ectoparasites of Panama* (R. L. Wenzel and V. T. Tipton, eds.), pp. 725–751. Field Museum of Natural History, Chicago.

Hershkovitz, P., 1969, The evolution of mammals on Southern continents. VI. The Recent mammals of the Neotropical Region: A zoogeographic and ecological review, *Q. Rev. Biol.* **44**(1):1–70.

Hershkovitz, P., 1970, Notes on Tertiary platyrrhine monkeys and description of a new genus from the late Miocene of Colombia, *Folia Primatol.* **12**:1–37.

Hershkovitz, P., 1972, The Recent mammals of the Neotropical Region: A zoogeographic and ecological review, in: *Evolution, Mammals and Southern Continents* (A. Keast, F. C. Erk, and B. Glass, eds.), pp. 311–431, State University of New York Press, Albany.

Hershkovitz, P., 1974, A new genus of late Oligocene monkey (Cebidae, Platyrrhini), with notes on postorbital closure and platyrrhine evolution, *Folia Primatol.* **21**:1–35.

Hirschfeld, S. E., and Marshall, L. G., 1976, Revised faunal list of the *La Venta Fauna* (Friasian–Miocene) of Colombia, South America, *J. Paleontol.* **50**(3):433–436.

Hoffstetter, R., 1968, Un gisement de mammifères déséadiens (Oligocène inférieur) en Bolivie, *C. R. Acad. Sci. Paris, Sér. D* **267**:1095–1097.

Hoffstetter, R., 1969, Un primate de l'Oligocène inférieur sud-Américain: *Branisella boliviana* gen. et sp. nov., *C. R. Acad. Sci. Paris, Sér. D* **269**:434–437.

Hoffstetter, R., 1970*a*, Vertebrados Cenozoicos de Colombia, *Actas IV Cong. Latinoamer. Zool., Caracas, 1968,* **2**:931–954.

Hoffstetter, R., 1970*b*, Vertebrados Cenozoicos del Ecuador, *Actas IV Cong. Latinoamer. Zool., Caracas, 1968,* **2**:955–969.

Hoffstetter, R., 1970*c*, Vertebrados Cenozoicos y mammiferos Cretacicos del Peru, *Actas IV Cong. Latinoamer. Zool., Caracas, 1968,* **2**:971–983.

Hoffstetter, R., 1970*d*, *Colombitherium tolimense,* pyrothérien nouveau de la formation Guandalay (Colombie), *Ann. Paléontol. (Verts.)* **56**(2):149–169.

Hoffstetter, R., 1970*e*, Radiation initiale des mammifères placentaires et biogéographie, *C. R. Acad Sci. Paris, Sér. D* **270**:3027–3030.

Hoffstetter, R., 1970*f*, L'histoire biogéographique des marsupiaux et la dichotomie marsupiaux-placentaires, *C. R. Acad. Sci. Paris, Sér. D* **271**:388–391.

Hoffstetter, R., 1970*g*, Les Paléomammalogistes Français et l'Amérique Latine (résumé d'un exposé avec projections), *Bull. Acad. Soc. Lorraines Sci.* **9**(1):233–243.

Hoffstetter, R., 1971, Le peuplement mammalien de l'Amérique du Sud. Rôle des continents austraux comme centres d'origine, de diversification et de dispersion pour certains groupes mammaliens, *An. Acad. Brasil. Ciên. (Suppl.)* **43**:125–144.

Hoffstetter, R., 1972, Données et hypothèses concernant l'origine et l'histoire biogéographique des marsupiaux, *C. R. Acad. Sci. Paris, Sér. D* **274**:2635–2638.

Hoffstetter, R., 1973. Origine, compréhension et signification des taxons de rang supérieur: Quelques enseignements tirés de l'histoire des mammifères, *Ann. Paléontol. (Verts.)* **59**(2):1–35.

Hoffstetter, R., 1974, Phylogeny and geographical deployment of the primates, *J. Hum. Evol.* **3**:327–350.

Hoffstetter, R., 1975, El origen de los Caviomorpha y el problema de los Hystricognathi (Rodentia), *Actas Prim. Cong. Arg. Paleontol. Bioestrat., Univ. Nac. Tucumán, Aug., 1974,* **2**:505–528.

Hoffstetter, R., 1977, Un gisement de mammifères Miocènes à Quebrada Honda (Sud Bolivien), *C. R. Acad. Sci. Paris, Sér. D* **284**:1517–1520.

Hoffstetter, R., and Lavocat, R., 1970, Découverte dans le Déséadien de Bolivie de genres pentalophodontes appuyant les affinités Africaines des rongeurs caviomorphes, *C. R. Acad. Sci. Paris, Sér. D* **271**:172–175.

Hoffstetter, R., Martinez, C., Muñoz-Reyes, J., and Tomasi, P., 1971, Le gisement d'Ayo Ayo (Bolivie), une succession stratigraphique Pliocène-Pléistocène datée par des mammifères, *C. R. Acad. Sci. Paris, Sér. D* **273**:2472–2475.

Hoffstetter, R., Martinez, C., and Tomasi, P., 1972, Nouveaux gisements de mammifères Néogènes dans les couches rouges de l'Altiplano Bolivien, *C. R. Acad. Sci. Paris, Sér. D* **275**:739–742.

Irving, E. M., 1975, Structural evolution of the northernmost Andes, Colombia, *U.S. Geol. Surv. Prof. Pap.* **846**:i–iv, 1–47.

Kirsch, J. A. W., 1977, The comparative serology of Marsupialia, and a classification of marsupials, *Aust. J. Zool., Suppl. Ser.,* **No. 52**:1–152.

Krause, D. W., and Baird, D., 1979, Late Cretaceous mammals east of the North American Western Interior seaway, *J. Paleontol.* **53**(3):562–565.

Kumar, N., and Embley, R., 1977, Evolution and origin of the Ceará rise: An aseismic rise in the western equatorial Atlantic, *Bull. Geol. Soc. Am.* **88**:683–694.

Kumar, N., and Gamboa, L. A. P., 1979, Evolution of the São Paulo Plateau (southeastern Brazilian margin) and implications for the early history of the South Atlantic, *Bull. Geol. Soc. Am.* **90**:281–293.

Ladd, J. W., 1974, South Atlantic sea-floor spreading and Caribbean tectonics, Ph.D. Thesis, Columbia University, New York.

Ladd, J. W., 1976, Relative motion of South America with respect to North America and Caribbean tectonics, *Bull. Geol. Soc. Am.* **87**:969–976.

Landry, S. O., Jr., 1957, The interrelationships of the New and Old World hystricomorph rodents. *Univ. Calif. Publs. Zool.* **56**(1):1–118.

Lavocat, R., 1971, Affinités systématiques des caviomorphes et des phiomorphes et origine Africaine des caviomorphes, *An Acad. Brasil. Ciên. (Suppl.)* **43**:515–522.

Lavocat, R., 1973, Les rongeurs du Miocène d'Afrique orientale. I. Miocène Inférieur, *Mém. Trav. École Pratique des Hautes Etude Inst. Montpellier* **1**:i–vi, 1–284.

Lavocat, R., 1974, What is an hystricomorph?, in: *The Biology of Hystricomorph Rodents* (I. W. Rowlands and B. J. Weir), *Symp. Zool. Soc. Lond.*, No. 34, pp. 7–20 [see also discussion, pp. 55–60], Academic Press, London.

Lavocat, R., 1977, Sur l'origine des faunes sud-américaines de mammifères du Mésozoïque terminal et du Cénozoïque ancien, *C. R. Acad. Sci. Paris, Sér. D* **285**:1423–1426.

Le Pichon, X., and Hayes, D. E., 1971, Marginal offsets, fracture zones, and the early opening of the South Atlantic, *J. Geophys. Res.* **76**:6283–6293.

Le Pichon, X., Melguen, M., and Sibuet, J.-C., 1978, A schematic model of the evolution of the South Atlantic, in: *Proceedings of the Joint Oceanographic Assembly, Edinburgh,* Plenum Press, New York.

Loomis, F. B., 1914, *The Deseado Formation of Patagonia,* Rumford Press, Concord, N.H.

Lund, P. W., 1841, *Blik paa Brasiliens dyreveren för sidste jordomvaeltning,* Bianco Luno, Copenhagen.

McKenna, M. C., 1956, Survival of primitive notoungulates and condylarths into the Miocene of Colombia, *Am. J. Sci.* **254**:736–743.

McKenna, M. C., 1973, Sweepstakes, filters, corridors, Noah's arks, and beached Viking funeral ships in palaeogeography, in: *Implications of Continental Drift to the Earth Sciences* (D. H. Tarling and S. K. Runcorn, eds.), Vol. 1, pp. 295–308, Academic Press, London.

McKenna, M. C., 1975, Toward a phylogenetic classification of the Mammalia, in: *Phylogeny of the Primates, a Multidisciplinary Approach* (W. P. Luckett and F. S. Szalay, eds.), pp. 21–46, Plenum Press, New York.

Mankinen, E. A., and Dalrymple, G. B., 1979, Revised geomagnetic polarity time scale for the interval 0–5 m. y. b. p., *J. Geophys. Res.* **84**(B2):615–626.

Marshall, L. G., 1978, Evolution of the Borhyaenidae, extinct South American predaceous marsupials, *Univ. Calif. Publs. Geol. Sci.* **117**:i–iv, 1–89.

Marshall, L. G., 1979, A model for paleobiogeography of South American cricetine rodents. *Paleobiology,* **5**:126–132.

Marshall, L. G., and Hecht, M. K., 1978, Mammalian faunal dynamics and the Great American Interchange: An alternative interpretation, *Paleobiology* **4**:203–206.

Marshall, L. G., Pascual, R., Curtis, G. H., and Drake, R. E., 1977, South American geochronology: Radiometric time scale for middle to late Tertiary mammal-bearing horizons in Patagonia, *Science* **195**:1325–1328.

Marshall, L. G., Butler, R. F., Drake, R. E., Curtis, G. H., and Tedford, R. H., 1979, Calibration of the Great American Interchange, *Science* **204**:272–279.

Marshall, L. G., Berta, A., Hoffstetter, R., and Pascual, R., in press *a*, Geochronology of the continental mammal bearing Quaternary of South America, in: *Vertebrate Paleontology as a Discipline in Geochronology* (M. O. Woodburne, ed.), University of California Press, Berkeley.

Marshall, L. G., Hoffstetter, R., and Pascual, R., in press *b*, Geochronology of the continental mammal-bearing Tertiary of South America, in: *Vertebrate Paleontology as a Discipline in Geochronology* (M. O. Woodburne, ed.), University of California Press, Berkeley.

Matthew, W. D., 1915, Climate and Evolution, *Ann. N.Y. Acad. Sci.* **24**:171-318.

Matthew, W. D., and Granger, W., 1925, Fauna and correlation of the Gashato Formation of Mongolia, *Am. Mus. Novit.* **No. 189**:1-12.

Newell, N.D., 1949, Geology of the Lake Titicaca region, Peru and Bolivia, *Mem. Geol. Soc. Am.* **36**:1-111.

Owen, R., 1842, Description of the skeleton of an extinct gigantic sloth, *Mylodon robustus,* Owen [etc.], R. and J. E. Taylor, London.

Pascual, R., 1965, Un Nuevo Condylarthra (Mammalia) de Edad Casamayorense de Paso de los Indios (Chubut, Argentina). Breves consideraciones sobre la Edad Casamayorense, *Ameghiniana* **4**:57-65.

Patterson, B., 1942, Two Tertiary mammals from northern South America, *Am. Mus. Novit.* **No. 1173**:1-7.

Patterson, B., 1952, Un Nuevo y extraordinario marsupial deseadano, *Rev. Mus. Municipal Cien. Nat. Tradic., Mar del Plata* **1**:39-44.

Patterson, B., 1958, Affinities of the Patagonian fossil mammal *Necrolestes, Breviora* **No. 94**:1-14.

Patterson, B., 1977, A primitive pyrothere (Mammalia, Notoungulata) from the early Tertiary of northwestern Venezuela, *Fieldiana, Geol.* **33**(22):397-422.

Patterson, B., and Pascual, R., 1968*a*, The fossil mammal fauna of South America, *Q. Rev. Biol.* **43**(4):409-451.

Patterson, B., and Pascual, R., 1968*b*, New echimyid rodents from the Oligocene of Patagonia, and a synopsis of the family, *Brevior* **No. 301**:1-14.

Paula Couto, C. de, 1952*a*, Fossil mammals from the beginning of the Cenozoic of Brazil. Marsupialia: Didelphidae, *Am. Mus. Novit.* **No. 1567**:1-26.

Paula Couto, C. de, 1952*b*, Fossil mammals from the beginning of the Cenozoic in Brazil. Condylarthra, Litopterna, Xenungulata, and Astrapotheria, *Bull. Am. Mus. Nat. Hist.* **99**:355-394.

Paula Couto, C. de, 1956*a*, Une chauve-souris fossile des argiles feuilletées Pléistocènes de Tremembé, État de São Paulo (Brésil). *Act. IV Cong. Internat. Quatern., Rome-Pise, 1953* **1**:343-347.

Paula Couto, C. de, 1956*b*, Resumo de memórias de Lund sôbre as cavernas de Lagoa Santa e seu conteúdo animal, traduzido de H. C. Orsted, *Univ. Brasil, Museu Nac. Publs. avulsas* **No. 16**:1-14.

Paula Couto, C. de, 1958, Idade geológica das bacias Cenozóicas do vale do Paraíba e de Itaboraí, *Bol. Mus. Nac., Rio de Janeiro, Brasil, Geol.* **No. 25**:1-18.

Paula Couto, C. de, 1961, Marsupiais fósseis do Paleoceno do Brasil, *An. Acad. Bras. Ciên.* **33**(3-4):321-333.

Paula Couto, C. de, 1962, Didelphideos fosiles del Paleoceno de Brasil, *Rev. Mus. Arg. Ciên. Nat. (Zool.)* **8**(12):135-166.

Paula Couto, C. de, 1963, Um Trigonostylopidae do Paleoceno do Brasil, *An. Acad. Bras. Ciên.* **35**(3):339-351.

Paula Couto, C. de, 1967, Contribuição à paleontologia do Estado do Pará. Um sirênio na Formação Pirabas. Atas do Simpósio sôbre a Biota Amazônica, *Geociên.* **1**:345-357.

Paula Couto, C. de, 1970*a*, Novo notoungulado no Riochiquense de Itaboraí, *Iheringia, Geol.* **No. 3**:77-86.

Paula Couto, C. de, 1970*b*, Evolução de communidades, modificações faunisticas e integrações biocenóticas do vertebrados cenozóicos do Brasil, *Actas IV Cong. Latinoamer. Zool., Caracas* **2**:907-930.

Paula Couto, C. de, 1974, Marsupial dispersion and continental drift, *An. Acad. Brasil. Ciên.* **46**(1):103-126.

Paula Couto, C. de, 1977, Fossil mammals from the Cenozoic of Acre, Brazil. I—Astrapotheria, *An. xxviii Cong. Brasil. Geol.,* pp. 237-249.

Paula Couto, C. de, 1978*a*, Ungulados fósseis do Riochiquense de Itaboraí R. J., Brasil. II—Condylarthra e Litopterna, *An. Acad. Brasil. Ciên.* **50**(2):209-218.

Paula Couto, C. de, 1978*b*, Ungulados fósseis do Riochiquense de Itaboraí, Estado do Rio de

Janeiro, Brasil, III—Notoungulata e Trigonostylopoidea, *An. Acad. Brasil. Ciên.* **50** (2):220–226.

Paula Couto, C. de, and Mezzalira, S., 1971, Nova Conceituação Geocronológica de Tremembé, Estado de São Paulo, Brasil, *An Acad. Brasil. Ciên. (Suppl.)* **43**:473–488.

Pitman, W. C., III, and Talwani, M., 1972, Sea-floor spreading in the North Atlantic, *Bull. Geol. Soc. Am.* **83**:619–646.

Portugal, J. A., 1974, Mesozoic and Cenozoic stratigraphy and tectonic events of Puno-Santa Lucia area, Department of Puno, Peru, *Bull. Am. Assoc. Petrol. Geol.* **58**(6):982–999.

Rabinowitz, P. D., 1976, A geophysical study of the continental margin of southern Africa, *Bull. Geol. Soc. Am.* **27**:1643–1653.

Repetto, F., 1977, Un mamífero nuevo en el Terciario del Ecuador (Azuay-Cañar), *Técnologica (Esc. Polit. Litoral, Guayaquil)* **1**(2):33–38.

Reyment, R. A., 1974, Application des méthodes paléobiologiques à la théorie de la dérive des continents, illustrée par l'Atlantique sud, *Rev. Géog. Phys. Géol. Dynam., Sér. 2* **16**(1):61–70.

Reyment, R. A., and Tait, E. A., 1972, Biostratigraphical dating of the early history of the South Atlantic Ocean, *Philos. Trans. Roy. Soc. London, Biol. Sci.* **264**(858):55–95.

Rigby, J. K., Jr., 1980, Swain Quarry of the Fort Union Formation, middle Paleocene (Torrejonian), Carbon County, Wyoming: Geologic setting and mammalian fauna, *Evolutionary Monographs* **3**:i–vi, 1–179.

Rímoli, R., 1977, Una nueva especie de monos (Cebidae: Saimirinae: *Saimiri*) de la Hispaniola. *Cuadernos del Centro Dominicano de investigaciones antropologicas (CENDIA), Univ. Autonoma de Santo Domingo,* **242**(1):1–14.

Rose, K. D., 1978, A new Paleocene epoicotheriid (Mammalia), with comments on the Palaeanodonta, *J. Paleontol.* **52**(3):658–674.

Rosenberger, A. L., 1977, *Xenothrix* and ceboid phylogeny, *J. Hum. Evol.* **6**:461–481.

Scientific Staff, 1974, Southwestern Atlantic, *Geotimes,* **November**:16–18.

Sibuet, J.-C., and Mascle, J., 1978, Plate kinematic implications of Atlantic equatorial fracture zone trends, *J. Geophys. Res.* **83**(B7):3401–3421.

Sigé, B., 1972, La faunule de mammifères du Crétacé supérieur de Laguna Umayo (Andes peruviennes), *Bull. Mus. Nat. d'Hist. Nat., Sér. 3, No. 99, Sci. Terr.,* **No. 19**:375–408.

Simons, E. L., 1972, *Primate Evolution, An Introduction to Man's Place in Nature,* Macmillan, New York.

Simpson, G. G., 1932, The supposed association of dinosaurs with mammals of Tertiary type in Patagonia, *Am. Mus. Novit.* **No. 566**:1–20.

Simpson, G. G., 1933, Structure and affinities of *Trigonostylops, Am. Mus. Novit.* **No. 608**:1–28.

Simpson, G. G., 1934, Provisional classification of extinct South American hoofed mammals, *Am. Mus. Novit.* **No. 750**:1–21.

Simpson, G. G., 1935, Descriptions of the oldest known South American mammals, from the Río Chico Formation, *Am. Mus. Novit.* **No. 793**:1–25.

Simpson, G. G., 1940, Review of the mammal-bearing Tertiary of South America, *Proc. Am. Philos. Soc.* **83**(5):649–709.

Simpson, G. G., 1942, Early Cenozoic mammals of South America, *Proc. 8th Am. Sci. Congr.* **4**:303–332.

Simpson, G. G., 1943, Notes on the mammal-bearing Tertiary of South America, *Proc. Am. Philos. Soc.* **86**:403–404.

Simpson, G. G., 1948, The beginning of the age of Mammals in South America. Part 1. Introduction. Systematics: Marsupialia, Edentata, Condylarthra, Litopterna and Notioprogonia, *Bull. Am. Mus. Nat. Hist.* **91**:1–232.

Simpson, G. G., 1967, The beginning of the age of Mammals in South America. Part 2. Systematics: Notoungulata, concluded (Typotheria, Hegetotheria, Toxodonta, Notoungulata incertae sedis); Astrapotheria; Trigonostylopoidea; Pyrotheria; Xenungulata; Mammalia incertae sedis, *Bull. Am. Mus. Nat. Hist.* **137**:1–260.

Simpson, G. G., 1970*a*, The Argyrolagidae, extinct South American marsupials, *Bull. Mus. Comp. Zool.* **139**(1):1–86.

Simpson, G. G., 1970*b*, Mammals from the early Cenozoic of Chubut, Argentina, *Breviora,* **No. 360**:1-13.

Simpson, G. G., 1970*c*, Additions to knowledge of the Argyrolagidae (Mammalia, Marsupialia) from the late Cenozoic of Argentina, *Breviora,* **No. 361**:1-9.

Simpson, G. G., 1970*d*, Addition to knowledge of *Groeberia* (Mammalia, Marsupialia) from the mid-Cenozoic of Argentina, *Breviora,* **No. 362**:1-17.

Simpson, G. G., 1971, The evolution of marsupials in South America, *An. Acad. Brasil. Ciên. (Suppl.)* **43**:103-118.

Simpson, G. G., 1978, Early mammals in South America: Fact, controversy, and mystery, *Proc. Am. Philos. Soc.* **122**(5):318-328.

Simpson, G. G., Minoprio, J. L., and Patterson, B., 1962, The mammalian fauna of the Divisadero Largo Formation, Mendoza, Argentina, *Bull. Mus. Comp. Zool.,* **127**(4):239-293.

Sinclair, W. J., 1906, Mammalia of the Santa Cruz beds: Marsupialia, *Rept. Princeton Univ. Exped. Patagonia* **4**(3):333-460.

Smith, P. J., 1977, Origin of the Rio Grande rise, *Nature* **269**:651-652.

Steiger, R. H., and Jäger, E., 1977, Subcommision on geochronology: Convention on the use of decay constants in geo- and cosmochronology, *Earth Planet. Sci. Lett.* **36**:359-362.

Stirton, R. A., 1947*a*, A rodent and a peccary from the Cenozoic of Colombia, *Compil. Est. Geol. Of. Colombia, (1946),* **7**:317-324.

Stirton, R. A., 1947*b*, The first lower Oligocene vertebrate fauna from northern South America, *Compil. Est. Geol. Of. Colombia, (1946),* **7**:327-341.

Stirton, R. A., 1951, Ceboid monkeys from the Miocene of Colombia, *Univ. Calif. Publs. Geol. Sci.* **28**:315-356.

Stirton, R. A., 1953, Vertebrate paleontology and continental stratigraphy in Colombia, *Bull. Geol. Soc. Am.* **64**:603-622.

Stirton, R. A., and Savage, D. E., 1950, A new monkey from the La Venta Miocene of Colombia, *Min. Minas Petrol., Serv. Geol. Nac., Compil. Est. Geol. Of. Colombia,* **7**:347-356.

Tedford, R. H., 1974, Marsupials and the new paleogeography, in: *Paleogeographic Provinces and Provinciality* (C. A. Ross, ed.), *Soc. Econ. Paleontol. and Mineral. Spec. Publ.* **No. 21**:109-126.

Van der Hammen, T., 1958, Estratigrafía del Terciario y Maestrichtiano continentales y tectogénesis de los Andes Colombianos, Colombia, *Serv. Geol. Nac. Bol. Geol.* **6**:67-128.

Van Houten, F. B., 1976, Late Cenozoic volcaniclastic deposits, Andean foredeep, Colombia, *Bull. Geol. Soc. Am.* **87**:481-495.

Van Valen, L., 1978, The beginning of the Age of Mammals, *Evol. Theory* **4**(2):45-80.

Wahlert, J. H., 1973, *Protoptychus,* a hystricomorphous rodent from the late Eocene of North America, *Breviora* **No. 419**:1-14.

Wardlaw, N. C., and Nicholls, G. D., 1972, Cretaceous evaporites of Brazil and West Africa and their bearing on the theory of continent separation, *Proc. 24th Intl. Geol. Cong.* **Sect. 6**:43-55.

Webb, S. D., 1976, Mammalian faunal dynamics of the Great American Interchange, *Paleobiology* **2**(3):220-234.

Webb, S. D., 1978*a*, Mammalian faunal dynamics of the Great American Interchange: Reply to an alternative interpretation, *Paleobiology* **4**(2):206-209.

Webb, S. D., 1978*b*, History of savanna vertebrates in the New World. Part II: South America and the Great Interchange, *Annu. Rev. Ecol. Syst.* **9**:393-426.

Wellman, S. S., 1970, Stratigraphy and petrology of the nonmarine Honda Group (Miocene), Upper Magdalena Valley, Colombia, *Bull. Geol. Soc. Am.* **81**:2353-2374.

Williams, E. E., and Koopman, K. F., 1952, West Indian fossil monkeys, *Am. Mus. Novit.* **No. 1546**:1-16.

Wood, A. E., 1972, An Eocene hystricognathous rodent from Texas: Its significance in interpretations of continental drift, *Science* **175**:1250-1251.

Wood, A. E., 1973, Eocene rodents, Pruett Formation, southwest Texas; their pertinence to the origin of the South American Caviomorpha, *Texas Mem. Mus., Pearce-Sellards Series,* **No. 20**:1-40.

Wood, A. E., 1974, The evolution of the Old World and New World hystricomorphs, in: *The Biology of Hystricomorph Rodents* (I. W. Rowlands and B. J. Weir, eds.), *Symp. Zool. Soc. Lond.*, No. 34, pp. 21–54 [see also discussion, pp. 55–60], Academic Press, London.

Wood, A. E., and Patterson, B., 1959, The rodents of the Deseadan Oligocene of Patagonia and the beginnings of South American rodent evolution, *Bull. Mus. Comp. Zool.* **120**(3):281–428.

Paleobiogeographic Models of Platyrrhine Origins II

The Origin of the Caviomorph Rodents from a Source in Middle America

3

A Clue to the Area of Origin of the Platyrrhine Primates

A. E. WOOD

Introduction

It has long been recognized that the mammalian fauna of South America consists of three distinct parts—an ancient fraction, that reached South America in an unknown manner and from an unknown source or sources, but at least as early as the Paleocene; a modern fraction, that arrived on the continent from North and Middle America at the end of the Pliocene or the beginning of the Pleistocene, when the Panama land bridge was formed; and a third group, of intermediate age, that arrived in South America by rafting

A. E. WOOD • Department of Biology, Amherst College, Amherst, Massachusetts 01002. Present address: 20 Hereford Avenue, Cape May Court House, New Jersey 08210.

some time between the Mustersan, of probable middle Eocene age, and the Deseadan of early Oligocene age (Simpson, 1950, pp. 368–376). This last fraction includes the caviomorph rodents and the platyrrhine primates, both restricted to South America and the Antilles until Middle and South America were connected at the end of the Tertiary. Both have been postulated as having come either from Africa or from Middle America.

In both South America and Africa, the rodents and primates are first known from the early Oligocene, perhaps 37 million years before the present (mybp). In each case, the African (Fayum) forms were already specialized in their own direction (loss of P^2_2 in the primates and retention of DM^4_4 in the rodents), to such an extent that there is no possibility that they could have given rise to the South American forms, whatever may have been true of their unknown Eocene ancestors. The most similar living relatives of both the caviomorph rodents and the platyrrhine primates are African.

This has led to disagreement between those who feel that these resemblances among the living forms indicate the correct ancestral relationships, and those who feel that the present geographic relationships with North America and the clear North American origin of the Pleistocene invaders of South America are signposts pointing to a North or Middle American origin for the South American rodents and primates. The recent development of the theory of plate tectonics has been interpreted by the former group as suggesting that the South Atlantic was by no means so important a barrier in the Eocene as it is at present. It was, however, a much wider expanse of ocean than that which would have been involved in an invasion through the Antilles.

The crucial question for this volume is the origin of the South American primates, a subject of which I am profoundly ignorant. However, the consensus is that the caviomorph rodents and the platyrrhine primates had the same geographic origin, whatever it may have been, and I feel quite competent to discuss the geographic and taxonomic source of the infraorder Caviomorpha, the South American rodents.

Hystricognath Classification

The rodent suborder Hystricognathi is characterized primarily by a lateral shift of the angular process of the lower jaw, so that it does not arise in the vertical plane that contains the lower incisor, but lateral to that plane. Among rodents, this character occurs only in members of this suborder, and is the only morphologic feature known to be universally distinctive of the Hystricognathi. This lateral shift of the angular process is probably related to a strengthening of the pterygoid muscles. Initially, the shifting was very slight, which is what would be expected for the initial stages of any new trend. In the course of evolution, the strengthening continued, increasing the amount of hystricognathy, and leading to the forward expansion of the other end of the pterygoid muscle, resulting in the enlargement of the pterygoid fossa and, ultimately, in its extension through the floor of the braincase, as in the Caviomorpha, Hystricomorpha, and Bathyergomorpha. A feature sometimes present, that developed independently of the changes in the pterygoid muscle, was the forward movement of the origin of the *masseter medialis*, penetrating the infraorbital foramen and spreading out on the face, resulting in the tremendous enlargement of that foramen (hystricomorphy).

At present, the suborder Hystricognathi has a predominantly southern distribution, being particularly abundant in South America, with a lesser development in Africa, a small number of forms in Asia south of the Himalayas, and with a handful of late immigrants in North America, Europe, and northern Asia. The suborder includes the infraorder Caviomorpha, South American rodents; the infraorder Hystricomorpha,* including the Hystricidae or Old World porcupines, and the African superfamily Thryonomyoidea, formerly abundant but now reduced to the genera *Thryonomys* and *Petromus*; and the infraorder Bathyergomorpha, which Stromer (1926, p. 134), followed by Lavocat (1973, pp. 109–133), thought were African derivatives of what I include in the Thryonomyoidea.† An earlier, much more primitive, and presumably ancestral infraorder, the Franimorpha, is known from the latest Paleocene through the early Oligocene of North and Middle America. Unless it be accepted that hystricognathy originated two or more times independently, the franimorphs must have had an important place in hystricognath evolution.

Hystricognaths spread from southern regions to the north in late Tertiary and Pleistocene times, the most successful of the migrants being the North American genus *Erethizon,* which has spread over all the forested regions of North America. The predominantly southern distribution of the hystricognaths has been explained either as parallelisms in the evolution of the South American and African forms (presumably from an already at least incipiently hystricognath franimorph ancestry), or as indications of early trans-Atlantic migrations between Africa and South America. No evidence has been brought forward as yet by the supporters of either point of view that seems, to supporters of the other proposition, to carry much weight. Whatever the source of the Caviomorpha, the absence of franimorphs from the Eocene of Asia is presumably either an accident of collecting or a lack of recognition of the incipient hystricognathy in representatives of this infraorder.

*Lavocat (1973, p. 160) has considered the Hystricidae and the group of African rodents that he called the Phiomorpha to belong to the same infraorder, for which he used the term Phiomorpha. I agree that they should be united at the infraordinal level. However, the name Hystricomorpha has been widely used for over a century for the same group of animals and should be preferred as older and better known.

†Patterson and Wood (in press) present evidence that the Bathyergidae were not derived from Lavocat's Phiomorpha but rather from forms referable to the north Asiatic Oligocene Tsaganomyinae, which in turn were derived from the North American Eocene–Oligocene Cylindrodontidae.

The development of hystricomorphy occurred several times, significantly later than the development of hystricognathy, although it had been acquired by several North American franimorphs in the middle to late Eocene, and was also present in the earliest (early Oligocene) members of both the Caviomorpha and Hystricomorpha. It also evolved, completely independently, in several lineages of the other rodent suborder, the Sciurognathi (Theridomyoidea, Anomaluridae, Ctenodactylidae, Pedetidae, Dipodoidea). Other features have been considered characteristic of hystricognaths, but none is universally present in the suborder (Wood, 1975, Fig. 1), and the dates of their development are difficult or impossible to determine.

Hystricognathy is incipiently present in the earliest described rodent jaw* (Wood, 1962, p. 139; 1974, p. 17), from the latest Paleocene of Wyoming, and this presumably indicates a fundamental dichotomy between the hystricognathous and sciurognathous rodents. In the latter, the angular process is in the same vertical plane as the alveolus of the incisor, and the pterygoid muscle is less important. Accepting the existence of the Hystricognathi as a valid taxon, with a single origin, it is surely significant that all the demonstrable Eocene members of the suborder are from North or Middle America. In the early Oligocene, hystricognaths appear, more or less simultaneously, in Egypt and in Bolivia and Patagonia.

The Primitive Caviomorphs

When dealing with the evolution of mammals, it seems to me axiomatic that (1) brachydont (low-crowned) cheek teeth are more primitive than higher crowned ones; tooth evolution begins with cuspidate forms, and progresses toward crested ones, so that forms in which the cusps are clearly evident are more primitive than those where the cusps have been absorbed into crests; (2) it is primitive for the deciduous teeth to be replaced in the normal manner, so that DM^4_4 are replaced by P^4_4 shortly after M^1_1 erupt; (3) and retrogressive evolution, where advanced structures revert to a condition completely indistinguishable from the primitive condition, is a rare occurrence. It has also been axiomatic with me, for the last forty-odd years, that parallelism is an exceedingly important feature of rodent evolution, more important than in all the rest of the Mammalia combined.

Accepting these principles, the most primitive of the caviomorphs *must* be the brachydont ones, next the mesodont ones, with the hypsodont ones the least primitive. On this basis, the most primitive Deseadan genera are the octodontids *Platypittamys* and an undescribed genus (*Migraveramus*; Patterson

*I interpret the remark by Dawson (1977, p. 197) that the incipient "hystricognathous condition of *Reithroparamys* . . . was 'incipient' only *a posteriori*," as being in agreement with my interpretation.

and Wood, in press), and *Protosteiromys,* an erethizontid. These three are brachydont, the cusps can be recognized as distinct portions of the crests, and (as in all Deseadan rodents) there is normal replacement of the deciduous teeth. The lower cheek teeth of all three genera are clearly four-crested,* as are the uppers of *Platypittamys* (Fig. 1A; Wood and Patterson, 1959, Figs. 31-32). The uppers of *Protosteiromys* (those of *Migraveramus* are unknown) show the initial stages in the development of a fifth crest that arose by subdivision of the posterior one of the four basic crests seen in the upper teeth of *Platypittamys* and in the lower teeth of all three brachydont genera.

Only slightly more advanced are the Deseadan echimyids—*Deseadomys, Sallamys* and *Xylechimys.* These have higher crowned cheek teeth, being mesodont rather than brachydont, but the cheek-tooth patterns are almost as primitive as those of the octodontids and erethizontids. The upper molars are four-crested (Wood and Patterson, 1959, Fig. 4; Lavocat, 1976, Pl. 1, Fig. 5), but the lowers are in various stages of a reduction from a four-crested to a three-crested pattern (Wood and Patterson, 1959, Figs. 5, 7; Patterson and Pascual, 1968, Fig. 2; Lavocat, 1976, Pl. 1, Fig. 6).

All other known Deseadan rodents are considerably more advanced in their progressive height of crown (unilaterally or bilaterally hypsodont) and complete absorption of the cusps into the crests. The upper molars have generally acquired the fifth crest whose initial stages are seen in *Protosteiromys*; the lowers have remained four-crested.

If one accepts as primitive those Deseadan caviomorphs that appear to be the most primitive, the ancestors of the Caviomorpha must have been animals in which the cheek teeth were low-crowned, with cusps clearly present, but uniting into four transverse crests in both upper and lower teeth. Since there are no known African rodents that show the slightest suggestion of this tooth pattern, this conclusion means that there is no support for an African origin of the Caviomorpha.

The Deseadan caviomorphs were generally hystricomorphous, although in one genus (*Platypittamys*) the infraorbital foramen was small (for a hystricomorph) and there is no evidence that the masseter muscle passed through it (Wood, 1949, p. 13). This suggests that hystricomorphy was not fully developed in the pre-Deseadan immigrants to South America.

Possible Ancestors of the Primitive Caviomorphs

Hystricognathous Eocene rodents are, so far, known only from North and Middle America. In Europe the advanced Eocene rodents (the sciurog-

*The only basis that I can see for the identification of any part of the lower molars of Deseadan caviomorphs as a remnant of a fifth crest (Lavocat, 1976, e.g., pp. 69-70) is the preconceived opinion that these forms had had ancestors with five-crested molars.

nathous Theridomyoidea) had developed cheek-tooth patterns based on five transverse crests, the fifth (the latest to evolve—the mesoloph in the upper teeth, the mesolophid in the lowers) being in the middle of the tooth. Because this type of structure was very familiar to them, many European students of fossil rodents have erroneously concluded that these five-crested rodents were ancestral to all later rodents with five-crested teeth, no matter what type of angular process or zygomasseteric structure was present (e.g., Schlosser, 1884; Stehlin and Schaub, 1951; Schaub, 1958; Lavocat, 1976), although there is no evidence that any theridomyoid ever migrated from Europe or that any ever gave rise to non-theridomyoid rodents. If the theridomyoids gave rise to the hystricognathous African phiomyids, it is conclusive evidence that hystricognathy arose at least twice, since there is no possibility that the theridomyoids were ancestral to any of the New World franimorphs.

Eocene caviomorph ancestors are as unknown in Africa as they are in South America. The most primitive of the Oligocene phiomyids from Egypt, *Phiomys* (Wood, 1968, Figs. 1–5), clearly demonstrates that these forms were descended from hystricognath ancestors with five-crested teeth including the mesoloph and mesolophid. Asian Eocene rodents are not so well known as

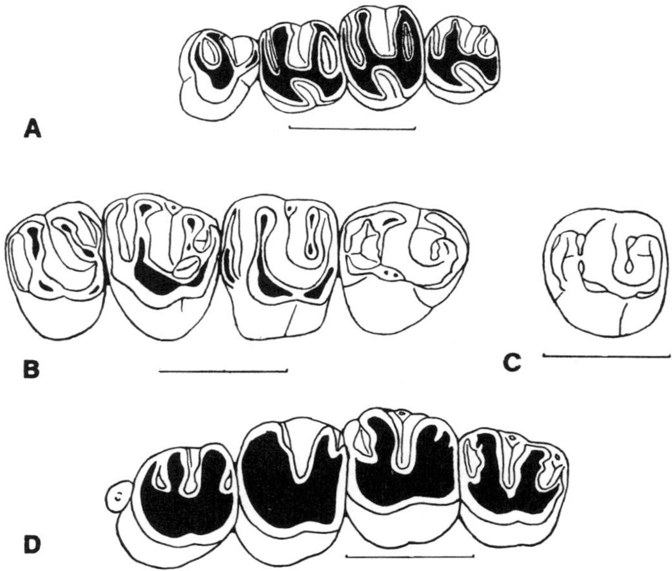

Fig. 1. Upper cheek teeth of a primitive caviomorph and of North American Eocene franimorph rodents. Black areas are dentine. Scale bars are 2 mm long. (A) *Platypittamys brachyodon,* Deseadan Oligocene of Patagonia, LP^4–M^3. After Wood, 1949, Fig. 3A. (B) *Mysops boskeyi,* mid-Eocene Pruett Formation, Texas, LP^4–M^3. Composite dentition redrawn from Wood, 1973, Fig. 3C,D,H,J. (C) *Pareumys troxelli,* late Eocene of Utah, LM^2. Am. Mus. Nat. Hist. No. 2021. (D) *Protoptychus hatcheri,* late Eocene of Utah, LP^3–M^3. After Wahlert, 1973, Fig. 1A.

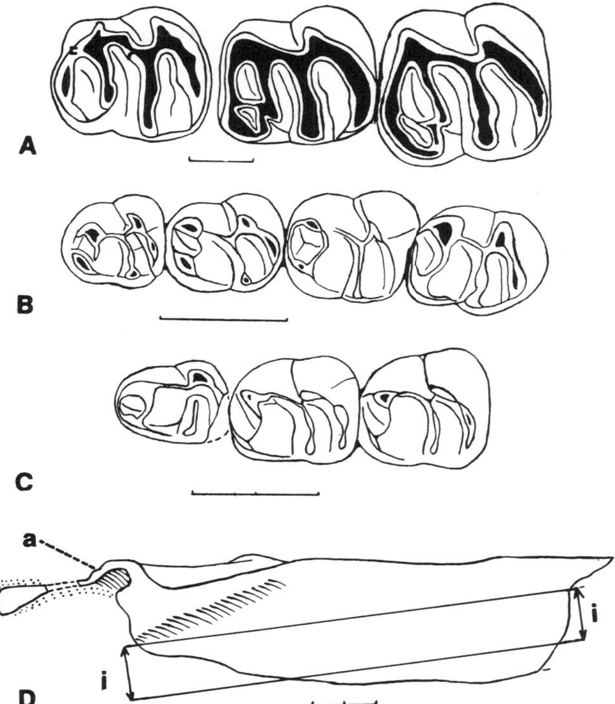

Fig. 2. Lower cheek teeth of a primitive caviomorph and of North American Eocene franimorph rodents, and ventral view of a franimorph lower jaw. Black areas are dentine. Scale bars are 2 mm long. (A) Undescribed octodontid, Deseadan Oligocene of Bolivia, RP_4–M_2. (B) *Mysops boskeyi,* mid-Eocene Pruett Formation, Texas, RP_4–M_3. Composite dentition redrawn from Wood, 1973, Fig. 4A,D,F,N. (C) *Pareumys troxelli,* late Eocene of Utah, LDM_4–M_2 reversed, Am. Mus. Nat. Hist. No. 2021. (D) Ventral view of lower jaw of *Cylindrodon fontis,* showing the angle (a) arising lateral to the course of the incisor (i).

those of Europe or North America, but none are known that could have been ancestral to the caviomorphs. Sahni and Srivastava (1976) have described middle Eocene rodents from India, including one isolated tooth questionably referred to the franimorph genus *Franimys*, but otherwise Asian franimorphs are unknown. All of the Indian genera are cuspidate, without development of transverse crests.

In North America, however, the four-crested pattern is very widespread in the middle and later Eocene, occurring in some of the more advanced members of the Paramyidae and Reithroparamyidae, in all of the Sciuravidae, Protoptychidae and Cylindrodontidae, as well as in the late Eocene of Oligocene Ischyromyidae. All of these families include genera that had brachydont cheek teeth in which the individual cusps clearly were distinct entities on the crests of which they are a part. In all of these forms, there is the normal mammalian replacement of deciduous teeth. It has recently been demonstrated (Black and Stephens, 1973, pp. 2–4; Wahlert, 1973, p. 14; Wood, 1962, p. 117; 1972, p. 1251; 1973, pp. 27–33) that a considerable number of these rodents, from both North and Middle America, were at least incipiently hystricognathous. Such forms have been placed in the infraorder Franimorpha (Wood, 1975). At least some of these genera, including among others *Protoptychus* (Wahlert, 1973, p. 7) and *Prolapsus* (unpublished data) were also hystricomorphous.

A few years ago, in reviewing the southernmost early Oligocene rodent fauna of North America (Wood, 1974), I pointed out a number of similarities in tooth structure between the Jaywilsonomyinae, a predominantly Middle American subfamily of cylindrodonts, and the Deseadan caviomorphs (Wood, 1974, p. 53). These include: suggestive similarities in the upper molars, very similar upper premolars, and fundamentally identical patterns of four-crested lower molars. *Jaywilsonomys* could not have been ancestral to the caviomorphs; like the known phiomyids, it is too late for that. But such middle to late Eocene cylindrodonts as *Mysops* (Figs. 1B, 2B; Wood, 1973, Figs. 3–4) and *Pareumys* (Figs. 1C, 2C) possess all the dental features needed in a caviomorph ancestor.

In 1974, I considered the resemblances of the Jaywilsonomyinae to the caviomorphs to be parallelisms, because I believed the cylindrodonts to have been sciurognathous, although there were no known specimens from North America with an intact angular process. Recently I found a single specimen of the early Oligocene *Cylindrodon* (American Museum of Natural History No. 94501) with the angular process preserved (Fig. 2D). This specimen is clearly subhystricognathous, so that the Cylindrodontidae must be transferred to the infraorder Franimorpha of the Hystricognathi. Incidentally, this makes sense of the similarities of the cylindrodonts to the Mongolian Oligocene Tsaganomyinae and of the latter to the African Miocene and later Bathyergoidea. A four-crested tooth pattern very similar to that of the cylindrodonts is found in the North American franimorph *Protoptychus* (Fig. 1D), in which both the upper teeth and the lowers (so far undescribed) are four-crested with visible remnants of the ancestral cusps. As in most North American Eocene rodents, there is no trace of a mesoloph or mesolophid. Figure 1 clearly demonstrates the close similarities between the upper molars of *Protoptychus* and those of the Deseadan octodont *Platypittamys* (Fig. 1A,D), and the similar but more primitive upper molar of the late Eocene cylindrodont *Pareumys* is shown in Fig. 1C. In the lower teeth, the basic pattern of four-crested molars seen in Deseadan octodontids (Fig. 2A) is present in North American Eocene cylindrodonts (Fig. 2B,C; Wood, 1973, Fig. 4). Only very minor modification would be required to derive the tooth pattern of one of the brachydont caviomorphs from teeth such as these. There is, so far as I am aware, not a single feature of the cheek teeth (essentially the only parts known) that would prevent *Mysops boskeyi* of the mid-Eocene of the Big Bend area of Texas from being an actual ancestor of the Caviomorpha, given my assumptions as to which of the Deseadan caviomorphs have the most primitive cheek teeth.

Since *Mysops* was a cylindrodont, it presumably was at least subhystricognathous.

Middle America as the Source Area for Caviomorphs

The problem of determining the source area for the caviomorphs is complicated by the fact that no potential immediate ancestors are known from Africa. Obviously, none of the North American hystricognaths could have been immediately ancestral to the caviomorphs—they lived too far to the north. For both the platyrrhines and the caviomorphs, demonstration of the possibility of derivation from a Middle American ancestry has been rendered difficult by the almost complete absence of early Tertiary fossil mammals from Mexico and Central America. However, as defined by Ferrusquia (1978, p. 198), the paleogeographic Middle America extends north at least to the latitude of the northern boundary of Mexico, and includes western and southwestern Texas. This adds considerably to the Middle American fauna, including both rodents and primates. As demonstrated by Wood (1974, pp. 102–104), the early Tertiary rodents of southern California and west Texas clearly were the northern representatives of a very diverse southern rodent fauna, quite different from the better known faunas farther north in the western United States. Moreover, hystricognath rodents are now known from the Eocene of Mexico and southwest Texas, and advanced primitive primates were present in southwest Texas, in contrast to the complete absence of both groups from the known Eocene of Africa.

There is often a tendency to think of Middle America as being too small an area to have been a potential source of much evolutionary change. However, the area of Middle America (including only Mexico and Central America) is more than half that of Europe excluding the USSR, and the latitudinal extent is over 25°, about the same as the latitudinal distance from Gibraltar to the Shetlands or to Bergen, Norway.

In spite of the large size of Middle America, only four genera and five species of pre-Pliocene rodents have been described from Mexico and Central America, represented by only a handful of specimens. The fact that half of these rodents were hystricognaths suggests that this area was an important one for the early Tertiary development of the Hystricognathi. In contrast, no Eocene hystricognaths have been reported from any part of the Old World.

If there were a strong east–west equatorial current flowing between Middle America and South America in the Eocene, as Lavocat has suggested (1976, p. 80), it is entirely possible that rodents were unable to make a direct crossing from Panama to Colombia. However, as I have pointed out (Wood, 1977, p. 104), there must have been a west–east current flowing along the northern side of the Greater Antilles, no matter what type of geography is postulated for the area. There has been extensive Tertiary movement between South America and the Antillean block, but even if this movement has amounted to 1200 km since the middle or late Eocene, the Antilles would have extended far enough to the east, at that time, to have permitted rodents that had reached the eastern Antilles to have been rafted to South America, even if there were a strong east–west current between the two areas. It is true that there are no Eocene caviomorphs known from the Antilles, but this is not surprising since no continental Eocene deposits have been reported from the area.

In summary, it seems quite clear to me that there is *at least* equal paleogeographic and paleoceanic evidence for a Middle American as for an African origin of the Caviomorpha, and that the Eocene rodents of Middle America are far and away, on morphologic grounds, the most likely ancestors of the Caviomorpha.

Why the Phiomyidae Cannot Be Caviomorph Ancestors

As I have shown above, the primitive caviomorphs possessed four-crested cheek teeth, almost without exception. And, in the one exception, the upper teeth of *Protosteiromys,* the extra crest is developing behind the metacone, rather than in front of it as is the case with the mesoloph present in the phiomyids. The phiomyids are in process of losing the mesolophid, but this is taking place after the loss of the distinctness of the individual cusps.

But the clearest evidence is in the matter of the replacement of the deciduous teeth. In all Deseadan caviomorphs, the replacement is of the normal mammalian pattern. In the most primitive of the phiomyids (*Phiomys;* Wood, 1968, Figs. 1–5), the replacement of the deciduous teeth by P^4_4 is greatly delayed, so that the deciduous teeth remained in use for most of the animals' lifetime. In most Oligocene and all later thryonomyoid genera, replacement of the deciduous teeth never occurs. This clearly demonstrates that even the earliest of the Thryonomyoidea were evolving in a direction very different from that followed by the caviomorphs, and that even *Phiomys* was far too specialized to be considered as having any close relationship to the Caviomorpha (Wood and Patterson, 1970, pp. 632–633).

The diversity of the African Oligocene primates indicates that they had been in Africa for a significant period of time. The lack of diversity of the rodents, on the other hand, indicates that they had not had a long history in Africa before the Oligocene.* There is evidence for only a limited amount of adaptive radiation of African rodents by the end of the early Oligocene, and all known genera had brachydont or at most mesodont cheek teeth. In contrast, in South America there is no evidence for any great diversity or notable

*Lavocat (1973, p. 257) suggested that Africa was invaded by unknown paramyid rodents probably in the Paleocene, but certainly no later than the beginning of the Eocene. This is highly unlikely, considering the lack of diversity of Fayum rodents, in view of the rapidity with which rodents have always diversified when they invaded a continent they had not previously occupied.

adaptive radiation of the primates in the early Oligocene, whereas the rodents had evolved to the point where there were hypsodont as well as brachydont forms, and where there is enough diversity to warrant the known fossils being distributed among at least seven families. That is, the present evidence supports the postulate that, when the Oligocene began, the primates had been in Africa some millions of years longer than they had been in South America, but that the reverse was true in the case of the rodents.

I should like to document the point about the much greater abundance and diversity of the rodents in the early Oligocene of South America than of Africa. The Egyptian Fayum (Jebel el Qatrani Formation) covers essentially the entire Oligocene (Simons, 1968, p. 4). The Lower Fossil Wood Zone (Quarries A, B, E) is early Oligocene; Quarry G is mid-Oligocene; the Upper Fossil Wood Zone (Quarry I) is middle to late Oligocene; and the series is capped by a basalt dated at about 25 mybp, or the end of the Oligocene. Yale Quarry I, where modern collecting methods were used and especial attention was given to the small forms, has yielded 386 specimens, of which only 18 (4.7%) were rodents (Patterson and Wood, in press). In contrast, Cabeza Blanca, Loomis' Deseado locality in Patagonia, with approximately the same areal extent as Quarry I, yielded 283 specimens from surface collecting. Of these, 102 or 36% were rodents. The collecting methods used at that time were certainly not such as to lead to a bias in favor of small forms. In the Lower Fossil Wood zone, the rodents again make up only a very small percentage of the specimens that have been collected. Quarry G is the only Fayum locality where the rodents are a major proportion of the individuals collected, and here the vast majority of the identified specimens belong to a single species, *Metaphiomys schaubi.* That is, the Oligocene African rodents do not seem to have been anywhere near so numerous or taxonomically diverse as those of the Deseadan.

If these comparisons mean anything, they indicate, I believe, that the rodents reached Africa in the latest Eocene or earliest Oligocene, most probably from Asia Minor, although their immediate ancestors are completely unknown. Because the Eocene rodents of Europe are relatively well known, that continent presumably was not the source of the phiomyids. Similarly, the South American evidence would seem to show that the rodents had been in that continent some millions of years, and may even have arrived immediately after the Mustersan, now generally considered Middle Eocene. The only described fossils from any part of the world that could have been the *immediate* ancestors of the Caviomorpha are Middle American members of the Cylindrodontidae.

Conclusion

The ancestors of both the platyrrhine primates and the caviomorph rodents reached South America in the latter part of the Eocene. No unquestioned ancestors of either are currently recognized. However, the franimorph population of North and Middle America is the only known Eocene rodent population from any part of the world that had the structural potential to have given rise to the Caviomorpha. The rodents from western United States were too far removed from South America to have been directly involved in the migration. Recent discoveries in Middle America, however, show that there was a diverse rodent fauna in this area, with a very significant proportion being hystricognathous. Such mid-Eocene cylindrodonts as *Mysops boskeyi* from southwestern Texas might even have been the actual ancestors of the caviomorphs. There is no morphological possibility that any of the known African Phiomyidae could have occupied such a position, and the evidence indicates that the phiomyids had not been in Africa very long when the Oligocene began. The currents in the Eocene Gulf of Mexico would presumably have favored the raft transportation of rodents from Middle America through the Antilles and thence to South America, with much shorter (both in time and distance) ocean voyages than would have been required if the caviomorph ancestors had come from Africa. Finally, the area of Middle America was amply sufficient to have provided a land mass upon which the evolution to an incipient caviomorph could have taken place.

I therefore conclude that, while we cannot at this moment prove that the caviomorph rodents reached South America from Middle America, such a source is the most reasonable on the basis of current knowledge, since it and North America alone included in their mid-Eocene rodent populations known forms that might have been the caviomorph ancestors, and since it was at least as feasible for rodents to have reached South America from Middle America via the Antilles as for them to have reached it by floating across the entire South Atlantic. And, as I indicated previously, I feel sure that the platyrrhine primates reached South America from the same source as did the rodents.

Note Added in Proof

In a recent paper, Hussain *et al.* (1978) have described a family Chapattimyidae from the Middle Eocene of Pakistan, which they consider to have been ancestral to the African Oligocene Thryonomyoidea, among others. However, all known lower jaws referred to as chapattimyids are sciurognathous; the incisor enamel (contrary to the statement by Hussain *et al.*) is pauciserial; and the cheek-tooth pattern clearly unites the chapattimyids with the Ctenodactylidae rather than with any hystricognaths [Hussain, S. T., DeBruijn, H., and Leinders, J. M., 1978, Middle Eocene rodents from the Kala Chitta Range (Punjab, Pakistan), *Proc. Konink. Nederland. Akad. Wetensch., Ser. B,* **81**:74–112].

References

Black, C. C., and Stephens, J. J. III., 1973, Rodents from the Paleogene of Guanajuato, Mexico, *Occasional Papers, The Museum, Texas Tech. Univ.* **No. 14**:1–10.

Dawson, M. R., 1977, Late Eocene rodent radiations: North America, Europe and Asia, *Géobios, Mém. Spéc.* **1**:195-209.

Ferrusquia-Villafranca, I., 1978, Distribution of Cenozoic vertebrate faunas in Middle America and problems of migration between North and South America, in: *Conexiones Terrestres entre Norte y Sudamerica* (I. Ferrusquia-Villafranca, ed.), pp. 193-321, Univ. Nac. Auton. México, Inst. Geología, Bol. **101.**

Lavocat, R., 1973, Les Rongeurs du Miocène d'Afrique Orientale. I. Miocène Inférieur, *Ecole Pratique des Hautes Etudes, Institut de Montpellier, Mémoires* **1**: i-iv, 1-284, separate folder of stereophotos.

Lavocat, R., 1976, Rongeurs caviomorphes de l'Oligocène de Bolivie. II. Rongeurs du Bassin Déséadien de Salla-Luribay, *Palaeovertebrata* **7**:15-90.

Patterson, B., and Pascual, R., 1968, New echimyid rodents from the Oligocene of Patagonia, and a synopsis of the family, *Breviora* **No. 301**:1-14.

Patterson, B., and Wood, A. E., 1981, Rodents from the Deseadan Oligocene of Bolivia and the relationships of the Caviomorpha, *Bull. Mus. Comp. Zool.* **149**.

Sahni, A., and Srivastava, V. C., 1976, Eocene rodents and associated reptiles from the Subathu Formation of northwestern India, *J. Paleontol.* **50**:922-928.

Schaub, S., 1958, Simplicidentata (= Rodentia), in: *Traité de Paléontologie* (J. Piveteau, ed.), Vol. 6, No. 2, pp. 659-818.

Schlosser, M., 1884, Die Nager des europäischen Tertiärs nebst Betrachtungen über die Organisation und die geschichtliche Entwicklung der Nager überhaupt, *Palaeontographica* **31**:1-143.

Simons, E. L., 1968, African Oligocene Mammals: Introduction, history of study, and faunal succession. Part I of: *Early Cenozoic Mammalian Faunas, Fayum Province, Egypt, Bull. Peabody Mus. Nat. Hist.* **28**:1-21.

Simpson, G. G., 1950, History of the fauna of Latin America, *Am. Sci.* **38**:361-389.

Stehlin, H. G., and Schaub, S., 1951, Die Trigonodontie der simplicidentaten Nager, *Schweiz. Paläontol. Abh.* **67**:1-385.

Stromer, E., 1926, Reste Land- und Süsswasser-bewohnender Wirbeltiere aus dem Diamantenfeldern Deutsch-Südwestafrikas, in: *Die Diamantenwüste Südwestafrikas,* Vol. 2, by E. Kaiser, pp. 107-153, Dietrich Reimer, Berlin.

Wahlert, J. H., 1973, *Protoptychus*, a hystricomorphous rodent from the late Eocene of North America, *Breviora* **No. 419**:1-14.

Wood, A. E., 1949, A new Oligocene rodent genus from Patagonia, *Am. Mus. Novitates* **No. 1435**:1-54.

Wood, A. E., 1962, The early Tertiary rodents of the Family Paramyidae. *Trans. Am. Phil. Soc., N. S.* **52**:1-261.

Wood, A. E., 1968, The African Oligocene Rodentia. Part II of: *Early Cenozoic Mammalian Faunas, Fayum Province, Egypt. Bull. Peabody Mus. Nat. Hist.* **28**:23-105.

Wood, A. E., 1972, An Eocene hystricognathous rodent from Texas: Its significance in interpretations of continental drift, *Science* **175**:1250-1251.

Wood, A. E., 1973, Eocene rodents, Pruett Formation, southwest Texas; their pertinence to the origin of the South American caviomorphs, *Texas Mem. Mus., Pearce-Sellards Series,* **No. 20**:1-40.

Wood, A. E., 1974, Early Tertiary vertebrate faunas, Vieja Group, Trans-Pecos Texas: Rodentia, *Bull. Texas Mem. Mus.* **21**:i-iv, 1-112.

Wood, A. E., 1975, The problem of the hystricognathous rodents, in: *Studies on Cenozoic Paleontology and Stratigraphy in Honor of Claude W. Hibbard, Univ. Mich. Papers Paleontol.* **12**:75-80.

Wood, A. E., 1977, The Rodentia as clues to Cenozoic migrations between the Americas and Europe and Africa, in: *Paleontology and Plate Tectonics* (R. M. West, ed.), *Milwaukee Publ. Mus. Spec. Publ. Biol. Geol.* **2**:95-109.

Wood, A. E., and Patterson, B., 1959, The rodents of the Deseadan Oligocene of Patagonia and the beginnings of South American rodent evolution, *Bull. Mus. Comp. Zool.* **120**:279-428.

Wood, A. E., and Patterson, B., 1970, Relationships among hystricognathous and hystricomorphous rodents, *Mammalia* **34**:628-639.

The Implications of Rodent Paleontology and Biogeography to the Geographical Sources and Origin of Platyrrhine Primates

4

R. LAVOCAT

Introduction

It is a normal practice in science to infer from what is better and more completely known in order to discover the structure and the meaning of that which is less well or only partly known. The similarity of facts known on both sides of a controversy suggests that the best documented side be taken as a good model for the reconstruction of the structures still unknown on the other side. Since the amount of information regarding the evolutionary history of rodents is much more complete than that of the primates, one can validly and usefully argue with reasonable probability from the rodents to the primates.

Following this view, one has first to recall the essential evidence about rodents and appreciate the scientific weight of this evidence. In 1953, Schaub, as a consequence of his research (following that of H. G. Stehlin), created the

R. LAVOCAT • Ecole Pratique des Hautes Etudes, Laboratoire de Paléontologie des Vertébrès, Faculte des Sciences, Place E. Bataillon 34000—Montpellier, France.

suborder Pentalophodonta. This suborder includes all the rodents with a dental structure based, in the opinion of these two authors, upon the *Theridomys* plan:

1. The Theridomorpha flourished in Europe from the Eocene to the late Oligocene; up until now they have been recovered only from Europe.
2. The Hystricidae (true porcupines) are now living in Africa, Asia, and the meridional part of Mediterranean Europe.
3. The Phiomorpha are typically African, known in 1953 by only a few forms, the status of which was not clear to Schaub. Some mention of their occurrence in the Miocene of India has been made. However, more recent studies of the Oligocene and Miocene faunas of Africa have shown the strong homogeneity and major importance of this African group.
4. The Caviomorpha lived, in South America only, from the lower Oligocene to the Pliocene but sent some migrants into North America during the Pliocene.

Notwithstanding dental similarities, Schaub's opinion on placing the Castoridae among the Pentalophodonta, can no longer be retained with certainty. One must also reject the inclusion of the Eupetauridae, Spalacidae, Rhizomyidae, and Pellegriniidae for several reasons not explained in this text.

Schaub has explicitly stated that the creation of this suborder did not imply that the Theridomorpha were the ancestors of the other Pentalophodonta. However, I think that there are now good arguments showing that this group can be very near to the ancestral group, but not precisely. The Theridomorpha are generally supposed to be the local descendants of some paramyids (until now not clearly recognized). As a result of some observed dental similarities, when first discovered, they have been placed near the South American Caviomorpha: witness for example the name of *Echimys* given by de Laizer to an Auvergne form. For similar reasons based on dental similarities, several reputable paleontologists thought that the African Phiomorpha were the direct descendants of the Theridomorpha. A. E. Wood has been and remains a strong opponent of the idea that there are structural similarities between the teeth of the Caviomorpha and the Theridomorpha.

The dental similarities being what they are, some osteological characters, if only a few, can also be documented as shared by the Theridomorpha with the other Pentalophodonta; for example, hystricomorph infraorbital structure, an advanced character which has been evolving within the theridomorph group itself. There are several other anatomical characters, including the sciurognath structure of the mandible, the anterior shutting of the pterygoid fossa, and the erinaceomorph structure of the middle ear. These are primitive and clearly distinguish this group from the Phiomorpha as well as from the Caviomorpha.

On the other hand, if one compares the anatomical structures of the Caviomorpha to those of the Phiomorpha, leaving aside the dental evidence for a further discussion of Wood's position, the similarities are astonishing.

They bear on all the structures or combinations of structures of the osteology of the skull, and also on the myological structures, as well as on very elaborate fetal structures. Of course, one or another of the osteological structures is also found in some of the other groups of rodents. But it is only in the Caviomorpha and Phiomorpha that all of these structures can be found in identical association with one another.

The earliest authors on this topic, for example Tullberg (1899), had clearly recognized this fact. After having thoroughly studied the two known African forms and many South American forms, Tullberg admitted a very close systematic connection between the two groups. Following the same view, many authors have included the two African genera respectively into two different caviomorph families. Having done so, one must take note of the rather difficult biogeographic problems that such an arrangement creates. W. D. Matthew, strongly favoring the stability of the continental masses as well as the continental ways of paleogeographical dispersal, admitted a remarkable exception for the rodents. He properly suggested that, notwithstanding the enormous oceanic width to be crossed, the Caviomorpha were direct offshoots from the African rodent ancestors of *Thryonomys* and *Petromus*.

Nevertheless, in 1950 Wood published an important paper with conclusions to which I subscribed at the time. A thorough analysis of the migration possibilities indicated that, on the basis of the current scientific knowledge of the day, no real possibility of migration could be accepted and no explanation could be given other than extreme parallelism (even if this unavoidable conclusion was highly surprising). Later, the discovery in the Deseado formation of Patagonia of the complete skeleton of the rodent *Platypittamys* suggested to Wood that the fundamental plan of the caviomorph teeth was tetra- and not pentalophodent. This conclusion could have far-reaching implications if it could definitely be proven. Further discussion of this point will be presented later.

The Zoological Evidence

Recently much new information has come to light concerning the Caviomorpha and Phiomorpha due to their intense study by several European and American workers. Wood (1968) has published on the Oligocene fauna of the Fayum, and documented the most ancient phiomorphid fauna known; I have also published an extensive study of the East African Lower Miocene rodents. The presence in this magnificent collection of the complete skulls of seven genera of the Phiomorpha previously entirely unknown and a very large series of dentitions gives a particular weight to the osteological comparisons. All of these studies resulted in new elements that could be added to the list of common characters previously known, among which are the following: the identical position of the lacrimal bone and openings; the great size of the sphenopalatine foramen; the position of the temporal bone; the structure of the external and middle part of the bony ear; the arterial circulation type; the structure of the fetal membranes, shown by Mossmann and Luckett (1968) to be of a very specialized advanced type; the identity of the muscular anatomy of the Caviomorpha and Phiomorpha (Woods, 1972); the close systematic relationships of the parasites (nematod) of some caviomorphs, with those of African *Thryonomys* (Durette-Desset, 1971); and the remarkable dental similarities between teeth of either the Miocene or Oligocene phiomorphs of Africa with homologous teeth of South American Caviomorpha.

Wood (1975) denied the value of the arguments taken from fetal membranes, saying that, since the fetal membranes of rodents so clearly different as the Ctenodactylidae, the Sciuridae, and the Anomaluridae are most similar, the similarity between the fetal membranes of the Caviomorpha and Phiomorpha has no real meaning. But one can easily reply with Luckett's (1971) argument that the reason for the similarity between the fetal membranes of Ctenodactylidae and the others is due to the fact that they are all of a primitive type. So, in the latter case, the only assumed relationship is really with the primitive type itself. On the contrary, the structure shown in the Caviomorpha and the Phiomorpha is a very sophisticated one and has evolved so the identity of structure between these two groups retains all of its systematic value.

So, zoologically speaking, there exist several strong arguments favoring extremely close relationships between the groups from the two different continents. Presently, it can be stated that if there does appear to be a reasonable possibility of migration across the South Atlantic Ocean between the two continents at a convenient time in the Eocene, the zoological arguments would strongly favor such a solution. Moreover, if it ever should be demonstrated that the Caviomorpha have a suitable structural ancestor in North America and thus could have originated there, we will then be faced with the problem of explaining the vast accumulation of arguments favoring close relationships between the African and South American rodents. It is impossible merely to forget these facts.

Two zoological arguments have been proposed by Wood (1974) against an African origin and in favor of a North American origin of the Caviomorpha. First, in his opinion, the similarity of structure between the teeth of the Caviomorpha on one side and the Theridomorpha and Phiomorpha on the other is a fallacious one. This point has already been discussed by Lavocat (1974). It is perhaps not possible, given the present state of knowledge, to produce a demonstration in itself absolutely apodictic for either of the opinions. However, in such a situation, one cannot neglect the fact that since all the other anatomical structures studied are intimately correlated, the probability that the structure of the dentition would entirely disagree with all the other structures seems highly unlikely. Clearly, the most parsimonious hypothesis should be the one adopted.

Concerning the problem of tooth structure, some new arguments can be added to those previously presented. For all paleontologists, including Wood, the Theridomorpha and Phiomorpha have teeth showing pentalophodont structures. Regarding the Phiomorpha, Wood, who does recognize their pentalophodont structure, suggests that the teeth of *Phiomys* from the Fayum Oligocene result from a structure either *tetra-* or *pentalophodont.* So, following his own logic, if he should demonstrate that the caviomorph teeth have a tetralophodont origin, he could not, however, demonstrate that their ancestor is not the same as the ancestor of the Phiomorpha and that this ancestor did not live in Africa.

Following Wood's view once again, the structure of the teeth of *Platypittamys,* assumed to be ancestral to that of Caviomorpha, is necessarily the most primitive, since these teeth are the most brachyodont. But the argument is not irrefutable. True, a brachyodont tooth is more primitive than a hypsodont tooth of the same lineage, and, as a consequence, its structure is the more primitive of the two. But there is no proof that a brachyodont tooth shows a more primitive structure than a mesodont or hypsodont one of another lineage. This is certainly the case with most of the other contemporaneous forms where the pentalophodont morphology is quite evident. A lineage can, at the same time, retain strong brachyodonty and evolve rapidly in dental design. So, if one compares different lineages, it is easy to understand that the brachyodont species can have a more advanced design of the tooth crown than the hypsodont species. With the same level of hypsodonty, two Miocene genera of Phiomorpha are extremely similar. But one shows a mesoloph, while the other does not. And I must add that it remains open for discussion as to which Caviomorpha is really the most primitive. Bugge has shown (1974) that, among the Caviomorpha, the Erethizontidae show a more primitive structure of the arterial circulation than any of the others. If so, following Wood's argument, there is good reason to say that the brachyodont teeth of *Coendu* are primitive. Indeed, Wood speaks of these teeth as primitive, but he offers them as examples of tetralophodont teeth, while *Coendu* shows an objectively pentalophodont structure.

The second zoological argument put forth by Wood is taken from the presence in the North American Eocene of hystricognathous rodents; moreover, *Protoptychus,* from the Upper Eocene, is not only hystricognathous, but also hystricomorphous (wide infraorbital foramen). These two characters shared with the Caviomorpha are certainly important (but the fact that the Phiomorpha also share these characters with the Caviomorpha has necessarily the same importance), and they demonstrate that the *zoological* probability of a North American origin of the Caviomorpha is by no means nonexistent. Wood himself remarks that the hystricomorph structure arose independently in several rodent suborders; so, in itself, the value of this character, *taken alone,* is low and mostly negative.

At any rate, it is evident that a balance between the weight of two common characters on one side, and the weight of a great number of common characters (practically all, down to the family level) on the other side must be established. Certainly it can not be logical to say that the presence of two characters in common is decisive proof, while the presence of 10 or 20 characters in common proves nothing. The known facts, although incomplete, show, in my opinion, that some relationship between the Caviomorpha and the North American rodents is perhaps not impossible zoologically speaking, but is not highly probable. In my opinion, the known facts support neither the existence of close relationships nor of the direction of these eventual relationships (whether in migration from North to South America or from South to North America)!

The other aspect of the problem is a paleogeographical one. Normally, zoologically close relationships such as those existing between Caviomorpha and Phiomorpha would be explained by direct connections between the two groups. For example, the two groups may be parts of the same living community, first united, then afterwards separated into the said two groups. Or, the origin of one of the two groups could have resulted from the dispersal of some migrants from the main ancestral population, thus giving rise in the new locale to a daughter population.

No paleontologist can properly propose the first hypothesis, since not one of them can admit that the rodents could have developed at a time when the two continents were closely enough connected to give to this group the possibility of overlapping group boundaries. The conclusions drawn from plate tectonics, as well as the demonstratively strong faunal differences at the time of the first presence of rodents in South America speak clearly against this hypothesis. Moreover, the faunal comparisons show quite clearly that migration towards South America has been subjected to highly variable selective conditions, not compatible with a land connection. Only a raft crossing, with its aleatory conditions, favoring small animals, preferably arboreal in their habits, easily taking their place on the rafting trees, can explain all the characteristics of this migration. Only this process agrees with all the available data.

So, one must examine, taking full advantage of the plate tectonic reconstructions whether the assumed process could have existed between North and South America, or between Africa and South America; and, if, eventually, each way is shown possible, one must then select which explanation is the more probable. I will hereafter compare the results of the zoological and the paleogeographical data sets, attempting to document the significance of the eventual agreement of all the evidence.

The Paleogeographical Evidence

Before the revolution in plate tectonics most scholars argued and some continue to argue that the distance between South America and Africa was much too considerable to permit a raft-crossing. I will show here that nothing is less certain; all the paleogeographical data resulting from the plate tectonic synthesis strongly favor the idea of migration from Africa to South America. I recently published (Lavocat, 1977) paleogeographical maps, deduced from the data given by Francheteau (1973), which show the relative positions of the three continental blocks between a 65 and 45 Myr. B. P. time interval during which the migration certainly happened. As recently as 1975, Wood has taken his argument from the position that as early as the Cretaceous the South Atlantic was already 3000 km wide, and states, "Lavocat . . . assumes, unjustifiably in my opinion, that in the late Eocene the western tip of Africa and the eastern tip of Brazil were on the same meridian" (p. 77). What actually seems to be unjustifiable is, properly, the affirmation that my opinion is unjustifiable! The comparison of my reconstruction with that of Smith and Briden (1977), based upon recent research results, shows a rather good agreement. In fact, around the end of the Cenomanian, the South Atlantic was already 3000 km wide . . . at Capetown latitude! But the position of the rotational pole at 21.5° North and 14° West, given by Francheteau, implies that, for the same angular rotation, up to the Turonian, the distance was expanding very rapidly at the Capetown latitude, since this region was then near the equator of rotation. Conversely, the expansion was very slow in the Natal-Recife region which was rather near to the pole of rotation. I think that the level of certainty of the position estimated in the Eocene by me is really good. And with the distance of no more than 1000 km approximately, and probably less, between the African and South American plates, in the Middle Eocene, the raft crossing is not at all inconceivable.

The situation is much less clear on the North American side. Here indeed, the Panamanian isthmus and the Antilles present strong problems not yet clearly solved. As shown by Smith and Briden (1977), following many others, a reconstruction of the Isthmus of Panama and of the Antilles in their same relative position to the North American plate as now results in a significant overlapping of these parts with the adjacent African and South American plates, prior to the separation of the North American plate. (There is no need to say that such an overlapping is a major impossibility.) Several solutions have been advanced by excellent authors, but it does not seem, up to now, that any of these are really satisfactory. What is certain is that not only is there no direct tectonic proof of any terrestrial connection between South and North America through Central America in the early Tertiary, but that every tectonic reconstruction will have to take into account the fact that paleontological evidence argues strongly for a total separation of the South and North American plates during most of the Tertiary.

The hypothesis of a short term connection at the beginning of the Tertiary, as elaborated by many paleontologists, is itself only a mere deduction from the idea that the South American fauna could not come from a land mass other than North America. To now retain this assertion without additional proof is only to assume as a principle what should have to be proven as a conclusion. Starting from the fact that certainly there were no terrestrial connections between North and South America for most of the length of the Tertiary following the earliest Paleocene, it seems hard to find a paleotectonic reconstruction in congruence with this paleontological certitude and with the idea of a land connection at a moment when the distance between the two plates appears to have been the greatest. At any rate, every paleontologist, including Wood, agrees that, at the time of the rodent migration, there was a channel or strait connecting the Atlantic and Pacific. Through this strait a continuous current certainly crossed, the possible speed of which has been evaluated at several knots. Even if little else is known about Central American events at that time, it seems that the probability should be at least in favor of a more westward relative position, in the plate complex, than there is now. If so, while the current crossing the African–Brazilian oceanic gap was thereafter moving along South America's northern coast (providing a good opportunity for a landing of the rafts crossing from Africa), the same current was probably reaching Central America and flowing by the strait in a northerly direction. Such a current flow would mean that rafts from North or Central America would have to run *against* the current to reach South America. Even if it were possible to cross the strait given enough length of coast on the other shore and allowing for the necessity of drift when crossing, it would have been impossible, without a motor, to go against the current. I doubt if strong winds blowing against the flow of the current could be expected to be much of a factor in this nearly equatorial region. On the contrary, if rafting could hardly be expected from North or Central America towards South America, there appears a real possibility of making use of currents to go by rafts from South to North or Central America. As a consequence, suppose that if, in the Upper Eocene and Lower Oligocene of Central and North America, rodents should be found showing strong affinities, without any doubt, with the Caviomorpha, then one could very seriously wonder if these rodents did not result from a migration started in South America.

Instead, one may almost completely exclude that the relationships between South American and African fauna at that time could be explained by a transoceanic migration starting from South America towards Africa. It is indeed very clear that, even in the Early Eocene, the equatorial current of the African–Brazilian strait must have been a one-way current from Africa to South America. From South America to Africa, the returning current, far in the South, was crossing such a width of ocean that here I agree with Wood that migration under such conditions was really impossible. So, a hypothesis proposing a North American origin of the Caviomorpha would run into biogeographic difficulties explaining the close relationships between the Caviomorpha and the Phiomorpha.

Thus, the data of plate tectonics and paleophysical geography as well as of paleontology and zoology run in the same direction to furnish a coherent explanation of the origin of the Caviomorpha from Africa.

Looking at the African rodent fauna itself, comparisons made with the Theridomorpha of Europe provide good reasons to propose that Europe is the ancestral region for African rodents sometime between the Paleocene and the Lower Eocene. The tectonics of the peri-Mediterranean region is a matter under heavy discussion. I should not say that the possibility of a land connection at the convenient time is positively established, but I think that this possibility must remain at least a working hypothesis. Africa has certainly received at least a part of its mammalian fauna from the outside, and there are several reasons favoring the hypothesis that Europe should be considered a good candidate for the area of origin of a part of this fauna. There also is agreement that more must be known of the (possible) connection of Africa with Southeastern Asia, a connection which seems the best explanation for the presence of anthracotheres in the Oligocene deposits of the Fayum region of Africa. But I do not think that one should exclude earlier contacts between Europe and Africa, before the time when the Indian plate became firmly connected with Asia.

Conclusion

As a result of the zoological affinities between the Phiomorpha and Caviomorpha, one has to acknowledge that the affinities between these two groups are such that they can be explained only by close direct phylogenetic relationships between the two groups. Evidence from plate tectonics show that the possibility of migration needed to explain this strong zoological affinity did really exist at the appropriate point in geologic time. Some zoological factors suggest, and the plate tectonics as well as the paleophysical geography clearly confirm, that the migration possibilities were *only* in a direction from Africa towards South America. As a consequence, the ancestral group of the Phiomorpha should be regarded as the ancestors of the Caviomorpha, and not the reverse. The study of plate tectonics and of marine currents shows that migration was possible from Africa to South America, but extremely difficult, if not altogether impossible, from North America towards South America (but probably possible from South toward North America). There are also good reasons to locate the origin of at least a part of the African placentals, including that of the rodents, in the earliest Tertiary of Europe.

What is the bearing of these conclusions on the question of South American primate origins? The oldest record of primates in South America appears there in the same time as do rodents. Without being a decisive proof, this suggests that the two groups may have had a common destiny, since it is now known that the possibilities of migration were rather limited. In sum, the identity of effects and of some circumstances suggests an identity of causes and other circumstances. If strong phylogenetic affinities can be demonstrated between the primates of Africa and South America, there is a good probability that the explanation is the same as for rodents, namely a migration from Africa. Regarding the hypothesis of crossing the Atlantic by rafts, small size coupled with arboreal behavior was evidently a strong advantage, and these are common to both the primates and the rodents. It is a normal assumption to reason that if perhaps rafts with rodents sometimes bore primates, there is a high probability that the rafts with primates also had rodents as passengers. The arguments derived from plate tectonics and paleophysical geography are not in themselves especially connected more with rodents than with primates. But probably, if the zoological arguments presented for the rodents were not so cogent, much less attention would have been paid to the plate tectonics and physical geography. (That is certainly true for myself.) As long as the indications given by the rodents favoring a European early Eocene origin remain intact, they suggest a similar origin for the African, and therefore, South American primates. Eventually, the presence in North America of some peculiar forms of fossil primates showing affinities with South American groups, or with primitive African primates could be explained by a migration from South to North America better than by a North-to-South migration pathway.

In conclusion, I think that the abundance and perfection of the "documentation" offered by the rodents, not only in Africa but also in South America (Salla fauna, Lavocat, 1976), establishes the need for future study of the rodents of these southern continents and the bearing of their unique phylogenetic interrelationships as strong support for the hypothesis of an African origin of the Platyrrhini.

References

Bugge, J., 1974, The cephalic arteries of hystricomorph rodents, in: *The Biology of Hystricomorph Rodents* (I. W. Rowlands and B. J. Weir, eds.), *Symposia of the Zoological Society of London,* No. 34, pp. 61–78, Academic Press, London.

Durette-Desset, M. C., 1971, Essai de classification des nématodes héligmosomes. Corrélations avec la paléogéographie des hôtes, *Mém. Mus. Natl. Hist. Nat., N.S.* **69**:1–126.

Francheteau, J., 1973, Plate tectonic model of the opening of the Atlantic Ocean south of the Azores, in: *Implications of Continental Drift to the Earth Sciences,* Vol. 1 (D. H. Tarling and S. K. Runcorn, eds.), pp. 197–202, Academic Press, London.

Lavocat, R., 1974, What is an hystricomorph?, in: *The Biology of Hystricomorph Rodents* (I. W. Rowlands and B. J. Weir, eds.), *Symposia of the Zoological Society of London,* No. 34, pp. 7–20, Discussion, pp. 57–60, Academic Press, London.

Lavocat, R., 1976, Rongeurs du Bassin Déséadien de Salla-Luribay, in: Rongeurs Caviomorphes de l'Oligocène de Bolivie, *Palaeovertebrata* **7** (3):15–90.

Lavocat, R., 1977, Sur l'origine des faunes sud-américaines de Mammifères du Mésozoique terminal et du Cénozoique ancien, *C. R. Acad. Sci., Paris, Sér. D* **285**:1423–1426.

Luckett, W. P., 1971, The development of the chorio-allantoic placenta of the African scaly-tailed squirrels (Family Anomaluridae), *Am. J. Anat.* **130**:159–178.

Mossman, H. W., and Luckett, W. P., 1968, Phylogenetic relationships of the African mole rat *Bathyergus janetta,* as indicated by the fetal membranes, *Am. Zool.* **8**:806.

Schaub, S., 1953, Remarks on the distribution and classification of the "Hystricomorpha," *Verh. Naturforsch. Ges., Basel* **64**:389–400.

Smith, A. G., and Briden, J. C., 1977, *Mesozoic and Cenozoic Paleocontinental Maps,* Cambridge University Press, Cambridge.

Tullberg, T., 1899, Ueber das System der Nagethiere, eine phylogenetische Studie, *Nov. Acta Reg. Soc. Upsala, Ser. 3* **18**:1–154.

Wood, A. E., 1950, Porcupines, paleogeography and parallelism, *Evolution* **4**:87–98.

Wood., A. E., 1968, The African Oligocene Rodentia, in: *Early Cenozoic Mammalian Faunas, Fayum Province, Egypt,* Part II, *Bull. Peabody Mus. Nat. Hist.* **28**:23–105.

Wood, A. E., 1974, The evolution of the Old World and New World hystricomorphs, in: *The Biology of Hystricomorph Rodents* (I. W. Rowlands and B. J. Weir, eds.), *Symposia of the Zoological Society of London,* No., 34, pp. 21–60, Discussion, 57–60, Academic Press, London.

Wood, A. E., 1975, The problem of the hystricognathous rodents, *Univ. Michigan, Papers on Paleontology* **12**:75–80.

Woods, C. A., 1972, Comparative myology of jaw, hyoid and pectoral appendicular regions of New and Old World hystricomorph rodents, *Bull. Am. Mus. Nat. Hist.* **47**:117–198.

Origin and Deployment of New World Monkeys Emphasizing the Southern Continents Route

5

R. HOFFSTETTER

Comprehension of the Platyrrhini

The name Platyrrhini has been erected to designate the extant neotropical monkeys (including marmosets), which zoologists agree to group into a superfamily Ceboidea. Living in the tropical forests of South and Central America, they are distinguished from the recent Old World monkeys by the presence of three premolars (instead of two), their ring-shaped ectotympanic (not developed into a bony tube), their inflated auditory bulla, several details of cranial architecture (notably the parietal–jugal contact, which must have been constant in the group at its origin), placentation (no cytotrophoblastic shell, trabecular condition of the placental disk: see Luckett, 1975) and also by platyrrhiny [which, as redefined by Hofer (1976), is diagnostic for the group]. Practically all fossil primates known from South America and the West Indies are referred to this group: late Pleistocene or Holocene of Brazil (five recent

R. HOFFSTETTER • Institut de Paléontologie, Muséum National d'Histoire Naturelle, 75005 Paris, France.

genera), Hispaniola (*Saimiri:* see Rimoli, 1977) and Jamaica (*Xenothrix*); middle Miocene of Colombia (*Neosaimiri, Cebupithecia, Stirtonia*); early Miocene (*Homunculus*) and latest Oligocene (*Dolichocebus, Tremacebus*) of Patagonia, and also early Oligocene of Bolivia (*Branisella*). However, Hershkovitz (1974*a*) has expressed doubts as to the inclusion of *Branisella* within the Platyrrhini and even within the Haplorhini. According to him, the small single-rooted P[2] of this fossil would indicate a trend toward the disappearance of this tooth, and

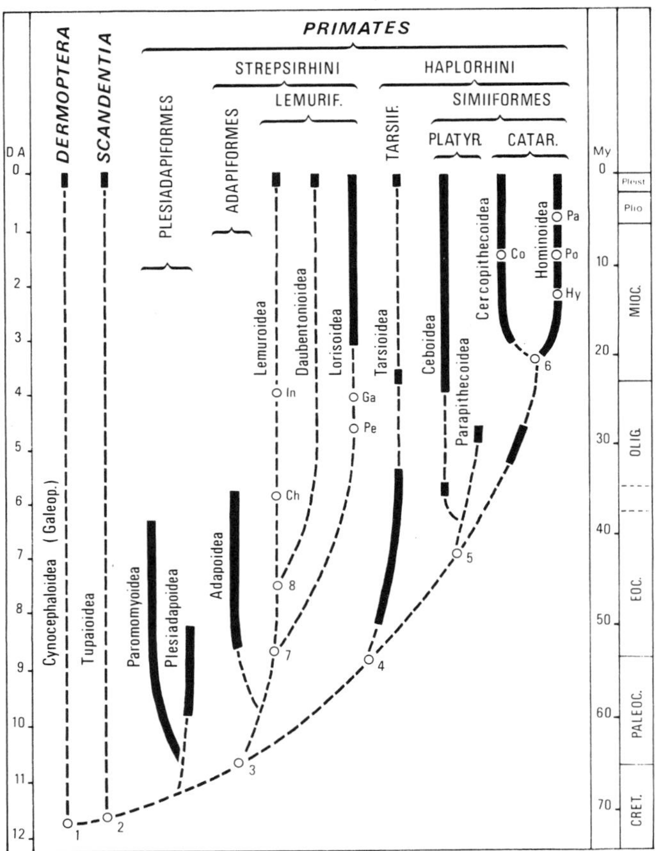

Fig. 1. Phylogeny of the Archonta, at superfamily level (after Hoffstetter 1977*c*). The initial diagram is constructed from the antigenic distances (DA) of Goodman (1975, Table 3, Fig. 1). Besides the main dichotomies (points 1 to 8), some branching points in the superfamilies are placed as indicators based on biochemical data. An attempt at a chronological calibration (origin of the Primates placed at −70 m.y.) allows a representation (in solid lines) of the stratigraphic extension of fossil representatives. The confrontation is satisfying. However, some questions are worth discussing, notably the interpretation of the Catarrhini from the Fayum [according to me a radiation, with mosaic evolution, preceding the individualization of undoubted Hominoidea (see Hoffstetter, 1977*c*, Fig. 3)].

the quadritubercular pattern of its upper molars (provided with a hypocone) would prevent this genus from being placed among the ancestors of the marmosets (whose triangular molars lack a hypocone). In my opinion (Hoffstetter,1969), *Branisella* has retained the ancestral characters of the Haplorhini in both respects (these characters are also found in the earliest Tarsiiformes: Omomyidae), and the Callitrichidae have secondarily lost their hypocones (a loss also observed in ancient Tarsiiformes such as *Chumashius* and modern ones such as *Tarsius*). This question will be returned to later.

The distinction between New World monkeys and those of the Old World appears less clear-cut if the oldest African fossils (Oligocene of the Fayum) are taken into consideration. Some of them retained characters (three premolars, ring-shaped ectotympanic) which today are restricted to neotropical monkeys. The resemblance is especially clear for the Parapithecidae to the extent that Simons (1967) has called them the "primitive New World monkey-like family." It is possible to go further and envision the inclusion of the Parapithecidae in the Platyrrhini *s.l.* (Hoffstetter, 1977*a,c*). It should be recognized, however, that the observed common characters are essentially symplesiomorphies (sharing of primitive characters) and therefore are not really demonstrative. It is even possible that the Platyrrhini are a paraphyletic group (Hennigian terminology), notably, if the still unknown common stem of the Simiiformes is included therein. But the proposed grouping accords well with a phylogenetic diagram and a biogeographical history which can be convincingly defended. The same grouping has the advantage of making a useful diagnosis of the Platyrrhini possible; these would include all monkeys with three premolars, whereas the Catarrhini would be characterized by the loss of P^2_2.

Problems Posed by the Phylogenetic Relationships and the Place of Origin of the Platyrrhini

Comprehension of the Platyrrhini depends closely on our information of the phylogenetic relationships and the geographical origin of the neotropical monkeys (Ceboidea). Paleontological evidence does not allow one to envision a South American origin for them, since no primates or potential ancestors for them have been found in the Paleocene and Eocene localities of this continent. Actually, the Ceboidea appears suddenly ca. 35 million years (m.y.) ago at the beginning of the Oligocene (*Branisella,* from the Deseadan of Bolivia) at the same time as the caviomorph rodents. South America was then isolated by marine barriers which the immigrants, the founders of the Ceboidea and the Caviomorpha, had to cross by rafting. There are but two possible geographical origins for the Ceboidea (as well as for the Caviomorpha). First, they could have come from North America (Matthew's hypothesis, long held as a dogma) where they could have their roots either in the nothrarctine Adapidae

(Gingerich, 1973, 1976)*, or in the omomyid Tarsiiformes (many authors, including Szalay, 1975*b*), or else in monkeys having differentiated in nuclear Central America (McKenna, 1967). Second, they could have arrived from Africa across a much less expansive South Atlantic, derived directly from Simiiformes (Parapithecidae or their immediate ancestors). This latter hypothesis seems the most likely to me (for example, see Hoffstetter, 1972, 1974*b*, 1977*b*), but it still requires further documentation.

These several problems concerning Platyrrhini origins are closely connected. Discussions and choices make it necessary to consider the successive dichotomies which have affected the order Primates. The beginning of the history of the group is poorly known. Many omit the Plesiadapiformes (= Paromomyiformes, incl. *Purgatorius*) which represent an early radiation, having apparently split off from the common stem of the Primates at a very early date. This group became extinct before the end of the Eocene, with no direct line to the history of the monkeys, mostly Platyrrhini, unless one envisions a Plesiadapiformes to Tarsiiformes (Gingerich, 1976) and Tarsiiformes to Platyrrhini (Szalay, 1975*b*) phylogenetic scheme.

The other groups of the Primates include both fossil and recent representatives, so that their phylogenetic relationships can be studied not only on the basis of bones and teeth, but also from soft anatomy, karyology, biochemistry (immunology, amino acid sequences, DNA) and even from ecology and ethology.

The Strepsirhini–Haplorhini Dichotomy

Leaving aside the Plesiadapiformes, there is today an almost general consensus to divide the Primates into two suborders, Strepsirhini (Adapiformes and Lemuriformes, incl. Lorisoidea) and Haplorhini (Tarsiiformes and Simiiformes). Both groups are already well individualized in Laurasia at the beginning of the Eocene with Adapidae and Omomyidae as respective representatives. Their Paleocene ancestors are still unknown, but various evidence goes to prove that they are two sister-groups issuing from the same dichotomy. The evidence brought forth by Pocock in 1918 is still valid; as far as the snout is concerned, the Strepsirhini retain the ancestral condition (laterally split nostril, rhinarium and philtrum with a naked and wet skin, frenum immobilizing the upper lip); on the other hand, the Haplorhini share apomorphous characters of their own (entire nostrils, a nose with hairy undifferentiated skin, no philtrum, and a much reduced or absent frenum). Luckett's very significant Hennigian diagram (1975) based on placentation and

*Gingerich and Schoeninger (1977) alter this hypothesis in deriving the Platyrrhini directly from an "Old World adapid" (with a true hypocone, thus adapine); but, curiously, in the same paper, they continue to defend a migration across the Panamanian straits (!).

fetal membranes shows that the Haplorhini share six synapomorphies, while the Strepsirhini have remained at the corresponding plesiomorphous stages. According to Starck (1975) the development of the chondrocranium confirms the validity of this subdivision. Biochemical studies, mainly immunodiffusion (Goodman, 1975; Dene *et al.*, 1976) and the amino acid sequences in the hemoglobin chains (Goodman, 1976) reveal the same dichotomy.* Although the DNA of the tarsier has been studied little, the works of Hoyer and Roberts (1967) already provide some concordant indications: the Strepsirhini–Haplorhini dichotomy should indeed have occured prior to the Tarsiiformes–Simiiformes branching. This dichotomy apparently corresponds to an orientation towards two activity rhythms: nocturnal for the Strepsirhini (in which the eye of the few diurnal forms retains a tapetum lucidum†) and diurnal for the Haplorhini (the few nocturnal genera of which, *Aotus* and *Tarsius,* still have a fovea centralis, but have not acquired a tapetum lucidum).

All of these observations are contrary to Gingerich (1973, 1976) who has the Simiiformes derived from the Lemuriformes *s.l.* (and the Tarsiiformes from the Plesiadapiformes). His views have been criticized by Hoffstetter (1974*a,b*), Hershkovitz (1974*b*), and Szalay (1975*a,* 1977) but have been favorably received by Russell (1977). As a matter of fact, *Apidium* is indeed a monkey, close to the Ceboidea, and the attachments of its tympanic ring agree with the Tarsiiformes and the juvenile stages of the Platyrrhini and probably correspond to the primitive condition of the Haplorhini (Hershkovitz, 1974*b*). *Cercamonius* is an adapid and the disposition of its P_3 alveoli betray at most a convergence with the Catarrhini. *Amphipithecus* may be a primate of doubtful subordinal position, but our uncertainty results from the poorness of the presently known material, not from the so-called "intermediate" position of this fossil between the Adapidae and the Simiiformes. As to the vertical and spatulate incisors, the presence of which in some Adapidae and in the Simiiformes is stressed by Gingerich, it would be necessary to specify whether it is a symplesiomorphy [the likelier hypothesis (cf. Szalay, 1977)], a synapomorphy (as assumed by Gingerich), or a convergence. Fusion of the

*Cronin and Sarich (1975) and Sarich and Cronin (1976), using another immunological method (microcomplement fixation), end up with a different cladogram, in which Dermoptera, Tupaiidae, Strepsirhini, Tarsiiformes and Simiiformes apparently have the same antiquity and are rooted in the same initial stock (which can only be a first approximation). Doubtless, this method does not allow one to distinguish the successional order of remote dichotomies, and thus to precisely say whether the nearest relatives of the Simiiformes are the Lemuriformes or the Tarsiiformes [it should be noticed that Sarich (1970, p. 225, Fig. 2) also gave the same antiquity to the Lorisoidea]. Immunodiffusion (Goodman) shows that the five above-mentioned groups are the result of four dichotomies, the last one separating Tarsiiformes and Simiiformes, which agrees with all other observations.

†However, the Adapidae, or at least the larger ones, may have been diurnal, either secondarily (like some Malagasyan Lemuroidea), or initially (in which case nocturnality would have appeared only with the Lemuriformes proper).

mandibular symphysis was doubtlessly acquired independently by some Adapidae and the Simiiformes.

It should be noted that the aim of Gingerich's reasoning is to demonstrate the impossibility of deriving the Simiiformes from the Eocene Omomyidae (a point on which I fully agree), while the question is whether the last (certainly pre-Eocene) common ancestor of the Tarsiiformes and the Simiiformes was older or more recent than that of the Lemuriformes and the Simiiformes. To sum up, and this is a fundamental point, I consider that, taking into account all known facts, the Haplorhini are a natural group and that any hypothesis which derives the Simiiformes directly from the Strepsirhini should therefore be discarded.

The Simiiformes, Sister-Group of the Tarsiiformes

Modern studies show that the Simiiformes (Platyrrhini and Catarrhini) constitute a natural, monophyletic group, contrary to Matthew's opinion, according to which neotropical monkeys and those of the Old World were two parallel or convergent groups, independently rooted in the Laurasian "prosimians" (some authors, among them Gingerich, still admit it). As a matter of fact, if both extant and fossil forms are considered, osteological and dental characters do not allow a separation of the two groups of monkeys (see above). Moreover, Luckett (1975), on the basis of placentation and fetal membranes, has pointed out four synapomorphies in the Simiiformes, which distinguish them from the Tarsiiformes.

Similarly, cladograms by Goodman (immunodiffusion) and by Sarich and Cronin (microcomplement fixation) show that the Platyrrhini–Catarrhini branching, definitely after the splitting of the Tarsiiformes, is preceded by a long common stem. If the individualization of the Primates is placed at −70 m.y. in Goodman's cladogram, the origin of the Simiiformes would take place at ca. −55 m.y., and their splitting at ca. −42 m.y., that is, in the later Eocene. Lastly, the amino acid sequences of several proteins and the data about DNA lead to concordant results (cf. Goodman, 1975, 1976). One is thus led to admit that the Simiiformes are monophyletic and represent the sister-group of the Tarsiiformes.

The distribution of fossils suggests that this dichotomy is the result of geographic segregation north and south of the Tethys. The Tarsiiformes appear suddenly at the beginning of the Eocene in Laurasia and remain confined there. The Simiiformes are known only from the early Oligocene onward, both in Africa (six genera, various families and superfamilies in the Oligocene of the Fayum) and in South America (Bolivia: one Oligocene genus, *Branisella*); they reach Eurasia on the one hand (Catarrhini, from Africa), and Central America and some islands of West Indies on the other (Platyrrhini, from South America) only later.

Differentiation of the two infraorders took place prior to the Eocene, as suggested by biochemists' cladograms. Moreover, it can be observed that as early as the Eocene the Laurasian Tarsiiformes (Omomyidae) exhibit progressive characters (ossified external auditory tube, shape of dental arch, specialized incisors, and modified tarsus) which exclude them from the ancestry of the Simiiformes, thus being necessarily differentiated in the same epoch.

Place of Origin of the Simiiformes

Africa and South America are the only lands known to have sheltered monkeys before the Miocene. It is thus legitimate to look for the cradle of the group there. As noted earlier, a South American origin is highly unlikely, not to say impossible, since the Ceboidea, as well as the caviomorph rodents, appear as immigrants from elsewhere. On the contrary, our lack of information on possible pre-Oligocene Simiiformes in Africa has a different meaning. As early as the Oligocene, the Fayum monkeys are already very diversified and other forms certainly lived during the same epoch in the intertropical forest zone where pre-Miocene fossil-bearing localities have not yet been discovered. The Eocene localities of Egypt, Libya and Mali exhibit very peculiar facies (with *Moeritherium,* palaeophid snakes and sometimes cetaceans) which do not fit the ecological requirements of primates. The Eocene locality in the western Sahara which has yielded *Azibius* and Hyracoidea, has only been subjected to rapid collecting. It remains very likely that the Fayum monkeys are but a marginal sample of an ancient simiiform fauna which differentiated in Africa. In other words, it seems that the primitive Haplorhini, if they were not born in Africa, entered it very early (at the same time as the Creodonta, Suiformes, and also Embrithopoda) to differentiate there into Simiiformes while their Laurasian relatives gave rise to the Tarsiiformes.

Ancestral Morphotype of the Simiiformes

At this point, it is legitimate to push hypotheses further by trying to reconstruct the characters of the early Simiiformes. Still very close to (perhaps indistinguishable from) the first Tarsiiformes, they must have exhibited the various plesiomorphous characters retained by one or another of extant or fossil monkeys.

They were probably small, tree-living animals (of a size comparable to that of *Branisella, Apidium moustafai, Callicebus,* etc.) diurnal, microsmatic, with a diet of insects and fruits. They undoubtedly possessed the diagnostic characters of the Haplorhini concerning the nose, placentation, and fetal membranes.* As compared with the Strepsirhini their braincase was relatively large and their snout shortened (but the snout could become secondarily longer, notably in baboons). The auditory bulla was inflated (it remained so in the Platyrrhini, and, to a lesser extent, in some Cercopithecidae). The extrabullar ring-shaped ectotympanic showed the condition found in *Apidium, Aegyptopithecus* and the Ceboidea. Carotid circulation was characterized by the disappearance of the mesial entocarotid, the reduction of the stapedial artery, and predomination by the promontory artery (main branch of the lateral entocarotid) in brain irrigation. The orbito-temporal fenestra was still widely open (as in the ancient Tarsiiformes and *Tremacebus*). Jacobson's organ was functional [since it is still so in *Tarsius* and several Platyrrhini (Starck, 1975)]. The dental formula, 2133/2133, was that of the Parapithecidae and the Cebidae. The incisors were subvertical and spatulate; the canines were of moderate size. The cheek teeth still possessed well-developed styles; P^2 was a small single-rooted tooth (as in several Omomyidae and in *Branisella*); P^3 and P^4 were transversally broadened and bicuspid (a trend already observed in the Tarsiiformes and also found in the oldest monkeys from Africa and South America); the upper molars, subtrapezoidal, broader than long, with a hypocone lingually shifted relative to the protocone (all those characters are known in the Omomyidae and are retained by *Apidium* and *Branisella*), still had well-individualized paraconules and metaconules (like the Omomyidae and the Parapithecidae). The lower premolars were single-rooted; P_2 was small but higher than the following premolars; the lower molars had four main cusps with, sometimes, a small hypoconulid, but these teeth (even M_3) did not show any marked elongation. The humerus possessed an entepicondylian foramen [which is present in *Tarsius* and has been retained by the Cebidae and by some Fayum monkeys (Simons, 1967, p. 16)]. The carpus comprised nine bones, including a central. In the leg, the tibia and fibula were separate.

To sum up, those early Simiiformes were essentially distinguished from their forebears (primitive Haplorhini) and from collateral forms (Tarsiiformes) by pneumatization of the petromastoid [already acquired in *Apidium* (see Szalay, 1975a, Fig. 15D)], enlargement of the carotid foramen (development of the promontory artery), complete fusion of the mandibular symphysis, molars with more bulbous cusps and without a vestibular cingulum, wholly intraorbital lacrymal bone and foramen, development of four apomorphous characters concerning placentation and fetal membranes (Luckett, 1975), and doubtless a more progressive stage of the brain. On these points, the more conservative Tarsiiformes had remained at a less advanced (plesiomorphous) stage, but it should be remembered that, on a few other points (see above), the Eocene Omomyidae showed an advance (apomorphy) as compared with monkeys.

*Moreover, according to Luckett (1975, p. 179) "a double discoidal placenta characterizes most Ceboidea and Cercopithecoidea, and this doubtlessly represents a primitive condition in Anthropoidea" [= Simiiformes].

Dichotomies of the Simiiformes in Africa

Early in the Tertiary (in the Eocene at the latest), the African Simiiformes had split into two branches, a conservative one (Parapithecidae) and a definitely more "progressive" one (Catarrhini *s.s.*). Both are represented in the Oligocene localities of the Fayum.

Curiously, the Parapithecidae (*Apidium* and *Parapithecus*) are known only from the upper levels of the Fayum, which correspond to ca. −30 to −28 m.y. (Simons, 1972), but they represent the most common primates in the fauna. Clearly, they had already differentiated well before this date (presumably more to the south), for they cannot be derived from the Catarrhini *s.s.*, which had already lost their P^2_2 in the early Oligocene. On the contrary, the Parapithecidae have retained three premolars, and their ectotympanic still has the primitive ringlike shape (a feature found in *Apidium,* not yet observed, but likely, in *Parapithecus*). In both genera, the upper premolars (including P^2) are transversally broadened; they possess three roots (two vestibular and one lingual observed in *Apidium moustafai*); their crown bears an eocone (= amphicone) and a protocone; the styles are distinct; moreover, a strong "paraconule," developed on all three premolars, appears as a characteristic of the family. The upper molars still have a fairly primitive pattern in *Apidium* (trapezoidal shape, hypocone shifted lingually relative to the protocone, paraconule and metaconule present, although the former is already somewhat reduced). *Parapithecus* shows a more advanced evolutionary stage; the hypocone is in line behind the protocone resulting in a square pattern of the four main cusps; moreover M_3 shows a definite reduction and the canines are enlarged (cf. Simons, 1974).

In their known characters, those Parapithecidae are undoubtedly very close to the neotropical Ceboidea. It is likely (see later) that they are two sister branches born in Africa as a result of a dichotomy dating back to the Eocene.

The Parapithecidae must have reached a certain range extent in Africa, but they are still known only from the Fayum. They do not seem to have survived after the Oligocene and have probably been eliminated by competition from the Cercopithecidae. However, Simons (1972 and before) assumes that the latter could be direct descendants of the Parapithecidae. We shall see later that this hypothesis meets with serious difficulties.

The other branch of the African Simiiformes, Catarrhini *s.s.,* is known as early as the Lower Fossil Wood Zone of the Fayum (ca. −32 to −30 m.y. according to Simons, 1972) with *Oligopithecus.* This branch is characterized by the loss of P^2_2, the gradual flattening of the auditory bulla and the development of the ectotympanic into a bony tube (a delayed development, since it has not yet been realized in *Aegyptopithecus* from the Fayum). P_3 has two roots obliquely positioned one behind the other and its crown forms an oblique ridge on which the upper canine honed itself (this character is observed even on the earliest Catarrhini *s.s.,* such as *Oligopithecus;* the transition to the single-rooted bicuspid P_3 of the Hominidae would thus be secondary and relatively late).

Luckett's cladograms based on placentation and fetal membranes (this volume), like those by Baba, Goodman, Cronin, and Sarich based on immunology (this volume) show that the Catarrhini have split into two branches, Cercopithecoidea and Hominoidea; the dichotomy seems to be a fairly late one (ca. −20 m.y. according to Goodman, Cronin and Sarich), which is in agreement with the absence of the Cercopithecidae in localities earlier than the Miocene.

Origin of the Ceboidea

It is likely that if remains of parapithecids had been found on a continent inhabited by the Ceboidea, one would not have hesitated to unite both groups, perhaps in the same superfamily. The observable dental and osteological characters are indeed in perfect agreement. It would be interesting to know whether there is also an agreement in the details of cranial architecture (notably the pterion) and in the postcranial skeleton. Other characters are unfortunately unrecognizable on fossils; in particular platyrrhiny, which, according to Hofer (1976), concerns only the internarium (alar cartilages and soft tissues), and not the septum nasi, cannot be observed on dry skulls.

It is highly probable that both groups are derived from a common, immediate ancestor, probably a primitive African monkey, related closely to parapithecids but still possessing a small single-rooted P^2. In the late Eocene, the founders of the Ceboidea succeeded in crossing the Atlantic by rafting. The phylogeny would then be as shown on Fig. 2A. As I have already suggested (Hoffstetter, 1977*a,c*), its adoption would lead to uniting the Parapithecidae and the Ceboidea into the same taxon (Platyrrhini *s.l.*), which would include all monkeys with three premolars and would be the sister-group of the Catarrhini *s.s.* (monkeys with two premolars).

The major problem remains the crossing of the Atlantic; the possibility of which is rejected by some authors. I have already discussed that point since 1970. Let us recall that in the late Eocene the Atlantic was notably smaller than today, and crossing by rafting was made easier by the equatorial current and the trade winds, both of east-west direction, since they are linked to the rotation of the Earth (see the maps by Berggren and Hollister, 1974). It should also be remembered that, whatever their geographic origin, the founders of the Ceboidea had to cross a marine barrier since South America was then completely isolated. [See Addendum (1); see also Tarling in this volume.]

The preceding phylogenetic and biogeographic scheme seems the likeliest to me. However, other hypotheses have been put forward, three of which, recently formulated, deserve to be discussed.

1. Simons' hypothesis (1972 and before), already mentioned, assumes that the Miocene Cercopithecidae would be directly derived from the Oligocene Parapithecidae. It is not compatible, in Fig. 2, with diagram A

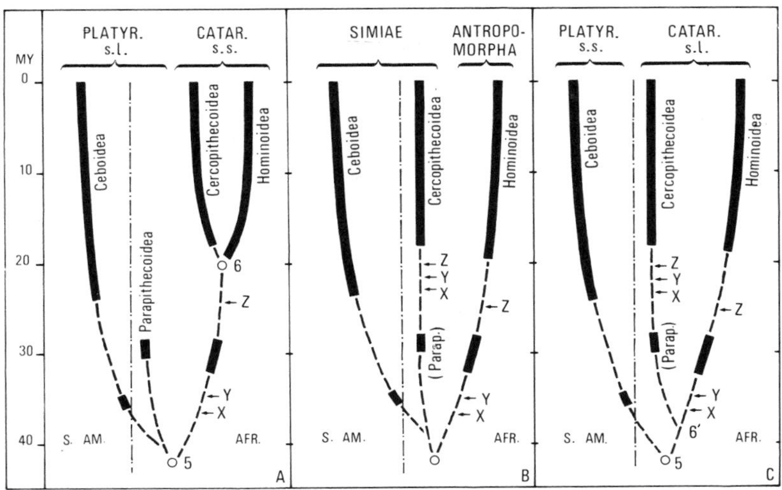

Fig. 2. Three hypotheses on the phylogenetic relationships of the superfamilies of the Simiiformes. Diagram A (adopted here) is in agreement with cladograms by biochemists (Goodman, Sarich, Cronin and others, this volume) and antomists (Luckett, this volume). This is not true of diagrams B and C [in which the Cercopithecidae are derived from the Parapithecidae, following the hypothesis of Simons (1974) and Kay (1977)]. See text for further explanation.

(adopted here), but it agrees with B and C. But diagram B would make the Hominoidea appear as the sister-group of the Cercopithecoidea + Ceboidea group, which is in contradiction with the cladograms by Luckett, Goodman, Sarich, etc. On the other hand, according to diagrams B and C, the two dichotomies responsible for the separation of the three extant superfamilies would have followed one another very closely in time (about −40 m.y.), while immunologists place them around −40 and −20 m.y., respectively. Lastly, in both cases, the Hominoidea and the Cercopithecidae would have acquired independently several apomorphous characters such as the loss of P^2_2, specialized hone-shaped P_3, and ectotympanic developed into a bony tube (characters X, Y, and Z in Fig. 2).

Simons' hypothesis has been criticized by Delson (1975). However, on the basis of the crests and wear facets of the molars, Kay (1977) also favors deriving the Cercopithecidae "from a *Parapithecus*-related stock" (*Apidium* being excluded from it). For my part, I think that divergent evolution of the premolars (enlarging or loss of P_2, notably) is at least as significant. In the interpretation adopted here (Fig. 2A) the Cercopithecidae would have replaced the Parapithecidae ecologically (hence the dental convergences observed by Kay), but they would not be their descendants.

2. Szalay (1975*b*) has put forward quite another hypothesis by inverting the order of migrations [this presumably in an attempt to explain the conservative character of the Ceboidea, as compared with the Catarrhini; upon this point, see Addendum (3)]. For him, the North American Tarsiiformes

(Omomyidae) would be at the origin of the South American Ceboidea; some of the latter would have then crossed the Atlantic from west to east to give rise to the Catarrhini *s.l.* in the Old World. This view meets with serious difficulties. It requires two marine crossings, both very hazardous: the first one from North America to South America, across an arm of the sea swept from east to west by the equatorial current; the second one across the Atlantic from west to east, necessarily along the counter-current that is at a fairly high latitude, or by an extremely long southern route under a climate unfavorable to primates (see maps by Berggren and Hollister, 1974). Besides, there is a considerable gap between the Omomyidae (the only Haplorhini known in the Tertiary of North America) and the Ceboidea (the only South American Primates): I have already discussed (Hoffstetter, 1974*b*, p. 346) the evidence which makes a differentiation of the Simiiformes in nuclear Central America (as envisioned by myself in 1954 and more explicitly by McKenna, 1967) improbable [see Addendum (3)]. Lastly, Szalay's hypothesis does not agree with the diversity exhibited as early as the Oligocene by the African Simiiformes, while everything seems to indicate that primates did not enter South America until shortly before the Oligocene.

3. Hershkovitz (1974*a*) has envisioned the possibility that the Simiiformes could have differentiated in a still undivided African-South American land mass. Disjunction of the two continents would have brought about, by geographic segregation, a parallel evolution leading to the Platyrrhini *s.s.* (Ceboidea) on the one hand, and to the Catarrhini *s.l.* on the other. However, it is known that Africa and South America have been completely separated at least since the Turonian and that before that time the Atlantic "rift" functioned as a biogeographic barrier. These chronological data and what we know of the antiquity of the various groups compel us to discard Hershkovitz's hypothesis.

To sum up, the hypothesis defended here, and schematized in Figs. 1 and 2A, appears the likeliest. It takes into account the various data from comparative anatomy, biochemistry, paleontology, and chorology.

Deployment of the Platyrrhini

According to the view defended here, the Platyrrhini originated in Africa, probably in the Eocene, as the sister-group of the Catarrhini *s.s.*; they have given rise locally to the Parapithecidae, which probably lived in the intertropical forest belt. It is not until the Middle Oligocene (ca. −30 to −27 m.y.), thus later than the Catarrhini, that this family reached the Fayum region. This expansion was successful (they were then the most abundant monkeys) but short-lived; the family became extinct in Africa at the end of the Oligocene.

It is probably in the later Eocene that a small population of primitive Platyrrhini succeeded in crossing the Atlantic by rafting. One could wonder why they were the only monkeys to have achieved that feat since the Catarrhini *s.s.* were also present in Africa, but as previously stated, until the Oligocene, the Parapithecidae (thus Platyrrhini *s.l.*) were less varied but more numerous than the Catarrhini, which gave them a statistical advantage.

The platyrrhine monkeys found in South America a land of opportunity, where they could achieve a radiation which they had been prevented from doing elsewhere. They had arrived there at an evolutionary stage close to that illustrated by *Branisella* (from the Deseadan, i.e., ca. −35 m.y.), in which the small, single-rooted P^2 (this character being a unique instance in the known Simiiformes) is indicative of a more primitive stage than *Apidium.* But the molars of *Branisella* have already lost the paraconule (in this respect, *Apidium* and *Parapithecus* are less advanced); its premolars also lack the large paraconule of the Parapithecidae (but it may have been secondarily acquired, i.e., apomorphous, in the latter). It seems legitimate to consider the whole group of extant and fossil neotropical monkeys (incl. *Branisella*) as one superfamily, the Ceboidea.

Most of them are diurnal; it is however remarkable that some have become adapted to nocturnal life [the extant *Aotus*; probably *Tremacebus,* of the latest Oligocene (cf. Hershkovitz, 1974*a*)], but the eye of *Aotus* shows that its ancestors were diurnal. This way of life, unknown in Old World monkeys, is doubtlessly explainable by the absence of the Strepsirhini in South America (conversely, some Malagasy Lemuroidea, in the absence of monkeys, have become adapted to diurnal life).

South American monkeys have remained closely linked to the forest; no savanna form, comparable to the African baboons, is known among them. They do not comprise any animal comparable in size to the large apes and man. On the other hand, they include essentially insectivorous dwarf forms (marmosets, *Callicebus,* etc.), some of them with claws which represent a biological type unknown among the Catarrhini.

Most South American monkeys are quadrupedal; some are able to walk on their hind legs (but of course their occasional bipedality is much less elaborate than that of man); others practice brachiation (in a less advanced stage than gibbons), but they are the only group of monkeys to have "invented" the prehensile tail, especially advantageous for life in the forest (*Ateles, Brachyteles, Lagothrix, Alouatta, Cebus*); only one (*Cacajao*) has an atrophied tail, but it still remains arboreal. The diet is varied, more or less exclusive or eclectic: insectivorous, microcarnivorous, frugivorous, and even phyllophagous.

South American monkeys, today restricted geographically to the tropical forests of South and Central America, have extended as far as Patagonia (50° S.) at the end of the Oligocene (Colhuehuapian) and the beginning of the Miocene (Santacruzian) probably owing to a temporary warming up of the climate. By rafting, they have reached at least two of the West Indian islands, Hispaniola (where a remarkably large *Saimiri* has just been reported by Rimoli, 1977) and Jamaica [where a particular, now extinct genus, *Xenothrix,* is known by a mandible from the Holocene (see Hershkovitz, 1974*a*; Rosenberger, 1977)]. Lastly, in the Pleistocene or shortly before, the emersion of the Panamanian isthmus has allowed them to colonize the warm forests of Central America up to 23° N. lat. However, the marmosets did not go farther than the Panamanian territory; unfortunately, we do not know any fossil that could provide more precise information on the dates and possible stages of this migration.

The phylogeny of the Ceboidea still poses some problems. Neontologists are agreed to distinguish two families, the Cebidae (monkeys) and the Callitrichidae (marmosets). The genus *Callimico* has sometimes been placed in a family of its own, but its ethology, and its biochemical and anatomical characters lead one to consider it as a marmoset whose much reduced M^3_3 have not yet completely disappeared. This is the generally used classification. But it should be recalled that Hershkovitz (1974*a*) goes so far as to divide the Ceboidea into 5 families (2 of which are extinct) and 14 subfamilies (4 of which are exclusively fossil), *Branisella* (which he considers as *incertae sedis*) not being included.

The cladograms established by immunologists (Goodman, Sarich, Cronin, and others) for the Ceboidea are rather contradictory: thus, for Cronin and Sarich (1975), *Aotus* would be the end product of a particular branch, having separated very early from all other Ceboidea (incl. marmosets); on the contrary, for Baba *et al.* (1975) it would be the result of a late differentiation, its nearest relative being *Callimico*. Present results should therefore be viewed with caution, for they require additional data and discussion.

Anatomically, despite Hershkovitz's contrary opinion, there is nothing to preclude a derivation of the various extant and fossil Cebidae from a stock which includes *Branisella*. They differ from it by more advanced, quadrangular upper molars, with the hypocone in line behind the protocone (thus of a type comparable to that of *Parapithecus*) and by their transversally broadened, two-rooted P^2 (which had thus followed with some delay the same evolutionary trend as P^3 and P^4). The radiation of the family is exuberant. For the 11 genera and 25 species recognized today, zoologists thus admit 7 subfamilies (Saimirinae, Aotinae, Callicebinae, Alouattinae, Pitheciinae, Cebinae, and Atelinae, according to Napier), which are, on the whole, confirmed by immunology (compare this exuberant radiation with the simple subdivision of the Cercopithecidae, the 15 extant genera of which constitute only two subfamilies, Cercopithecinae and Colobinae, also confirmed by immunology). If fossils are also considered, other groups should be added: Homunculinae, Xenothricinae, Tremacebinae, Stirtoninae, Cebupithecinae; Hershkovitz has even proposed families of their own for the former two. Perhaps a familial or subfamilial taxon should also be erected for *Branisella,* which there is no reason to exclude from the Ceboidea, but which may represent an offshoot. Anyway, this betrays the variety of ecological niches offered to monkeys by the American tropical forests.

The Callitrichidae (5 genera and 15 species, all extant with no fossils known before the Holocene) pose a more difficult problem. They do possess various apparently primitive characters, but the interpretation of these is not obvious, for some may result from evolutionary reversals.

Thus, the triangular molars, without a hypocone, probably represent a secondary condition, linked to an insectivorous diet; the same simplified morphology is also to be found in the Tarsiiformes *Chumashius* (later Eocene) and *Tarsius* (extant), whereas the earliest known Tarsiiformes and Simiiformes (and the oldest primate, *Purgatorius*) possess a hypocone. It should be added that various biochemical studies are in favor of a relatively late individualization of marmosets [ca. −15 m.y. according to Cronin and Sarich (1975)]. Moreover, other characters of the latter, notably the loss of M_3, and that of the epicondylian foramen of the humerus, are certainly secondary. The marmosets indeed seem to be monkeys whose evolution is directed towards dwarfism and an insectivorous diet. Even the claws (which in them occur on all digits except the hallux) may be the result of reversed evolution, in relation to a mode of locomotion comparable to that of the squirrel.*

On immunological evidence, Cronin and Sarich (1975) consider the Callitrichidae (incl. *Callimico*) as a natural monophyletic group, having separated 15 m.y. ago, the radiation of which, however, would not have begun until about −10 m.y. This is in fairly good agreement with anatomical and ecoethological observations. But the cladograms by Sarich and Cronin would lead one to consider this group as a simple subfamily of the Cebidae (from Goodman's results, it cannot be excluded that it may be an artificial grouping including various branches originating in the Cebidae).

Conclusion

Although there is still a considerable amount of speculation, the history of monkeys as a whole, and of the Platyrrhini in particular, is already fairly well established. It cannot be understood except within a mobilist paleogeography, and, in this way, it supports modern conceptions. It is obvious that the southern continents have played an important, too long underrated, part in the evolution of terrestrial faunas. For the problem discussed here, a special effort must bear on the search for mammal-bearing localities prior to the Oligocene in Africa; there lies the key to getting rid of any ambiguity as to the origin of the Simiiformes. It should also be recalled that there is an extraordinary resemblance between the chronological and chorological characteristics of the history of catarrhine and platyrrhine monkeys, on the one hand, and of phiomorph and caviomorph rodents on the other; any evidence acquired on

*On this point, see Le Gros Clark (1936). The claws of marmosets have a histological structure intermediate between that of a true claw (falcula) and that of a nail (ungula). They may be either claws evolving towards nails or a secondary reversal. Le Gros Clark favors the former interpretation, but the latter one cannot be definitively discarded: thus Napier alludes to the "claw-like nails" of the Callitrichidae and Le Gros Clark himself (1971) remains undecided.

one of these groups throws light on the history of the other. Thus, parasitological studies (Durette-Desset, 1971; Quentin, 1973) bring forth serious evidence in favor of the African origin of the Caviomorpha, and in this way make such an origin more probable for the Platyrrhini (for which parasites have not yet brought any valid indications on this problem).

Another point deserves to be stressed. Some biogeographers claim that, in the distribution of an animal group, more advanced forms always occupy a peripherical position. This has even been considered as a biogeographic law. Actually, an area remote from the cradle often serves as a refuge for primitive forms. As far as monkeys are concerned, the Ceboidea are seen to retain a number of ancestral characters, today lost by most Catarrhini (but some of which are curiously present also in the Hominidae). As a matter of fact, the radiation of the Ceboidea, independent from that of the African monkeys, has merely been different; it essentially depends on the characteristics of the founders (here a plesiomorphous branch of the Simiiformes)* and also on the physical and biological environment in which this radiation took place; one does not see how remoteness or nearness of the cradle of the group could play any part.

Addendum

Only after this paper was written did I have the opportunity to read Simons' paper (1976), in which this author rejects the possibility of an African origin of the Platyrrhini. According to him, the South American and African monkeys would be independently derived from a Laurasian stock (possibly *Teilhardina*). This paleobiogeographic pattern is still also defended by Gingerich and Schoeninger (1977), who, however, favor an adapoid stock. The evidence put forward by these authors should be discussed here.

1. *Possibility of a mammalian migration across the South Atlantic during the Eocene.* This is rejected by Simons on the grounds of the absence of terrestrial mammals on the islands of Hawaii and Tristan da Cunha. Actually, the case of South America is quite different: the probability of a raft becoming stranded on an island or a continent depends on the distance to be covered, on currents, and on winds, and also on the length of the coastline of the receiving land; moreover, the implantation of possible immigrants is made easier by the number and variety of available ecological niches. An exemplary instance of insular settlement is provided by Madagascar (despite Simons' contrary opinion); its unbalanced mammalian fauna is obviously the result of successive settlements, probably from the Eocene until a recent date. Another example is

*An advanced evolutionary stage is not necessarily a decisive advantage to colonize a territory. Thus, of the two groups of Proboscidea present in North America, those which invaded South America in the Pleistocene are the mastodonts (plesiomorphic branch), not the elephants (apomorphic branch), which did not go beyond Central America.

that of the Greater West Indies, where the Insectivora, the Megalonychidae, the Caviomorpha, as well as *Xenothrix,* are the descendants of immigrants which arrived by rafting, despite Simons' opinion that *Xenothrix* would have been introduced by man! Of course the route to South America from Africa was more hazardous; as a result, two groups only (Ceboidea and Caviomorpha) could achieve this feat, some 40 m.y. ago, while all other "attempts" failed.

Darlington (1957, pp. 14–15) stresses the importance of "floating islands" for the distribution of animals and provides testimonies as to their existence, their formation and their performances. He notes that, according to Powers, "a raft . . . , with trees 30 feet high, evidently tied together by the roots of living plants, was seen in the Atlantic off the coast of North America in 1892, and is known to have drifted at least 1000 miles."

According to Simons, the biological requirements of the primates exclude for them the possibility of a long rafting. As a matter of fact, we do not know what these requirements were for the primates of the Eocene. Petter *et al.* (1977, pp. 88–90) have observed that *Cheirogaleus,* a small lemuriform from Madagascar, accumulates fatty reserves during the favorable season; it can then become lethargic when conditions become unfavorable; its temperature falls and it can stay thus during several months without drinking or eating. According to Petter, the Cheirogaleidae would represent a very primitive biological type. This ability to survive may have been inherited from ancient primates, and may also have been present in the early Simiiformes. However that may be, the founders of the Ceboidea must have crossed a sea one way or another to reach South America.

A contradiction can be mentioned on the part of Gingerich (1975, Fig. 8, p. 77), who denies the possibility of a transatlantic migration while suggesting a colonization of Madagascar by lemurs from the Indian plate [a hypothesis which Simons (1976, p. 38) does not reject].

2. *Faunal exchanges between North America and South America at the beginning of the Tertiary.* For Gingerich and Simons, the occurrence of notungulates in North America and eastern Asia in the early Eocene is very significant. Actually, this distribution helps to demonstrate that mammalian exchanges took place between the Americas, but *before* the Eocene, since *Arctostylops* and *Palaeostylops* constitute a particular family unknown in South America; such exchanges are also evidenced as early as the late Cretaceous by marsupials and condylarths and suggested by other groups. But one cannot consider this as evidence for or against the existence of *later* exchanges concerning monkeys and caviomorph rodents.

A detailed discussion on the basis of paleogeography or plate tectonics would still be premature: the map chosen by Gingerich and Schoeninger [1977, Fig. 7, p. 497 (from Malfait and Dinkelman, 1972)] is but one hypothesis among others. Moreover, it is surprising that Gingerich should still consider this part of the discussion of such importance while at the same time he is rooting the Ceboidea in the Eurasian Adapinae, no longer in the North American Notharctinae!

3. *Ancestral stock of the Ceboidea.* I have already stressed the improbability of an independent differentiation of American and African monkeys from a "prosimian" stock, which would necessitate an unlikely accumulation of parallelisms, while no decisive character would have appeared to enable a separation of the two groups (see above). Furthermore, the adapoid stocks envisioned by Gingerich (*Pelycodus, Notharctus, Cercamonius,* not to mention poorly known genera, such as *Hoanghonius* and *Amphipithecus*) are in contradiction with biochemical and odontological data. A relationship between the Tarsioidea and the Simiiformes is much more likely, but it is not possible to assert that it is really an ancestor–descendant relationship. Anyway, the gap between *Teilhardina* (which appears to be already well engaged into the tarsioid direction) and *Branisella* (the oldest known ceboid) is considerable. McKenna (1967) did envision the differentiation in nuclear Central America of (unknown) Simiiformes which would then have reached South America: I have already underlined the improbability of this hypothesis, since, during the climatic warming-up which took place around the Oligocene-Miocene limit, the expansion of *Ekgmowechashala* up to South Dakota demonstrates the existence farther south of tarsioids, not of Simiiformes (while in South America, thanks to the same warming-up, Patagonia was reached by *monkeys*).

To sum up, in accordance with Simons' observation (1967), it does appear that, morphologically and probably phylogenetically, the African Parapithecidae are the group most closely related to the Ceboidea; all data being considered, this leads us to admit an African origin for the latter.

References

Baba, M. L., Goodman, M., Dene, H., and Moore, G. W., 1975, Origins of the Ceboidea viewed from an immunological perspective, *J. Hum. Evol.* **4**:89–102.

Berggren, W. A., and Hollister, C. D., 1974, Paleogeography, paleobiogeography and the history of circulation in the Atlantic Ocean, in: *Studies in Paleo-oceanography* (W. W. Hay, ed.), *Society of Economic Paleontologists and Mineralogists, Spec. Publ.* **20**:126–186.

Cronin, J. E., and Sarich, V. M., 1975, Molecular systematics of the New World monkeys, *J. Hum. Evol.* **4**:357–375.

Darlington, P. J., Jr., 1957, *Zoogeography,* John Wiley, New York.

Delson, E., 1975, Toward the origin of the Old World monkeys, *Coll. Intl. C.N.R.S. (Paris, 1973)* **No. 218**:839–850.

Dene, H. T., Goodman, M., and Prychodko, W., 1976, Immunodiffusion evidence on the phylogeny of the Primates, in: *Molecular Anthropology* (M. Goodman and R. E. Tashian, eds.), pp. 171–190, Plenum Press, New York.

Durette-Desset, M. C., 1971, Essai de classification des Nématodes Héligosomes. Corrélations avec la paléobiogéographie des hôtes, *Mém. Mus. Natl. Hist. Nat. Paris, Sér. A* **69**:1–126.

Gingerich, P. D., 1973, Anatomy of the temporal bone in the Oligocene anthropoid *Apidium* and the origin of Anthropoidea, *Folia Primatol.* **19**:329–337.

Gingerich, P. D., 1975, Dentition of *Adapis parisiensis* and the evolution of lemuriform primates, in: *Lemur Biology* (I. Tattersall and R. W. Sussman, eds.), pp. 65–80, Plenum Press, New York.

Gingerich, P. D., 1976, Cranial anatomy and evolution of Early Tertiary Plesiadapidae (Mammalia, Primates), *Univ. Mich. Mus. Paleontol. Papers Paleontol.* **15**:1–142.

Gingerich, P. D., 1977, Radiation of Eocene Adapidae in Europe, *Géobios, Mém. Spéc.* **1:**165–182.
Gingerich, P. D., and Schoeninger, M., 1977, The fossil record and primate phylogeny, *J. Hum. Evol.* **6:**483–505.
Goodman, M., 1975, Protein sequence and immunological specificity, in: *Phylogeny of the Primates* (W. P. Luckett and F. S. Szalay, eds.), pp. 219–248, Plenum Press, New York.
Goodman, M., 1976, Toward a genealogic description of the Primates, in: *Molecular Anthropology* (M. Goodman and R. E. Tashian, eds.), pp. 321–353, Plenum Press, New York.
Hershkovitz, P., 1974*a*, A new genus of Late Oligocene monkey (Cebidae, Platyrrhini), with notes on postorbital closure and platyrrhine evolution, *Folia Primatol.* **21:**1–35.
Herschkovitz, P., 1974*b*, The ectotympanic bone and origin of higher primates, *Folia Primatol.* **22:**237–242.
Hofer, H. O., 1976, Preliminary study of the comparative anatomy of the external nose of South American monkeys, *Folia Primatol.* **25:**193–214.
Hoffstetter, R., 1954, Les mammifères fossiles de l'Amérique du Sud et la biogéographie, *Rev. Gen. Sci.* **51**(11–12):348–378.
Hoffstetter, R., 1969, Un Primate de l'Oligocène inférieur sud-américain: *Branisella boliviana* gen. et sp. nov. *C.R. Acad. Sci. Paris, Sér. D* **269:**434–437.
Hoffstetter, R., 1972, Relationships, origins, and history of the ceboid monkeys and caviomorph rodents: A modern reinterpretation, in: *Evolutionary Biology*, Vol. 6 (Th. Dobzhansky, M. K. Hecht, and W. C. Steere, eds.), pp. 323–347, Appleton-Century-Crofts, New York.
Hoffstetter, R., 1974*a*, *Apidium* et l'origine des Simiiformes (= Anthropoidea), *C.R. Acad. Sci. Paris, Sér. D.* **278:**1715–1717.
Hoffstetter, R., 1974*b*, Phylogeny and geographic deployment of the Primates, *J. Hum. Evol.* **3:**327–350.
Hoffstetter, R., 1977*a*, Origine et principales dichotomies des Primates Simiiformes (= Anthropoidea), *C.R. Acad. Sci. Paris, Sér. D* **284:**2095–2098.
Hoffstetter, R., 1977*b*, Primates: Filogenia e historia biogeográfica, *Studia Geol.* **13:**211–253.
Hoffstetter, R., 1977*c*, Phylogénie des Primates: Confrontation des résultats obtenus par les diverses voies d'approche du problème, *Bull. Mém. Soc. Anthropol. Paris* **4**(13):927–946.
Hoyer, B. H., and Roberts, R. B., 1967, Studies of nucleic acid interactions using DNA-agar, in: *Molecular genetics* (H. Taylor, ed.) Part II, pp. 425–479, Academic Press, New York.
Kay, R. F., 1977, The evolution of molar occlusion in the Cercopithecidae and early catarrhines, *Am. J. Phys. Anthropol.* **46**(2):327–352.
Le Gros Clark, W. E., 1936, The problem of the claw in Primates, *Proc. Zool. Soc. London* **1936:**1–24.
Le Gros Clark, W. E., 1971, The antecedents of Man, 3rd edn. Edinburgh University Press, Edinburgh.
Luckett, W. P., 1975, Ontogeny of the fetal membranes and placenta, in: *Phylogeny of the Primates* (W. P. Luckett and F. S. Szalay, eds.), pp. 157–182, Plenum Press, New York.
Malfait, B. T., and Dinkelman, M. G., 1972, Circum-Caribbean tectonic and igneous activity and the evolution of the Caribbean plate, *Geol. Soc. Am. Bull* **83:**251–272.
McKenna, M. C., 1967, Classification, range and deployment of the prosimian primates, *Coll. Intl. C.N.R.S. (Paris, 1965)* **163:**603–610.
Petter, J.-J., Albignac, R., and Rumpler, Y., 1977, Mammifères lémuriens, in: *Faune de Madagascar,* Publ. **44** O.R.S.T.O.M.-C.N.R.S., Paris.
Pocock, R. I., 1918, On the external characters of the lemurs and of *Tarsius, Proc. Zool. Soc. London* **1918:**19–53.
Quentin, J. C., 1973, Affinités entre les oxyures parasites des rongeurs Hystricidés, Erethizontidés et Dinomyidés. Intérêt paléobiogéographique, *C. R. Acad. Sci. Paris, Sér. D* **276:**2015–2017.
Rimoli, R. O., 1977, Una nueva especie de monos (Cebidae: Saimirinae: *Saimiri*) de la Hispaniola, *Cuadernos del Cendia; Univ. Autónoma Santo Domingo* **242**(1):1–14.
Rosenberger, A. L., 1977, *Xenothrix* and ceboid phylogeny, *J. Hum. Evol.* **6**(5):461–481.
Russell, D. E., 1977, L'origine des Primates, *La Recherche* **8**(82):842–850.
Sarich, V. M., 1970, Primate systematics with special reference to Old World monkeys: A protein perspective, in: *Old World Monkeys* (J. R. and P. H. Napier, eds.), pp. 175–225, Academic Press, London.

Sarich, V. M., and Cronin, J. E., 1976, Molecular systematics of the Primates, in: *Molecular Anthropology* (M. Goodman and R. E. Tashian, eds.), pp. 139–157, Plenum Press, New York.

Simons, E. L., 1967, New evidence on the anatomy of the earliest catarrhine primates, in: *Neue Ergebnisse der Primatologie* (D. Starck, R. Schneider, and H. J. Kuhn, eds.), pp. 15–18, Fischer, Stuttgart.

Simons, E. L., 1972, *Primate Evolution: An Introduction to Man's Place in Nature,* Macmillan, New York.

Simons, E. L., 1974, *Parapithecus grangeri* (Parapithecidae, Old World higher primates): New species from the Oligocene of Egypt and the initial differentiation of Ceropithecoidea, *Postilla* **166:**1–12.

Simons, E. L., 1976, The fossil record of primate phylogeny, in: *Molecular Anthropology* (M. Goodman and R. E. Tashian, eds.), pp. 35–62, Plenum Press, New York.

Starck, D., 1975, The development of the chondrocranium in Primates, in: *Phylogeny of the Primates* (W. P. Luckett and F. S. Szalay, eds.), pp. 127–155, Plenum Press, New York.

Szalay, F. S., 1975*a*, Phylogeny of primate higher taxa: The basicranial evidence, in: *Phylogeny of the Primates* (W. P. Luckett and F. S. Szalay, eds.), pp. 91–125, Plenum Press, New York.

Szalay, F. S., 1975*b*, Phylogeny, adaptations and dispersal of the tarsiiform primates, in: *Phylogeny of the Primates* (W. P. Luckett and F. S. Szalay, eds.), pp. 357–404, Plenum Press, New York.

Szalay, F. S., 1977, Constructing primate phylogenies: A search for testable hypotheses with maximum empirical contents, *J. Hum. Evol.* **6:**3–18.

Phylogenetic Models of Platyrrhine Origins

III

Eocene Adapidae, Paleobiogeography, and the Origin of South American Platyrrhini

6

P. D. GINGERICH

Introduction

The origin of South American platyrrhine monkeys or Ceboidea is among the most interesting problems in primatology. This problem is basically an historical one, and geological evidence has special importance for any solution. Fossil primates, mammalian faunas, and paleogeography have a direct bearing on the origin of South American monkeys. Fortunately, much has been learned in the past twenty years about the fossil record of primate evolution. Several recent discoveries are particularly important for understanding the origin of higher primates. Furthermore, new evidence about climatic history and faunal migration during the early Cenozoic provides an improved background for interpreting the primate fossil record. Much remains to be learned, but the evidence available at present is sufficient to suggest a reasonably detailed hypothesis of ceboid origins.

South American Faunas

Paleocene and Eocene mammalian faunas of South America (Riochican to Mustersan) include a diverse group of Marsupialia, edentates of the order

P. D. GINGERICH • Museum of Paleontology, University of Michigan, Ann Arbor, Michigan 48109.

or suborder Xenarthra, and a variety of ungulates representing the orders Condylarthra, Notoungulata, Litopterna, Trigonostylopoidea, Xenungulata, and Astrapotheria (Patterson and Pascual, 1972). The major Cenozoic faunal events in South America are summarized in Fig. 1.

Marsupials, edentates, condylarths, and a notoungulate are all known from the late Paleocene and early Eocene of North America (Jepsen and Woodburne, 1969; Rose, 1978). Thus some faunal exchange between North

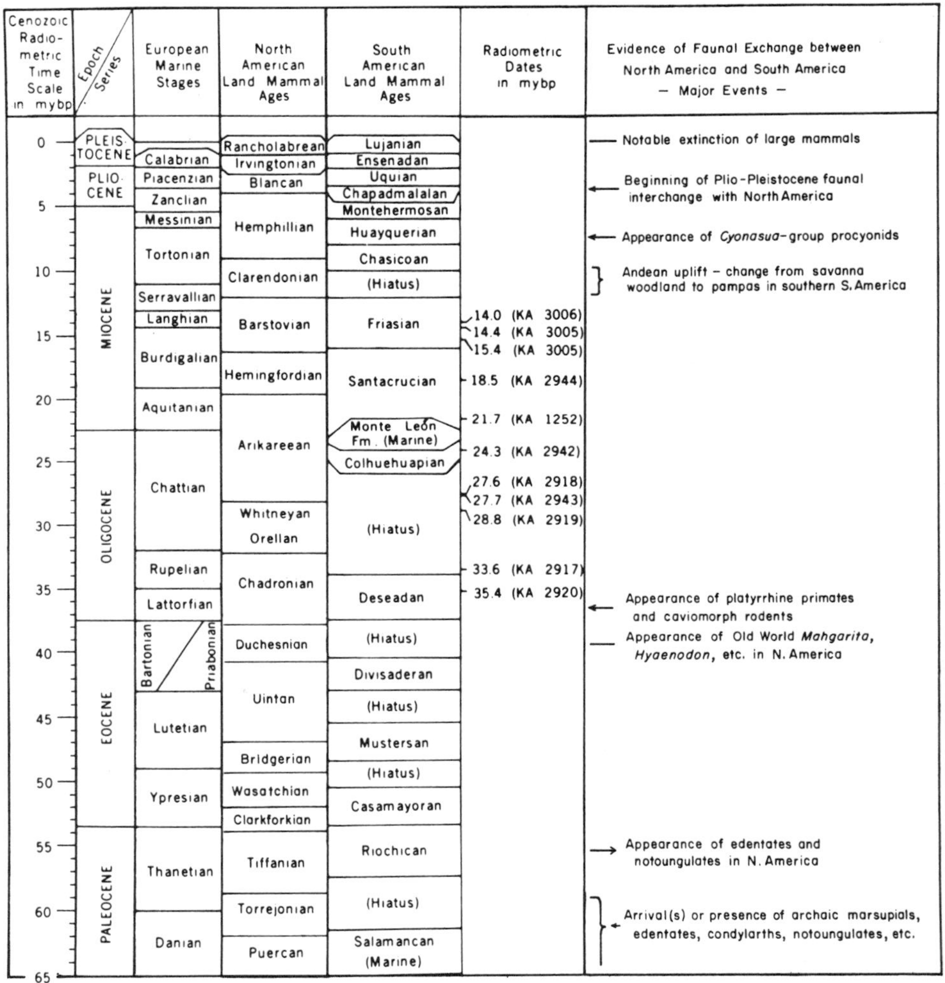

Fig. 1. Faunal succession and radiometric time scale for Cenozoic mammalian evolution in South America compared to sequences in North America and Europe. Major faunal events with a bearing on faunal migrations are indicated in the right-hand column. Data principally from Marshall *et al.* (1977) and Patterson and Pascual (1972), with additions from Wilson and Szalay (1976), Rose (1978), and others.

America and South America must have occurred during the Paleocene, filtered by a discontinuous land connection and/or the intermediate zone of tropical climate. This evidence contradicts statements by some recent authors that Paleocene and Eocene faunal migration between North America and South America was improbable or impossible, based on the Eocene position of South America relative to North America published by Frakes and Kemp (1972). Frakes and Kemp's Eocene reconstruction has been widely cited in discussing the origin of South American primate and rodent faunas, but it was constructed for another purpose and not tested against known faunal distributions before being published. A more reliable reconstruction of continental positions during the Eocene, taken from the recent book by Smith and Briden (1977), is shown in Fig. 4. Here the connection between North America and South America more closely resembles the filtered route suggested by the known distribution of Paleocene and Eocene land mammal faunas.

A major event in the history of South American mammalian faunas was the appearance of both platyrrhine primates and caviomorph rodents in the early Oligocene (Deseadan), dated at about 35–36 million years (m.y.) before present (Marshall *et al.*, 1977). The principal evidence of primates in this fauna is the type specimen of *Branisella boliviana* described by Hoffstetter (1969). Additional remains of primates from the Deseadan of Bolivia are fragmentary and all appear to represent *Branisella* as well. In contrast, the early caviomorph rodents known from the Deseadan are a diverse group including representatives of all five major suborders *Erethizontoidea, Chinchilloidea, Octodontoidea, Cavioidea,* and *Hydrochoeroidea* (Hartenberger, 1975). This diversity suggests that caviomorph rodents began radiating elsewhere before several different lines reached South America or, more probably, that they reached South America in the late Eocene. If primates arrived with rodents as part of the same faunal immigration, then primates too may have entered South America in the late Eocene. The late Eocene in South America, the "Divisaderan," is very poorly known and the Divisadero Largo fauna itself represents a peculiar facies difficult to date or relate to the mainstream of mammalian evolution (Simpson *et al.*, 1962). Thus there is no real evidence that rodents and primates were absent, and there is some slight evidence favoring their entry into South America during the late Eocene. Early and middle Eocene mammalian faunas (Casamayoran and Mustersan) are well known, include abundant microfauna, but lack primates or rodents, and it is therefore very unlikely that primates and rodents entered South America before the late Eocene.

Another filtered interchange occurred in the late Miocene with the appearance of the procyonid *Cyonasua* in South America in the Huayquerian. Subsequently, in the Montehermosan or Chapadmalalan, a land bridge between North America and South America was established through Central America and the great American mammalian interchange began (Webb, 1976). The documented occurrences of faunal interchange between North America and South America in the early Tertiary and again in the late Tertiary and Quaternary, suggests that some limited faunal interchange in the middle Tertiary was at least a possibility.

Branisella, Apidium, Aegyptopithecus, and the Origin of Simiiform Primates

Assuming that the earliest Platyrrhini and Caviomorpha entered South America in the late Eocene or earliest Oligocene, we can consider their relationship to primates and rodents in the late Eocene and early Oligocene elsewhere in the world. Deseadan primates and rodents are often compared with the Fayum Oligocene rodents and primates of northern Africa (Hoffstetter, 1972; Lavocat, 1974; and others). Fayum primates and rodents are too young geologically to have given rise to Deseadan elements of these orders in the South American fauna, but they show such similarity in structural grade that some reasonably close relationship is indicated. I am not sufficiently familiar with Eocene and Oligocene rodents to discuss the origin of Caviomorpha, but I have studied the original specimens of virtually all fossil primates relevant to the origin of Simiiformes (higher primates or "Anthropoidea"). I shall attempt to outline the nature of the paleontological evidence bearing on the origin of higher primates as simply as possible.

Branisella boliviana is known principally from the holotype maxillary fragment (Hoffstetter, 1969). In size and dental morphology this species corresponds closely to the living squirrel monkey (Fig. 2). The molars of *Branisella* in the holotype are somewhat worn, but they show the same trigon cusp and crest relationships, with a small hypocone on the internal cingulum, as seen in the living squirrel monkey. Virtually all of the fossil primates known from South America are similar to living genera and species of Cebidae, and it appears that living cebids do not differ greatly in general structure from their South American ancestors in the Oligocene.

At least five genera of primates are known from the Fayum Oligocene of Egypt. These fall naturally into three groups: (1) the adapoid *Oligopithecus,* (2) the parapithecoids *Apidium* and *Simonsius,* and (3) the hominoids *Propliopithecus* and *Aegyptopithecus* (Simons, 1965, 1972; Gingerich, 1978*a*). *Oligopithecus* is known only from a single mandible that resembles the Eocene adapid *Hoanghonius* from China (Gingerich, 1977*c*). The two genera that are best known anatomically and contribute most to our understanding of the morphology of Fayum anthropoids are *Apidium* and *Aegyptopithecus.* Cranially and postcranially *Apidium* and *Aegyptopithecus* resemble South American Cebidae to a remarkable degree (Simons, 1959, 1969, 1972; Gingerich, 1973; Conroy, 1976; Fleagle, 1978; Fleagle *et al.,* 1975; Fleagle and Simons, 1978). Thus, Oligocene *Branisella, Apidium,* and *Aegyptopithecus* taken together present a reasonably unified picture of the anatomy of a truly primitive simiiform primate. Among living primates, primitive Oligocene Simiiformes most closely resemble cebids and not callitrichids, tarsiids, or lemurids.

Fig. 2. Comparison of upper cheek teeth of primates related to the origin of South American primates, all drawn at same scale: (A) little worn left P^4M^{1-3} of the extant squirrel monkey *Saimiri sciureus;* (B) moderately worn left P^4M^{1-2} of the holotype of Oligocene *Branisella boliviana;* (C) little worn left M^{2-3} of the middle Eocene adapid *Periconodon huerzeleri.* Note close resemblance in overall size, and detailed similiarity of trigon and hypocone cusps and crests in *Branisella* and *Saimiri. Periconodon* differs from these two principally in having a distinct pericone on the lingual cingulum, but otherwise it apparently represented an Eocene primate very similar in body size and dental adaptation to *Branisella* or even *Saimiri. Branisella* specimen is in the Muséum National d'Histoire Naturelle, Paris, and the *Periconodon* is in the Naturhistorisches Museum, Basel (Bchs. 640).

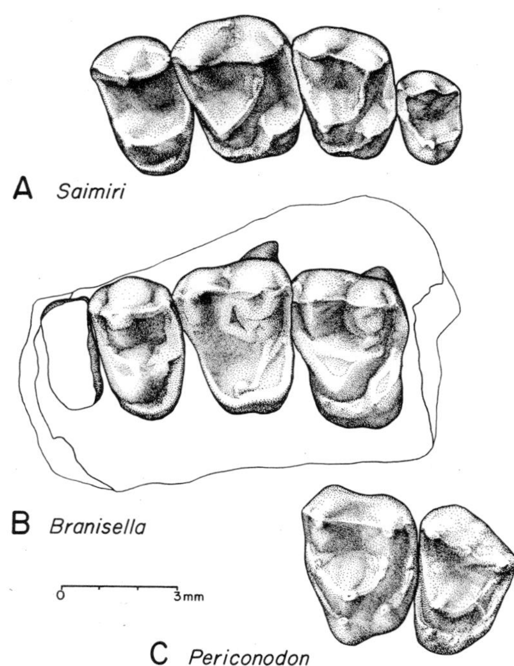

Two large families of primates of modern aspect are known from the Eocene: the tarsiiform Omomyidae and the lemuriform Adapidae. Anatomical characteristics seen in Oligocene higher primates are listed in Table 1 for comparison with the characteristics of Eocene omomyid and adapid primates possibly ancestral to the simiiform radiation.

Paleontology and comparative anatomy furnish two complementary approaches to understanding the adaptations and evolutionary history of primates. In a group like the primates for which the fossil record is reasonably well known, it is possible to outline the phylogenetic history of the group based on hard parts preserved in the fossil record (Gingerich and Schoeninger, 1977; Gingerich, 1978*b*). Interpreting the distribution of anatomical traits of living members in light of this phylogeny yields information about the probable evolutionary pathways of other hard parts and of soft anatomical characteristics not preserved in fossils. Many possible phylogenetic trees showing the relationships of primates can be suggested based on the comparative anatomy of living animals, but only one of these can reflect the actual historical pathway followed. Reversals, parallelism, and convergence are three well-documented evolutionary processes that cannot be detected by comparative study alone. For this reason, direct historical information about the actual stages of primate evolution is essential for reconstructing the evolutionary phylogeny of primates. In terms of the general question addressed in this chapter, the origin of higher primates, this reduces to two more specific questions: (1) What were the most primitive higher primates like?; and (2) what were possible precursors at an earlier stage like, and to which of these are primitive anthropoids most similar? In other words the general problem of the origin of higher primates focuses on the question of whether Oligocene Simiiformes more closely resemble Eocene Omomyidae or Eocene Adapidae.

Table 1 lists 16 anatomical characteristics preserved as hard parts in Eocene Omomyidae and Adapidae, and in Oligocene simiiform primates. Four of these are indeterminate, being shared equally by all, by Eocene lower primates but not Oligocene anthropoids, or by Oligocene anthropoids but not Eocene lower primates. Of the remaining twelve characteristics, eleven are similarities shared by Eocene Adapidae and Oligocene Simiiformes but not Eocene Omomyidae. Only one of the twelve diagnostic characteristics, relative brain size estimated by the encephalization quotient, favors Eocene Omomyidae as the ancestors of higher primates.

Kay (1975) has shown that insectivorous and folivorous primates differ in body size, with the former usually being smaller than 500 g and the latter being greater than 500 g in body mass. This size threshold at about 500 g may appropriately be called "Kay's threshold." Omomyids radiated on the insectivorous side of Kay's threshold, whereas adapids radiated at larger body size on the folivorous side of the threshold (Fleagle, 1978; note that 500 g corresponds to an M_2 length of about 3.2 mm, or ln M_2 length = 1.2, Gingerich, 1977*a*). The late Eocene and Oligocene radiation of simiiform primates was also on the folivorous side of Kay's threshold (Fleagle, 1978).

The dental formula of omomyids and adapids is variable and by itself does not suggest special affinity of either group to early simiiform primates. On the other hand, virtually all other dental characteristics distinguish Adapidae and Simiiformes from Omomyidae. The mandibular symphysis of omomyids is never fused. Fusion occurred independently at least five times in adapids. There also appears to be a trend toward fusion in progressively smaller adapids through the course of the Eocene. Thus by the late Eocene even *Mahgarita stevensi* with a body weight estimated at about 1 kg had a solidly fused mandibular symphysis (Wilson and Szalay, 1976) like that of early Simiiformes. As discussed elsewhere (Gingerich, 1977*b*), the anterior dentition of adapids and anthropoids differs from that of omomyids in having vertically implanted, spatulate incisors with the lower central incisors smaller than the lateral ones. Omomyids, on the other hand, typically have enlarged central incisors and reduced lateral incisors and canines, with the central incisors forming an almost bird-like beak (Fig. 3). Adapids have projecting, interlocking canines honed by an anterior premolar as in primitive Simiiformes (Ginerich, 1975). In addition, the canine teeth of some adapids appear to be sexually dimorphic (Stehlin, 1912; Gregory, 1920; Gingerich, 1979*b*) like those of primitive simiiform primates. Canine dimorphism has never been documented in Omomyidae, and in most omomyid genera, the canines are greatly reduced in size relative to the central incisors (Fig. 4).

The earliest Omomyidae and Adapidae have molars that are very similar in morphology, the only diagnostic differences in the dentition being in the

Table 1. Characteristics of Primitive Oligocene Simiiform Primates Compared to Those of Eocene Omomyidae and Adapidae[a]

Morphological characteristics	Eocene Omomyidae	Oligocene Simiiformes	Eocene Adapidae
Body size			
Radiation[d]	Below Kay's threshold	Above Kay's threshold	Above Kay's threshold
Dentition			
Dental formula[b]	2.1.4.3 to 2.1.2.3 or 1.1.3.3	2.1.3.3 or 2.1.2.3	2.1.4.3 to 2.1.3.3
Mandibular symphysis[d]	Unfused	Fused	Unfused to fused
Incisor form[d]	Pointed	Spatulate	Spatulate
Incisor size[d]	$I_1 \geqslant I_2$	$I_1 < I_2$	$I_1 < I_2$
Canine occlusion[d]	Limited	Interlocking	Interlocking
Canine dimorphism[d]	Absent	Present	Present
Canine/premolar hone[d]	Absent	Present	Present
Molar form[d]	Tritubercular	Quadrate	Tritubercular to Quadrate
Position of hypocone[b]	On basal cingulum	On basal cingulum	On postproto- or basal cingulum
Encephalization quotient[c]	EQ = 0.42 to 0.97	EQ = 0.85	EQ = 0.39 or 0.41
Postorbital closure[b]	None	Partial–complete	None
Ectotympanic[d]	Tubular	Free(?)–fused anulus	Free anulus
Stapedial artery[b]	Small	Lost	Large or small
Postcranium			
Calcaneum/navicular[d]	Elongated	Short	Short
Tibia/fibula[d]	Fused	Unfused	Unfused

[a] Where there is variation and a clear direction of evolution documented by the fossil record, the trend is written as primitive-to-derived. Variation is indicated wherever it is known. The complete dental, cranial, or postcranial anatomy is not known for any genus, and future discoveries may show that presently known genera do not adequately represent Omomyidae, Adapidae, or primitive Simiiformes. See text for discussion.
[b] Characteristics in which Adapidae and Omomyidae share equal similarity or dissimilarity with primitive Simiiformes.
[c] Characteristics in which Omomyidae are more similar than Adapidae to primitive Simiiformes.
[d] Characteristics in which Adapidae are more similar than Omomyidae to primitive Simiiformes.

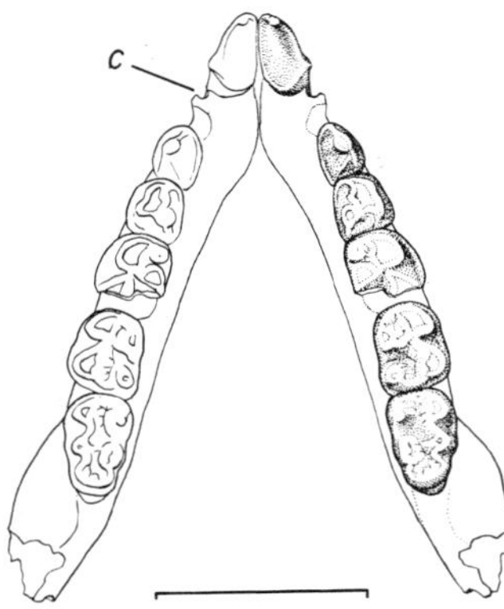

Fig. 3. Reconstruction of left and right mandibles of *Microchoerus erinaceus* showing the lower dentition in occlusal view. Note the large pointed central incisors (I_1), forming an almost birdlike beak. There are no second incisors (I_2) in *Michrochoerus,* and the lower canine (labelled C in the figure) is greatly reduced in size relative to I_1 or P_2. Specimen is in the British Museum of Natural History, London (M30345 and 30347). Scale bar is 1 cm.

morphology of the premolars and anterior dentition. Most omomyid genera retain a paraconid on the lower molars and a basically tritubercular molar structure (like that of *Tarsius*). Adapids, on the other hand, lost the paraconid on the lower molars relatively early and their molar structure is more quadrate than tritubercular. Oligocene simiiform primates have quadrate molars like those of adapids rather than omomyids. This is why genera like late Eocene *Amphipithecus* and *Pondaungia,* and early Oligocene *Oligopithecus* are difficult to classify. They have the molar structure of both Adapidae and Simiiformes (Szalay, 1970, 1972; Simons, 1971; Gingerich, 1977*c*). The hypocone in most representatives of all three groups, Omomyidae, Adapidae, and Simiiformes, is a so-called "true" hypocone on the basal cingulum.

In cranial structure, the relative size of the brain can be measured using Jerison's (1973) encephalization quotient (EQ). Radinsky (1977) calculated EQ values of .42, .79, and .97 for the omomyids *Tetonius, Necrolemur,* and *Rooneyia,* respectively. He gives EQ values of .41 and .39 for *Smilodectes* and *Adapis,* respectively. *Aegyptopithecus* had an EQ of about .85 (Gingerich, 1977*a*), so in relative brain size omomyids are closer to *Aegyptopithecus* than adapids are. Postorbital closure separates primitive Simiiformes from both Omomyidae and Adapidae, and thus does not indicate any affinity with one family or the other.

The structure of the ectotympanic in Omomyidae is tubular as it is in *Tarsius.* Adapidae have a free ectotympanic within the auditory bulla like that of living Malagasy lemurs. The ectotympanic of both *Aegyptopithecus* and *Apidium* was ringlike, and it undoubtedly filled much of the lateral wall of the auditory bulla like it does in living Ceboidea. It is possible that this primitive anthropoid condition could be derived from the tubular ectotympanic of an omomyid, but I am not aware of any other examples of loss of the tubular extension of the ectotympanic in primate evolution. In addition, the squamosal of *Apidium* has a small cup-shaped depression that received the distal end of the ectotympanic anulus. The anulus itself is not preserved, but the presence of a distinct depression where its free end articulated with the squamosal suggests that the anulus was not solidly fused to the auditory bulla

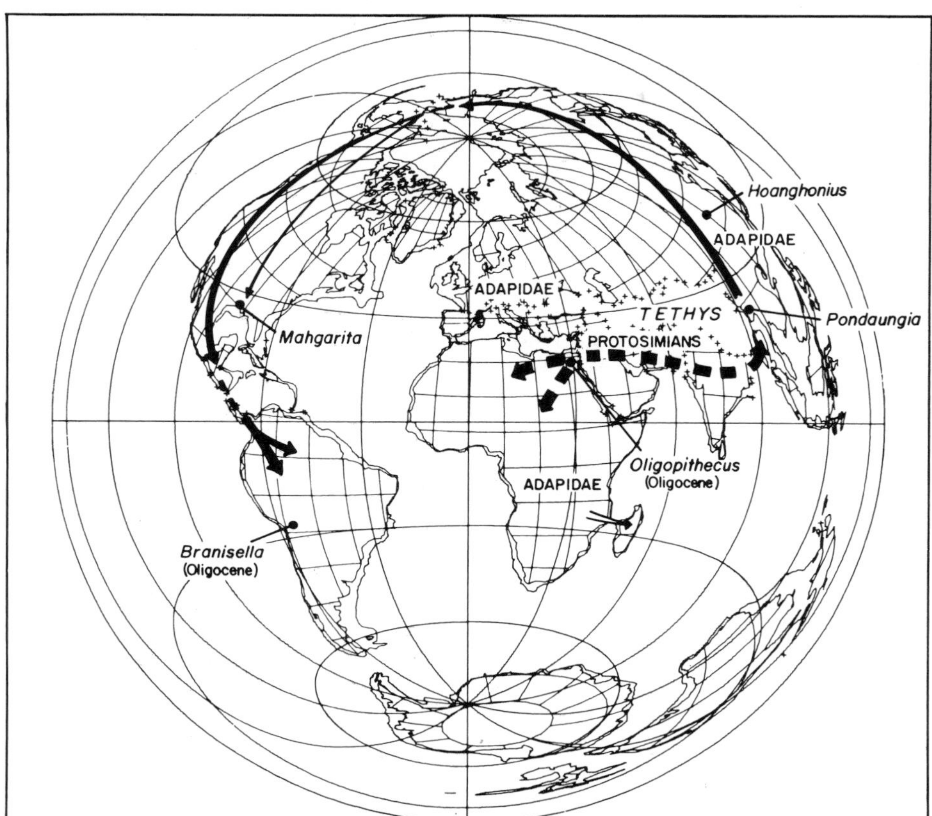

Fig. 4. Late Eocene paleocontinental map showing the position of South America relative to other continents. The geographic distribution of late Eocene adapid primates included Europe (*Adapis,* etc.), Asia (*Hoanghonius*), North America (*Mahgarita*), and almost certainly Africa and south Asia. By the middle or late Eocene the lemur fauna of Madagascar was probably isolated after derivation from African adapids. Late Eocene *Pondaungia* and *Amphipithecus* (both from the same general area of Burma) and early Oligocene *Oligopithecus* are transitional adapid-simiiform primates linking higher primates to an adapid origin. Early Oligocene *Branisella* is the earliest record of Ceboidea in South America. Note that the Burmese localities yielding *Pondaungia* and *Amphipithecus* were north of Tethys and part of Laurasia in the late Eocene. The evidence available at present favors origin of Simiiformes from an advanced adapid "protosimian" stock in south Asia or Africa or both. Part of the protosimian stock radiated in Africa, giving rise to the earliest Hominoidea by the middle and late Oligocene (*Propliopithecus, Aegyptopithecus*). Plausibly another part of the protosimian stock accompanied *Mahgarita, Hyaenodon,* and other Asian mammals across the Bering route into southern North America in the late Eocene. The protosimian stock then crossed from North America into South America by island-hopping via the route of present Central America or the West Indies. This hypothesis is shown by solid lines superimposed on the map. It is also possible that the adapid-protosimian stock ancestral to Ceboidea crossed the South Atlantic directly from Africa to South America. Base map is a Lambert equal-area projection from Smith and Briden (1977).

in *Apidium* (Gingerich, 1973). Hershkovitz (1974) has shown that the distal end of the ectotympanic sometimes does not fuse to the squamosal in *Tarsius* and in ceboids, but it is always solidly fused to the auditory bulla. The distal portion of the ectotympanic in *Tarsius* and ceboids is broad and flat, and it does not fit into a cup-like depression like that seen in *Apidium*. Obviously, more complete specimens of *Apidium* are required to determine the detailed relationship of the ectotympanic to the auditory bulla and squamosal, but evidence at hand indicates that neither *Apidium* nor *Aegyptopithecus* had an omomyid-like tubular ectotympanic fused to the auditory bulla. The facet for the distal articulation of the ectotympanic with the squamosal in *Apidium* suggests that the primitive simiiform configuration may have included a partially free ectotympanic anulus more similar to that of adapids.

Simiiform primates differ from all Eocene lower primates in lacking a stapedial branch of the internal carotid artery. Omomyids and at least some adapids have a relatively reduced stapedial and enlarged promontory branch of the internal carotid artery (Gingerich, 1973). Carotid circulation does not indicate any special similarity of either omomyids or adapids to early higher primates.

The postcranial skeleton of omomyids, adapids, and primitive Simiiformes is not yet sufficiently well described to permit a detailed comparison, but two aspects of hind limb anatomy deserve mention. The calcaneum is known in a number of different genera of omomyids, including *Hemiacodon, Teilhardina, ?Tetonius, Necrolemur, Nannopithex,* and *Arapahovius,* and in every case it is relatively elongated compared to generalized primates (Szalay, 1976; Savage and Waters, 1978). The tibia and fibula have been described in two omomyids, *Necrolemur* and *Nannopithex* (Schlosser, 1907; Weigelt, 1933; see also Simons, 1961; Le Gros Clark, 1962), and in *Necrolemur* at least the fibula appears to be reduced in size and fused to the tibia. The conformation of the fibula in *Nannopithex* is less certain (Simons, 1961). Calcaneal elongation and fibular fusion are resemblances of omomyids to living *Tarsius,* but they distinguish this group postcranially from both Adapidae and from Simiiformes.

Primitive Oligocene simiiform primates resemble Eocene Adapidae much more than they do Eocene Omomyidae. The most parsimonious interpretation of this evidence is that higher primates evolved from Adapidae and not from Omomyidae. It is generally accepted that living lemurs are derived from Eocene Adapidae and the living *Tarsius* from Eocene Omomyidae. Consequently, anthropoid primates and lemurs are probably more closely related to each other than either is to *Tarsius.* The implications for comparative anatomy are several. Anatomical characteristics such as the reduced rhinarium and nasal fossa (Cave, 1973), and the hemochorial placenta (Luckett, 1975) shared by *Tarsius* and Simiiformes but not Lemuriformes may be parallel evolutionary acquisitions (or possibly retained primitive states). The reliability of phylogenetic distances inferred from immunology and protein sequences (Goodman, 1975) appears somewhat questionable when these distances span a total temporal separation on the order of 80–100 million years

(40–50 m.y. in each lineage compared). I doubt that placing lemurs and lorises slightly closer to anthropoids than *Tarsius* is would significantly decrease the parsimony of the immunological or protein sequence result.

There is disagreement regarding the major phyletic relationships of Tarsiiformes, Lemuriformes, and Simiiformes, with different results depending on whether one attempts to trace phyletic groups through the fossil record or to infer history from the comparative anatomy of living forms. This means on the one hand that our evidence regarding primate phylogeny is still far from complete, and on the other hand that we need to take a more critical look at different methods being used to reconstruct primate history. Parallelisms and reversals are common evolutionary phenomena. For this reason I tend to trust a phylogeny based on closely spaced historical records preserved as fossils rather than one based on selected comparisons of animals living, so to speak, 40 or 50 m.y. after the fact.

Paleobiogeography

The approximate distribution of continental land masses during the late Eocene, when simiiform primates evolved from their adapid ancestors, is shown in Fig. 4. Superimposed on early Cenozoic paleogeography was a series of major worldwide climatic changes documented paleobotanically on the continents (Wolfe and Hopkins, 1967; Wolfe, 1978) and isotopically in the oceans (Savin *et al.,* 1975). The late Paleocene was generally a time of climatic cooling, followed by a definite warming trend at the end of the epoch that continued into the Eocene. After several fluctuations in the Eocene, there was a sharp drop in temperature worldwide at the end of the Eocene corresponding to Stehlin's (1909) "*grande coupure*" in European mammalian faunas. Climate strongly affects the distribution of mammalian faunas, and there is evidence that high latitutde land bridges like the Bering route between Asia and North America were effectively opened or closed during the early Cenozoic by changes in climate as well as sea level.

Modern primates, more than most other orders of mammals, are sensitive to climate. Thus it is probably no accident that the introduction of Omomyidae and Adapidae into Europe and North America coincided with early Eocene climatic warming, and the reduction in diversity of both families on northern continents also coincided with climatic cooling. The *grande coupure* marks the final exit of both Eocene families from Europe. Simpson (1947) made an extensive analysis of mammalian faunal similarity between North America and Eurasia. He showed that the greatest faunal interchange between North America and Eurasia took place during the late Eocene just before the *grande coupure.*

The major faunal interchange between North America and Eurasia in the late Eocene assumes special importance in explaining the distribution of

Adapidae at this time. The principal radiation of adapids documented to date was in the Eocene of Europe, but finds in the poorly known early Cenozoic faunas of Asia *(Hoanghonius)* and Africa *(Oligopithecus)* suggest that major radiations of Adapidae may have taken place there as well. The notharctine adapid radiation in North America apparently became extinct early in the late Eocene, but one adapine genus, *Mahgarita,* is known from the late Eocene of Texas (Wilson and Szalay, 1976). It presumably reached North America as part of the late Eocene invasion from Asia that included the Old World creodont *Hyaenodon,* anthracotheres, etc. (Gingerich, 1979*a*). As a result, Adapidae apparently enjoyed a virtually worldwide distribution in the late Eocene.

Hoffstetter (1972, 1974) has advanced the hypothesis that tarsiiform primates radiated north of Tethys in Laurasia while their "sister-group" the simiiform primates radiated south of Tethys, initially in Africa and then in South America. It is true that omomyids are unknown outside of Laurasia, but it is difficult to see how Simiiformes could be derived from Tarsiiformes given this geographical exclusivity. A more reasonable hypothesis, I think, is that higher primates were derived from a group that shared a similar geographical distribution. For this reason, and all of the anatomical reasons discussed above, Adapidae as a group are a better candidate for simiiform ancestry than tarsiiform Omomyidae.

The most likely area of origin of higher primates, based on present evidence, appears to be Africa and/or South Asia. This is the region labelled "Protosimians" in Fig. 4, which lies between the known distribution of *Amphipithecus* and *Pondaungia* in Burma and *Oligopithecus* in Africa. All three of these genera have the distinction of being ambiguous adapid-simiiform intermediates at the time when simiiform primates were first differentiating.

The remaining problem is how the ancestors of Ceboidea reached South America if they originated in Africa or South Asia. There are two possibilities: (1) they crossed the Bering land bridge from Asia during a warm interval in the late Eocene and, with *Mahgarita,* colonized the southern part of North America, then crossed one of two possible volcanic island arcs bordering the Carribean Plate (see Gingerich and Schoeninger, 1977) and entered South America (Fig. 4); or (2) they crossed the South Atlantic directly, either by rafting or by island-hopping across the Walvis-Rio Grande rise (see Tarling, this volume). Of these two hypotheses, I favor the former because of the difficulty primates would have crossing large tracts of open ocean on rafts. It was no doubt necessary to cross some ocean by island-hopping in either case, but this would be minimized in crossing from Central America to South America. Unfortunately, there is little evidence available as yet to test either hypothesized route.

A third possibility deserves mention, although I do not yet think the evidence is sufficient to warrant serious consideration. The new mandible of *Pondaungia* (Fig. 5) recently described by Ba Maw, *et al.* (1979) from the late Eocene of Burma bears a surprising resemblance to *Notharctus.* Pilgrim (1927)

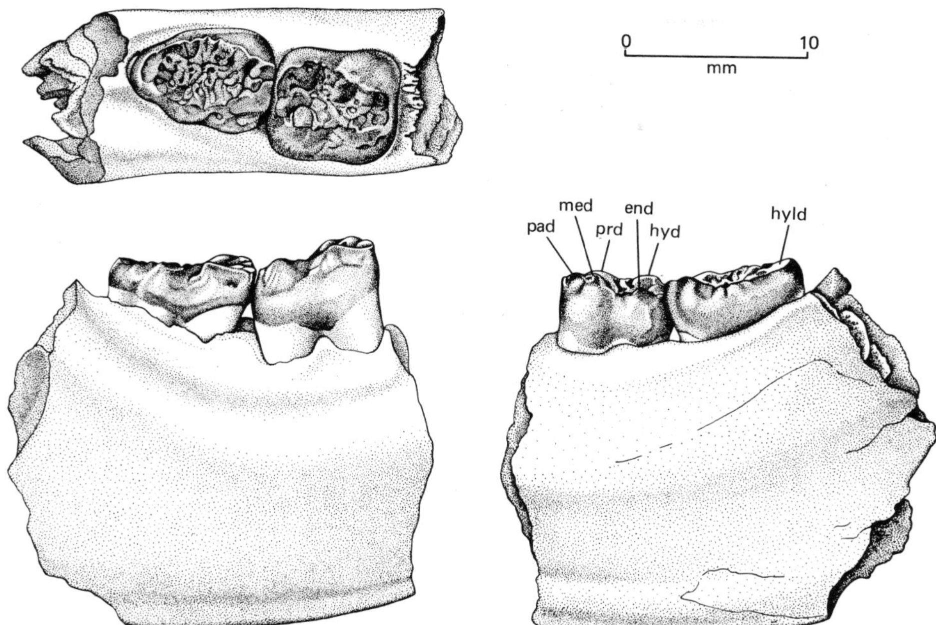

Fig. 5. Recently discovered right mandible of *Pondaungia* sp. with M_{2-3} from the late Eocene of Burma. Specimen resembles the adapid *Notharctus robustior* in size, placement of trigonid cusps and crests, development of talonid cusps and crests, and to some extent in enamel crenulation. The molars are relatively broader and flatter in *Pondaungia* indicating an anthropoid-like frugivorous adaptation similar to that of *Aegyptopithecus,* whereas *Notharctus* has more crested cheek teeth indicating a predominently folivorous dietary adaptation. In spite of this difference in adaptation, among Eocene primates the molars of *Pondaungia* are most similar in structure to those of notharctine Adapidae, as Pilgrim (1927) stated over fifty years ago based on less well-preserved specimens. Figure reproduced from Ba Maw *et al.* (1979).

also compared *Pondaungia* extensively with *Pelycodus* and *Notharctus.* The dentition of the new specimen differs in being adapted for a more frugivorous diet, whereas *Notharctus* has more crested folivorous cheek teeth, but the basic plan of trigonid and talonid cusps is very similar. If *Pondaungia* is a notharctine, it is possible that higher primates originated in southern North America and subsequently migrated in the late Eocene from North America into South America and also from North America across the Bering route into south Asia and ultimately Africa. This is not a serious hypothesis at present, but it is a possibility.

Much has been learned in the past twenty years about the evolution and geographical development of primates, and continued recovery of new fossil specimens at the current rate will undoubtedly contribute in the next twenty years to a better understanding of the origin of South American primates. The most critical tests of phylogenetic hypotheses are new fossils.

ACKNOWLEDGMENTS

I would like to thank many museum curators for generous access to the fossil specimens discussed here, especially Drs. Elwyn L. Simons, Duke University Primate Center, Durham; D. E. Savage, University of California, Berkeley; M. C. McKenna, American Museum of Natural History, New York; R. Hoffstetter and D. E. Russell, Muséum National d'Histoire Naturelle, Paris; R. A. Reyment, Paleontologiska Institutionen, Uppsala; B. Engesser and J. Hürzeler, Naturhistorisches Museum, Basel; R. Wild and E. Heizmann, Staatliches Museum für Naturkunde, Stuttgart (Ludwigsburg); Darwish Alfar and Baher El-Kashab, Cairo Geological Museum, Cairo; and M. V. A. Sastry and A. K. Dutta, Geological Survey of India, Calcutta. I also thank Drs. R. L. Ciochon and D. E. Savage for permitting me to reproduce a figure of one of their most important new specimens from Burma in Fig. 5. Ms. Karen Payne drew the specimens in Figs. 2 and 3. Ms. Karna Steelquist assisted with photography and Ms. Anita Benson typed the manuscript.

This research has been supported principally by a NATO Postdoctoral Fellowship at the Université de Montpellier (1975), grants from the Smithsonian Foreign Currency Program for museum research in Egypt and India, and NSF grant DEB 77-13465 for research on Paleocene and Eocene faunas in North America.

References

Ba Maw, Ciochon, R. L., and Savage, D. E., 1979, Late Eocene of Burma yields earliest anthropoid primate, *Pondaungia cotteri*, *Nature (London)* **282**:65–67.

Cave, A. J. E., 1973, The primate nasal fossa, *Biol. J. Linnean Soc.* **5**:377–387.

Conroy, G. C., 1976, Primate postcranial remains from the Oligocene of Egypt, *Contrib. Primatol.* **8**:1–134.

Fleagle, J. G., 1978, Size distribution of living and fossil primate faunas, *Paleobiology* **4**:67–76.

Fleagle, J. G., and Simons, E. L., 1978, Humeral morphology of the earliest apes, *Nature (London)* **276**:705–707.

Fleagle, J. G., Simons, E. L., and Conroy, G. C., 1975, Ape limb bone from the Oligocene of Egypt, *Science* **189**:135–137.

Frakes, L. A., and Kemp, E. M., 1972, Influence of continental positions on Tertiary climates, *Nature (London)* **240**:97–100.

Gingerich, P. D., 1973, Anatomy of the temporal bone in the Oligocene anthropoid *Apidium* and the origin of Anthropoidea, *Folia Primatol.* **19**:329–337.

Gingerich, P. D., 1975, A new genus of Adapidae (Mammalia, Primates) from the Late Eocene of southern France, and its significance for the origin of higher primates, *Contrib. Mus. Paleontol. Univ. Mich.* **24**:163–170.

Gingerich, P. D., 1977*a*, Correlation of tooth size and body size in living hominoid primates, with a note on relative brain size in *Aegyptopithecus* and *Proconsul*, *Am. J. Phys. Anthropol.* **47**:395–398.

Gingerich, P. D., 1977*b*, Dental variation in early Eocene *Teilhardina belgica* with notes on the anterior dentition of some early Tarsiiformes, *Folia Primatol.* **28**:144–153.

Gingerich, P. D., 1977*c*, Radiation of Eocene Adapidae in Europe, *Géobios, Mém. Spéc.* **1**:165-182.

Gingerich, P. D., 1978*a*, The Stuttgart collection of Oligocene primates from the Fayum Province of Egypt, *Paläontol. Z.* **52**:82-92.

Gingerich, P. D., 1978*b*, Phylogeny reconstruction and the phylogenetic position of *Tarsius*, in: *Recent Advances in Primatology*, Vol. 3, *Evolution* (D. J. Chivers and K. A. Joysey, eds.), pp. 249-255, Academic Press, New York.

Gingerich, P. D., 1979*a*, Phylogeny of Middle Eocene Adapidae (Mammalia, Primates) in North America: *Smilodectes* and *Notharctus*, *J. Paleontol.* **53**:153-163.

Gingerich, P. D., 1979*b*, Sexual dimorphism in Eocene Adapidae: Implications for primate phylogeny and evolution, *Am. J. Phys. Anthropol.* **50**:442.

Gingerich, P. D., and Schoeninger, M. J., 1977, The fossil record and primate phylogeny, *J. Hum. Evol.* **6**:482-505.

Goodman, M., 1975, Protein sequence and immunological specificity, their role in phylogenetic studies of primates, in: *Phylogeny of the Primates: A Multidisciplinary Approach* (W. P. Luckett and F. S. Szalay, eds.), pp. 219-248, Plenum Press, New York.

Gregory, W. K., 1920, On the structure and relations of *Notharctus*, an American Eocene primate, *Mem. Am. Mus. Nat. Hist.* **3**:49-243.

Hartenberger, J.-L., 1975, Nouvelles decouvertes de rongeurs dans le Deseadien (Oligocène inférieur) de Salla Luribay (Bolivie), *C. R. Acad. Sci. Paris, Sér. D* **280**:427-430.

Hershkovitz, P., 1974, The ectotympanic bone and origin of higher primates, *Folia Primatol.* **22**:237-242.

Hoffstetter, R., 1969, Un primate de l'Oligocène inférieur sud-américain: *Branisella boliviana* gen. et sp. nov, *C. R. Acad. Sci. Paris, Sér D.* **269**:434-437.

Hoffstetter, R., 1972, Relationships, origins, and history of the ceboid monkeys and caviomorph rodents: A modern reinterpretation, in: *Evolutionary Biology*, Vol. 6 (Th. Dobzhansky, M. K. Hecht, and W. C. Steere, eds.), pp. 323-347, Appleton-Century-Crofts, New York.

Hoffstetter, R., 1974, Phylogeny and geographical deployment of the primates, *J. Hum. Evol.* **3**:327-350.

Jepsen, G. L., and Woodburne, M. O., 1969, Paleocene hyracothere from Polecat Bench Formation, Wyoming, *Science* **164**:543-547.

Jerison, H. J., 1973, *Evolution of the Brain and Intelligence*, Academic Press, New York.

Kay, R. F., 1975, The functional adaptations of primate molar teeth, *Am. J. Phys. Anthropol.* **43**:195-216.

Lavocat, R., 1974, The interrelationships between the African and South American rodents and their bearing on the problem of the origin of South American monkeys, *J. Hum. Evol.* **3**:323-326.

Le Gros Clark, W. E., 1962, *The Antecedents of Man*, Edinburgh University Press, Edinburgh.

Luckett, W. P., 1975, Ontogeny of the fetal membranes and placenta, their bearing on primate phylogeny, in: *Phylogeny of the Primates, a Multidisciplinary Approach* (W. P. Luckett and F. S. Szalay, eds.), pp. 157-182, Plenum Press, New York.

Marshall, L. G., Pascual, R., Curtis, G. H., and Drake, R. E., 1977, South American geochronology: Radiometric time scale for middle to late Tertiary mammal-bearing horizons in Patagonia, *Science* **195**:1325-1328.

Patterson, B., and Pascual, R., 1972, The fossil mammal fauna of South America, in: *Evolution, Mammals, and Southern Continents* (A. Keast, F. C. Erk, and B. Glass, eds.), pp. 247-309, State University of New York Press, Albany.

Pilgrim, G. E., 1927, A *Sivapithecus* palate and other primate fossils from India, *Mem. Geol. Surv. India* **14**:1-26.

Radinsky, L., 1977, Early primate brains: Facts and fiction, *J. Hum. Evol.* **6**:79-86.

Rose, K. D., 1978, A new Paleocene epoicitheriid (Mammalia), with comments on the Palaeanodonta, *J. Paleontol.* **52**:658-674.

Savage, D. E., and Waters, B. T., 1978, A new omomyid primate from the Wasatch formation of southern Wyoming, *Folia Primatol.* **30**:1-29.

Savin, S. M., Douglas, R. G., and Stehli, F. G., 1975, Tertiary marine paleotemperatures, *Bull. Geol. Soc. Am.* **86**:1499–1510.

Schlosser, M., 1907, Beitrag zur Osteologie und systematischen Stellung der Gattung *Necrolemur,* sowie zur Stammesgeschichte der Primaten überhaupt, *Neues Jahrb. Mineral. Geol. Paläontol. Festbd.* **1807–1907:**197–226.

Simons, E. L., 1959, An anthropoid frontal bone from the Fayum Oligocene of Egypt: The oldest skull fragment of a higher primate, *Am. Mus. Novit.* **No. 1976**:1–16.

Simons, E. L., 1961, Notes on Eocene tarsioids and a revision of some Necrolemurinae, *Bull. Br. Mus. Nat. Hist. (Geol.)* **5**:45–69.

Simons, E. L., 1965, New fossil apes from Egypt and the initial differentiation of Hominoidea, *Nature (London)* **205**:135–139.

Simons, E. L., 1969, Recent advances in paleoanthropology, *Yearb. Phys. Anthropol.* **1967**:14–23.

Simons, E. L., 1971, Relationships of *Amphipithecus* and *Oligopithecus, Nature (London)* **232**:489–491.

Simons, E. L., 1972, *Primate Evolution, An Introduction to Man's Place in Nature,* Macmillan, New York.

Simpson, G. G., 1947, Holarctic mammalian faunas and continental relationships during the Cenozoic, *Bull. Geol. Soc. Am.* **58**:613–688.

Simpson, G. G., Minoprio, J. L., and Patterson, B., 1962, The mammalian fauna of the Divisadero Largo Formation, Mendoza, Argentina, *Bull. Mus. Comp. Zool.* **127**:237–293.

Smith, A. G., and Briden, J. C., 1977, *Mesozoic and Cenozoic Paleocontinental Maps,* Cambridge University Press, Cambridge.

Stehlin, H. G., 1909, Remarques sur les faunules de mammifères des couches éocènes et oligocènes du bassin de Paris, *Bull. Soc. Géol. Fr.* **9**:488–520.

Stehlin, H. G., 1912, Die Säugetiere des schweizerischen Eocaens—*Adapis, Abh. Schweiz. Paläont. Ges.* **38**:1165–1298.

Szalay, F. S., 1970, Late Eocene *Amphipithecus* and the origins of catarrhine primates, *Nature, (London)* **227**:355–357.

Szalay, F. S., 1972, *Amphipithecus* revisited, *Nature (London)* **236**:170–180.

Szalay, F. S., 1976, Systematics of the Omomyidae (Tarsiiformes, Primates): Taxonomy, phylogeny, and adaptations, *Bull. Am. Mus. Nat. Hist.* **156**:163–449.

Webb, S. D., 1976, Mammalian faunal dynamics of the great American interchange, *Paleobiology* **2**:220–234.

Weigelt, J., 1933, Neue primaten aus der mitteleozänen (oberlutetischen) Braunkohle des Geiseltals, *Nova Acta Leopold.* **1**:97–156.

Wilson, J. A., and Szalay, F. S., 1976, New adapid primate of European affinities from Texas, *Folia Primatol.* **25**:294–312.

Wolfe, J. A., 1978, A paleobotanical interpretation of Tertiary climates in the northern hemisphere, *Am. Sci.* **66**:694–703.

Wolfe, J. A., and Hopkins, D. M., 1967, Climatic changes recorded by Tertiary land floras in northwestern North America, in: *Tertiary Correlations and Climatic Changes in the Pacific* (K. Hatai, ed.), *11th Pacific Sci. Symp.* **25**:67–76.

On the Tarsiiform Origins of Anthropoidea

7

A. L. ROSENBERGER and F. S. SZALAY

Introduction

Many systematic papers of the past decade have employed a cladistic approach. Reasons for this lie largely in the superficially rigorous appearance of this method and, therefore, the impression of a powerful tool. An unvarying commitment to the operational underpinnings of cladism has led to the acceptance of a systematic methodology which perforce only recognizes, and can only search for, sister-group relationships (e.g., Engelman and Wiley, 1977). The notion of ancestor–descendant relationships, or to put it differently, that phena *transform* from antecedent phena, has been simply set aside. This fundamental aspect of real evolutionary history has come to be ignored *at the expense of a method* because hypotheses of descent are claimed to be unfalsifiable, whereas sister-group relationships are thought to be easily refutable.

More recently, the testability of ancestor–descendant relationships and the problems posed by a simplistic methodology which ignores the theoretical and empirical foundations of evolutionary biology have been treated by Bock (1977*b*), Szalay (1977*a*) and Naylor (manuscript). All of these authors make the point that a rigorous application of cladistics (considerably removed from Hennig's original formulation of phylogenetic systematics), as advocated by Cracraft (1974, 1978), Engelman and Wiley (1977), Tattersall and Eldredge (1977), Nelson (1973), and many others, in fact deprives a systematic inquiry of the most precise (hence most easily corroborated) and therefore the most

A. L. ROSENBERGER • Department of Anthropology, University of Illinois at Chicago Circle, Box 4348, Chicago, Illinois 60680. F. S. SZALAY • Department of Anthropology, Hunter College, New York, New York 10021.

easily falsifiable of phylogenetic hypotheses—ancestry and descent. Discovery of single characters or fossils in a new stratigraphic position supplies ready and immediate tests of an ancestor-descendant hypothesis (Szalay, 1977*a*; Naylor, manuscript), either corroborating or forbidding it. As Naylor (manuscript) aptly puts it, "In this event, the hypothesis has been falsified and must be replaced by a *less precise* and *less readily tested* hypothesis of sister-group relationships." The discovery of a single autapomorphy in a postulated ancestral species immediately falsifies that hypothesis, and automatically relegates an explicit, boldly stated (ancestor–descendant) hypothesis to a more general one of lower order (a sister-group hypothesis). On the other hand, as Naylor further argues, the relative levels of primitiveness of characters, an empirically recognized consequence and attribute of evolution, offer irrelevant tests of sister-group relationships and are thus ignored by cladistics. But in postulating ancestor–descendant hypotheses, the recognition of additional primitive states can provide a strong corroboration of a transformation sequence, given that other features are not contradictory.

It should also be noted that character analysis, the cornerstone of phylogenetic studies, and particularly cladistics, relies heavily on inferences founded on vertical and horizontal comparisons (see Bock, 1977*a*), even though the stratigraphic position of the relevant morphologies must often be ignored because of the incompleteness of the record. These comparisons are transformational (in a sense ancestor–descendant) by nature and are required and justifiable simply because evolution is "descent with modification," whether this occurs during the history of an intact lineage or gene pool or at the time of splitting. Thus a polarity inference based on vertical comparisons is eminently falsifiable and equivalent, if not higher, in order than one based on horizontal comparisons.

Clearly, the synthesis resulting from a rigorously cladistic approach will be much less satisfactory biologically, both in terms of characters and organisms, than one which employs transformational hypotheses. With these foregoing remarks we proceed to employ and test ancestor–descendant hypotheses *and* sister-group hypotheses, preferring to allow the developing evidence determine which of these is more appropriate. Inasmuch as we treat even living animals as taxa and as characters (see especially Simpson, 1963), our operational concept of ancestry is an abstract one, realizing fully that our results can only be a first approximation of history. The nature of the problem dictates it.

The Problem

The radiation of Eocene euprimates (including strepsirhines and haplorhines) produced two rather successful groups of species that inhabited parts of North America and Eurasia, the Omomyidae and Adapidae [taxonomic and vernacular terms follow Szalay and Delson (1979)]. Each have been implicated as possible ancestors of the anthropoids. More specifically, some students have suggested direct phyletic ties of either omomyids or adapids with the South American platyrrhine primates. While these hypotheses may be tested and refuted, we believe that fruitful pursuit of that question, searching for the true origins of the platyrrhines, is beyond today's power of resolution. The New World monkeys are still poorly known paleontologically and serious studies of their morphological diversity and genealogical interrelationships are only now being undertaken. Furthermore, the evidence seems to point to a common ancestry of platyrrhines and catarrhines, requiring more broadly based comparisons. The emphasis of this contribution, therefore, is the bearing of omomyids and the extant tarsiiforms upon anthropoid origins, which is a phylogenetic question rather than a purely taxonomic one. As such, we will not restrict our examination to omomyids alone.

The morphology and relationships of omomyids have been considered by several workers during the past three decades. Essentially all are in agreement that the lineage leading to the living *Tarsius* stems directly from an omomyid or some closely allied group, hence the widespread recognition of the taxon Tarsiiformes. But the higher relationships of tarsiiforms remain in dispute. A balance of cranial, dental, soft tissue and molecular evidence (see reviews in Luckett and Szalay, 1975, 1978) support the hypothesis that tarsiiforms and anthropoids are sister-taxa, together comprising the Haplorhini. This implies that higher primates are descendant of some species that would perhaps be classified as tarsiiform, or perhaps omomyid. Cartmill and Kay (1978) offer a bold refinement of this hypothesis, specifying the *Tarsius* lineage as the actual sister-group of Anthropoidea. A divergent view, one that we regard as highly unlikely, is that tarsiiforms are actually more closely related to the predominantly Paleocene plesiadapiform primates (e.g., Gingerich, 1975*b,* 1976, 1978; Schwartz, 1978; Schwartz *et al.*, 1978; Krishtalka and Schwartz, 1979). This view usually entails the corollary proposition that adapids gave rise to anthropoids. A variety of arguments specifically countermanding much of the rationale for this twin hypothesis has been provided by Szalay (1977*b*), Szalay and Delson (1979) and Cartmill and Kay (1978); see also Archibald (1977). Our chief purpose here is to review what we consider to be the most compelling morphological and phylogenetic arguments favoring a derivation of anthropoids from tarsiiforms such as omomyids and denying their origin from *bona fide* tarsiids and adapids.

Omomyids and Anthropoids

The overall construction of the omomyid cranium compares favorably with anthropoids, as Le Gros Clark (1963) and others have ably demonstrated, particularly in details of the facial skeleton and some of the derived aspects of the ear region. Compared to the proportions encountered among strepsirhines, the facial skull is reduced in length, though not especially abbreviated as in some modern anthropoids (e.g., colobines, callitrichines), and is hafted low on the neurocranium. The nasal fossa is apparently diminished in overall size but is dorsoventrally deep. The internal architecture of the nasal capsule is still unknown but the closely approximated orbits (Kay and Cartmill, 1977), which fuse posteriorly to form an apical interorbital septum in all the known omomyid skulls (Cartmill, 1975; Cartmill and Kay, 1978), suggest that its posterior, olfactory components were correspondingly reduced. This is characteristic of the microsmatic *Tarsius* and the anthropoids (see Haines, 1950), both of which have a reduced system of olfactory scrolls and also lack a moist rhinarium (Pocock, 1918; Cave, 1967). Perhaps correlated with this is the expansion of the occipital lobes, which come to moderately overlap the cerebellum in *Tetonius* and *Necrolemur,* and the relative reduction of the olfactory bulbs noted in *Necrolemur* and *Rooneyia* (Radinsky, 1970). The fossilized brain of *Rooneyia* also lacks a coronolateral sulcus, common to strepsirhines but replaced in anthropoids by the central sulcus, possibly signaling a close relationship of omomyids with anthropoids (Radinsky, 1970).

Generally this organization stands in sharp contrast with that observable or inferrable for living strepsirhines and for adapids and plesiadapiforms: a long, shallow face, widely spaced orbits, deep olfactory recess situated between the orbital fossae, well-developed nasal turbinates, naked rhinarium, large olfactory bulbs, and small occipital lobes. Cartmill and Kay (1978) and Kay and Cartmill (1977) propose that the relatively large size of the infraorbital foramen of *Tetonius* and *Rooneyia* suggests that omomyids "... retained well-developed vibrissae and perhaps even a naked rhinarium" (1978, p. 212). While mosaicism and primitive retentions would not be surprising, particularly since some callitrichines (Hershkovitz, 1977) present a fairly rich complement of vibrissal tactile hairs (but far fewer than visually oriented carnivorans), much of the evidence appears to reflect an advanced dominance of the visual system. Furthermore, because the foramen is or tends to be multiple in haplorhines (e.g., most anthropoids, *Tarsius, Tetonius, Washakius,* and *Rooneyia*) rather than single as in most of the species examined by Cartmill and Kay, we suspect that a variety of functions, biological roles, and conditioning factors are being sampled by their measure. The apparent reduction in *Tarsius,* for example, may reflect an encroachment of the hypertrophic eyeballs and orbits upon the posterior maxilla and the infraorbital canal. We therefore regard such implied similarities to strepsirhines to be of dubious sytematic value.

The morphology of the omomyid ear region is well preserved in *Rooneyia* and *Necrolemur,* yielding much information. But there remains vexatious questions concerning the homologies and morphocline polarity of the tubular external auditory meatus (see Szalay, 1972, 1976; Archibald, 1977), possessed by all known tarsiiforms, plesiadapiforms, some strepsirhines, and modern catarrhines. Also, the functional significance of this and other characteristics, and the several pattern variations, is unclear. Still, tarsiiforms present a number of features whose polarities can be inferred with reasonable assurance, and they also present a transformation series which illustrates the manner in which the anthropoid auditory region may have evolved. Detailed recent discussions of this region can be found in Cartmill (1975), Szalay (1976), Archibald (1977), and Szalay and Delson (1979).

Omomyids typically present an enlarged promontory artery, which enters the bulla posteromedially, and a relatively small stapedial artery branching from it. The location of the carotid foramen and the arterial proportions contrast with the presumed primitive eutherian and primate pattern (Gregory, 1920; Szalay, 1975*b*; Bugge, 1974; Archibald, 1977) and are thus derived conditions. Anthropoids present an essentially similar pattern, with the stapedial atrophying in adults. Although Cartmill and Kay (1978) caution that such features are liable to undergo convergence when lineages are independently evolving enlarged forebrains, it seems to us that the detailed similarities of haplorhine intrabullar circulation strongly increases the likelihood that these derived conditions are indeed homologous. This is especially so if the reduction of the stapedial correlates with the development of an anastomosis between the meningeal and the maxillary arteries to supply the dura in place of the stapedial, as is known in *Tarsius* and anthropoids (Hill, 1953; Bugge, 1974). If so, Cartmill's (1978) view that this meningeal provision is possibly a haplorhine synapomorphy cannot be easily reconciled with his notion that *excessively* reduced stapedial arteries occur convergently in *Rooneyia* and *Tarsius*.

The general pattern of bullar inflation may also differentiate tarsiiforms and anthropoids from adapids, most lemuriforms and plesiadapiforms. In the latter groups, the tympanic cavity and hypotympanic sinus combine to form an expansive diverticulum ventral to the promontorium and middle ear. The bullar capsule is a spherical, inflated pocket. The major volume of the hypotympanic sinus is situated medial to the promontorium and extends posterolaterally around it. In haplorhines, on the other hand, the inflated bulla is diagonalized and the hypotympanic sinus extends anteromedially before the promontorium. The posterior recess of the hypotympanic cavity is much narrowed (related to the migration of the carotid foramen, see above) but the anterior portion is much enlarged. Thus the petrosal bone and the hypotympanic sinus of anthropoids becomes pneumatized elliptically in the very area enlarged by tarsiiforms, and almost to its anterior apex, quite unlike other primates. This is undoubtedly a derived morphology. While *Rooneyia* is probably relatively primitive in retaining a capacious tympanic cavity and a laterally expansive, deep subtympanic recess (Cartmill, 1975; Szalay and Wilson, 1976), *Necrolemur* presents a narrowed tympanum (as does *Tarsius*) and a dorsoventrally reduced subtympanic space, much as in anthropoids.

The anterior dentitions of omomyids and adapids are still poorly known.

Nevertheless, they have figured prominently in recent debates on higher primate phylogeny. Gingerich (e.g., 1975*b,* 1978) and Schwartz (1978; see also Schwartz *et al.,* 1978; Krishtalka and Schwartz, 1979) have stressed the morphological similarity of tarsiiform and plesiadapiform lower incisors. They interpret this as evidence that tarsiiforms and plesiadapiforms comprise a monophyletic group that shared only a relatively remote common ancestry with adapiforms and anthropoids, which are presumed to represent another monophyletic group. Based upon the dental development and eruption patterns of a number of living primates and a small sample of nonadult fossils, Schwartz has explicitly argued that the enlarged anterior teeth (upper and lower) of plesiadapiforms and tarsiiforms are homologous *bona fide canine* teeth (see also Le Gros Clark, 1963), hence a synapomorphy linking these taxa phyletically. Cartmill and Kay (1978) strongly refuted his argument, exposing its shaky biological foundation and its failure to accord with the facts of *Tarsius*' dental development. We fully endorse their criticism.

Arguing in a more traditional vein, Gingerich (e.g., 1976, 1977) has presented a dichotomous picture of incisor evolution in the primates, exemplified by enlarged apically pointed lower incisors appearing in plesiadapiforms and tarsiiforms on the one hand, and vertical spatulate incisors appearing in adapiforms and anthropoids on the other. He strongly implied that each pattern is independently derived relative to some unspecified ancestral primate condition (e.g., 1976; Fig. 42), but elsewhere (1977) indicated that the similarity which he finds between the primitive omomyid *Teilhardina* and plesiadapids represent ancestral retentions.

The morphological evidence of the anterior dentition advanced by Gingerich at first seems compelling, but his analysis can be questioned at several levels. There is indeed a striking similarity between the lower incisors of certain plesiadapiforms (*Plesiadapis*) and some omomyids (*Trogolemur*). But as Szalay (1976) demonstrated, the morphology and proportions of omomyid antemolar teeth are variable within the group, and incisor hypertrophy, one of the most common independent trends among mammals, might well be expected to occur convergently (see also Szalay, 1977*b*; Cartmill and Kay, 1978). From another perspective, we consider the differences between plesiadapiform, omomyid, and anthropoid anterior teeth to have been generally overemphasized in the literature. All three groups share in common rather high-crowned, buccolingually thick and mesiodistally narrow upper and lower medial incisors (see Fig. 1a,b). This basic pattern (although not the exaggerated form of *Plesiadapis* or *Phenacolemur*) may well be primitive for the order, regardless of the number and conformation of peripheral crown conules seen in the uppers of derived plesiadapids (some of which however may also be ancestral for primates). This interpretation is supported by the distribution of large, procumbent, cylindriform lowers and correlatively robust, tall uppers in closely related archontans such as the tupaiids, mixodectids and microsyopsids (see Szalay, 1969, 1976; Gingerich, 1976; Szalay and Drawhorn, 1980). We envision various lineages modifying this primitive pat-

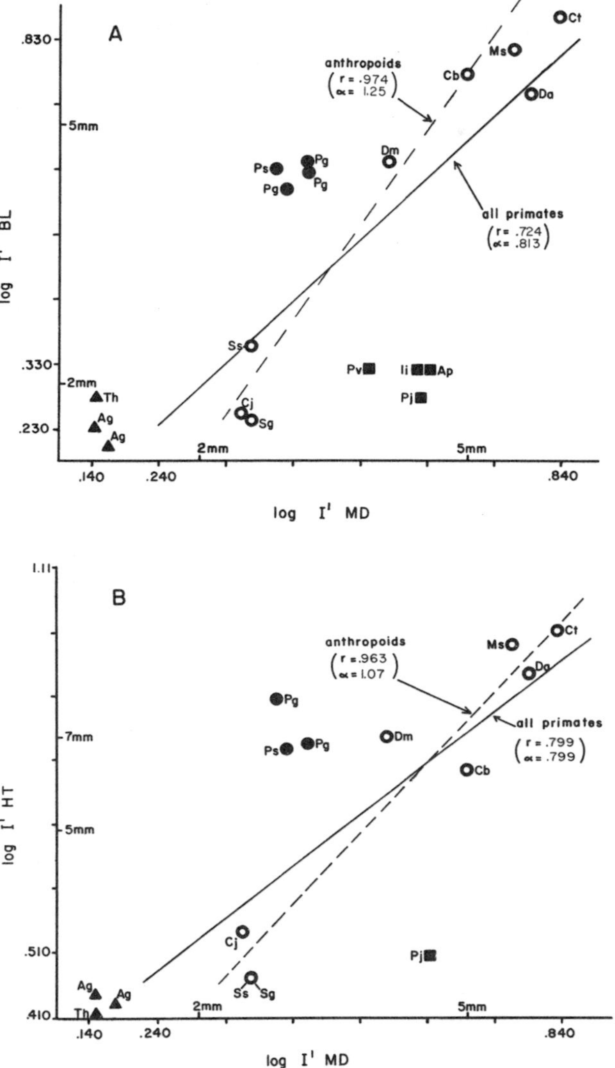

Fig. 1. Bivariate plots of (A) mediodistal length versus buccolingual breadth and (B) mediodistal length versus crown height (apical margin to cementoenamel junction) of the upper medial incisor. Regression based on sample means. Nonanthropoid fossils are represented by individual specimens. Conventions and sources: (triangles, omomyids) Ag, *Arapahovius gazini,* Th, *Tetonius homunculus* (Rosenberger, 1979); (rings, anthropoids) Cb, *Colobus badius,* Ct, *Cercocebus torquatus,* Ms, *Macaca sylvanus* (Delson, 1973); Da, *Dryopithecus africanus* (Andrews, 1978); Cj, *Callithrix jacchus,* Sg, *Saguinus geoffroyi,* Ss, *Saimiri sciureus* (Rosenberger, 1979); (squares, strepsirhines) Ap, *Adapis parisiensis* (Rosenberger, 1979); Ii, *Indri indri,* Pv, *Propithecus verreauxi* (Gingerich and Ryan, 1979); Pj, *Pelycodus jarrovii* (Gregory, 1920); (circles, plesiadapids); Pg, *Plesiadapis gidleyi,* Ps, *Plesiadapis sp.* (?*gidleyi*) (Rosenberger, 1979). Strepsirhines contrast haplorhines and plesiadapids in their gracile, broad, low-crowned I[1]s.

tern according to the selectional demands of the differing diets and varying harvesting roles encountered in an arboreal milieu. The secondary spatulateness of anthropoid I_1s (and I_2s) is correlated with a major reorganization of their antemolar dentition, no less dramatic than the development of a toothcomb in living strepsirhines. Thus, irrespective of the real size and proportions of the ancestral primates' incisor teeth, the genetic substrate determining at least $I\frac{1}{1}$ shape may have been inherited essentially unchanged by tarsiiforms and anthropoids from a euprimate common ancestor, which derived its morphology from some plesiadapiform.

A possible implication of this polarity hypothesis is that the much heralded similitude of adapid and anthropoid antemolar teeth, the spatulate incisors, is synapomorphic as Gingerich (e.g., 1976) indicated. We regard this as highly unlikely for there are important morphological and occlusal features distinguishing the two. For example, the upper incisors of the anthropoid morphotype, like all modern anthropoids, were robust and heteromorphic, I^1 being quite high-crowned and I^2 rather conical (Rosenberger, 1979). Adapids such as *Pelycodus, Notharctus, Europolemur* and *Adapis,* in contrast, have low-crowned, buccolingually thin and transversely broad $I^{1,2}$ (Figs. 1 and 2). The breadth of I^1 is acutely enhanced by a strong mesial process, while I^2 has a relatively simple, triangular shape. This gracile morphology is quite distinct from the postulated primate and anthropoid morphotypes and is clearly apomorphic. Furthermore, the pattern closely resembles all the essential details seen among the tooth-combed strepsirhines that have not drastically reduced the uppers (e.g., indriids). This very possibly represents a shared derived feature of known adapids and living strepsirhines.

The mandibular incisors of some adapids are also weakly developed, although their crowns are known from only a few specimens (see especially *Europolemur klatti*). Gregory (1920) described at least three kinds of lower

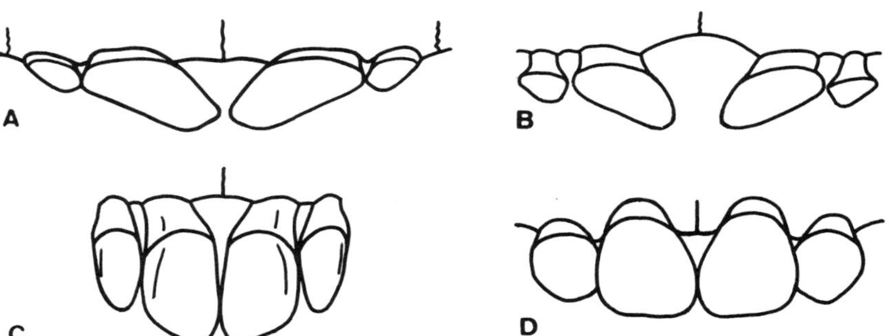

Fig. 2. Anterior views of the upper incisors of (A) *Adapis parisiensis,* (B) *Propithecus verreauxi,* (C) a compound of *Tetonius homunculus* ($I^{1,2}$) and *Arapahovius gazini* (I^1), and (D) *Saimiri sciureus.* Not to same scale. Note the comparatively high-crowned incisors of haplorhines and the low-crowned shape of the strepsirhines; the strong medial process or angulation of the I^1 crown in the strepsirhines, and their diminutive I^2s.

incisors: what he postulated to be the most primitive form found in *Pelycodus*, "... of small size, not chisel-shaped, not strongly procumbent ..." (p. 229) but "semi-erect"; the more spatulate variety seen in *Notharctus;* and the progressively derived pattern of *Adapis parisiensis.* He suggested that the apices of *Notharctus* $I_{1,2}$ would have presented "bluntly pointed" or "rounded truncate" tips rather than the chisellike edge which typifies those of anthropoids. In *Adapis* the apically broadened lower incisors are ranked closely with the canine to form a cropping mechanism. Gregory (1920), Gingerich (1975*a*), Szalay and Seligsohn (1977) and others have indicated that this is clearly a highly derived condition. It also involves a unique, derived occlusion of C_1 palatal to C^1. This articulation differs from *Notharctus* (Gregory, 1920) and virtually all anthropoids (Rosenberger, 1979), where C_1 occludes mesial to C^1 and into the precanine diastema. Furthermore, C_1 is exceptionally robust and low crowned in the known *Adapis,* quite unlike notharctine or protoadapin adapines such as *Mahgarita* and the presumed anthropoid ancestral condition. Thus, vertical spatulate incisors are not at all characteristic of adapids and bear only superficial resemblance to anthropoids. That genus which exhibits the most detailed similarities, *Adapis,* appears to also present the most derived incisor mechanism, evolving in a direction altogether different from that taken by anthropoids.

As a final point we wish to mention those omomyids which do conform with the projected ancestral anthropoid pattern. For example, the upper and lower median incisors of *Ourayia* are very similar to modern anthropoids. I^1 is robust, moderately high crowned and transversely broad at the apical margin (Fig. 3). I_1 is similarly shaped and vertical in orientation (Fig. 4). Upper central incisors assigned to *Tetonius* by Szalay (1976), and especially others allocated to the newly described *Arapahovius* (Savage and Waters, 1978), are also highly similar to anthropoid I^1s (Fig. 2). Simple metrical comparisons bear this out (Figs. 1a,b) and dramatically contrast a common omomyid-anthropoid pattern with that found in adapiforms and strepsirhines. These similarities themselves do not offer indications of descent but underscore the important point that the variability exhibited among omomyids includes a detailed transformation

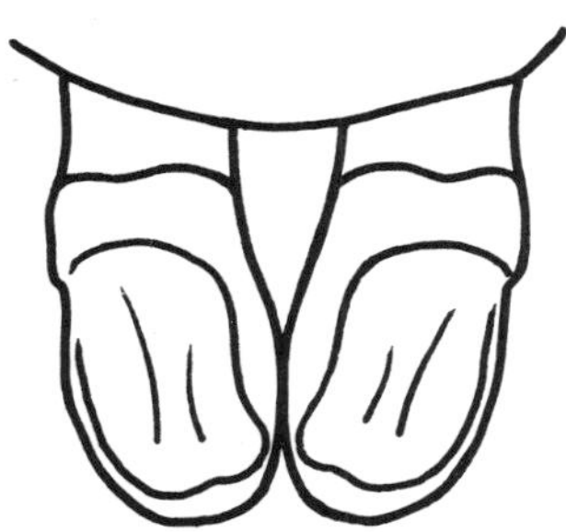

Fig. 3. Lingual view of upper medial incisors of *Ourayia uintensis,* showing their relatively broad apical margin and anthropoid- or haplorhine-like crown shape. Compare with Fig. 2.

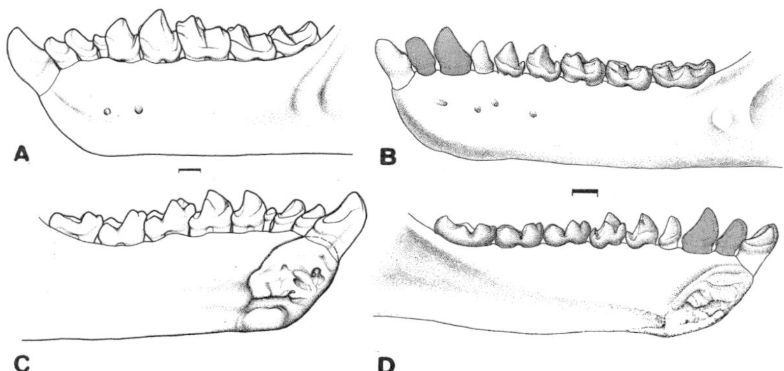

Fig. 4. Lateral (A,B) and medial (C,D) views of *Tetonius homunculus* (left) and *Ourayia uintensis* (right) lower jaw rami and toothrows. (From Szalay, 1976, courtesy of American Museum of Natural History). Scales represent 1 mm. Although the I_1 of *Tetonius* is robust it is not disproportionately enlarged in height or thickness relative to the posterior premolars and molars. The proportions of anterior and cheek teeth in *Ourayia* may be even more similar to that of ancestral anthropoids.

series ranging from what might be near the ancestral primate morphology (e.g., *Teilhardina*) to ones approaching the more derived condition found in anthropoids.

The cheek teeth of omomyids are exceptionally diverse when compared with the Adapidae (Szalay, 1976). Consequently, many genera present derived morphologies and tooth proportions that would exclude them from having direct bearing on anthropoid ancestry. In many features, however,

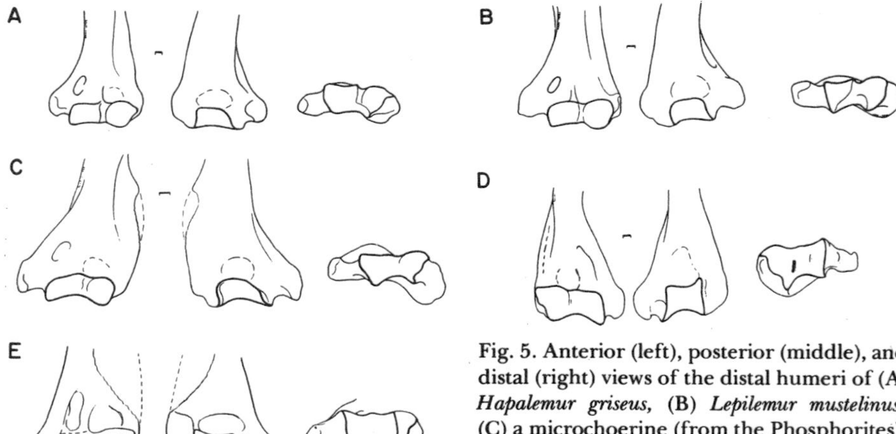

Fig. 5. Anterior (left), posterior (middle), and distal (right) views of the distal humeri of (A) *Hapalemur griseus*, (B) *Lepilemur mustelinus*, (C) a microchoerine (from the Phosphorites), (D) *Saimiri sciureus*, and (E) a Fayum anthropoid (YPM 24939). Scales represent 1 mm. All but D are from the left side.

including conule and cingulum development, overall tooth shape and occlusion, anaptomorphines such as *Teilhardina* and *Anemorhysis* are clearly primitive and are probably close to the ancestral euprimate molar morphology. More advanced forms such as *Omomys* and *Chumashius* may well represent conditions intermediate between an ancestral omomyid pattern and an anthropoid or platyrrhine condition (see Szalay, 1976). Kay (1977) has recently discussed how a *Hemiacodon*-like molar occlusion may be reasonably transformed into a primitive catarrhine pattern. Clearly, there was a sufficient pool of appropriate variation from which higher primate molar and premolar teeth may be derived (see Kay, this volume). But the great primitive as well as convergent similarity of many adapid and omomyid cheek teeth, reflecting their common euprimate ancestry, diminish the significance of a transformational approach in this case.

One area of morphological analysis which still is inadequately studied is the postcranium. Part of the reason is due to their poor representation in the fossil record (Szalay, 1976). However, a recent comparative study by Szalay and Dagosta (in press) on the distal humerus of Paleogene and living primates uncovered some interesting examples of special similarity between omomyids and primitive anthropoids (Fig. 5). The omomyid sample is relatively good: there are at least three different kinds of North American and at least two different European omomyids represented. It was found that the medial half of the trochlea curves distally in a highly diagnostic manner in omomyids, Fayum anthropoids, and a number of platyrrhine genera which are supposed to have retained this feature from a primitive anthropoid stage. The functional significance of this subtle but recognizable character is not clear. We suspect, however, that this similarity is a shared, homologously derived one, as it does not occur among known adapids and lemuriforms.

Tarsius and Anthropoids

In their recent, cladistically slanted contribution on the affinities of *Tarsius,* Cartmill and Kay (1978) denied Simons' (1972) hypothesis of a close relationship with microchoerine omomyids (as did Szalay, 1975*b*) and proposed that "Derived features shared by *Tarsius* and anthropoids, but not by any omomyids, are more numerous (p. 211)." Subsequent to the statement of their hypothesis of *Tarsius*-anthropoid monophyly, they characterized the morphology of omomyids in such ways as to cast doubt, but not dismiss, their "haplorhine" status. However, we find little disagreement in the character analyses presented above and that produced by Cartmill and Kay. Both point to the inescapable conclusion that omomyids are members of the monophyletic Haplorhini. Perhaps we are less struck by the potentially primitive tarsiiform characters exhibited by omomyids. The numerous cranial features shared between omomyids and anthropoids does not allow the genealogical association of omomyids with the Strepsirhini even if the latter are to be defined by shared patristic (and not primitive) features.

The hypothesis advocated by Cartmill and Kay (1978) implies that anthropoids shared an immediate common ancestor with tarsiids *sensu stricto* rather than with some more primitive omomyid or haplorhine. The features itemized in support of this hypothesis are: the presence of an intrabullar transverse septum partitioning the hypotympanic sinus from the tympanic cavity; loss of the subtympanic recess beneath the ectotympanic; verticalized, anterolateral placement of the internal carotid canal; and presence of an alisphenoid contribution to the postorbital septum. As discussed below, we consider these features to be either convergent similarities or indications of broader relationships within the haplorhines.

The restricted communication between hypotympanic and tympanic cavities in platyrrhines, such as *Cebus,* is due to the approximation of the promontorium and promontory canal with the lateral wall and ventral floor of the bulla rather than to the interposition of a transverse septum. This is at least partly conditioned by the lack of a subtympanic recess and a limited hypertrophy of the hypotympanic chamber, a function of the negative allometry between the bullar cavities and body size (see Cartmill, 1975). The carotid canal of platyrrhines (representing the ancestral anthropoid state) retains its primitive haplorhine position. It courses laterad across the promontorium from a medial entry into the bulla at the carotid foramen, which lies opposite the ectotympanic meatus. The primitive anthropoid condition, therefore, is distinctly less derived than the pattern shown by *Tarsius,* where the internal carotid ascends vertically into the bulla. In smaller platyrrhines, such as callitrichines, the hypotympanic sinus and the tympanum may be relatively larger in volume, but the partitioning of these cavities is still effected by a lamina of the *internally* pneumatized petrosal bone, connecting the bullar floor proper to the dorsal aspect of the carotid canal. This is perhaps the more important point in terms of potential homologies with *Tarsius,* for the trabeculate petrosal is a striking and fundamental synapomorphy of anthropoids (e.g., Gregory, 1920; Szalay, 1975*a*).

A combination of several factors distinguish *Tarsius* from the patterns just described and from other haplorhines. The internal carotid artery enters the bulla centrally and the canal is directed vertically through to the endocranium. The canal is situated anterolateral to the promontorium, not upon it, and anterior to the external auditory meatus. The hypotympanic sinus is markedly inflated, ventrad and anteriad. The enveloping petrosal is but a thin, single sheet of bone. This arrangement could produce a vast continuous chamber connecting tympanic and hypotympanic cavities. It does not because the vertical promontory canal is connected to the medial wall of the bulla by a transverse septum extending dorsoventrally between the promontorium and the ventrally displaced bullar floor. Recourse to the more primitive haplorhine patterns exhibited by omomyids and platyrrhines suggests that the development of the septum may be related to verticalization of the carotid canal. This is borne out by von Kampen's (1905) ontogenetic studies which show that the canal is initially more horizontal, with the carotid foramen opening medially into the bullar wall. It reorients as the vertical diameter of the bullar chamber increases, dragging behind it tissues that later ossify to become the transverse septum. Thus the pneumatic foramen between tympanic and hypotympanic cavities lies lateral to the canal and promontorium somewhat as in anthropoids. However, neither the common presence nor the similar position of this foramen, actually representing the persistently primitive communication between bullar chambers, is sufficient evidence of synapomorphy since different morphologies and conditioning factors appear to be involved.

Similarly, these differences between platyrrhine and *Tarsius* bullae indicate that the central location of the carotid foramen and vertical orientation of the promontory canal of *Tarsius* are autapomorphic. The similar position of this foramen in catarrhines and larger platyrrhines like *Alouatta, Lagothrix* and *Cacajao,* relates to their large body size and/or transversely expanded petrosals. Despite this, no platyrrhine shows a verticalized promontory canal or a transverse septum eminating from its medial (i.e., dorsal) aspect.

The absence of a subtympanic recess, which is associated with a narrowing of the tympanic cavity, also need not be a synapomorphy of anthropoids and *Tarsius.* While *Rooneyia* does present a considerable ventrolateral inflation of the tympanic cavity beneath the ectotympanic, which is probably primitive for haplorhines and primates (Cartmill, 1975), the recess is also reduced in *Necrolemur.* Convergence is therefore a possible explanation of this *Tarsius*-anthropoid similarity, as Cartmill (1975) had previously advocated. However, we cannot exclude the possibility that this derived feature represents a phylogenetic marker within haplorhines, perhaps linking anthropoids with a tarsiiform lineage that does not include *Rooneyia* or linking *Necrolemur* and *Tarsius* to one another (see Szalay, 1976). This issue can only be resolved with additional data on this and other characters.

We also regard it unlikely that *Tarsius*' advanced degree of postorbital closure is synapomorphous with the postorbital septum of anthropoids. The participation of an alisphenoid element in the formation of the *Tarsius* partition (Le Gros Clark, 1963; Hershkovitz, 1977) may well be correlated with hypertrophic eyeballs, as is the expansion of the frontal bone (Starck, 1975; Cartmill and Kay, 1978) to which the sphenoid is sutured. Similar postorbital flanges occur in large-eyed strepsirhines such as *Loris,* and an incipient postorbital septum derived from the frontal is probably primitive for haplorhines (see *Rooneyia, Necrolemur*). Furthermore, the sutural details of the ancestral anthropoid postorbital septum are still in doubt because platyrrhines and catarrhines typically exhibit alternative mosaics (Ashley-Montagu, 1933).

As discussed above, we regard the proportions of the stapedial and promontory arteries of haplorhines, particularly the enlargement of the promontory artery in the protohaplorhine, as homologous. Because *Necrolemur* has a relatively larger stapedial than *Tarsius,* Cartmill and Kay (1978) regarded its morphology to represent the primitive haplorhine condition, implying that the *extreme* reduction in *Tarsius* and anthropoids is synapomorphic. We fail to see how this is a more parsimonious interpretation of this taxonomic distribution, especially when the great similarity of other aspects of the *Necrolemur* and *Tarsius* ear regions are taken into account. We prefer to regard the *extreme* reduction of the stapedial of *Tarsius* and of anthropoids to be convergent.

Finally, we advance one other criticism of the tarsier–anthropoid hypothesis. *Tarsius* is the only living representative of its immediate lineage. Other known tarsiiforms cannot be allied with *Tarsius* with much certainty [see Szalay (1975*b*) for a critique of the microchoerine-*Tarsius* link; see also Cartmill and Kay (1978)]. Therefore, the characters of *Tarsius* are the diagnostic features of the tarsiid morphotype. Nearly every important aspect of its craniodental and postcranial anatomy [see also Luckett (1975) on placentation] are thought to be derived relative to the conditions found elsewhere among tarsiiforms or haplorhines. To list but a few (see Szalay and Delson, 1979): hypertrophic eyeballs and orbits; markedly flexed basicranial axis; verticalized carotid canal; hypertrophied bulla; markedly reduced petromastoid and loss or absence of a pneumatized condition of the mastoid; loss of I_2; upper and lower incisor proportions and occlusion; low intermembral index; accessory toilet-claw. The extinct tarsiiforms known by cranial material are far less derived, as are many taxa based on dentitions. All these are therefore better suited as *structural ancestors of anthropoids.* It follows that the direct genealogical ties of such an "omomyid," cladistic semantics aside, is more relevant to anthropoid ancestry, whereas the *Tarsius* lineage has little direct bearing on the question of anthropoid origins.

The imposition of a rigorous cladistic philosophy, a search for the sister-group, is often less important than a search for a structural morphotype and a suite of character transformations within the context of an acceptable phylogeny. The exclusive use of the former approach, we think, has led to somewhat irrelevant, if not erroneous, results in this case. On neontological grounds, *Tarsius* has long been regarded by many as the immediate collateral relative of anthropoids, but this has limited bearing on *descent.* Even were our character analysis proven faulty, and that of Cartmill and Kay (1978) upheld, we would still refrain from recognizing *Tarsius* as the nearest actual relative of anthropoids for it would severely limit our phylogenetic and anagenetic hypothesis.

Conclusion

In light of the phyletic analyses presented above, we think it likely that the "protoanthropoid" species was derived from a haplorhine primate that might well be regarded as an omomyid. A more precise inference is possible only in terms of dismissing several genera and lineages of omomyids from possible ancestral status. This in turn increases the likelihood that forms such as the omomyids represent the nearest morphological approximation to such an ancestral lineage. But, thus far, they are known only by fragmentary dentitions. Crania of *Necrolemur* and *Tarsius* (to a much lesser degree) suggest possible, though highly tenuous, indicators of anthropoid affinities, but when other aspects of their anatomy are taken into account (Szalay, 1976) their lineages are also barred from direct ancestral status. Given our limited sample of the early tarsiiform radiation and our inadequate knowledge of the morphology of omomyids, postulating any known taxon as an anthropoid ancestor would be premature.

Even with a reasonably confident assessment of the monophyletic status of the anthropoids (Szalay, 1975*a*) and their position within the Haplorhini, a zoogeographic explanation of their current distribution, or for that matter that of the hystricognathous rodents, does not become much simpler. The tarsiiforms were apparently widely distributed across Laurasia during the Paleogene, and some survivors certainly had at least a relic distribution within Eurasia during the Neogene. Since the Bolivian *Branisella* does appear to have direct platyrrhine affinities (Hoffstetter, 1969; Rosenberger, 1977, 1979), contrary to Hershkovitz's (1977) remarks, the differentiation of anthropoids must have occurred prior to the earliest Oligocene. The moderately diverse Fayum faunas of Africa's late Oliogcene attest to this (Simons, 1972; Szalay and Delson, 1979). Polyphyly aside, the major points arguing for an independent derivation of catarrhine and platyrrhine stocks from the north are: (1) the essential absence of recognizable platyrrhines and catarrhines in Laurasian Paleogene faunas, implying an *in situ* Gondwanan evolution of each group; (2) the presence of an Atlantic barrier making distant overwater dispersal unlikely (see especially Simons, 1976) and overland dispersal impossible; (3) the lack of suitable, primitive forms on either continent that might be ancestral to collateral relatives on the other. At the moment, all of these reasons are essentially negative evidence, having neither falsifying nor corroborative powers.

Advocates of a vicariance explanatory model of anthropoid zoogeography (e.g., Brundin, 1966; Hershkovitz, 1977), attributing their disjunction to sea-floor spreading between Africa and South America directly or indirectly, have correctly recognized the implications of anthropoid monophyly but incorrectly assumed that platyrrhines and catarrhines were strictly endemic to South America and Africa, respectively. While the evidence for Cenozoic faunal endemism is good for South America, (e.g., Patterson and Pascual, 1972), the constitution of Paleogene African faunas indicates otherwise (Cooke, 1972). Whether or not the Burmese *Pondaungia* proves to be an Eocene anthropoid (Ba Maw *et al.*, 1979), there is no reason to expect that catarrhines will not be found outside Africa in Laurasia during the Eocene and early Oligocene. For similar reasons, one cannot assuredly argue that archaic catarrhines gave rise to platyrrhines after island-hopping across a then narrower South Atlantic ocean. Certainly, the biological evidence counters Hoffstetter's (1977) proposal that parapithecids were directly ancestral to platyrrhines. They are far too derived in their morphology (see Szalay and Delson, 1979; Kay, this volume) as are all *bona fide* catarrhines (Rosenberger, 1980). This leads us to the conclusion that it was an unknown early anthropoid which was ancestral to both infraorders, and that that species need not have inhabited either of these southern continents. This hypothesis does not, of course, escape difficulties. It too relies on negative evidence inasmuch as it needs a "home-land" and actual organisms which would serve as a putative ancestral phenon.

In sum, the evidence is still too incomplete to seriously contemplate the probabilities of a trans-South Atlantic dispersal by anthropoid primates during the early Cenozoic, even via archipelagos. The probability that this in fact occurred is a function of the availability of a suitable route, which can only be determined geophysically (see Tarling, this volume). Without relevant fossils, our estimation of the likelihood that primates followed this or a pair of parallel north–south dispersal routes across the Tethys, one into Africa and one into South America, should be based on the degree to which other Cenozoic animals of similar geographical situation are known to be in sister-group or ancestral–descendant relationship with one another. Either of the alternatives outlined for anthropoids is compatible with the evidence, and neither can be falsified by it. Our hesitancy to favor one or the other explanation reflects our own division on this issue. In either case, neither of us would think that the weight of the evidence would tilt the balance much toward either side.

References

Andrews, P. J., 1978, A revision of the Miocene Hominoidea of East Africa, *Bull. Br. Mus. (Nat. Hist.), Geol.* **30**:85–224.

Archibald, J. D., 1977, Ectotympanic bone and internal carotid circulation of eutherians in reference to anthropoid origins, *J. Hum. Evol.* **6**:609–622.

Ashley-Montagu, M. F. A., 1933, The anthropological significance of the pterion in the primates, *Am. J. Phys. Anthropol.* **18**:160–336.

Ba Maw, Ciochon, R. L., and Savage, D. E., 1979, Late Eocene of Burma yields earliest anthropoid primate, *Pondaungia cotteri, Nature (London)* **282**:65–67.

Bock, W. J., 1977*a,* Adaptation and the comparative method, in: *Major Patterns in Vertebrate Evolution* (M. K. Hecht, P. G. Goody, and B. M. Hecht, eds.), pp. 57–82, Plenum Press, New York.

Bock, W. J., 1977*b,* Foundations and Methods of Evolutionary Classification, in: *Major Patterns in Vertebrate Evolution* (M. K. Hecht, P. C. Goody, and B. M. Hecht, eds.), pp. 851–896, Plenum Press, New York.

Brundin, L., 1966, Transantarctic relationships and their significance as evidenced by chironomid midges, with a monograph of the subfamilies Podonominae and Aphroteniinae and the austral Heptagyiae, *Kl. Sven. Vetenskaps akad. Handl.*, Sér 4 **11**:1–472.

Bugge, J., 1974, The cephalic arterial system in insectivores, primates, rodents and lagomorphs, with special reference to the systematic classification, *Acta Ana.* **82**:1–160.

Cartmill, M., 1975, Strepsirhine basicranial structures and the affinities of the Cheirogaleidae, in: *Phylogeny of the Primates* (W. P. Luckett and F. S. Szalay, eds.), pp. 313–353, Plenum Press, New York.

Cartmill, M., 1978, The orbital mosaic in prosimians and the use of variable traits in systematics, *Folia Primatol* **30**:81–114.

Cartmill, M., and Kay, R. F., 1978, Craniodental morphology, tarsier affinities, and primate suborders, in: *Recent Advances in Primatology*, Vol. 3, *Evolution* (D. J. Chivers and K. A. Joysey, eds.), pp. 205–214, Academic Press, London.

Cave, A. J. E., 1967, Observations on the platyrrhine nasal fossa, *Am. J. Phys. Anthropol.* **26**:277–288.

Cooke, H. B. S., 1972, The fossil mammal fauna of Africa, in: *Evolution, Mammals and Southern Continents* (A. Keast, F. C. Erk, and B. Glass, eds.), pp. 89–139, S.U.N.Y. Press, Albany.

Cracraft, J., 1974, Phylogenetic models and classifications, *Syst. Zool.* **23**:71–90.

Cracraft, J., 1978, Science, philosophy and systematics, *Syst. Zool.* **27**:213–216.

Delson, E., 1973, Fossil colobine monkeys of the circum-Mediteranean region and the evolutionary history of the Cercopithecidae (Primates, Mammalia), University Microfilms, Ann Arbor.

Englemann, G. F., and Wiley, E. O., 1977, The place of ancestor-descendant relationships in phylogeny reconstruction, *Syst. Zool.* **26**:1–11.

Gingerich, P. D., 1975*a*, Dentition of *Adapis parisiensis* and the evolution of lemuriform primates, in: *Lemur Biology* (I. Tattersall and R. W. Sussman, eds.), pp. 65–80, Plenum Press, New York.

Gingerich, P. D., 1975*b*, Systematic position of *Plesiadapis, Nature* **253**:111–113.

Gingerich, P. D., 1976, Cranial anatomy and evolution of early Tertiary Plesiadapidae (Mammalia, Primates), *Mus. Paleontol. Univ. Mich. Papers Paleontol.* **15**:1–141.

Gingerich, P. D., 1977, Dental variation in early Eocene *Teilhardina belgica,* with notes on the anterior dentition of some early tarsiiformes, *Folia Primatol.* **28**:144–153.

Gingerich, P. D., 1978, Phylogeny reconstruction and the phylogenetic position of *Tarsius,* in: *Recent Advances in Primatology,* Vol. 3, *Evolution* (D. J. Chivers and K. A. Joysey, eds.), pp. 249–255, Academic Press, London.

Gingerich, P. D., and Ryan, A. S., 1979, Dental and cranial variation in living Indriidae, *Primates* **20**:141–159.

Gregory, W. K., 1920, On the structure and relations of *Notharctus,* an American Eocene primate, *Mem. Am. Mus. Nat. Hist.,* **3**:49–243.

Haines, R. W., 1950, The interorbital septum in mammals, *J. Linn. Soc. London* **41**:585–607.

Hershkovitz, P., 1977, *Living New World Monkeys (Platyrrhini) with an Introduction to Primates,* Vol. 1, University of Chicago Press, Chicago.

Hill, W. C. O., 1953, The blood-vascular system of *Tarsius, Proc. Zool. Soc. London* **123**:655–694.

Hoffstetter, R., 1969, Un primate de l'Oligocène inférieur sud-américain: *Branisella boliviana* gen. et sp. nov, *C. R. Acad. Sci. Paris, Sér. D* **269**:434–437.

Hoffstetter, R., 1977, Origine et principales dichotomies des Primates Simiiformes (= Anthropoidea), *C. R. Acad. Sci. Paris, Sér. D* **284**:2095–2098.

Kay, R. F., 1977, The evolution of molar occlusion in the Cercopithecidae and early catarrhines, *Am. J. Phys. Anthropol.* **46**:327–352.

Kay, R. F., and Cartmill, M., 1977, Cranial anatomy and adaptations of *Palaechthon nacimienti* and other Paromomyidae (Plesiadapoidea, ?Primates), with a description of a new genus and species, *J. Hum. Evol.* **6**:19–53.

Krishtalka, L., and Schwartz, J. H., 1979, Phylogenetic relationships of plesiadapiform-tarsiiform primates, *Ann. Carnegie Mus. Nat. Hist.* **47**:515–540.

Le Gros Clark, W. E., 1963, *The Antecedents of Man,* Harper and Row, New York.

Luckett, W. P., 1975, Ontogeny of the fetal membranes and placenta: Their bearing on primate phylogeny, in: *Phylogeny of the Primates: A Multidisciplinary Approach* (W. P. Luckett and F. S. Szalay, eds.), pp. 157–182, Plenum Press, New York.

Luckett, W. P., and Szalay, F. S., 1975, *Phylogeny of the Primates: A Multidisciplinary Approach,* Plenum Press, New York.

Luckett, W. P., and Szalay, F. S., 1978, Clades versus grades in primate phylogeny, in: *Recent*

Advances in Primatology, Vol. 3, *Evolution* (D. J. Chivers and K. A. Joysey, eds.), pp. 227–237, Academic Press, London.

Naylor, B. G., manuscript, A paleontological view of cladistic analysis, ancestors and descendants.

Nelson, G. J., 1973, Classification as an expression of phylogenetic relationships, *Syst. Zool.* **22**:344–359.

Patterson, B., and Pascual, R., 1972, The fossil mammal fauna of South America, in: *Evolution, Mammals and Southern Continents* (A. Keast, F. C. Erk, and B. Glass, eds.), pp. 227–309, S.U.N.Y. Press, Albany.

Pocock, R. I., 1918, On the external characters of the lemurs and of *Tarsius, Proc. Zool. Soc. London* **1918**:19–53.

Radinsky, L. B., 1970, The fossil evidence of prosimian brain evolution, in: *The Primate Brain* (C. R. Noback and W. Montagna, eds.), pp. 209–224, Appleton, New York.

Rosenberger, A. L., 1977, *Xenothrix* and ceboid phylogeny, *J. Hum. Evol.* **6**:461–481.

Rosenberger, A. L., 1979, Phylogeny, evolution and classification of New World monkeys (Platyrrhini, Primates), Thesis, C.U.N.Y., New York.

Rosenberger, A. L., 1980, Gradistic views and adaptive radiation of the platyrrhine primates, *Z. Morphol. Anthropol.* **71**:157–163.

Savage, D. E., and Waters, B., 1978, A new omomyid primate from the Wasatch formation of southern Wyoming, *Folia Primatol.* **30**:1–29.

Schwartz, J. H., 1978, If *Tarsius* is not a prosimian, is it a haplorhine? in: *Recent Advances in Primatology,* Vol. 3, *Evolution* (D. J. Chivers and K. A. Joysey, eds.), pp. 195–202, Academic Press, London.

Schwartz, J. H., Tattersall, I., and Eldredge, N., 1978, Phylogeny and classification of the primates revisited, *Yearb. Phys. Anthropol.* **21**:95–133.

Simons, E. L., 1972, *Primate Evolution: An Introduction to Man's Place in Nature,* Macmillan, New York.

Simons, E. L., 1976, The fossil record of primate phylogeny, in: *Molecular Anthropology* (M. Goodman, R. E. Tashian, and J. H. Tashian, eds.), pp. 35–62, Plenum Press, New York.

Simpson, G. G., 1963, The meaning of taxonomic statements, in: *Classification and Human Evolution* (S. L. Washburn, ed.), pp. 1–31, Aldine, Chicago.

Starck, D., 1975, The development of the chondocranium in primates, in: *Phylogeny of the Primates: A Multidisciplinary Approach* (W. P. Luckett and F. S. Szalay, eds.), pp. 127–155, Plenum Press, New York.

Szalay, F. S., 1969, Mixodectidae, Microsyopidae, and the insectivore-primate transition, *Bull. Am. Mus. Nat. Hist.* **140**:195–330.

Szalay, F. S., 1972, Cranial morphology of the early Tertiary *Phenacolemur* and its bearing on primate phylogeny, *Am. J. Phys. Anthropol.* **36**:59–76.

Szalay, F. S., 1975*a,* Haplorhine phylogeny and the status of the Anthropoidea, in: *Primate Functional Morphology and Evolution* (R. Tuttle, ed.), pp. 3–22, Mouton, The Hague.

Szalay, F. S., 1975*b,* Phylogeny of primate higher taxa, in: *Phylogeny of the Primates: A Multidisciplinary Approach* (W. P. Luckett and F. S. Szalay, eds.), pp. 91–125, Plenum Press, New York.

Szalay, F. S., 1976, Systematics of the Omomyidae (Tarsiiformes, Primates): Taxonomy, phylogeny and adaptations, *Bull. Am. Mus. Nat. Hist.* **156**:157–450.

Szalay, F. S., 1977*a,* Ancestors, descendants, sister groups and testing of phylogenetic hypotheses, *Syst. Zool.* **26**:12–18.

Szalay, F. S., 1977*b,* Constructing primate phylogenies: A search for testable hypotheses with maximum empirical content, *J. Hum. Evol.* **6**:3–18.

Szalay, F. S., and Dagosto, M., Locomotor adaptations as reflected on the humerus of Paleogene primates, *Folia Primatol.* (in press).

Szalay, F. S., and Delson, E., 1979, *Evolutionary History of Primates,* Academic Press, New York.

Szalay, F. S., and Drawhorn, G., 1980, Evolution and diversification of the Archonta in an arboreal milieu, in: *Comparative Biology and Evolutionary Relationships of Tree Shrews* (W. P. Luckett, ed.), p. 133–170, Plenum Press, New York.

Szalay, F. S., and Seligsohn, D., 1977, Why did the strepsirhine tooth comb evolve?, *Folia Primatol.* **27**:75–82.

Szalay, F. S., and Wilson, J. A., 1976, Basicranial morphology of the early Tertiary tarsiiform *Rooneyia* from Texas, *Folia Primatol.* **25**:288–293.

Tattersall, I., and Eldredge, N., 1977, Fact, theory and fantasy in human paleontology, *Am. Sci.* **65**:204–211.

Von Kampen, 1905, Die tympanalgegend des Säugetierschadde, *Gegenbaurs Morphol. Jahrb.* **34**:321–722.

Evidence from Dental Anatomy Studies

IV

Platyrrhine Origins

A Reappraisal of the Dental Evidence

8

R. F. KAY

Introduction

Primatologists are generally agreed that living Neotropical primates are monophyletic. However, opinion is deeply divided when it comes to the question of their origins. Currently, three families are considered by different authors to be ancestral to the group—Adapidae, Omomyidae, or Old World Anthropoidea (*via* either Omomyidae or Adapidae). Advocates of any of these groups for platyrrhine ancestry find it difficult to account, at the same time, for the adaptive similarity of Old and New World anthropoids and their present mutual geographic isolation. This had led to two very different opinions about platyrrhine origins (Fig. 1). One view is that separate stocks of omomyids or adapids gave rise to Old and New World anthropoids. Although this requires the assumption that the numerous adaptive and morphological similarities between Old and New World anthropoids were attained convergently it has considerable paleogeographic advantages (e.g., demonstrated land bridges between the northern continents, and plausible island-hopping routes between northern and southern continents at the appropriate times in the past [see Simons (1976) for a review]). The alternative is that Neotropical primates take their origin from an African anthropoid which found its way

R. F. KAY • Department of Anatomy, Duke University Medical Center, Durham, North Carolina 27710.

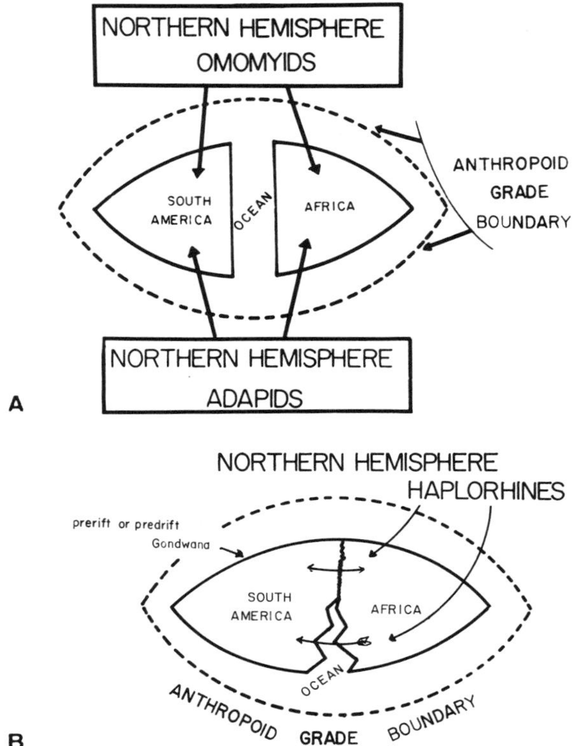

Fig. 1. Alternative views of platyrrhine origins. (A) The northern continent theories whereby either omomyids or adapids separately invaded the southern continents and there differentiated into the anthropoid grade of organization. (B) The southern continent theories whereby northern hemisphere haplorhines differentiated into anthropoids; one group entered Africa, and then found its way to South America.

across the South Atlantic ocean at a time when the intercontinental distance was a minimum of 700 miles. In spite of this obvious difficulty, the latter scheme has the advantage that the apparent derived features of Old and New World anthropoids need only have evolved once in a common anthropoid ancestor.

The purpose of this paper is to assess the dental evidence for the affinities of New World monkeys with adapids, omomyids, or Old World anthropoids. We will start with an analysis of the probable dental characteristics, or "morphotype," of the ancestral platyrrhine species. Once this is done, the morphology of adapids, omomyids, and African anthropoids will be reviewed to determine: (1) which dental features of platyrrhines are primitive retentions from the last common ancestor of all primates of modern aspect; (2) which features are convergent between the various groups; and (3) which features are shared by New World monkeys and any of the other taxa because they appeared in the last common ancestor of the two. It is assumed that only the shared derived features are useful indicators of phylogenetic relationship.

Mandibular and Dental Morphotype of the New World Monkey Ancestor

The living genus *Aotus* provides a useful model of the morphotype of the ancestral platyrrhine (Fig. 2). The mandible of the last common ancestor of all New World monkeys probably had a fused symphysis. The permanent dental formula was 2.1.3.3, above and below. Nearly equal-sized spatulate lower incisors had vertically implanted roots. The first upper incisor was larger than the second and both teeth were relatively procumbent. The upper and lower canines were not extremely large but would have projected slightly above the occlusal plane. The paracristid of P_2 formed a short shearing edge with the upper canine postparacrista. The P_2 metaconid was very small, as was the P^2 protocone. Both P2's were single-rooted. Upper and lower third and fourth premolars were two-cusped and very similar to one another (e.g., "homomorphic"). The lower third molars were smaller than the second molars.

The lower molars had a paraconid, at least on M_1; a hypoconulid was lacking on M_3. The lateral protocristid, supporting shearing facet 1 (Kay and Hiiemae, 1974) was oriented mediolaterally and contributed to the formation of the distal wall of the trigonid basin. The premetacristid formed a steep wall on the medial side of the trigonid.

The first and second upper molars had small, but discrete, paraconules and metaconules. Preparaconule and postparaconule cristae were present; the metaconule crests were greatly reduced. A "*Nannopithex*-fold" was absent. A small hypocone was situated on a well-developed posteromedial cingulum at least on M^1, and probably on M^2. A rudimentary posthypocrista was present but the prehypocrista and its shearing functions were probably limited or absent.

In some instances, recognition of the ancestral platyrrhine morphotype in terms of these dental characteristics presents no difficulty since all species of New World monkeys have a similar morphology. However, in most cases, some variation exists in the presence or degree of development of each character, so that further justification is needed to establish which of the various possible morphologies is the most primitive (refer to Figs. 2–4).

Symphyseal Fusion

The mandibular symphysis is fused among all known living and fossil Neotropical primate species and must have characterized their last common ancestor (Fig. 2A).

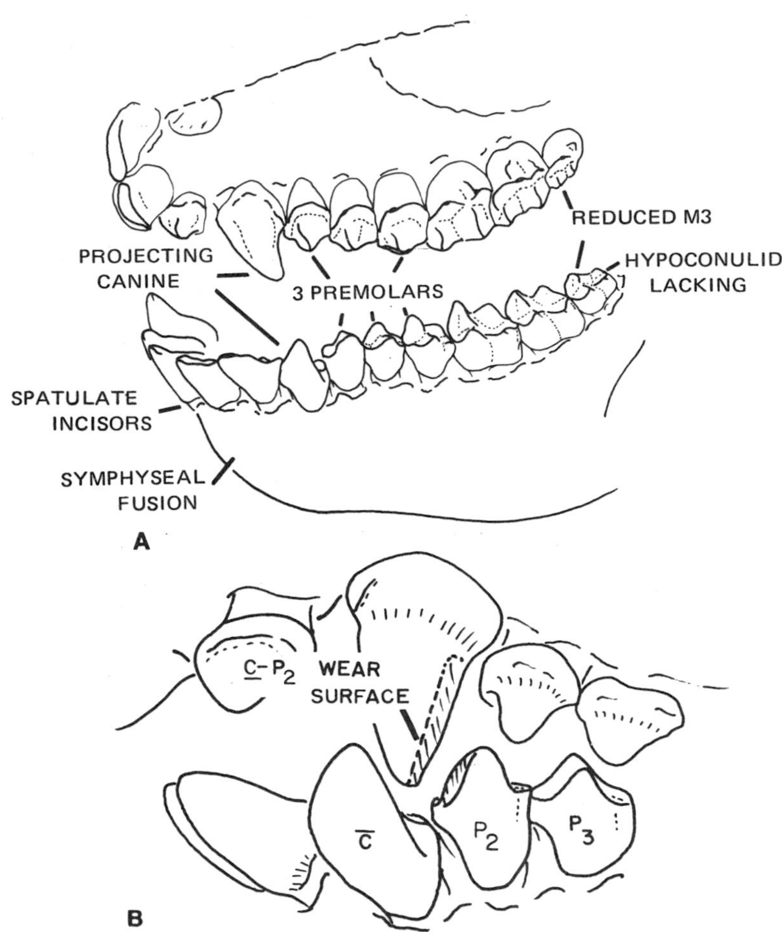

Fig. 2. Aspects of the morphology of the ancestral platyrrhine as exemplified by various extant species. (A) *Aotus trivirgatus:* Lower left mandibular corpus and upper right palate. Features of note include: symphyseal fusion; spatulate, vertically implanted incisors; projecting canines; 3 premolars; a single rooted, single cusped P^2; and a reduced M_3 lacking a hypoconulid. (B) *Aotus trivirgatus:* Lateral view of canines and premolars illustrating a wear surface along the leading edge of P_2 and the position of the matching canine wear surface along its lingual aspect. (The canine wear surface, outlined by dashed line, is not visible from the lateral view.)

Dental Formula

An upper and lower dental formula 2.1.3.3 is found universally among cebids and *Callimico;* other species, usually unified within the Callitrichidae, have two molars. Inasmuch as all other nonhuman primate groups commonly have three upper and lower molars, it is assumed that the platyrrhine ancestor had three molars and that the third molars have been lost in the callitrichids (Fig. 2A).

Incisor Size and Morphology

The lower incisors of most New World monkeys have spatulate crowns; incisor roots are oriented nearly normal to the occlusal plane and arranged transversely in an arc (Fig. 2A). In contrast, *Cebuella* and *Callithrix* have their pointed medial lower incisors set anteriorly to the lateral incisors; the lateral incisors are directly anterior to the canines. This, and related peculiarities of the symphyseal region, are probably related to a specialization for feeding on tree sap or gum and are unlikely to have characterized primitive New World monkeys (Rosenberger, 1977).

Canine Size

The canines of most New World monkey species are moderately large and project well beyond the occlusal plane. The lower canines of *Callicebus* are somewhat smaller (Kinzey, 1972). The upper canines were also small in the Oligocene genus *Tremacebus,* as inferred from the roots (Hershkovitz, 1974). The canines of the ancestral platyrrhine must have projected sufficiently for C to shear against the paracristid of the lower second premolar (Fig. 2A,B).

Premolar Morphology

All three lower premolars of New World monkeys have a protoconid and a metaconid; on P_3 and P_4 the metaconid is somewhat smaller than the protoconid, whereas the metaconid of P_2 is very small. Similarly, the upper premolars of platyrrhines are characterized by having small protocones and much larger paracones (some callitrichids are exceptions). The condition seen in most species is probably the primitive morphology. Similarly, single-rooted upper and lower P2's probably were found in the ancestral platyrrhine, inasmuch as it is seen almost universally in living and fossil species.

Third Molar Size and Structure

The ancestral platyrrhine went through a stage of third-molar reduction with the loss of the M_3 hypoconulid. Subsequently, in some lineages M_3 was lost altogether (e.g., callitrichids), whereas in other lineages it was reenlarged (e.g., *Alouatta*). This sequence of events explains the peculiar derived condition of the lower third molars of platyrrhines with large M_3's, in which a large "heel" on the back of the M_3 is produced by a "bowing-out" of the postcristid (posthypocristid and medial postentocristid). (M_3 enlargement in nonplatyrrhine species is achieved by the expansion of the hypoconulid, not the postcristid.)

Molar Structure

A major controversy with important implications for understanding platyrrhine phylogeny is over whether callitrichid molar morphology represents a retained primitive primate pattern or a "retrogressive" simplification. Wortman (1904) took the position that marmosets retain the primitive organization:

> Tritubercular upper molars furnish another character of considerable importance in determining the relationship of the marmosets to other groups. No primate of the Eocene is known to possess fully quadritubercular molars. . . . By far the greater number of species have simple tritubercular upper molars, and, with the exception of the marmosets and *Tarsius,* all the modern representatives of the Anthropoidea have four, fully developed, cusps. It follows, therefore, that these two groups are survivals from this early condition of the tritubercular stage of development of the molars, and that their detachment from the main axis could not have taken place later than the Eocene. The loss of the last molar of the marmosets, while unusual for a Primate, has clearly taken place since that time, as in the Eocene, all the known species have three fully developed molars [p. 220].

To this, Gregory (1916) responded:

> . . . as to the Hapalidae, I can only state here that nearly all the characters which, according to Dr. Wortman (1904, page 222), indicate derivation from some early "paleopithecine apes," are, in my judgement, either retrogressive or specialized characters which do not exclude Hapalidae from close alliance with the Cebidae [p. 262].

Hershkovitz (1977, p. 306), follows Wortman, and argues that the process of molar loss has no bearing on callitrichid molar morphology, whereas Gregory (1916), Remane (1960), and Rosenberger (1977), to name a few, take the view that molar simplification and molar loss form part of a common adaptive package, perhaps related to a secondary body-size reduction associated with adaptations for feeding on soft plant foods.

The structure of the molars of marmosets is not at all "primitive" in the sense intended by Wortman: there are major functional differences between the molars of callitrichids and those of Eocene primates with a "tribosphenic" molar pattern. These functional differences strongly imply that the callitrichid pattern represents a secondary approach to the "primtive," or ancestral primate condition and certainly is not a persistent relict of the likely ancestral morphological pattern.

Figure 3 illustrates the anatomy of the molars of *Tetonoides* (? = *Anemorhysis*) *tenuiculus,* an Eocene omomyid primate with the primitive "tribosphenic" molar pattern, *Callithrix argentatus,* a marmoset with the supposed "primitive" pattern, and *Callimico goeldii,* which much more closely approaches the probable ancestral condition for New World monkeys (e.g., Rosenberger, 1977).

On the M^1 of *Callimico,* as on the M^{1-2} of *Tetonoides* (and early Adapidae, as well; cf. Fig. 8), there are discrete metaconules and paraconules. Both species have preparaconule and postparaconule cristae supporting small Phase I shearing facets 1b and 3b [see Kay (1977) for a detailed review of this terminology]. These primitive features of the upper molars, which are almost certainly present on the M^1's of the last common ancestor of New World monkeys, are extremely reduced or lacking on the upper molars of *Callithrix.*

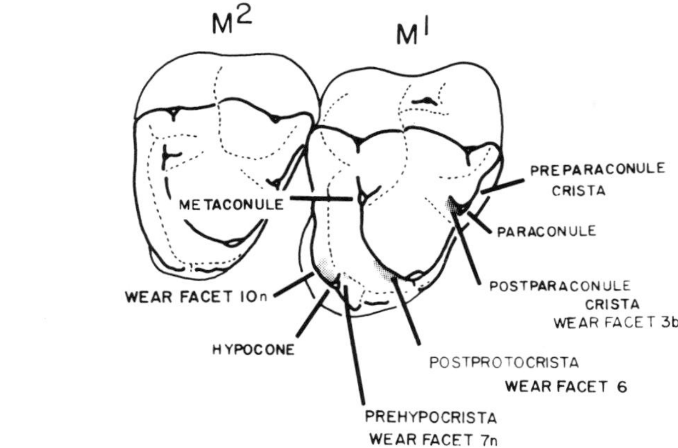

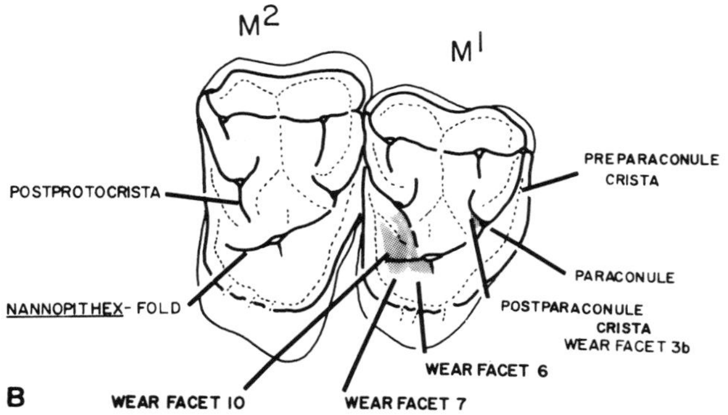

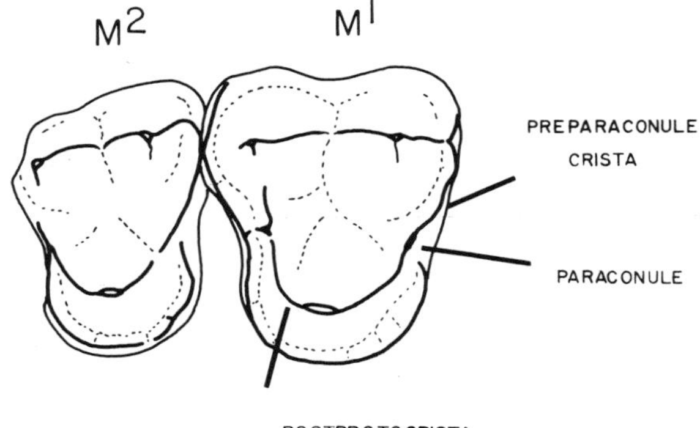

Fig. 3. The right upper first and second molars of *Callimico goeldii* (A), *Tetonoides (anemorphysis) tenuiculus* (B), and *Callithrix argentata. Callimico* and *Callithrix* are recent platyrrhines; *Tetonoides* is an early Eocene omomyid.

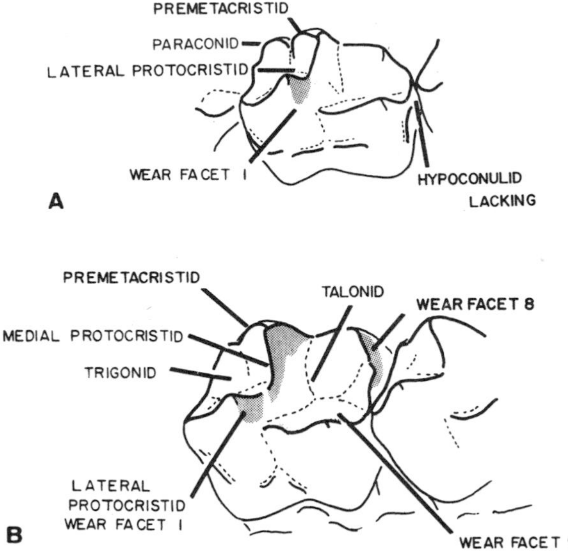

Fig. 4. Lateral views of left lower first and second molars of the recent platyrrhines *Saguinus midas* (A), *Saimiri sciureus* (B), and *Cebus apella* (C), and the middle Eocene omomyid *Omomys carteri* (D).

A paraconule is variably present but not a postparaconule crista. Another feature which early Eocene primates share with *Callimico,* but not with *Callithrix,* is a premetaconule crista. In short, the molars of *Callithrix* and other marmosets are quite dissimilar from those of the most primitive primates. Additional details of molar morphology suggest that callitrichids have specialized in their own unique divergent direction. For example, a complete cingulum wrapping around the protocone and a greatly reduced entoconid are derived features.

The upper molars of the last common ancestor of the platyrrhines were derived with respect to the morphology of early Eocene primates in several ways. Most importantly, there was a considerable modification in the system of crests leading posteriorly from the protocone. *Callimico* (Fig. 3) and all other New World monkeys have lost the "*Nannopithex*" system of wear surfaces and crushing areas. The *Nannopithex*-fold is a crest seen on Eocene primate molars (Fig. 3). This crest supports wear surfaces 6 and 7 where it wears against the postentocristid and the paraconid, respectively. The surface between the *Nannopithex*-fold and the short, poorly developed, postprotocrista wears against the trigonid to produce wear surface 10. In platyrrhines, the postprotocrista has subsumed the function of the *Nannopithex*-fold, supporting shearing wear facet 6; facet 7 has disappeared. A small hypocone supports a new shearing blade, 7n, and a new crushing surface 10n. As the hypocone enlarged, an additional shearing blade (8) was formed in some platyrrhines [stages in the evolution of this type of system are dealt with in detail by Kay (1977)].

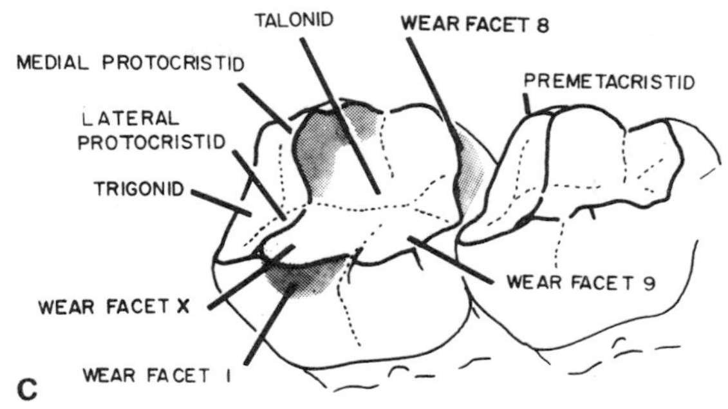

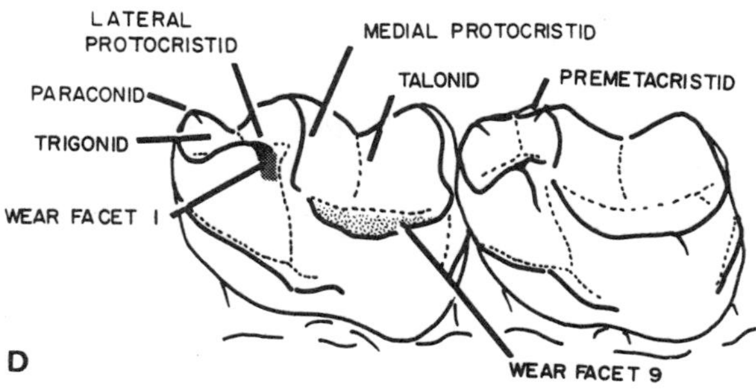

Fig. 4 (*Continued*)

The lower molars of the last common ancestor of platyrrhines must have been quite primitive, resembling the organization seen in early Eocene primates in many respects (Fig. 4). Although lower molars of most platyrrhines lack a discrete paraconid, this cusp is frequently present on the M_1 of *Saguinus,* and occasionally on that of *Saimiri* (on one of the five specimens I have examined). This cusp is present on the molars of all early Eocene primates and was almost certainly retained in the earliest New World monkeys. On the trigonid of all platyrrhines, a sharp crest, the premetacristid, leads anteriorly from the metaconid enclosing the medial margin of the trigonid (Fig. 4). A hypoconulid on M_{1-2} is a rare and specialized feature on the lower molars of early Eocene primates and is very unusual among New World monkeys. The platyrrhine ancestor almost certainly lacked this cusp.

The posterior wall of the trigonid exhibits two types of conformation among New World monkeys (Fig. 4). In the marmosets, *Callimico, Saimiri,* and *Alouatta,* the trigonid is raised above the talonid basin. The posterior wall of the trigonid is formed by the confluence of the medial protocristid and lateral protocristid. The lateral protocristid forms the leading edge of a Phase I shearing crest supporting wear facet 1; the medial protocristid acts as a guide for Phase I movement. The crushing and grinding action of the protocone is restricted to the talonid basin and its wear facet 9 does not occur on the posterior wall of the trigonid (see Fig. 4). In many advanced cebids, like *Cebus*

(Fig. 4C), the protocone is enlarged; its crushing and grinding surface overlaps the back of the trigonid (Maier, 1977) to form facet X. Consequently, the leading edge of shearing crest 1 has been pushed to the lateral edge of the lower molars, such that the lateral protocristid is no longer the leading edge of shearing blade 1.

As interpreted here, the restriction of Phase II wear to the molar basins, as seen in marmosets and cebids like *Saimiri,* is ancestral for platyrrhines. This conclusion is based on the fact that this configuration is found in some members of both platyrrhine families and is also characteristic of all Eocene primate families (e.g., *Omomys*, Fig. 4D), Paleocene primates, and most extant strepsirhines.

Among most New World monkeys, as among early Eocene primates, the postentocristid forms the posterior edge of the occlusal surface of M_1 and M_2. As the hypocone enlarged in many New World lineages, its anterior edge, the prehypocrista, came to occlude onto the posterior edge of the postentocristid producing wear surface 8 (Fig. 4C). With further enlargement, a small posterointernal basin was produced (e.g., *Callicebus,* not figured). Given that the hypocone of the ancestral platyrrhine species was likely to have been quite small, it is unlikely that a posterointernal basin was present in the ancestral platyrrhine.

The Case for Adapid Ancestry

An adapid ancestry for platyrrhine primates was originally advocated by Gidley (1923). Gingerich (1975) and Schwartz (e.g., Schwartz *et al.*, 1978) are its strongest current proponents. Gidley cited three characters of the masticatory apparatus shared by *Notharctus* (a North American Eocene adapid) and platyrrhines which might indicate a special relationship between the two: fusion of the mandibular symphysis, small, erect incisors, and projecting canines. Gingerich has elaborated this position, noting that adapids and anthropoids both have deep mandibular corpora that are midsagitally fused, and small, vertically implanted, spatulate lower incisors, with I_1 smaller than I_2. He adds that some progressive adapids share with anthropoids a shearing system developed between a projecting upper canine and the anteriormost lower premolar (Figs. 5, 6). However, the systematic distribution of these features among adapids suggests that they may not be useful indicators of a special relationship.

Symphyseal fusion is demonstrably labile in its occurrence. It has been evolved independently among many mammalian groups, including several carnivore lineages, elephants, marsupials, and indriid primates. Accordingly, one might expect the shared occurrence of this feature to have a low phyletic valence. In fact, several different groups of advanced adapids evolved mandibular fusion independently. *Notharctus* and *Adapis,* advanced members of two major adapid lineages, had fused mandibular symphyses even though their last common ancestor, *Pelycodus,* did not. Furthermore, *Cercamonius,* the

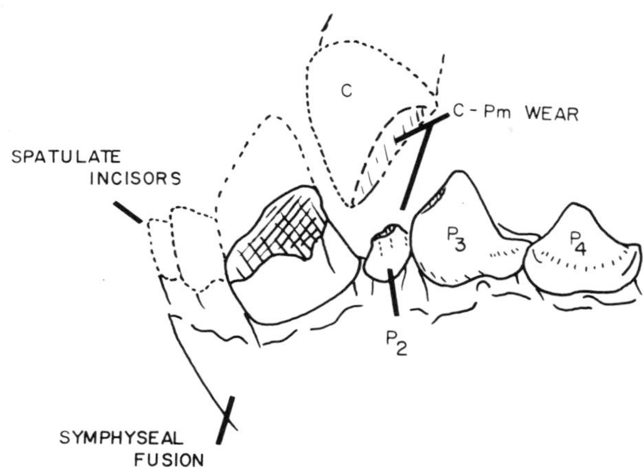

Fig. 5. *Leptadapis magnus,* a European Eocene adapid. A right jaw redrawn in mirror image after Gingerich's (1975) photograph of a specimen in the Musée National d'Histoire Naturelle, Paris. The incisor and canine morphology is reconstructed after the preserved parts of other specimens. Many Eocene and Oligocene adapids have symphyseal fusion, spatulate incisors, and projecting canines which wear against closely packed premolars. In *Leptadapis,* the upper canine wore against P_1 and P_2. The upper canine of the ancestors of *Leptadapis* probably contacted P_1 alone. Progressive reduction in the size of the P_2 and enlargement of the P_3 apparently led to a transfer of the canine-premolar shearing device distally while not disturbing this functional relationship.

European adapid which Gingerich nominates for a special relationship with Old World anthropoid primates apparently did not exhibit symphyseal fusion. Thus, the fact that symphyseal fusion occurs among platyrrhines and adapids does not lend very much support to the argument that the two groups are specially related.

Gingerich notes that projecting canines and small, vertically implanted, spatulate lower incisors, which probably characterized the ancestral platyrrhine, are found in adapids as well. However, as Gregory (1915) pointed out, these features might best be regarded as primitive retentions from the last common ancestor of all primates of modern aspect. The more primitive members of the Adapidae (*Pelycodus*) and the Omomyidae (*Teilhardina, Chumashius, Anaptomorphus,* and *Washakius*) had relatively large canines and small, subequal-sized, slightly procumbent lower incisors (as inferred from the morphology of the roots of these teeth, Figs. 6, 11, 12). Thus, while the specialized enlargement of the first incisor of some omomyids would eliminate those species from a platyrrhine ancestry, the occurrence of small, slightly procumbent incisors in an Eocene species does not indicate a special phylogenetic relationship with platyrrhines.

Gingerich notes that the large upper canine of some advanced adapids (e.g., *Leptadapis*) wears against the anterior lower premolar forming a shearing mechanism similar to that seen in anthropoids (Fig. 5). He suggests that a well-developed upper canine/lower premolar shearing contact constitutes a shared derived feature linking the two groups. The most obvious objection to this argument is that the upper canine of platyrrhines (Fig. 2B) and some

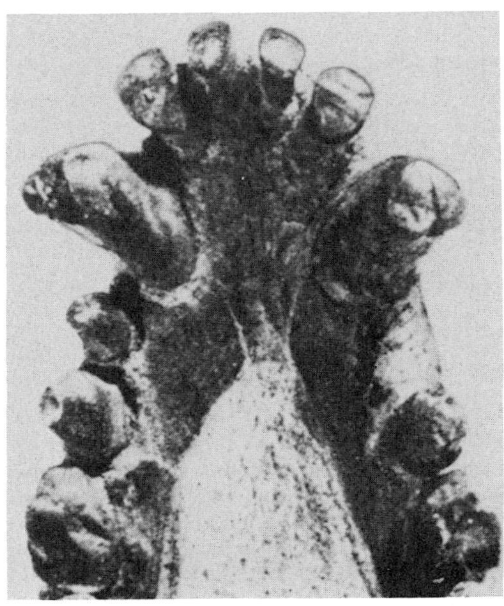

Fig. 6. The incisor structure of a specimen of *Notharctus "pugnax"* (American Museum of Natural History No. 11480). [Reproduced from Gregory (1920, Fig. 32).]

fossil catarrhines [such as *Parapithecus;* Delson (1975)] (Fig. 13B) contacts the lower second premolar, whereas the upper canine among other fossil and extant catarrhines contacts the lower third premolar. This seems to indicate that the canine–premolar shearing mechanism has evolved independently in several groups of anthropoids rather than being a derived feature of the ancestral anthropoid. Gingerich counters this argument by showing how the canine wear has shifted to more posterior premolars as the more anterior premolar was reduced and lost in adapid evolutionary history (Fig. 5). He argues that the loss of an anterior premolar could have occurred in various catarrhine lineages while the canine–premolar shearing mechanism remained functionally undisturbed.

An ingenious alternative to account for premolar reduction from three premolars to two premolars without disturbance of the upper canine–lower premolar shearing mechanism has been advanced by Osborn (1978). He suggests that the dP4 of the ancestral catarrhine was retained as a permanent, "M1," while the development of the permanent fourth premolar was suppressed. Simultaneously, the M3's were lost. This would lead to an adult post-canine dentition with two premolariform teeth and three molariform teeth. (DP4 of the ancestral catarrhines is assumed to have been molariform.) Thus, the extant catarrhine dental formula, which is apparently two premolars and three molars, would *actually* be composed of two permanent premolars (the first of which could retain its shearing function), one unreplaced, molariform, deciduous premolar and two true molars.

Even if either Gingerich's or Osborn's scenerios prove correct, the occurrence of an upper canine–lower premolar contact is not a unique condition shared by adapids and anthropoids alone. Among many anthropoids, the enamel on the anterolateral surface of the anterior lower premolar has become thickened. This surface serves as a "hone" to keep the posterior edge of the upper canine sharp (Zingeser, 1969, 1971; Every, 1970). There has been an uncritical tendency for recent workers to equate any canine–premolar contact with the occurrence of a specialized "honing mechanism." In fact, an occlusal contact between the upper canine and anterior lower premolar occurs among all primates which exhibit a combination of a large upper canine and tooth packing (the premolars set together without intervening spaces) in the lower jaw. Such a condition is commonly seen in strepsirhines as well as haplorhines (Cartmill and Kay, 1978; Zingeser, 1971). As noted below, (Fig. 7) it probably occurred in omomyids as well as adapids. Its presence is no more unique than an occlusal contact between, say, the back edge of P^3 and the front edge of P^4.

Schwartz *et al.* (1978) have suggested that the ancestral primate had five

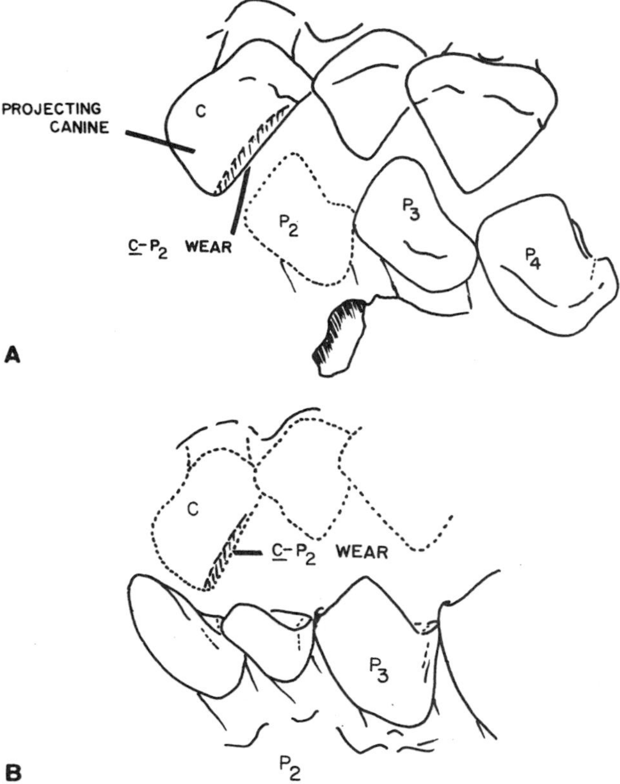

Fig. 7. Lateral views of the left lower canine and premolars of *Necrolemur antiquus* (A) from the collections at the Musée National d'Histoire Naturelle, Paris (No. QU11060 and No. QU10909), and the type of *Chlororhysis knightensis* (B), in the United States National Museum (No. 21901). The posteromedial surface of the upper canine of *Necrolemur* has a wear facet for the P_2 (the latter tooth is not preserved). The anterolateral surface of the P_2 of *Chlororhysis* has a facet for the upper canine (not preserved). Note the premolar packing in both species.

premolars, and that adapids (and strepsirhines generally) share with anthropoids a derived loss of the third tooth in the series, resulting in a total of four or fewer premolars. They argued that *Tarsius* and the earliest omomyids differ from all other primates in having retained five lower premolars. Cartmill and Kay (1978) have summarized the salient features of this hypothesis and the grounds for rejecting it. Among other things, Schwartz's scheme requires that the lower canine of *Tarsius* and omomyids occlude behind the upper one, and that the upper canine is lodged in the premaxilla.

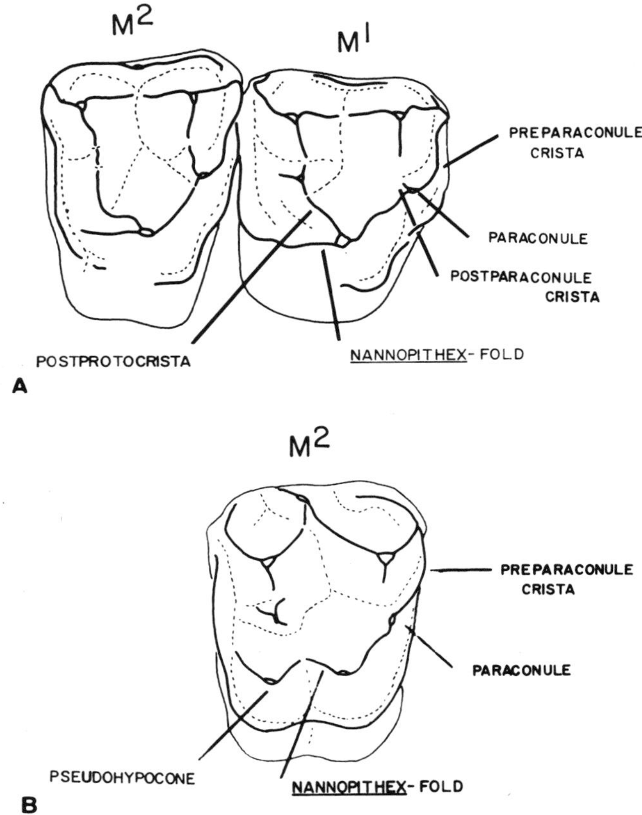

Fig. 8. Upper right molar morphology of the adapids *Pelycodus sp.* (A), Yale Peabody Museum No. 25012, from Yale Locality 290, Bighorn Basin, Wyoming; *Notharctus sp.* (B), a Yale Peabody Museum specimen; *Adapis parisiensis* (C), Museum of Comparative Zoology No. 8884. Early *Pelycodus* have a well-developed *Nannopithex*-fold but lack hypocones. *Notharctus,* a North American descendant of *Pelycodus,* has a hypocone formed by "splitting" of the protocone and the "migration" of the new cusp posteriorly along the *Nannopithex*-fold. *Adapis,* a European adapid, has a hypocone which was formed from an elaboration of the posteromedial cingulum. Note that the *Nannopithex*-fold does not contribute to the hypocone. The *Nannopithex*-fold is small in *Adapis;* some advanced European adapids apparently have lost the *Nannopithex*-fold completely [e.g., *Caenopithecus,* from a drawing by Stehlin (1916) and *Pronycticebus* (e.g., QU11056)].

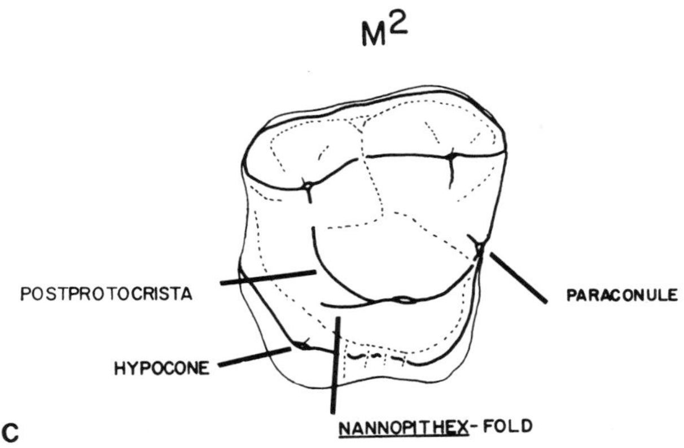

Fig. 8 (*Continued*)

The upper molars of some European adapids have lost the *Nannopithex*-fold and its system of wear surfaces, a possible derived feature linking them with anthropoids (Fig. 8). However, this crest is absent in some omomyids, (Fig. 9) indicating a high degree of parallelism in this feature.

The most vocal recent opponent of a possible adapid ancestor for platyrrhines is Hoffstetter (1974), who argues that the anatomy of the upper molars is sufficient to exclude notharctines (New World adapids) from the ancestry of New World monkeys. Hoffstetter has referred specifically to the configuration of the hypocone in three North American notharctids: *Notharctus, Smilodectes,* and advanced species of *Pelycodus.* In this group, the hypocone evolves by progressive splitting of the protocone with migration of the new cusp (a "pseudohypocone") posteriorly along the *Nannopithex*-fold (Gregory, 1920) (Fig. 8). Studies of the hypocone development in the platyrrhine *Alouatta* have shown that this cusp develops on the posterior cingulum as it does in most mammalian species (Tarrant and Swindler, 1973). This would apparently rule out advanced notharctines from the ancestry of platyrrhines.

The mode of hypocone development, however, cannot be used to rule out early notharctines such as *Pelycodus,* or European-derived adapines from platyrrhine ancestry. Early *Pelycodus* do not possess a hypocone, and European adapines have hypocones developed from the posterior cingulum in a manner similar to that seen in anthropoids (Fig. 8).

A progressive specialization seen in all European adapids which tends to rule out known forms from the ancestry of platyrrhines, is the progressive loss of the paraconid, and, more importantly, the complete absence of a premetacristid (Fig. 10A,B). The premetacristid in New World monkeys forms a distinct wall on the medial aspect of the trigonid (Fig. 4) whereas the trigonid of European adapids is open. The paraconid of adapids, where present, is linked to an extension of the paracristid; in New World monkeys it is connected to the premetacristid.

In summary of the above comments, possible shared derived features,

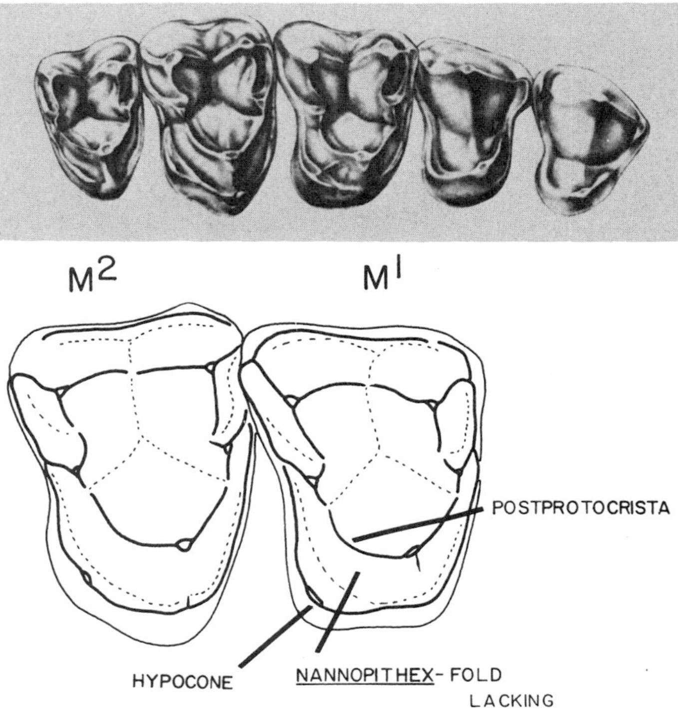

Fig. 9. Right P^3–M^3 of *Omomys* (middle Eocene) (top), based primarily on American Museum specimens [drawing from Szalay (1976) by permission], and a Yale Peabody Museum specimen, No. 13228-1 (bottom), illustrating the configuration of the upper molar cingulum with a variably present hypocone and the absence of a *Nannopithex*-fold.

such as the presence of symphyseal fusion, a canine–premolar shearing mechanism, and the loss of the *Nannopithex*-fold, suggest possible adapid–platyrrhine affinities. However an equally plausible alternative is that these characteristics evolved in parallel in the different groups. The common possession in anthropoids and adapids of two small, vertically implanted, spatulate incisors on each side of the lower jaw is plausibly interpreted as a retained primitive primate characteristic, since the available, fragmentary evidence suggests that these characteristics may also have been found among primitive omomyids. The possibility that adapids and anthropoids share a peculiar dental formula, as interpreted by Schwartz *et al.* (1978) appears unlikely for developmental reasons. Furthermore, one or more derived molar characteristics (pseudohypocone, progressive loss of the paraconid, or the absence of a premetacristid) tend to rule out advanced notharctines and adapines (including "cercamonines") from platyrrhine ancestry. On the other hand, I know of no dental characteristics which would rule out early unspecialized adapids, such as *Pelycodus* or *Donrussella*, from a possible role in platyrrhine ancestry.

The Case for Omomyid Ancestry

The case linking the Omomyidae (essentially as defined by Szalay, 1976) with the ancestry of anthropoids was first detailed by Wortman (1904). Additional elements have been added by Gregory (1916), Gazin (1958), Simons (1961), and Hoffstetter (1969).

Wortman (1904) advanced two arguments in favor of an omomyid ancestor. First, the structure of the molars of omomyids is very generalized for primates as a whole, one from which it is possible to derive all the more complex dental morphologies. Second, the dental formula for the omomyid lower jaw, and, presumably, that for the upper as well, is the same as in Cebidae, so that the latter could be descended from the former. (It should be noted that Wortman restricted these arguments to the Cebidae, arguing that marmosets were of more ancient origin.)

Gregory (1916) argued that European microchoerids (presently classified among the Omomyidae), although too specialized in many ways, exhibit characters not found in other primate groups which might be looked for in the immediate ancestors of anthropoids. Such characters are the mode of development of the hypocone, and the tendency to lose the paraconid.

Gazin (1958) has reviewed the evidence for an omomyid–platyrrhine relationship. He identified *Washakius* as the omomyine having the most features

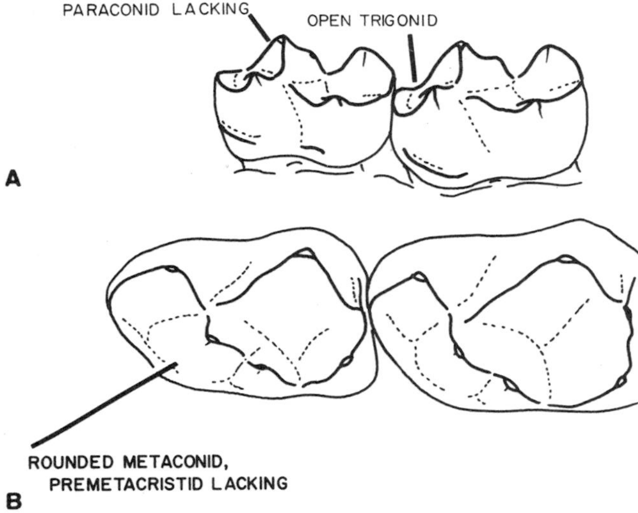

Fig. 10. *Adapis parisiensis.* (A) lateral view of the left M^{2-3} of QU10966, a specimen in the Musée National d'Histoire Naturelle, Paris, and (B) an occlusal view of a specimen in the Museum of Comparative Zoology, Harvard University. These specimens illustrate the open trigonid, the absence of a paraconid, and the smooth rounding of the anterior slope of the metaconid (premetacristid absent) characteristic of European adapids.

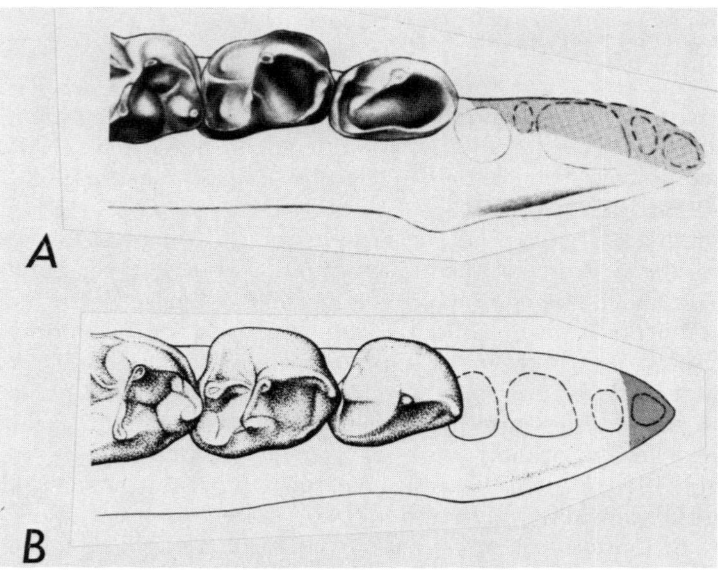

Fig. 11. The lower anterior teeth of several omomyids [reproduced from Szalay (1976) by permission]. (A) *Teilhardina belgica*, (P_{3-4}, part of M_1) early Eocene of Belgium. (B) *Loveina zephyri*, (P_{3-4}, part of M_1) early Eocene of Wyoming. (C) *Washakius insignis*, (canine–P_4, part of M_1) middle Eocene of Wyoming. (D) *Chlororhysis knightensis*, (canine–P_4) early Eocene of Wyoming.

reminiscent of platyrrhines: (1) the anterior part of the lower jaw is medially deflected toward the plane of the symphysis so that paired rami of the lower jaws are more U-shaped, as in platyrrhines, and unlike other omomyids; (2) as judged from the preserved tooth sockets and roots, the lower incisors were small, evidently about equal in size, and erect (Fig. 11C); and (3) the canine socket was twice the size of those of the incisors (Fig. 11C).

Simons (1961) argues that *Omomys* and its immediate forebears are the most likely to have had a direct relationship to the rise of South American monkeys. Several characteristics are advanced to support this position: (1) *Omomys* is small enough (Simons argued that the ancestry of platyrrhines should be sought in a small species, since some extant platyrrhines are small, and an evolutionary trend toward size decrease is very uncommon); (2) *Omomys* has a "suitably unspecialized" molar crown pattern; (3) *Omomys* has relatively small third molars; (4) it possesses a well-developed medial upper molar cingulum, with accessory cusps frequently developed on it, in a manner similar to that of the platyrrhine genus *Saimiri;* (5) it has the appropriate dental formula; (6) some omomyids (e.g., *Macrotarsius*) evolved deep jaws. Figures 7, 9, 11, 12 summarize some of these and other characteristics of omomyids.

Four of the characters of omomyids cited by Wortman, Gregory, Gazin, and Simons are simply primitive retentions from the last common ancestor of primates of modern aspect. For this reason, none provide evidence of a spe-

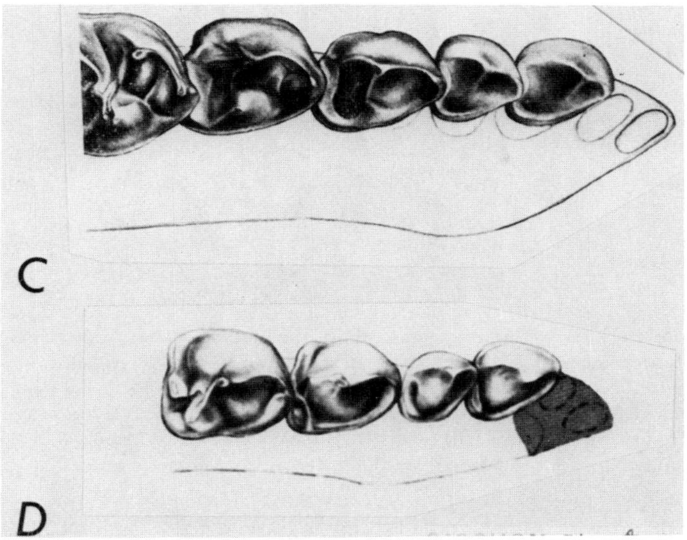

Fig. 11 (*Continued*)

cial omomyid–platyrrhine relationship: (1) small size; (2) the "generalized" molar structure evident among most omomyid genera; (3) the presence of two, small, vertically implanted incisors; and (4) projecting canines, larger than incisors. (If this incisor and/or canine configuration is not a primitive retention then various lineages of adapids and omomyids independently evolved small vertically implanted incisors and/or large canines, in which case these characteristics by themselves are not strong evidence for special relationship of any one omomyid with the platyrrhines.)

Three other derived characters that evolved convergently in omomyids, platyrrhines, and adapids, evolved also in adapids and cannot be weighted heavily in linking omomyids and platyrrhines: (1) a reduced dental formula by the loss of the first upper and lower premolars; (2) the development of a hypocone from the posterointernal cingulum; and (3) deepening of the horizontal mandibular rami.

Several characteristics of one or another advanced omomyids, cited by earlier workers as evidence of omomyid–anthropoid affinity, actually are crossing-specializations which rule out their possessors from an ancestral position for platyrrhines. The precocious loss of a paraconid and the elaboration of medial cingulum cusps in some advanced European and North American omomyid genera are not features to be expected in the immediate common ancestor of all platyrrhines, as reconstructed above.

The lower third molars are small in some omomyids. In platyrrhines, third molar reduction has been effected by the loss of the hypconulid, as well as an overall size reduction (Fig. 2A). Omomyid M_3's often are reduced in size but the hypoconulid is always present (e.g., Fig. 12).

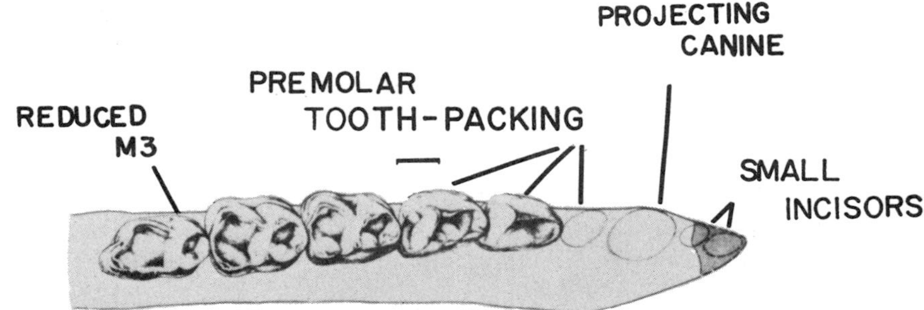

Fig. 12. An occlusal view of the lower dentition of the omomyid *Chumashius balchi* (late Eocene, California). Lower jaw reconstruction based on specimens in the Los Angeles County Museum, illustrates the reduced M_3, relatively large canine, and small incisors [drawing from Szalay (1976) by permission]. Gazin (1958) states that in the type (LACM 1391), the I_1 is "only slightly enlarged" and the I_2 "noticeably reduced"; he notes that a second jaw, No. 1390, shows less disparity between the incisors, and a slight increase in the size of the canine.

Several authors, most recently Hershkovitz (1977), have ruled out the omomyids from the ancestry of platyrrhines because the latest-occurring omomyids were specialized in directions other than that expected for a platyrrhine ancestor. Such specializations would include the excessive reduction in the size and/or number of premolars, canines, and incisors, and a tendency for the first incisor to be enlarged and procumbent. It is true that most known omomyids do exhibit "crossing specializations" sufficient to rule them out as platyrrhine ancestors. However, not all late-occurring genera exhibit such specializations. Thus, this does not completely exclude from consideration an omomyid–platyrrhine relationship.

Anthropoids and omomyids share certain derived characteristics not cited by earlier authors. The dental morphology of the genus *Chumashius* (Fig. 12) of the late Eocene of California is particularly suggestive of being in or near anthropoid ancestry. Five possible shared derived features link this genus with anthropoids: (1) In *Chumashius,* three well-developed lower premolars are closely packed together, perhaps a corollary of facial shortening; (2) no upper canines are preserved so it is not known that a shearing contact was present between the upper canine and P_2, but a combination of a well-developed lower premolar paracristid and the possibility of a slighty projecting canine (the lower canine root and socket are quite enlarged) is very similar to what might be expected in an ancestral platyrrhine. A similar upper canine–premolar shearing device evolved in *Necrolemur,* and some other omomyids (Fig. 10C); (3) *Chumashius* has lost the primitive *Nannopithex*-fold and evolved a postprotocrista, which is a configuration strikingly reminiscent of the platyrrhine condition; (4) the medial trigonid wall of the M_{2-3} of *Chumashius* has a well-developed premetacristid; and (5) the trigonids of *Chumashius* are low, and the talonid basins are expanded. Wear facet 1 is greatly reduced and pushed quite far laterally, reminiscent of primitive platyrrhines. At the same time, *Chusmashius* lacks the crossing-specializations seen in other late-occurring omomyids, such as excessive size or number of premolars, anterior incisor expansion, and canine size reduction. *Chumashius*, or its yet unknown relatives, certainly represent a plausible ancestral stock for platyrrhines. At the same time, it must be remembered that most of the characters supporting this hypothesis evolved among various adapids as well.

To summarize, many of the dental characteristics which have been used to support a special omomyid–platyrrhine relationship are either primitive retentions or are shared by advanced adapids suggesting convergent evolution. When these characters have been identified and eliminated, we are left with one uniquely shared derived character between platyrrhines and omomyids: lower third-molar size reduction. The characteristic third-molar size reduction of platyrrhines was accompanied by the loss of the hypoconulid (Fig. 2A), whereas M_3 reduction in omomyids was accomplished by an overall size decrease with the hypoconulid persisting. This difference in the mode of reduction seems to indicate that the M_3 size reduction of omomyids and platyrrhines was accomplished independently.

The Case for an Old World Anthropoid Ancestry

Recently, several authors have argued that the last common ancestor of platyrrhines had an African origin. Following the differentiation of the Anthropoidea in Africa, a considerable period of time elapsed during which anthropoid characteristics evolved. Finally, an African anthropoid was introduced into South America where it produced the platyrrhines.

Hoffstetter (1977) argues that the ancestral anthropoid stock entered Africa very early in primate history. In Africa, the stem group differentiated into a New World ancestral group (which includes the Oligocene Parapithecidae) and the common ancestor of cercopithecoids and hominoids as represented by the Fayum "pongids." A primitive parapithecid then found its way into South America in the late Eocene by rafting across the South Atlantic Ocean.

Hershkovitz (1977) subscribes to a variant of this hypothesis. He argues that an ancient "presimian" stock evolved in the common South America—Africa land mass prior to their being greatly separated by a developing South Atlantic Ocean. He contends, however, that parpithecids are not ancestral New World monkeys.

According to Hoffstetter (1977) the ancestral simian shared the following dental characteristics with stem haplorhines: (1) a dental formula of two incisors, one canine, three premolars, and three molars, above and below; (2) vertical, spatulate incisors; (3) moderately shearing canines; (4) a distinct upper molar sylar shelf; (5) an upper second premolar with a protocone; (6) upper third and fourth premolars with two cusps; (7) upper molars wider than long, with a hypocone situated lingual to the protocone; (8) well-developed molar paraconules and metaconules; and (9) single-cusped lower premolars. Shared derived characters of stem anthropoids, as visualized by Hoffstetter, include complete fusion of the mandibular symphysis and bulbous molar cusps. The characteristics which Hoffstetter visualizes for the African platyrrhine ancestor are mostly those of the stem anthropoids. However, he apparently believes that the paraconule was lost and the metaconule reduced in this group (to be reevolved secondarily, and/or secondarily enlarged in *Parapithecus*).

Most of the dental characteristics cited by Hoffstetter as supporting an anthropoid relationship with other haplorhines are simply stem characters of all primates of modern aspect, including adapids. These include traits 2–6 and 8 in the list enumerated above. Traits 1 and 7 are possibly shared derived features linking anthropoids and omomyids but, as noted above, are seen also in various advanced adapids. It is doubtful that character 9, single-cusped lower premolars, characterized the ancestral haplorhine, since virtually all omomyid genera possess lower fourth premolars with a paraconid and metaconid. In any case, all platyrrhines and catarrhines, living and fossil, have a two-cusped P_4, so that a single-cusped P_4 would be an unlikely ancestral configuration.

Two shared derived traits, which Hoffstetter considers as having characterized the ancestral simian, symphyseal fusion, and bulbous molar cusps, may well be simply convergent in New and Old World monkeys, inasmuch as these characters have been repeatedly independently evolved in various primate lineages (see above, p. 168). Thus, the dental evidence marshaled by Hoffstetter to support an anthropoid–stem haplorhine relationship is extremely weak. Furthermore, no reliable features are advanced to support the possibility of a special dental relationship between Old and New World anthropoids.

Hoffstetter contends that there are several crossing specializations which would exclude known parapithecids from the ancestry of the earliest New World monkeys. At the same time he feels that more primitive members of the family could be ancestral to platyrrhines. First, known parapithecids have two-rooted P^2's. Hoffstetter rightly notes that this is probably a derived feature of parapithecids. The P^2's of the Oligocene South American genera *Branisella* and *Tremacebus* are single-rooted (Hoffstetter, 1969; Hershkovitz, 1974); a single-rooted P^2 characterizes callitrichids, *Callimico,* and most Cebids (Hershkovitz, 1977); this condition is seen as well among all Eocene primates [see figured material in Szalay (1976) and Stehlin (1916)]. [In spite of this, Hershkovitz (1977) unaccountably has stated that a small, single-rooted P^2 is a specialized condition which would exclude *Branisella* from the direct ancestry of later platyrrhines.]

Second, Hoffstetter, following arguments developed by Delson (1975), points to the enlarged upper-molar conules of parapithecids as being derived for that group. He argues that the ancestral platyrrhines must have lacked molar paraconules (a derived feature) so that parapithecids are specializing in a direction opposite that expected of a platyrrhine ancestor. I agree with

Hoffstetter that parapithecids certainly show a derived enlargment of a paraconule and metaconule not to be expected in the platyrrhine ancestor. However, ancestral platyrrhines must certainly have had paraconules, inasmuch as this cusp is present on the distal end of the preprotocristae of M^{1-2}in *Saimiri* and *Callimico* (Fig. 3). I have seen this cusp as well on the M^1's of some *Callithrix*. It is also present on the M^1 of *Tremacebus* (Hershkovitz, 1974). Its absence in the early Oligocene platyrrhine *Branisella* (Hoffstetter, 1969) is probably a secondary specialization in that genus. A correlated crossing specialization of parapithecids is the presence of greatly enlarged paraconules on the premolars (see Simons, 1972, Fig. 76, p. 188). No New World monkeys have this feature.

In spite of these difficulties, Hoffstetter feels that an unknown, more primitive, parapithecid could be ancestral to platyrrhines. However, an additional derived feature of parapithecid molars indicates otherwise since it is shared by contemporary two-premolared African Oligocene primates, as well as all other more recent living and fossil catarrhines. The posterior wall of the molar trigonid of *Parapithecus* and *Apidium* is dominated by a large Phase II wear facet [facet X of Kay (1977) or facet 11 of Maier (1977)] (Fig. 13A). Wear facet 1, primitively associated with the lateral protocristid, is associated with a new crest running posteromedially from the protoconid. [Alternatively, the lateral protocristid and facet 1 may have been rotated to a more lateral position, and a new crest taken its place; Kay, (1977)]. Although a similar condition exists in some cebids, most notably *Cebus* (Fig. 4C), this peculiar specialization of the trigonid was almost certainly absent in the last common ancestor of platyrrhines, since the primitive configuration is still seen in all callitrichids, *Callimico,* and, the cebids *Saimiri* and *Alouatta* (Fig. 4). The presence of facet X among parapithecids indicates that this group shares a communality of origin with other catarrhines, and can have nothing to do with platyrrhine ancestry.

Interestingly, facet X is lacking in *Oligopithecus,* a two-premolared early Oligocene African species (Kay, 1977) (Fig. 13A). If *Oligopithecus* proves to have been a catarrhine, the presence in this genus of a more primitive type of molar organization is additional support for the contention that wear facet X evolved convergently among platyrrhines and would also provide evidence that premolar reduction among catarrhines must have occurred at least twice among catarrhines (once in *Oligopithecus,* and, later, in the common ancestor of extant catarrhines).

Two possible shared derived dental features of platyrrhines and catarrhines are the absence of a *Nannopithex*-fold (Fig. 13C) and the presence of an upper canine–lower premolar tooth contact. The *Nannopithex*-fold is apparently lacking in all anthropoids. However, the loss of the *Nannopithex*-fold has occurred also in omomyids (Fig. 9) and adapids (Fig. 8, caption) and does not necessarily indicate that platyrrhines should be derived from an African anthropoid. An upper canine–P_2 contact facet occurs in parapithecids (Fig. 13B) in a fashion similar to that seen in New World monkeys (Fig. 2B).

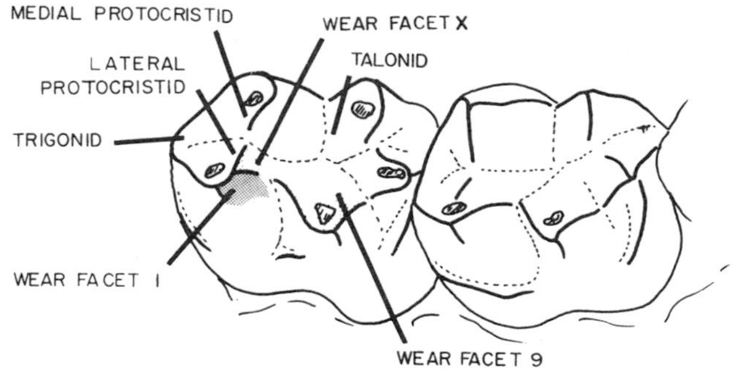

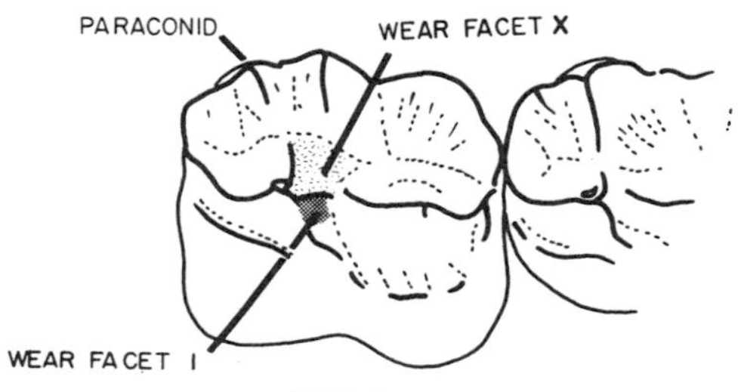

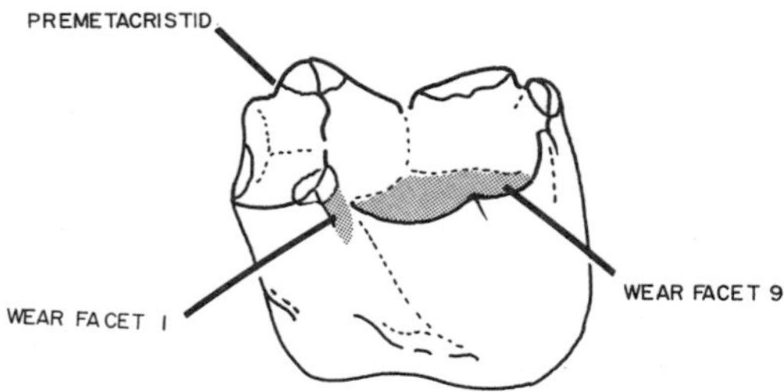

A

OLIGOPITHECUS

Fig. 13. Aspects of the dental anatomy of certain early Old World anthropoids. (A) Dorsolateral view of the molars of *Parapithecus fraasi* ($M_{1\text{-}2}$), from a cast of the type specimen, courtesy of P. D. Gingerich; *Oligopithecus savagei,* (M_2) from a cast of the type, courtesy of P. D. Gingerich; and *Pondaungia cotteri,* (M_{2-3}) in mirror image from a cast of courtesy of D. Savage and R. Ciochon. *Parapithecus* and *Pondaungia* share with all more recent Old World anthropoids the possession of a wear surface X, on the posterior and lateral side of the trigonid tooth. *Oligopithecus* lacks such a wear surface, resembling Eocene European and North American primates in that respect. Note the presence of paraconids on the $M_{2\text{-}3}$ of *Pondaungia.*

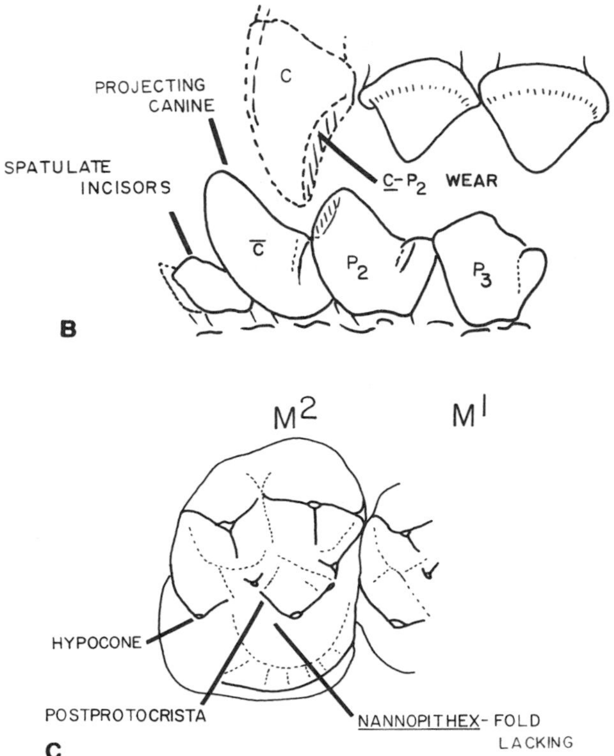

Fig. 13 (*Continued*). (B) Lateral view of the left lower anterior teeth of *Parapithecus fraasi* (type) with the upper teeth reconstructed from specimens of *Parapithecus grangeri*, and an isolated upper canine tentatively assigned to *Apidium*. The P_2 of *Parapithecus* has a wear facet indicative of an occlusal contact with the upper canine. (C) Occlusal view of *Aegyptopithecus zeuxis* (based on Cairo Geological Museum No. 40237), illustrating the absence of a *Nannopithex*-fold.

However, an analogous type of wear occurs in advanced adapids (Fig. 5) and omomyids (Fig. 7). Therefore, this condition provides no evidence for special relationship.

To summarize, all the dental characteristics thus far adduced to support a special relationship between catarrhines–platyrrhines are either primitive characters of primates of modern aspect, or are shared with one or more Eocene lower primates. I have been unable to suggest definitely any shared derived dental characters linking Old and New World anthropoids. The case for linking any known parapithecid with platyrrhine ancestry is not supportable. Furthermore, parapithecids share certain derived dental characters with other Old World anthropoids which make it extremely unlikely that a platyrrhine ancestor could be drawn from an as yet unknown more primitive member of that group.

Recently, an additional specimen of the enigmatic genus *Pondaungia* has been recovered in Burma from sediments dated at around 40 million years before present (Ba Maw *et al.* 1979). This specimen preserves M_2 and M_3 (Fig. 13A). Wear facet 1 may be recognized on the lateral margins of M_2 and M_3, and Phase II wear facet X is apparent on the back of the trigonid of M_2. This shared derived development of facet X definitely links *Pondaungia* with African Oligocene parapithecids and primitive apes.

Pondaungia is considerably more primitive than African anthropoids in having a well-developed paraconid on M_2. A well-developed premetacristid links up with the paraconid tc form a complete medial wall for the trigonid in a manner reminiscent of advanced omomyids and early anthropoids. Such a trigonid organization would appear to rule out known European adapids from the ancestry of *Pondaungia* since they have specialized, medially open trigonids (Fig. 10).

Discussion of the Dental Evidence

In this paper, I have endeavored to draw together all the aspects of dental anatomy which have been utilized heretofore to indicate a special affinity of platyrrhines with adapids, omomyids, or Old World anthropoids. In the process of reviewing the dental structure of all the relevant groups, I have identified several additional characteristics which might be considered relevant in this context. A total of 15 characters are discussed (Table 1).

The dental evidence linking platyrrhines to any of the groups considered is not compelling. The greatest number of dental similarities frequently cited as evidence for platyrrhine ancestry are either characters of the last common ancestor of all primates of modern aspect or have been evolved independently in both Omomyidae and Adapidae. Neither type of character can provide evidence as to special relationships among the groups considered. For example, the possession of two, spatulate, vertically implanted lower incisors has been demonstrated (or strongly inferred) for Adapidae, Omomyidae, and Old and New World Anthropoidea. Its common presence in any pair of these taxa would no more indicate that they are closely related than would the common possession of five digits on their hands and feet. For a different reason, the presence of three, rather than four, premolars, or the presence of an upper canine–lower anterior premolar occlusal contact, do not constitute strong evidence for relationship because they have occurred convergently in so many lineages of living and fossil primates, possibly as a result of selection for a shortened face.

When all these characters have been eliminated, we are left with single, seemingly trivial characters linking some adapids with anthropoids (symphyseal fusion) and some omomyids with anthropoids (M_3 size reduction). This story is further complicated because adapid species with symphyseal fusion can be ruled out of anthropoid ancestry owing to other crossing specializations of the dental structure, and M_3 size reduction apparently occurred by a different series of steps in omomyids and platyrrhines.

No dental characters have been identified indicating a special platyrrhine–catarrhine affinity. Oligocene parapithecids and Eocene *Pondaungia* have characters specially linking them with more advanced catarrhines.

The nondental anatomical evidence supports the hypothesis that omomyids share a last common ancestor with *Tarsius*. Both groups have a large promontory artery entering the auditory bulla posteromedially (Szalay, 1975) and an unpneumatized apical interorbital septum (Cartmill, 1975). This and the satisfactorily documented case that *Tarsius* is the sister group of Anthropoidea, suggest that platyrrhine ancestry should be sought among the omomyids.

Given our state of knowledge of primate fossils it is scarcely surprising that the dental evidence is uninformative or equivocal on the question of platyrrhine origins (or the origins of Anthropoidea generally). The principal obstacle standing in the way of a resolution to this question is that large gaps occur in the evidence at critical points in the fossil record. Most of our evidence comes from the Eocene of Europe and central North America. These

Table 1. Summary of the Dental Characteristics of the Ancestral Platyrrhine, Allocated according to Their Relevance for Indicating Platyrrhine Affinities[a]

Dental characteristics of the ancestral platyrrhine	A	B	C	D	E	F
I_2 subequal to I_2; both teeth spatulate, vertically implanted	√					
Canine projecting above the occlusal plane	√					
P^2 single rooted, single cusped	√					
Paraconule present	√					
Paraconid present	√					
Hypoconulid lacking on M_3						√
Posterior wall of trigonid lacks facet X	√					
Loss of premolar 1		√				
Lower premolars packed together with the lower anterior premolar wearing against the upper canine		√				
Metaconule crests reduced or absent		√				
Presence of a hypocone on a medial cingulum		√				
Premetacristid well developed as a medial wall for the trigonid		√				
Nannopithex-fold absent		√				
Symphyseal fusion				√		
M_3 reduced			√			

[a] Column A: Dental features of the platyrrhine ancestor which are retained from the last common ancestor of all primates of modern aspect. Only those characters are mentioned which have been used in the past either as indicating a special affinity of platyrrhines to some other group, or to rule out such a relationship. Column B: Characters of the platyrrhine ancestor which are known to have been evolved convergently in the Adapidae and Omomyidae. Column C: Possible unique shared derived characters of some omomyids and the ancestral platyrrhine. Column D: Possible unique shared derived characters of some adapids and the ancestral platyrrhine. Column E: Possible unique shared derived characters of some Old World anthropoids and the ancestral platyrrhine. Column F: Characters unique to platyrrhines.

areas were probably peripheral to the evolutionary events leading to anthropoid origins. A fossil record for primates is virtually lacking from most of the African and Central American Paleogene, from which we might reasonably expect to find the evidence of anthropoid origins. The comments of Simpson (1955) bear repeating in this context:

> [The order Primates] does not seem to have split into two (or a few) suborders, infraorders, or other major subdivisions from one or another of which the late forms arose. It seems, rather, to have split into a large number of divergent, progressing lineages, symbolizable especially as families or subfamilies, among which some five or six survived and separately gave rise to later Cenozoic (Oligocene-Recent) Primates. Not a single one of those ancestral lineages has been surely identified in the Paleocene or Eocene. (They may well have lived in regions now tropical, from which Eocene mammals are virtually unknown.) The known Paleocene-Eocene groups do, nevertheless, cast light on the origins of higher primates. Some may, indeed, prove to be ancestral groups or closely related to such. In any event they do, in sum, indicate . . . the levels of progression reached in the Paleocene and Eocene [p. 439].

Considering the paucity of good dental evidence (or anatomical evidence of any kind) linking platyrrhines with any of the groups considered here, it is surprising that many authors have so strongly (and often dogmatically) advocated one or another phylogenetic arrangement. This tendency has been exacerbated recently by the rise of phylogenetic systematics whose advocates call for a one-to-one correspondence between a phylogeny and a classification. Under this scheme the unwary nonspecialist will be led by the new classifications to underestimate the uncertainties in the relationships between major groups. Present indications are that there is an inverse relationship between the level of a newly erected primate taxon and the amount of evidence on which it is based.

Acknowledgments

I thank William Hylander and Alfred Rosenberger who read and commented on early stages of this manuscript. This research was supported by NSF grant BNS77-08939.

References

Ba Maw, Ciochon, R. L. and Savage, D. E., 1979, Late Eocene of Burma yields earliest anthropoid primate, *Pondaungia cotteri, Nature (London)* **282**:65–67.

Cartmill, M., 1975, Strepsirhine basicranial structures and the affinities of the Cheirogaleidae, in: *Phylogeny of the Primates: A Multidisciplinary Approach* (W. P. Luckett and F. S. Szalay, eds.), pp. 313–356, Plenum Press, New York.

Cartmill, M., and Kay, R. F., 1978, Carnio-dental morphology, tarsier affinities, and primate sub-orders, in: *Recent Advances in Primatology,* Vol. 3, *Evolution* (D. J. Chivers and K. A. Joysey, eds.), pp. 205-213, Academic Press, London.

Delson, E., 1975, Evolutionary history of the Cercopithecidae, in: *Contributions to Primatology,* Vol. 5 (F. S. Szalay, ed.), pp. 167-217, Karger, Basel.

Every, R. G., 1970, Sharpness of teeth in man and other primates, *Postilla* **143**:1-30.

Gazin, C. L., 1958, A review of the middle and upper Eocene primates of North America, *Smithsonian Misc. Coll.* **136**:1-112.

Gidley, J. W., 1923, Paleocene primates of the Fort Union, with discussion of relationships of Eocene primates, *Proc. U.S. Natl. Mus.* **63**:1-38.

Gingerich, P. D., 1975, A new genus of Adapidae (Mammalia, Primates) from the late Eocene of southern France, and its significance for the origin of higher primates, *Univ. Mich. Contrib. Mus. Paleontol.* **24**:163-170.

Gregory, W. K., 1915, On the relationship of the Eocene lemur *Notharctus* to the Adapidae and to other primates, *Bull. Geol. Soc. Am.* **26**:419-446.

Gregory, W. K., 1916, Studies on the evolution of the primates, *Bull. Am. Mus. Nat. Hist.* **35**:239-355.

Gregory, W. K., 1920, On the structure and relations of *Notharctus,* an American Eocene primate, *Mem. Am. Mus. Nat. Hist.* **3**(2):49-243.

Hershkovitz, P. 1974, A new genus of late Oligocene monkey (Cebidae, Platyrrhini) with notes on postorbital closure and platyrrhine evolution, *Folia Primatol.* **21**:1-35.

Hershkovitz, P., 1977, *Living New World Monkeys (Platyrrhini), with an Introduction to Primates,* Vol. 1, University of Chicago Press, Chicago.

Hoffstetter, R., 1969, Un primate de l'Oligocène inférieur Sud-Américain: *Branisella boliviana* gen. et sp. nov., *C. R. Acad. Sci. Paris, Sér. D* **269**:434-437.

Hoffstetter, R., 1974, Phylogeny and geographical deployment of the primates, *J. Hum. Evol.* **3**:327-350.

Hoffstetter, R., 1977, Phylogénie des primates, *Bull. Mém. Soc. Anthropol. Paris,* **4** (Sér. 13): 327-346.

Kay, R. F., 1977, The evolution of molar occlusion in the Cercopithecidae and early catarrhines, *Am. J. Phys. Anthropol.* **46**:327-352.

Kay, R. F. and Hiiemae, K. M., 1974, Jaw movement and tooth use in recent and fossil primates, *Am. J. Phys. Anthropol.* **40**:227-256.

Kinzey, W. G., 1972, Canine teeth of the monkey, *Callicebus moloch:* Lack of sexual dimorphism, *Primates* **13**:365-369.

Maier, W., 1977, Die bilophodonten molaren der Indriidae (Primates)—ein evolutionsmorphologischer modellfall, *Z. Morphol. Anthropol.* **68**:307-344.

Osborn, J. W., 1978, Morphogenetic gradients: Fields versus clones, in: *Development, Function and Evolution of Teeth* (P. M. Butler and K. A. Joysey, eds.), pp. 171-202, Academic Press, New York.

Remane, A., 1960, Zahne and Gebeiss, in: *Primatologia* **III**(2) (H. Hofer, A. H. Schultz, and D. Stark, eds.), pp. 637-846.

Rosenberger, A., 1977, *Xenothrix* and ceboid phylogeny, *J. Hum. Evol.* **6**:461-481.

Schwartz, J. H., Tattersall, I. and Eldridge, N., 1978, Phylogeny and classification of the primates revisited, *Yearb. Phys. Anthropol.* **22**:95-133.

Simons, E. L., 1961, The dentition of *Ourayia:* Its bearing on relationships of omomyid prosimians, *Postilla* **54**:1-20.

Simons, E. L., 1972, *Primate Evolution: An Introduction to Man's Place in Nature,* Macmillan, New York.

Simons, E. L., 1976, The fossil record of primate phylogeny, in: *Molecular Anthropology* (M. Goodman and R. Tashian, eds.), pp. 35-62, Plenum Press, New York.

Simpson, G. G., 1955, The Phenacolemuridae, new family of early primates, *Bull. Am. Mus. Nat. Hist.* **105**(5):415-441.

Stehlin, H. G., 1916, Die sügetiere des schweizerischen Eocänes: Kritischer Katalog der materialen, *Überhandlungen Schweizerishen Palaöntol. Ges.* **38**:1297-1552.

Szalay, F. S., 1975, Phylogeny of higher primate taxa: The basicranial evidence, in: *Phylogeny of the Primates* (P. Luckett and F. S. Szalay, eds.), pp. 91–126, Plenum Press, New York.

Szalay, F. S., 1976, Systematics of the Omomyidae (Tarsiiformes, Primates) taxonomy, phylogeny, and adaptations, *Bull. Am. Nat. Hist.* **156**(3):157–450.

Tarrant, L. H., and Swindler, D. R., 1973, Prenatal dental development in the black howler monkey (*Alouatta caraya*), *Am. J. Phys. Anthropol.* **38**:255–260.

Wortman, J. L., 1904, Studies of Eocene Mammalia in the Marsh collection, Peabody Museum. Part II. Primates, Suborder Anthropoidea, *Am. J. Sci.* **17**:204–250.

Zingeser, M. R., 1969, Cercopithecoid canine tooth honing mechanisms, *Am. J. Phys. Anthropol.* **31**:285–214.

Zingeser, M. R., 1971, The prevalence of canine tooth honing in primates (abstract), *Am. J. Phys. Anthropol.* **25**:300.

Dental Evolutionary Trends of Relevance to the Origin and Dispersion of the Platyrrhine Monkeys

9

F. J. ORLOSKY

Introduction

The evolutionary history of the platyrrhine primates has been the subject of considerable controversy. Comparative studies of extant Old and New World higher primates suggest that these forms and the platyrrhines are each monophyletic, but the nature and origin of common ancestral groups remains in doubt. It has been hypothesized that the Platyrrhini evolved from a lemuroid stock (Liedy, 1873; Loomis, 1911; Gregory, 1920, 1921; Gingerich, 1973, 1975, 1977), a tarsioid stock (Wortman, 1904; Gazin, 1958; Simons, 1961, 1972; Hoffstetter, 1969, 1974; Orlosky and Swindler, 1975; Szalay, 1976) or a monkey stock (Stirton, 1951; Olson, 1964; Hershkovitz, 1969; Sarich, 1970; Cronin and Sarich, 1975). Most of the above authors assumed that the South American primates were derived from some North American or Central American primate group. However, some have hypothesized that

F. J. ORLOSKY • Department of Anthropology, Northern Illinois University, DeKalb, Illinois 60115.

an African stock might have been ancestral to the platyrrhine primates (Sarich, 1970; Hoffstetter, 1972, 1974; Lavocat, 1974; Cronin and Sarich, 1975). Indeed, recent geological evidence (see Tarling, this volume) indicates that an immigration of primates into South America from Africa would be quite as feasible during the late Eocene to early Oligocene as a North American origin for these primates.

Problems of the phyletic history and the question of continental origins of the New World monkeys have been perpetuated by the limited late Paleogene primate fossils of Africa and South America. A reasonable understanding of what primates were distributed to Africa during the late Eocene and early Oligocene and a better awareness of early Oligocene South American primates would provide a superior basis on which to address the above controversies. For example, it is currently impossible to determine whether or not a fused mandibular symphysis is a primitive feature for the New World monkeys or a shared derived condition of all Neogene platyrrhines. Currently the earliest platyrrhine fossil evidence for such fusion is from the early Miocene. The discovery of a single complete primate mandible from the early Oligocene of South America would provide more conclusive basis for evaluating primitive conditions and probable ancestral stocks.

The present paper is a review of New World monkey relationships and of probable nonplatyrrhine affinities based on dental morphology. Interpretations of phyletic relationships within the platyrrhines are greatly aided by the use of dental features. The Ceboidea constitute a morphologically diverse superfamily of primates. The dental diversity of the ceboids allows for relatively easy distinctions between extant forms and for assessment of extinct ceboid relationships. In this paper, however, phyletic relationships within the platyrrhines will be discussed only to the extent that they reflect on general New World monkey dental trends. This paper is primarily an attempt to present a probable primitive morphotype for platyrrhine dentitions. The primitive model was constructed by ignoring obvious and probable derived dental conditions and retaining probable primitive features. The proposed morphotype allows for the derivation of all platyrrhine dental specializations in the most parsimonious manner. The limitation of such a dental model is that it might represent a primitive morphotype for a more recent common ancestor than were the original South American primates, i.e., the primitive morphotype for early Oligocene and Neogene South American primates might well represent a higher number of derived dental features compared to the first immigrants.

It is assumed in this paper that dental morphology evidenced by the earliest primates in South America provides a more probable assessment of primitive states than would a model inferred solely from extant forms. If several morphoclines for particular features are equally plausible, the hypothesized primitive morphology that best fits known extinct forms is assumed to indicate the most probable polarity. It is also assumed that morphological states observed in earlier members of a lineage have a higher probability of reflecting primitive conditions. Although fossil specimens might represent unique derived adaptations, these are necessary assumptions in light of the sparsity of fossil specimens and the extreme diversity in dental form exhibited by the extant platyrrhines.

Extinct Ceboid Dentitions

A traditional classification done to subfamily is presented in Fig. 1 in order to facilitate discussion of the various ceboids. The extinct taxa (asterisked) are included within the appropriate taxa as well as their approximate age. The classification follows Hershkovitz (1977) except for extinct ceboids and the classification of *Callimico*. Also, extinct taxa that are not relevant to the current discussions are omitted.

The Oligocene through Miocene fossil evidence for ceboid evolution is essentially limited to seven genera. The late Miocene ceboid fossil record consists of three monospecific genera. Since each is known from one individual, information regarding intrageneric variability is lacking. *Cebupithecia sarmientoi* clearly indicates an incipient pithecine dental pattern consisting of: procumbent incisors, buccally splayed upper canines, C_1 with a distinct triangular basal shape, and premolars and molars with low cusps and shallow occlusal basins. These dental features are shared derived traits in common with the Pithecinae. Possible primitive features include: lingual cingula more

Superfamily—Ceboidea
- Family—Callitrichidae
  - Subfamily—Callitrichinae
  - Subfamily—Callimiconinae
- Family—Cebidae
  - Subfamily—Aotinae
  - Subfamily—Cebinae
  - Subfamily—Pitheciinae
    - **Cebupithecia sarmientoi* (Late Miocene)
  - Subfamily—Saimiriinae
    - **Neosaimiri fieldsi* (Late Miocene)
  - Subfamily—Alouattinae
    - **Stirtonia tatacoensis* (Late Miocene)
  - Subfamily—Tremacebinae*
    - **Tremacebus harringtoni* (Late Oligocene) (=*Homunculus harringtoni*)
  - Subfamily—Homunculinae*
    - **Homunculus patagonicus* (Early Miocene) (=*Homunculus ameghinoi*)
    - **Dolichocebus gaimanensis* (Late Oligocene)
- Family—Uncertain
  - **Branisella boliviana* (Early Oligocene)

Fig. 1. Classification of extinct ceboids (*extinct forms).

pronounced on P^4 and molars, smaller distal fossae on P_3 and P_4, and hypocones located slightly more distolingual to protocones on upper molars.

Neosaimiri fieldsi is known from a single mandibular specimen with all teeth except I_1 and M_3 retained. The mandibular teeth closely resemble those of *Saimiri sciureus,* a relationship recognized by earlier investigators (Stirton, 1951; Hershkovitz, 1970). Possible primitive features include: more closely approximated molar cusps and more strongly developed molar cingula, I_2 absolutely and relatively smaller in mesiodistal and buccolingual diameters, I_2 with more pronounced mesial marginal elevation, and I_2 with greater labial inclination.

Stirtonia tatacoensis is known from one mandibular specimen with the preserved crowns of the canine, P_3, P_4, M_1, and right M_2 as well as from one isolated I_2 and two additional molars. The specimens share the following derived features with *Alouatta* spp.: premolars with oblique occlusal outline, M_1 and M_2 with buccal cusps located mesiobuccally to corresponding lingual cusps, molars with trigonid buccolingual diameters smaller than talonid breadths, and molars with the same hypocristid patterns as respective *Alouatta* molars. Derived characters, perhaps primitive for the Alouattinae, distinguish *S. tatacoensis* from *Alouatta* spp. General ceboid primitive conditions may be represented in incisor size which is absolutely and relatively smaller in mesiodistal diameter than in extant forms. *S. tatacoensis* probably represents a specialized folivore dentition, and it may represent a form that had reduced incisor size beyond that of extant Alouattinae. Alternatively, the more constricted I_2, also indicated by the alveoli of the mandibular specimen, can be considered a retention of a primitive feature. Primitive conditions might also be indicated by the presence of small paraconids on the premolars and the presence of lingually open premolar trigonids.

There is little dental evidence representing the late Oligocene and early Miocene ceboids. *Dolichocebus gaimanensis* is known from one nondentulous cranial specimen of upper Oligocene age (Kraglievich, 1951). The dentition of *Tremacebus harringtoni* is known from three worn and chipped maxillary molars of one specimen (Hershkovitz, 1974). Hypocones are well developed on M^1 and M^2. A lingual cingulum is present on these molars and is continuous with the apex of the hypocone and the distal marginal ridge. M^3 is reduced in size comparable to modern cebids and the root of P^2 indicates an enlarged tooth with modern cebid proportions.

The dentition of early Miocene *Homunculus patagonicus* is best known from P^2 and M^3 preserved in a facial fragment and a mandibular specimen with the worn crowns of all teeth represented except C_1. P^2 is bicuspid with a well-developed protocone and M^3 is reduced in size but of unique shape. Features of both preserved maxillary teeth appear to be derived. The mandibular dentition indicates the following possibly primitive features: I_2 slightly larger than I_1, I_2 positioned distolingual to I_1, incisors labially inclined, incisors relatively smaller than more recent cebids, premolars with lingually open trigonids, and molars without constricted trigonids.

The early Oligocene primates are known mostly from one specimen, *Branisella boliviana* (Hoffstetter, 1969). The preserved single root of P^2 indicates a somewhat small tooth that was expanded buccolingually and probably unicuspid. The root of P^3 implies a greater mesiodistal compression than P^4. The fourth premolar is bicuspid with a prominent lingual cingulum as well as parastyles and mesostyles. The molars have the following: prominent trigons with subequal paracones and metacones and prominent protocones, small parastyles and metastyles connected with major cusps by crests, low but well-developed hypocones continuous with broad lingual cingula, and low poorly developed metaconules. Many of these dental features can be considered to be primitive for the platyrrhines.

Primitive Platyrrhine Dental Features

The following discussions include inferred primitive platyrrhine dental features for each tooth group (see also Orlosky, 1973).

Incisors

Most extant platyrrhines (12 genera) have heteromorphic upper incisors with caniniform lateral incisors. In 5 genera, upper incisors are homomorphic caniniform. Upper central incisors are generally larger than lateral incisors (except *Alouatta villosa*). The mean mesiodistal ratios of I^2 to I^1 ranges from 102% for *A. villosa* through 83% for *Saimiri* to 64% for *Aotus*. Most extant platyrrhines (10 genera) have heteromorphic lower incisors with caniniform lateral incisors. Four genera have homomorphic caniniform incisors, whereas 2 genera are homomorphic incisiform. In all platyrrhines, I_2 is greater or equal to I_1 in crown diameters. The mean mesiodistal ratios of I_1 to I_2 ranges from 77% in *Cebus* to 82% in *Saimiri* to about 100% for the Callitrichidae.

Incisors are poorly represented in the platyrrhine fossil record. Tentative trends indicated by comparing late Miocene cebids to related extant forms imply within-lineage increase in incisor size. The positioning, orientation, and relatively small incisors of *H. patagonicus* were mentioned previously. These features of *H. patagonicus* may be derived as labially inclined incisors as in *Cebuella, Callithrix,* and the Pitheciinae. The lack of other dental specializations and the relatively early occurrence of this form lends credence to the possibility that *H. patagonicus* represents a primitive condition from which later vertical and more procumbent incisor orientations were derived.

A probable incisor primitive morphotype is: I^1 slightly larger than I^2, upper and lower lateral incisors caniniform, upper and lower central incisors with primarily flat incisal edges, I_2 slightly larger than I_1, I_2 positioned distolingual to I_1, and lower incisors with moderate labial inclination.

Canines

In most platyrrhines, canines are well developed with apices extending at least minimally beyond the occlusal plane. Canines have become reduced in some forms such as *Callithrix* and greatly increased in size such as in *Cebus* and the Pitheciinae. Well-developed canine size is indicated in the known dentitions of extinct New World monkeys and at least moderately developed canines must be considered a primitive feature for the platyrrhines. It should be noted, however, that canine size sexual dimorphism, markedly present in *Saimiri, Cebus,* and *Alouatta,* is not common among the platyrrhines and cannot be considered as a primitive phenomenon.

Premolars

Most platyrrhines have primarily unicuspid second mandibular premolars and bicuspid P^{3-4}_{3-4}'s. In the maxillary premolars, transverse ridges typically connect the two cusps which are also connected by peripheral ridges. No maxillary premolars are molarized beyond the two-cusp state with the exception of *Alouatta.* In the maxillary premolars, the mesiodistal diameter is always much less than the buccolingual diameter, e.g., 64% to 78% for platyrrhine second premolars and, with increased distal width diameters, 53% to 67% for the fourth upper premolar. Ceboid upper second premolars are equal to or greater than third premolars in length and width.

Platyrrhine second lower premolars are equal to or greater than P_3 in mesiodistal diameter and approximately equal in buccolingual dimensions. P_3 and P_4 typically have the mesial fossa closed lingually by crests passing from the metaconids to the mesial marginal ridges. Although P_4 is primarily bicuspid in extant ceboids, low distal cusps are frequently added in the cebids. Extreme molarization of the lower premolars does not occur among the platyrrhines.

Lower premolars of extinct platyrrhines conform to the above general morphology of extant forms except that *S. tatacoensis* and *H. patagonicus* exhibit lingually open mesial fossae on P_3 and P_4. The few known extinct platyrrhine maxillary premolars conform to the general bicuspid condition except for *B. boliviana.* This specimen deserves special comment since it has been assumed that P^2 of this species was extremely small, represents a condition where second premolars were in the process of disappearing, and that *B. boliviana,* therefore, could not represent an ancestor to more recent platyrrhines (Szalay, 1976; Hershkovitz, 1977). These evaluations are apparently based on the comparison of P^2 root dimensions with the basal crown diameters of P^3 presented by Hoffstetter (1969). The diameters of the second premolar were taken some distance from the crown of the tooth. The mesial portion of the P^2 alveolus is not preserved and estimates of mesiodistal cervical root size cannot be made. The actual crown of P^2 might actually have been considerably larger than currently believed. *B. boliviana* does indicate a P^2 that was expanded buccolingually as in modern platyrrhines, perhaps one half to two thirds the size of P^3 but not extremely reduced, and probably unicuspid.

The inferred primitive platyrrhine premolar morphology is: buccolingually expanded maxillary premolars, unicuspid P^2_2 smaller than P^3_3 but not greatly reduced, and P^{3-4}_{3-4} bicuspid without extensive molarization.

Molars

All cebids exhibit reduced M^3 diameters relative to M^2. With the exception of *Alouatta villosa,* all cebids have reduced M_3 diameters. Molar reduction is also present in the Callimiconinae and, in its extreme form of M^3_3 agenesis, in the Callitrichinae. The primitive morphotype must, of course, include three molars.

Maxillary molars typically have well-developed hypocones in the Cebidae, but these are usually absent or poorly developed off of the lingual cingulum in the Callitrichidae. In all platyrrhines, buccolingual molar diameters exceed mesiodistal dimensions. Mesiodistal to buccolingual diameter ratios of M^2 vary from about 60% for *Saguinus* through 71% for *Saimiri* to 90% for *Alouatta.*

Mandibular molars of platyrrhines are four cusped. Protoconids and metaconids are most frequently situated opposite one another and are connected by transverse crests. Trigonid widths very frequently are not constricted. With the exception of *Brachyteles arachnoides,* paraconids are not present and the mesial fossae are lingually closed.

Maxillary molars are poorly represented in the cebid fossil record. The molars of *T. harringtoni* are of modern form comparable to *Aotus* or *Callicebus*. The molars of *B. boliviana* resemble those of *Saimiri* except for the presence of small metaconules. The moderate stylar development on *B. boliviana* molars provides a plausible primitive morphotype from which enlarged and reduced style and buccal cingula conditions could be derived in various platyrrhines (Kinzey, 1973, 1974). Like *Saimiri, B. boliviana* has a low hypocone developed on the lingual cingulum and positioned distolingual to the protocone. Primitive maxillary molar platyrrhine morphology is best represented by *B. boliviana.* Previous protoceboid models have concluded that the ancestral molar condition resembled that of *Aotus* and *Callicebus* and that more advanced molar patterns evolved from such a "primitive" type (Gregory, 1920; Kinzey, 1974). The *Branisella-Saimiri* model, however, provides a more parsimonius explanation for the derived molar patterns of the Callitrichidae and the Cebidae with the reduction of the small hypocone in the former and enlargement in the latter.

Lower molars of extinct cebids conform to the most frequent general features of extant cebids. Except for the derived *Alouatta*-like *Stirtonia* specimen, trigonids are not markedly constricted, paraconids and hypoconulids are absent, and mesial fossae are lingually closed. These general features seem to be primitive for platyrrhine lower molars.

Nonplatyrrhine Comparisons

The absence of pre-Oligocene primates in South America suggests that the original immigration occurred no earlier than the middle Eocene. There are no known African primates from the Eocene. Comparisons with the African Oligocene Fayum primates are possible. The earliest of these, *Oligopithecus savagei* possesses but two lower premolars and P_3 is specialized into a quasisectorial tooth (Simons, 1962). *O. savagei* is obviously too specialized to provide an ancestral model for the ceboids, but its generalized molars indicate the possibility of common ancestry with the platyrrhines. The more recent *Apidium* spp. and *Parapithecus* spp. share a common dental formula with the platyrrhines, but they exhibit advanced dental features compared to the primitive ceboid morphotype. Both exhibit large hypocones and well-developed conules on upper molars. *Apidium* also has advanced P^2 morphology unequalled by most extant cebids and prominent centroconids of lower molars. Some similarities in dental morphology are observed when comparing these genera to selected extant cebids (Simons, 1972; Delson and Andrews, 1975), but these are indicative of parallel evolution rather than close phyletic relationships. The remaining Fayum primates are obviously advanced catarrhines and need not be discussed in this paper.

Since it is assumed that the earliest ceboids were derived from some middle to late Eocene primate, it is logical to compare the proposed prototype with the known possible ancestral groups, the omomyids and adapids of this time period. Detailed comparisons with all adapids and omomyids are beyond the scope of this paper, and the following comments cannot encompass all of the variations within each of these taxa.

The adapids are characterized as having vertically oriented, spatulate incisors (Simons, 1972; Gingerich, 1975, 1977). The degree of adapid I_2 incisal edge flattening discussed by Gingerich exceeds that proposed in the primitive platyrrhine morphotype and observed in extant ceboids. Also, the vertical lower incisor orientation and relative positioning are more advanced than proposed by the ceboid prototype.

As with extinct ceboids, incisors are not well known for the omomyids. It is implied by Gingerich (1977) that omomyids do not possess spatulate, vertically oriented incisors, and many omomyids do have narrow, procumbent, caniniform incisors (Szalay, 1976). Spatulate tendencies are described for *Arapahovius gazini* (Savage and Waters, 1978) and *Ourayia uintensis* (Simons, 1961). No known omomyid seems to approach the adapids in the vertical positioning of incisors, but some omomyids approached the platyrrhine morphotype more closely than is currently assumed.

The canines of most adapids appear to be medium to large in size and extend beyond the occlusal plane (Simons, 1972; Gingerich, 1975). In some genera, canine hypertrophy exceeds that of modern ceboids but adapids closely approximate the platyrrhine prototype in canine size. In contrast, most omomyids have canines that are relatively small and do not project beyond the occlusal plane. The exception seems to be *Omomys carteri* (Szalay, 1976). In this respect, the omomyids do not conform to the proposed ceboid morphotype.

The premolars and molars of the adapids are varied but generally exhibit greater molarization than the omomyids. The North American forms appear to be particularly specialized. The notharctines exhibit twinning of P^4 protocones, the development of hypocones off of the protocone, and specialized maxillary molars which superficially resemble *Aloutta* spp. The recently described *Mahgarita stevensi,* which resembles European adapids, exhibits a molarized P_4, a greatly expanded buccal half of P^4, extreme reduction of P^2_2, and the development of square rather than mesiodistally compressed maxillary molars (Wilson and Szalay, 1976). In these features, this genus does not conform to the platyrrhine morphotype. The lower molars of *M. stevensi* do exhibit advanced, ceboid features. European adapids in general have advanced square molars with large hypocones, a frequent emphasis on lower molar buccal crest development and at least some exhibit advanced molarization of P^4_4.

The premolars and molars of the omomyids [tribes: Omomyini and Washakini, Szalay (1976)] are more conservative than those of the adapids. Third and fourth maxillary premolars are most frequently bicuspid and lower premolars generally are not molarized beyond that expected in the platyrrhine morphotype. Second premolars, however, are relatively small and P^2 most frequently is not buccolingually expanded. The exception seems to be *Washakius insignis* as indicated by its P^2 alveolus (Gazin, 1958; Szalay, 1976). Most omomyids have retained mesiodistally compressed maxillary molars with low hypocones but with conule development exceeding that of hypothetical early platyrrhines. Lower molars most frequently have constricted trigonids that support paraconids and are lingually open. These are primitive molar features that do not conform to the slightly more advanced platyrrhine primitive morphotype.

Neither the adapids nor omomyids of the middle to late Eocene wholly conform to the proposed early Oligocene platyrrhine morphotype. It would appear that more evolutionary reversals would be required to derive such South American primates from the adapids. In the balance, therefore, the omomyids appear to be the more likely ancestral group. This may simply reflect the retention of more primitive features in omomyid dentitions. Indeed, it does not seem possible to demonstrate shared derived dental features for the omomyids and ceboids. On the other hand, such a generalized omomyid dentition appears to be the likely precursor to ceboid dentitions.

Continental Origins

The possible continental origins of the ancestral platyrrhines are from either North America or Africa. If the omomyid dental pattern most closely approximates early platyrrhine dental features, ancestry from these numerous and varied North American primates favors origin from that continent. Affinities with the known North American adapids seem unlikely. Although perhaps less probable than an African origin on the basis of continental positioning, a chance origin from North American omomyids seems quite possible.

The alternative possibility of an African origin necessitates the hypothesis of a broad primate distribution in this continent during the Eocene. If such primates were not widely distributed in Africa, the relative positions of Africa and South America would be meaningless. Since there is no fossil record for such primates, one could hypothesize primates approaching the early platyrrhine dental prototype as being derived from either early adapids or omomyids of Eurasian origin. It is possible that such primates did exist but, in the absence of any fossil evidence, possible African origins remain speculative.

Based on the existing empirical evidence, the chance origin of the platyrrhines from the North American omomyids still seems to be the most parsimonious explanation for the origins of the South American primates. Discoveries of new Oligocene South American fossils indicating that *B. boliviana* does not reflect primitive platyrrhine features or of mid-Paleogene African primates would necessitate revisions of the above statements.

ACKNOWLEDGMENTS

I would like to thank Dr. Daris R. Swindler for permitting me to examine his New World monkey dental cast collection which served as the basis for much of this study. I am grateful to Philip Hershkovitz for his permission to examine fossil and extant specimens stored at the Chicago Field Museum. I am also grateful to Dr. Robert Hoffstetter for providing me with a cast of *Branisella boliviana*.

References

Cronin, J. E. and Sarich, V. M., 1975, Molecular systematics of the New World monkeys, *J. Hum. Evol.* **4**:357–375.

Delson, E., and Andrews, P., 1975, Evolution and interrelationships of the catarrhine primates, in: *Phylogeny of the Primates* (W. P. Luckett, and F. S. Szalay, eds.), pp. 405–446, Plenum Press, New York.

Gazin, C. L., 1958, A review of the middle and upper Eocene primates of North America, *Smithsonian Misc. Coll.* **136**:1–112.

Gingerich, P. D., 1973, Anatomy of the temporal bone in the Oligocene anthropoid *Apidium* and the origin of Anthropoidea, *Folia Primatol.* **19**:329–337.

Gingerich, P. D., 1975, Dentition of *Adapis parisiensis* and the evolution of lemuriform primates, in: *Lemur Biology* (I. Tattersall, and R. W. Sussman, eds.), pp. 65–80, Plenum Press, New York.

Gingerich, P. D., 1977, New species of Eocene primates and the phylogeny of European Adapidae, *Folia Primatol.* **28**:60–80.

Gregory, W. K., 1920, Evolution of the human dentition. Part IV. The South American monkeys, *J. Dent. Res.* **2**:404–426.

Gregory, W. K., 1921, On the structure and relationships of *Notharctus,* an American Eocene primate, *Mem. Am. Mus. Nat. His.* **3**:49–243.

Hershkovitz, P., 1969, The evolution of mammals on Southern continents. VI. The Recent mammals of the Neotropical Region: A zoogeographic and ecological review, *Q. Rev. Biol.* **44**:1–70.

Hershkovitz, P., 1970. Notes on Tertiary platyrrhine monkeys and description of a new genus from the late Miocene of Colombia, *Folia Primatol.* **12**:1–37.

Hershkovitz, P., 1974, A new genus of late Oligocene monkey (Cebidae, Platyrrhini) with notes on postorbital closure and platyrrhine evolution, *Folia Primatol.* **21**:1–35.

Hershkovitz, P., 1977, *Living New World Monkeys (Platyrrhini) with an Introduction to Primates,* Vol. 1, University of Chicago Press, Chicago.

Hoffstetter, M. R., 1969, Un primate de l'Oligocène inférieur sud-américain: *Branisella boliviana* gen. et sp. nov., *C. R. Acad. Sci. Paris, Ser. D* **269**:434–437.

Hoffstetter, R., 1972, Relationships, origins, and history of the ceboid monkeys and caviomorph rodents: A modern reinterpretation, in: *Evolutionary Biology,* Vol. 6 (Th. Dobzhansky, M. K. Hecht, and W. C. Steere, eds.), pp. 323–347, Appleton-Century-Crofts, New York.

Hoffstetter, M. R., 1974, Phylogeny and geographical deployment of the primates, *J. Hum. Evol.* **3**:327–350.

Kinzey, W. G., 1973, Reduction of the cingulum in Ceboidea, in: *Craniofacial Biology of Primates,* Vol. 3 (M. A. Zingeser, ed.), pp. 101–127, Karger, Basel.

Kinzey, W. G., 1974, Ceboid models for the evolution of hominoid dentition, *J. Hum. Evol.* **3**:193–203.

Kraglievich, J. L., 1951, Contribuciones al conocimiento de los primates fósiles de la Patagonia. 1. Diagnosis previa de un nuevo primate fósil del Oligocene superior (Colhuehuapiano) de Gaimán, Chubut, *Com. Inst. Nac. Invest. Cienc. Nat. (Buenos Aires), Cienc. Zool.* **11**:55–82.

Lavocat, R., 1974, The interrelationships between the African and South American rodents and their bearing on the problem of the origin of South American monkeys, *J. Hum. Evol.* **3**:323–326.

Liedy, J., 1873, Contributions to the extinct vertebrate fauna of the western territories, *Report of U.S. Geological Survey Territory (Hayden)* **1**:1–358.

Loomis, F. B., 1911, The adaptations of the primates, *Am. Nat.* **45**:479–492.

Olson, E. C., 1964, The geology and mammalian faunas of the Tertiary and Pleistocene of South America, *Am. J. Phys. Anthropol.* **22**:217–226.

Orlosky, F. J., 1973, Comparative dental morphology of extant and extinct Cebidae, Ph.D. Thesis, University of Washington.

Orlosky, F. J., and Swindler, D. R., 1975, Origins of New World monkeys, *J. Hum. Evol.* **4**:77–83.

Sarich, V. M., 1970, Primate systematics with special reference to Old World monkeys, in: *Old World Monkeys: Evolution, Systematics, and Behavior* (J. R. Napier and P. H. Napier, eds.), pp. 175–226, Academic Press, New York.

Savage, D. E., and Waters, B. T. 1978, A new omomyid primate from the Wasatch formation of southern Wyoming, *Folia Primatol.* **30**:1–29.

Simons, E. L., 1961, The dentition of *Ourayia:* Its bearing on relationships of omomyid prosimians, *Postilla* **No. 54**:1–20.

Simons, E. L., 1962, Two new primate species from the African Oligocene, *Postilla* **No. 64**:1–12.

Simons, E. L., 1972, *Primate Evolution: An Introduction to Man's Place in Nature,* Macmillan, New York.

Stirton, R. A., 1951, Ceboid monkeys from the Miocene of Colombia, *Bull. Univ. Calif. Publ. Geol. Sci.* **28**:315–356.

Szalay, F. S., 1976, Systematics of the Omomyidae (Tarsiiformes, Primates) taxonomy, phylogeny and adaptations, *Bull. Am. Mus. of Nat. Hist.* **156**:157–450.

Wilson, J. A., and F. S. Szalay, 1976, New adapid primate of European affinities from Texas, *Folia Primatol.* **25**:294–312.

Wortman, J. L., 1903–1904, Studies of Eocene Mammalia in the Marsh Collection, Peabody Museum. Part II: Primates, *Am. J. Sci.* **15**:163–176, 399–414, 419–436, and **17**:23–33, 133–140, 203–214.

Implications of Enamel Prism Patterns for the Origin of the New World Monkeys

10

D. G. GANTT

Introduction

The origin and dispersal of the New World monkeys to the South American continent present an interesting dilemma, for paleogeographical evidence indicates that South America had already separated from Africa by the end of the Cretaceous. During the Tertiary, South America remained an island continent isolated from the rest of the world by oceanic barriers, from the Middle Paleocene [ca. 55 million years (m.y.) ago] to the Middle Pliocene (ca. 3.5 m.y. ago) (Hershkovitz, 1977; Hoffsteffer, 1972, 1974; Orlosky and Swindler, 1975). The source of origin of the New World monkeys both from a biogeographic and a phylogenetic perspective is thus uncertain and controversial. Two main hypotheses have been proposed:

1. The Platyrrhini as well as the Catarrhini were independently derived from Laurasian Paleogene primates, presumably an omomyid lower primate. Therefore, the close resemblance of New and Old World monkeys can be interpreted or explained as a result of convergence and/or parallelism from a lower primate ancestor.

D. G. GANTT • Department of Anthropology, Florida State University, Tallahassee, Florida 32306.

2. The Platyrrhini and the Catarrhini were derived from a common ancestral anthropoid stock probably African in origin with the dispersal of the platyrrhine ancestors occurring through direct faunal interchange (rafting) between the southern continents across the South Atlantic Ocean.

The two extant platyrrhine families are vastly different in morphology and behavior, especially when compared to the Old World monkeys. To complicate matters further, the fossil evidence of the Platyrrhini is exceedingly fragmentary and rare (Rosenberger, 1977, 1978*a*). Therefore, in order to shed new light on the problem of New World monkey origins, new techniques and methods of analysis are needed to aid in the assessment of the present paleomorphological and odontometric data.

Materials and Methods

Recently, comparative studies of dental histology have documented the potential of using structural and microstructural features of enamel and dentin as possible taxonomic indicators (Boyde, 1971; Gantt, 1977, 1979*b,c;* Lavelle *et al.,* 1977). Dental tissues of certain species are sufficiently distinct to be readily recognized and allow an unknown specimen or an isolated fossil specimen to be assigned to a particular taxon. Tooth structure, especially that of the enamel, *i.e.,* the enamel prism, would certainly be of great value as a taxonomic indicator when the material is too fragmentary or worn to otherwise be reliably described.

An attempt was made to apply these recent advances to the problem of the origin of the New World monkeys. Molars and premolars of extant Ceboidea were obtained for this study. In addition, comparisons of the enamel prism patterns were made to the patterns obtained from specimens of extant and extinct Hominoidea, Cercopithecoidea, Lemuriformes, and Tarsiiformes (see Table 1).

The specimens were prepared for the scanning electron microscope (SEM) and the following procedures were applied: (1) a few specimens were split by compression resulting in natural fracture planes, mainly in the longitudinal plane; (2) several specimens were sectioned both longitudinally and horizontally—the cut faces were then polished and etched with N hydrochloric acid (HCl) for 10–20 sec and then washed in water; (3) natural enamel surfaces were studied in a few species in which unerupted teeth were obtained; and finally, (4) the majority of specimens were studied by placing a highly polished facet on the mid-lateral crown which was then etched with a 0.074 M solution of phosphoric acid (H_3PO_4) for 60 sec and then washed in water [see Gantt (1979*a*) and Boyde *et al.* (1978) for further details]. The fractured surfaces and etched surfaces were used to study the enamel prisms in the longitudinal and transverse planes. All specimens were soaked in acetone and ethanol, allowed to air dry, and were then mounted on stubs and coated with gold/palladium by evaporation. They were then examined with a

Table 1. Primate Genera Used in This Study

Genera	Number of individuals	Number of molars and premolars
Cebidae		
Alouatta	3	10
Ateles	2	5
Aotus	1	3
Cebus	4	10
Callicebus	2	4
Saimiri	1	2
Callitrichidae		
Callithrix	3	5
Cebuella	7	20
Saguinus	1	2
Cercopithecidae		
Papio	5	20
Macaca	5	20
Pongidae		
Pongo	5	10
"Lower" Primates		
Lemur	2	5
Necrolemur	1	2
Adapis	2	4
Notharctus	4	10
Smilodectes	1	2

Cambridge S-4 scanning electron microscope at 10 and 15 kV in the secondary emission mode.

The Enamel Prism

The basic unit of the enamel covering of the tooth is the prism or rod. The enamel prism consists of hydroxyapatite crystals in an ordered arrangement (see Figs. 1 and 2). The prisms are a product of the ameloblasts and run obliquely from the dentinoenamel junction to the surface of the tooth. Studies of the structure of developing mammalian dental enamel by Boyde (1964, 1965, 1971, 1976) have documented that distinctive developmental prism patterns do exist (see Fig. 1). Boyde has documented at least three basic patterns and several variations on each pattern (see Fig. 1). However, the range and degree of variation is still an unsolved question.

Recently, Shellis and Poole (see Lavelle *et al.*, 1977, Chapter 5) have attempted to provide a survey of the calcified dental tissues of primates. Their study documented that the enamel prism patterns were an important character in assessing taxonomic affinities. Using Boyde's developmental patterns [only Patterns 1, 2B, and 3B (see Fig. 1) were considered], they were able to distinguish a number of primate species.

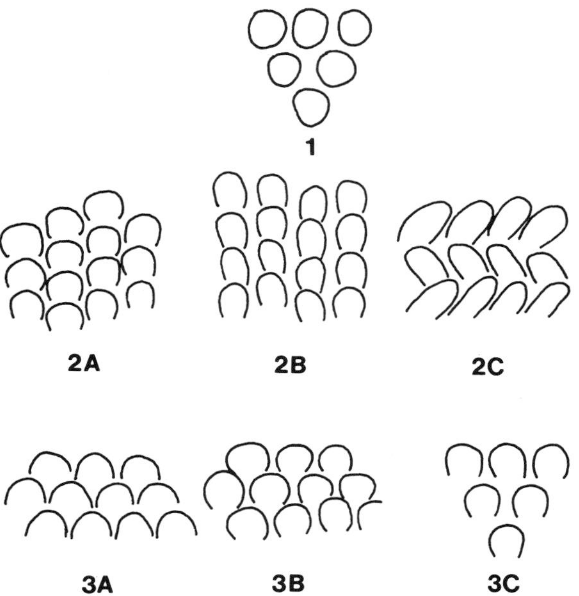

Fig. 1. The developmental patterns in enamel have been documented by Boyde (1964, 1965, and 1976). Pattern 1 prisms are most common in the Insectivora, Sirenia, and some of the Chiroptera. It is also found in some Lemuriformes and in human early cuspal enamel. In the developing enamel the Tomes' process (the secretory part of the ameloblast) lies in a pit in the floor of the developing enamel and forms the central portion of the prism pattern. In Pattern 1 prisms the pits are nearly straight cylinders with flat floors. The prisms are separated from one another by a continuous interprismatic region which formed the mutual walls of the pit at the formative surface and by the sides of adjacent Tomes' processes (Boyde, 1976, p. 341 and Figs. 7 and 10A). Pattern 1 prisms are the result of the ameloblast movement which is nearly perpendicular to the developing surface which causes the crystal orientation to be the same in the centers of the interprismatic and prismatic regions.

Pattern 2 prisms are found in most ungulates, marsupials, and rodents. In Pattern 2 prisms the ameloblasts are not perpendicular to the surface as they are in Pattern 1 prisms. The pits enter the surface obliquely so that the floor of a pit becomes tilted and forms the missing part of the wall (Boyde, 1976, p. 341 and Figs. 9 and 10B). In Pattern 2 prisms the floor is the cervical side of the pit with the ameloblast aligned in rows parallel with the longitudinal axis of the tooth. The crystallites in the floor of the pits grow parallel with the direction of movement of the ameloblast while the majority of the crystals in the inter-row sheets grow perpendicularly to the tops of the interpit walls which are in the general plane of the developing surface. In Pattern 2 prisms, where two ameloblasts make each prism, there are also two predominant crystal orientations—those in the prisms and those in the interprismatic sheets.

Pattern 3 prisms are most commonly found in human enamel being described as the "keyhole" enamel prism pattern (Gantt *et al.,* 1977). The Pattern 3 prism pits are inclined to the general plane of the ameloblast matrix. The pits are arranged in transverse rows, though the row axis is rotated in areas where the net cell movement has a lateral as well as cuspal component. The floor is the cervical side of the pit which merges with the interpit wall next to it instead of joining the next cervical wall as in Pattern 2 prisms (Boyde, 1976, p. 341 and Figs. 9 and 10C). There exists an extreme range of crystal orientations, but the orientation in the interprismatic (winged process or tail) regions of the prisms merges gradually into that in the prism body proper. The tail regions are what was the interpit wall and contain the crystals which grew mostly in relation to the tops of the interpit walls and thus normal to the general plane of the developing surface (Boyde, 1976, p. 342 and Fig. 15). The crystals in the prism body (head region) centers are parallel with the prism direction, so that there must be a span of orientations between the body and the tail at least as great as the angle of inclination of the prisms with respect to the developing surface. However, the crystals in the most cuspal side of the prism may diverge cuspally towards the next prism by some 5 or 10°; and those few which grow in the limited amount of new matrix added along the long cuspal wall of the pits (i.e., in the far cervical, fan-shaped part of the tails of the prisms) diverge cervically toward that surface, so that the total fan angle of orientations from

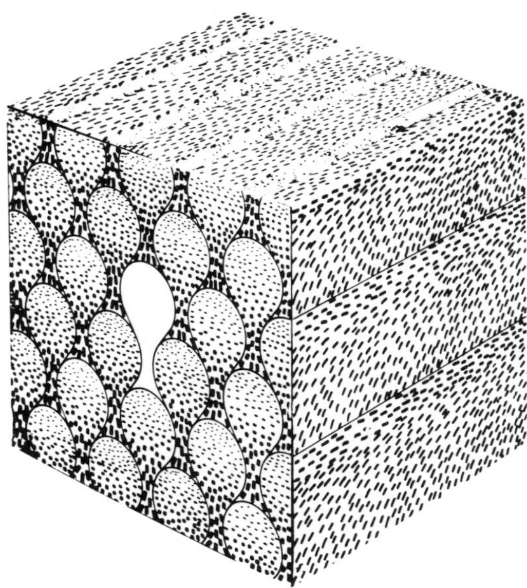

Fig. 2. A model of the enamel prism structure of human enamel showing the outline of the keyhole enamel prisms (developmental enamel prism Pattern 3B, see Fig. 1). The outline of keyholes on the front surface is a cut which is perpendicular to the direction of the prisms. The top view, which is a cut that is horizontal to the section, gives a banded appearance and is caused by cross-sectioned heads and tails which have different crystal arrangements. The side view, a vertical cut, shows the tilting of crystallites in the head to tail portions of the keyhole prisms [see Poole and Brooks (1961) for a more detailed explanation].

Prism Patterns of the Cebidae

The analysis of the enamel prism patterns of *Alouatta, Ateles, Cebus, Callicebus,* and *Aotus* documents that members of this group have a C-shaped prism pattern. The "heads" of the prisms were widely separated in *Ateles* and *Alouatta* by interprismatic enamel. The prism pattern in these forms more closely resembles Pattern 2A as is seen in *Papio* while *Cebus, Callicebus,* and *Aotus* have Pattern 2B prisms (see Figs. 3, 4, and 5).

---

top-of-head to bottom-of-tail of Pattern 3 prisms may exceed 60° (Boyde, 1976, p. 343, and see Fig. 2, this text). Each prism is contributed to by four ameloblasts.

Although, three basic patterns have been documented by Boyde, several variations have also been identified (see Fig. 1 and Boyde, 1964). The range and extent of variation has at present not been fully documented. In addition, there have been few attempts to correlate the prism patterns of the fully mineralized enamel surface to the developmental prism patterns documented by Boyde (1964, 1965, 1976).

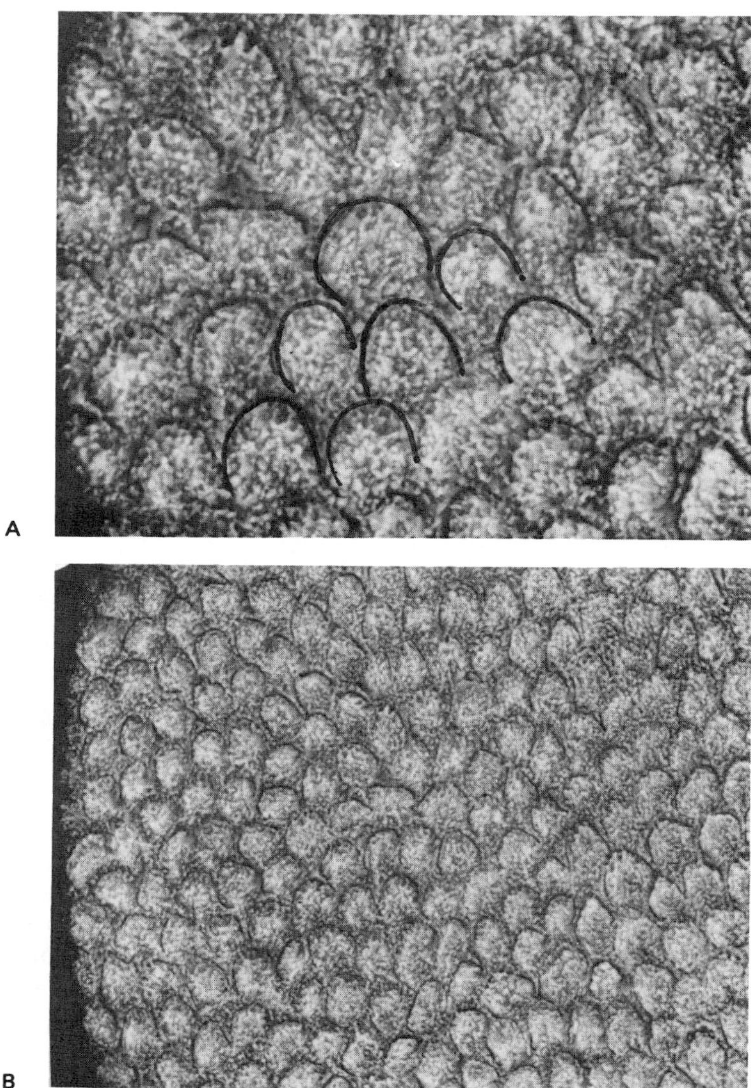

Fig. 3. Surface enamel prism patterns of an Old World monkey, *Papio*. The SEM microphotos were taken at 2000× (A) and 1000× (B) which documents an arcade-shaped prism pattern which is comparable to the developmental Pattern 2A (see Fig. 1).

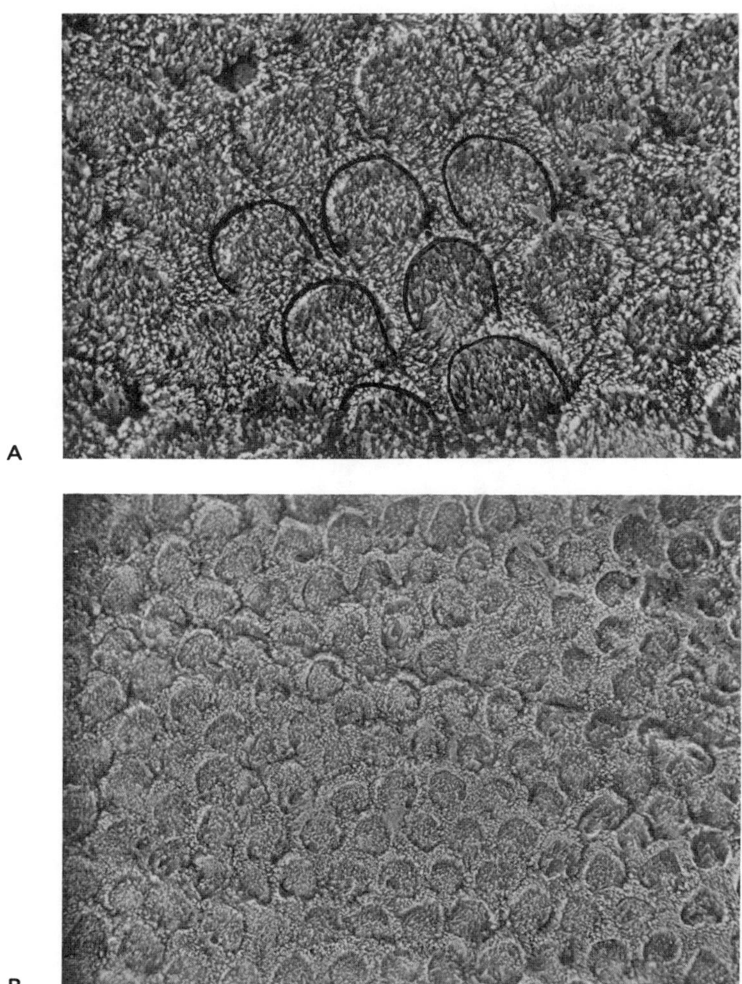

Fig. 4. Surface enamel prism patterns of *Alouatta*. The SEM microphotographs were taken at 2000× (A) and 1000× (B). These photos document that in *Alouatta* and *Ateles* the enamel prism pattern is C-shaped which is comparable to the developmental Pattern 2A (see Fig. 1).

A

B

Fig. 5. Surface enamel prism patterns of *Cebus*. The SEM microphotographs were taken at 2000× (A) and 1000× (B). These photos document that in *Cebus, Callicebus,* and *Aotus* the enamel prism pattern is C-shaped but differs from those observed in the larger cebids, *Alouatta* and *Ateles,* resembling developmental Pattern 2B (see Figs. 1, 4, and 10).

Of the cebids *Saimiri* most closely resembles the callitrichids. The enamel prisms were parallel throughout the thickness of the enamel and stood almost perpendicular to the surface. Shellis and Poole also noted that *Saimiri* appeared to be aligned closer to the callitrichids than to the cebids in a number of other dental characters (Lavelle *et al.*, 1977).

Prism Pattern of Callitrichidae

The analysis of several callitrichids, *Callithrix, Cebuella,* and *Saguinus,* documents that the prism pattern is circular or polygonal which resembles Pattern 1 of Boyde's classification (see Figs. 6 and 7). The prisms are separated by interprismatic regions about 1 μm thick. The prisms were largely straight and oriented at about 80° to the surface, but *Callithrix* did show some prism decussation (see Fig. 6; also see Fig. 58 of Lavelle *et al.*, 1977).

Also of interest is the evidence obtained from the study of the anterior dentition of *Cebuella.* Rosenberger (1978b) has reported the absence of enamel on the lingual surfaces of the anterior dentition in *Cebuella.* Rosenberger suggested that this loss of enamel was a specialization due to gum feeding. Analysis of several individuals' anterior dentition confirms that the lingual surface instead does have a covering of enamel a few micra thick which is rapidly worn away and is absent in adult animals (see Fig. 6).

Prism Patterns of Fossil Lower Primates and Their Relationship to the Ceboidea

An attempt was made to study as many fossil lower primates as were available. Two groups were evaluated, the Adapidae (*Adapis, Notharctus,* and *Smilodectes*) and the Tarsiidae (*Necrolemur*). The adapids are considered to share a tightly packed hexagonal-shaped prism which is similar to the developmental Pattern 1 of Boyde's scheme (see Fig. 8). The analysis of *Necrolemur* also revealed a prism pattern which is similar to Pattern 1; however, these prisms are circular to oval-shaped as in the callitrichids with wide interprismatic regions (see Figs. 9 and 6).

Summary and Conclusion

The analysis of the enamel prism patterns of the Ceboidea reveals a number of important relationships: (1) the prism patterns of the large cebids, *Ateles* and *Alouatta,* resemble those obtained for Old World monkeys, *Papio*

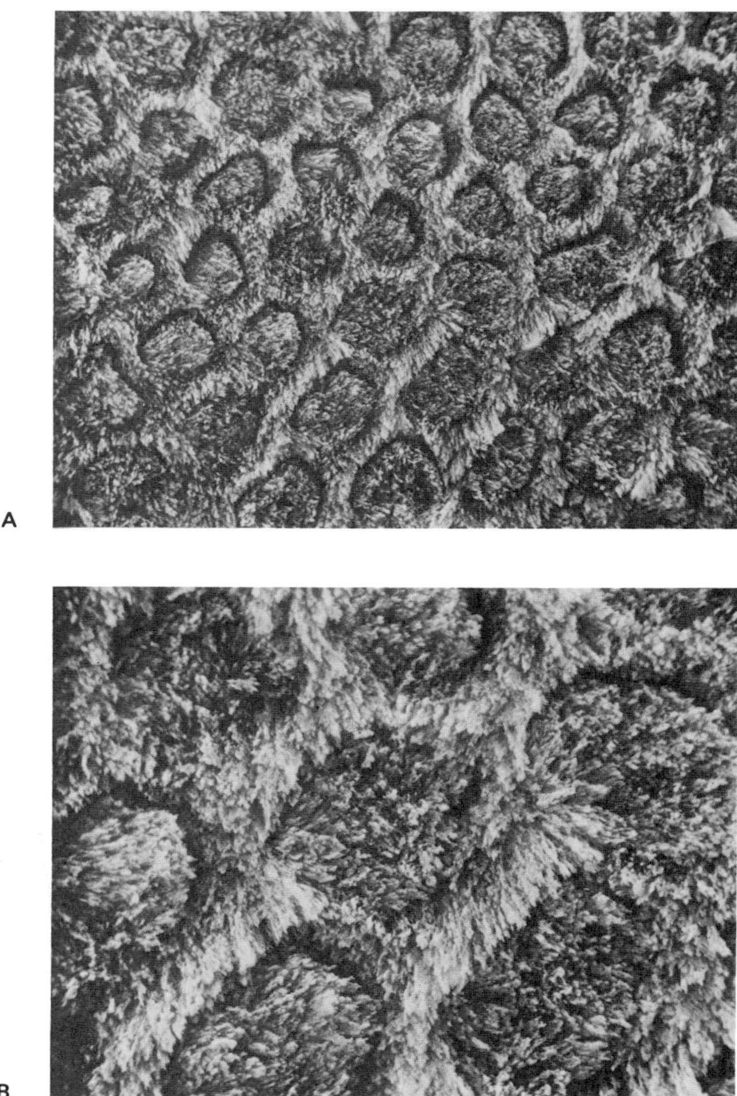

Fig. 6. Surface enamel prism patterns of *Cebuella*. The SEM microphotographs of longitudinal and transverse planes document that the prism pattern which is characteristic of the callitrichids is that of a circular or polygonal-shaped prism with wide interprismatic regions. This pattern is

and *Macaca:* developmental Pattern 2A; (2) the prism patterns for the smaller cebids, *Aotus, Callicebus, Cebus,* and *Saimiri,* differ from the above by having C-shaped prisms which are tightly packed: developmental Pattern 2B; (3) the callitrichids studied, *Callithrix, Cebuella,* and *Saguinus,* all show a circular or

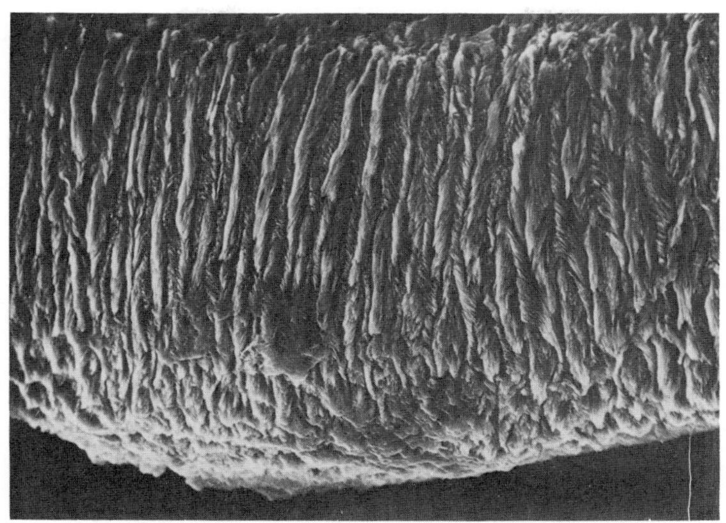

C

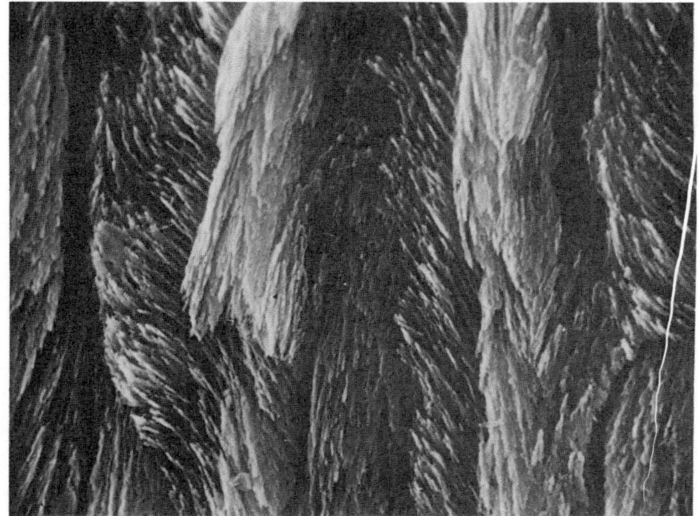

D

similar to developmental Pattern 1 (see Fig. 1). (A) *Callithrix*—circular pattern with wide interprismatic enamel, 2000×; (B) *Callithrix*—5000×; (C) *Callithrix*—longitudinal section, 700×; (D) *Callithrix*—5000×.

polygonal-shaped prism pattern with wide interprismatic regions: developmental Pattern 1; (4) the tarsiiform, *Necrolemur,* also revealed a prism pattern which is very similar to that obtained in the callitrichids, a circular or polygonal-shaped prism pattern with wide interprismatic regions: develop-

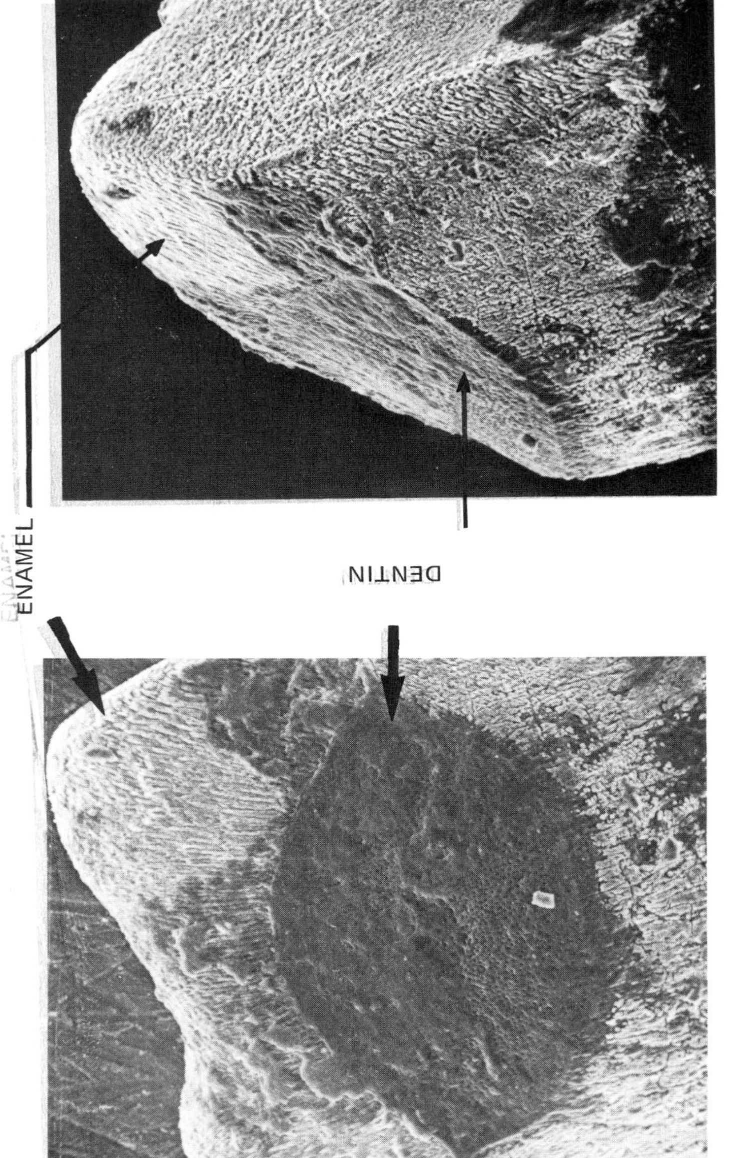

Fig. 7. Analysis of the surface enamel of the anterior dentition of *Cebuella*. The SEM microphotographs document that the lingual surface of the anterior dentition is covered by a thin layer of enamel which is rapidly worn away.

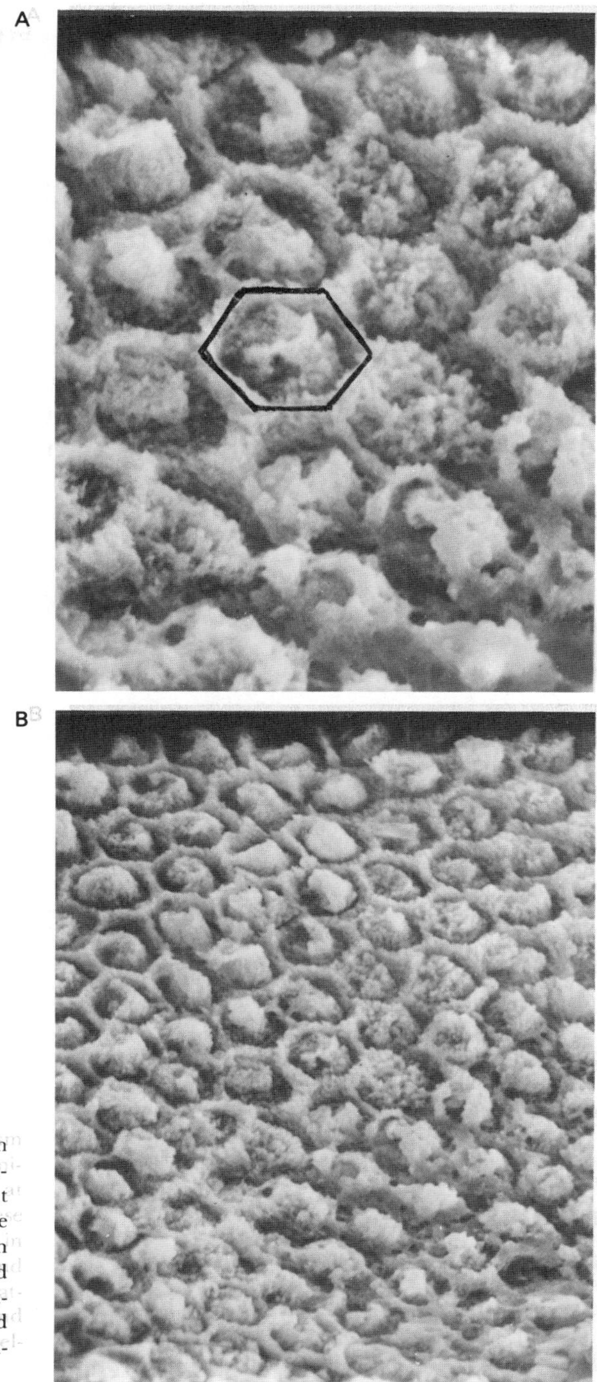

Fig. 8. Surface enamel prism pattern of *Adapis*. The SEM microphotographs were taken at 2000× (A) and 1000× (B). These photographs document that in adapids, *Adapis, Notharctus,* and *Smilodectes,* the enamel prism pattern is hexagonal-shaped and tightly packed, resembling developmental Pattern 1 (see Fig. 1).

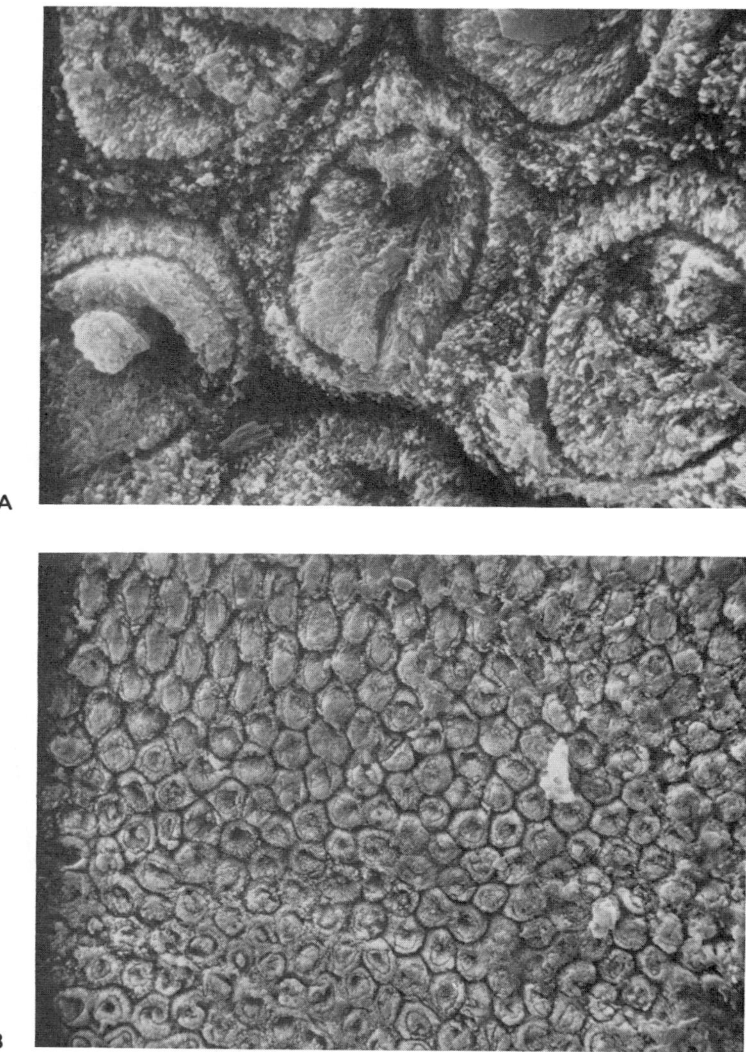

Fig. 9. Surface enamel prism pattern of *Necrolemur.* The SEM microphotographs were taken at 7000× (A) and 1000× (B). These photos document that in the fossil tarsiiform, *Necrolemur,* the enamel prism pattern is circular to oval-shaped with wide interprismatic regions, resembling developmental Pattern 1 (see Fig. 1).

mental Pattern 1; and (5) the adapids, *Adapis, Notharctus,* and *Smilodectes,* the prism pattern is a tightly packed, hexagonal-shaped prism, which is similar to developmental Pattern 1, although differing from that of the callitrichids and *Necrolemur.* Also, the analysis of *Lemur* (Fig. 10) suggests, that at least in this genus the prism pattern and enamel structure differs from that observed in the fossil adapids (see Lavelle *et al.* 1977, p. 248).

A

B

Fig. 10. Surface enamel prism patterns of the lower primate, *Lemur*. The SEM microphotographs were taken at 2000× (A) and 1000× (B). These photos document that in *Lemur* the enamel prism patterns are oval-shaped to C-shaped, that is, they tend to have open sides which would place the pattern in a transition from developmental Pattern 1 to Pattern 3 (see Fig. 1). Shellis and Poole's analysis of the enamel of *Lemur* also documented that "the enamel had an extremely unusual structure" (Lavelle *et al.*, 1977, p. 248, also see Figs. 69–72).

The relationships stated above suggest that the Platyrrhini and Catarrhini were derived from a common ancestral stock which show closer affinities to the omomyid lower primates than to the adapid lower primates. However, before definitive statements can be made regarding the exact phyletic relationships an intensive study of Paleogene and extant lower primates must be conducted to establish the full taxonomic potential of tooth structure, especially the enamel prism pattern.

ACKNOWLEDGMENTS

I wish to thank Mr. Bill Miller and Mr. Dan Cring for their assistance with the SEM analyses. I am also grateful to Drs. Ian Tattersall, Alan Walker, G. R. H. von Koenigswald, Fred Szalay, and Susan Ford for the loan of primate dental specimens. I am indebted to Dr. Alan Boyde for his comments and suggestions concerning the effects of acid etchants and procedures of analysis and also to Drs. David Poole and Peter Shellis for their comments and suggestions.

I wish to acknowledge an Alexander von Humboldt-Stiftung Research Fellowship from the German Federal Republic that allowed me to complete the study on this project in 1979, while on leave from Florida State University.

References

Boyde, A., 1964, *The Structure and Development of Mammalian Enamel.* Ph.D. Thesis, University of London.

Boyde, A., 1965, The structure of developing mammalian dental enamel, in: *Tooth Enamel-I* (R. W. Rearnhead and M. V. Stack, eds.), pp. 163–167, John Wright and Sons, Bristol.

Boyde, A., 1971, Comparative histology of mammalian teeth, in: *Dental Morphology and Evolution* (A. Dalberg, ed.), pp. 81–93, University of Chicago Press, Chicago.

Boyde, A., 1976, Amelogenesis and the structure of enamel, in: *Scientific Foundations of Dentistry* (B. Cohen and I. R. H. Kramer, eds.), pp. 335–352, William Heinemann, London.

Boyde, A., Jones, S. J., and Reynolds, P. S., 1978, Quantitative and qualitative studies of etching with acid and EDTA, *Scanning Electron Microscopy* **2**:991–1002.

Gantt, D. G., 1977, Enamel of primate teeth: Its thickness and structure with reference to functional and phyletic implications, Ph.D. Thesis, Washington University, St. Louis, University Microfilms, Ann Arbor.

Gantt, D. G., 1979*a*, A method of interpreting enamel prism patterns, *Scanning Electron Microscopy* **2**:975–981.

Gantt, D. G., 1979*b*, Taxonomic implications of primate dental tissue, *J. Biol. Buccale* **7**:149–156.

Gantt, D. G., 1979*c*, Comparative enamel histology of primate teeth, in: *Proceedings of the Third International Symposium on Tooth Enamel, J. Dent. Res.* **58** (Special Issue **B**):1002–1003.

Gantt, D. G., Pilbeam, D. R., Steward, G., 1977, Hominoid enamel prism patterns, *Science* **198**:1155–1157.

Hershkovitz, P., 1977, *Living New World Monkeys (Platyrrhini)*, Vol. I, University of Chicago Press, Chicago.

Hoffstetter, R., 1972, Relationships, origins, and history of the ceboid monkeys and caviomorph rodents: A modern reinterpretation, in: *Evolutionary Biology*, Vol. 6 (Th. Dobzhansky, M. K. Hecht, and W. C. Steere, eds.), pp. 323–347, Appleton-Century-Crofts, New York.

Hoffstetter, R., 1974, Phylogeny and geographic deployment of the primates, *J. Hum. Evol.* **3**:327–350.

Lavelle, C. L. B., Shellis, R. P., and Poole, D. F. G., 1977, *Evolutionary Changes to the Primate Skull and Dentition*, Chapter 5: The calcified dental tissues of Primates, pp. 197–279, Charles C. Thomas, Springfield.

Orlosky, F. J., and Swindler, D. R., 1975, Origin of New World monkeys, *J. Hum. Evol.* **3**:77–83.

Poole, D. F. G., and Brooks, A. W., 1961, The arrangement of crystallites in enamel prisms, *Arch. Oral Biol.* **5**:14–26.

Rosenberger, A. L., 1977, *Xenothrix* and ceboid phylogeny, *J. Hum. Evol.* **4**:461–481.

Rosenberger, A. L., 1978*a*, New data on *Branisella* and *Homunculus*, *Am. J. Phys. Anthropol.* **48**:431.

Rosenberger, A. L., 1978*b*, The loss of incisor enamel in marmosets, *J. Morphol.* **59**:207–208.

Evidence from Cranial Anatomy Studies

V

Nasal Structures in Old and New World Primates

11

W. MAIER

Introduction

The structures of the rostral parts of the nose are not well known in primates. Most past information has been derived from studies of serial sections of various fetal stages, but unfortunately postnatal stages are normally skinned and cleaned to such a degree that all cartilage disappears. The only comprehensive studies on this topic were carried out by Wen (1930), who investigated all major primate taxa, and by Schultz (1935), who contributed to the knowledge of the Hominoidea. Recently Hofer (1976, 1977) has directed his interest toward the structures of the external nose.

Largely unknown are the skeletal structures of the anterior nasal floor in older animals, since they are not accessible for gross anatomical dissection. These structures are primarily connected with the vomeronasal organ of Jacobson and with the primary choana, the nasopalatine duct of Steno. For parts of this study, it has been necessary to sacrifice juvenile and adult heads for serial sectioning; from these histological series cardboard models have been prepared which display much more details than any dissection.

The nasal cartilages are derivatives of the primordial cranium (Starck, 1975) that do not ossify for special functional reasons. Therefore, it is to be expected that these structures will be further differentiated postnatally. In most primitive primates (as in most mammals) the nasal cartilages support a protruding rhinarium. This exposed anterior end of an animal is usually equipped with a specialized skin with rhinoglyphics, glands, and numerous

W. MAIER • Klinikum der Johann Wolfgang Goethe-Universität, Zentrum der Morphologie, Dr. Senckenbergische Anatomie, Theodor-Stern-Kai 7, D-6000 Frankfurt 70, West Germany.

mechanoreceptors (Halata, 1975), but it also needs protection from mechanical impacts. The movements of the cartilages surrounding the external nares have not yet been extensively studied, however some preliminary suggestions regarding their functional meaning will be presented below.

Primates are said to be distinguishable by a tendency for reduction of the olfactory sense with a concomitant expansion of the visual system. Without any doubt, this conception is too simplified to be of much help, since many more factors have to be considered only for interpreting the adaptations of the facial skull. In order to provide a morphological background, some data of the insectivore-lower primate grade are presented first; then the description of the nasal structures of the platyrrhine and catarrhine higher primates may be better understood.

Nasal Structures in Insectivores

The primates must have originated from some Cretaceous mammals of insectivoran grade (McKenna, 1966; Van Valen, 1967; Kay and Cartmill, 1977), but it is far from clear as to which of the numerous groups of fossil Insectivora would be considered as the most likely ancestor. Szalay (1975) and others seem to favor the erinaceotan Adapisoricidae, but Maier (1979) has cast some doubts on this judgment.

Nasal structures of various fetal Insectivora have been described (Broom, 1900, 1915; Fawcett, 1918; De Beer, 1929; Roux, 1947), but adult specimens of some taxa (*Solenodon, Erinaceus, Setifer*) were only recently investigated (Maier *et al.*, in prep.). To demonstrate some basic structures of the nasal region, *Tupaia glis* has been chosen as a model of a pre-primate stage of differentiation although this taxon shows some slight reduction of the whole olfactory apparatus (Spatz, 1964). This author had modelled a wax-plate reconstruction of the cranium of a neonate *Tupaia;* Fig. 1 shows an enlarged view of the anterior nasal region newly drawn from that original model. The subsequent descriptions are based on this model and on sectional series of adult specimens, which are in general accordance. The only differences appear in some cartilage elements, which are separated by distinct sutures in the adults and function as syndesmal joints. Since the study of craniogenesis is an old morphological discipline with its own terminology, it is necessary to introduce a number of specialized terms.

The slit-like nasal apertures (nares) are directed sideward in *Tupaia,* and, therefore, a relatively wide internarial area can be seen which is covered by the rhinarium. For more details of the microscopic anatomy of the rhinarium of *Tupaia,* see Klauer (1976). This internarial space is always supported by the paired cupulae nasi which are the most anterior parts of the fetal nasal capsule. Therefore, the rostrum of *Tupaia* can be considered as "platyrrhine." This condition has erroneously been referred to as a "broad nasal septum," but, in fact, it depends on the formation of the cupulae nasi alone.

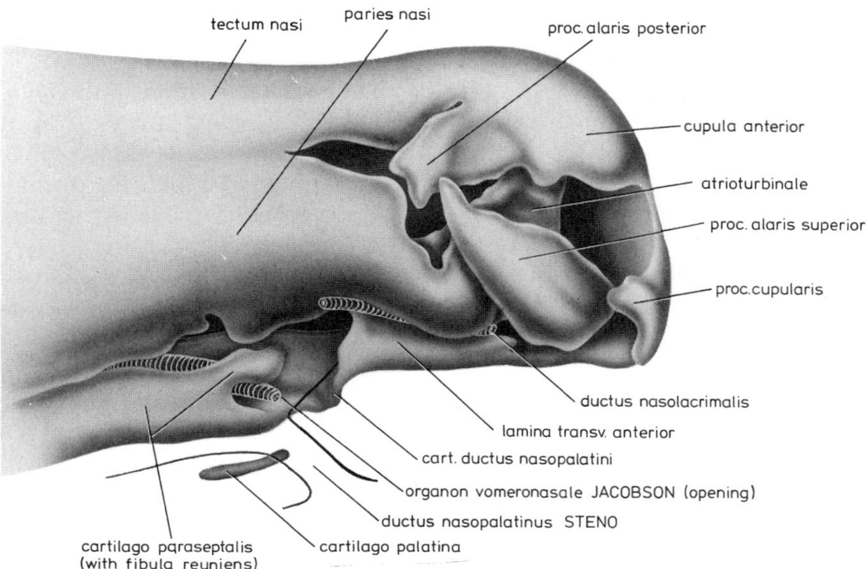

Fig. 1. Model of the anterior nasal cartilages of a late fetus of *Tupaia glis* (model of Spatz, 1964). The exoskeletal bones are not drawn. In the present stage, the process of differentiation of the cartilages has not yet reached its terminal phase. Note the complicated structure of the skeletal elements surrounding the nares. In *Tupaia,* the opening of the vomeronasal organ is situated at the nasal end of the nasopalatine canal.

Laterally, these cupulae show an incisura cupulae, which binds the nasal openings anteriorly. Ventrally, these openings are bordered on each side by a stiff processus cupularis and a moveable processus alaris superior, whereas the dorsal margin is rolled inward to form a marginoturbinale; inside the nose this continues into the atrioturbinale. The atrioturbinal cartilage supports a small concha which controls the air passage at the nasal entrance. For this reason, several bundles of anterior facial muscles insert into the outer lamina of the atrioturbinale and on the processus alaris superior (Seiler, 1976); the latter process is linked with the lower part of the atrioturbinale and thus can act as a lever. From dissected animals it would appear plausible that the atrioturbinal concha separates the air stream into an upper and a lower component. The upper component is directed to the ethmoideal conchae, which are covered by ordinary olfactory epithelium; the lower component flows through the lower nasal duct and there passes by the opening of the vomeronasal organ and the nasal aperture of the nasopalatine canal. However, more detailed experiments and observations are necessary to understand this area more properly.

The fenestra nasi superior is formed by a deep fissure which separates most parts of the cupula nasi as a moveable cartilago alaris major from the remaining lateral and dorsal nasal cartilages. The so-called processus alaris posterior probably corresponds to the lateral lamina of the wing cartilage.

The detachment of the wing cartilages may serve as an elastic buffer against mechanical impact to the rostrum.

The lateral wall of the nasal capsule bulges outside the marginoturbinale, and this prominentia supraconchalis is separated by a deep sulcus from the lamina transversalis anterior. This sulcus houses the ductus nasolacrimalis, which in *Tupaia* opens right in front of the marginoturbinale; this primitive condition prevails in lower primates, whereas the opening below the maxilloturbinale is a derived feature of all higher primates. The lamina transversalis connects the lower margin of the nasal septum with the lateral parts of the external nasal cartilages; it gives some mechanical support to the marginoturbinale by lying conformably on the premaxillary bone (omitted in Fig. 1).

The lamina transversalis and the accessory cartilages of the nasopalatine ducts and of the vomeronasal organs are normally classified as the nasal floor elements of the primordial nasal capsule. The structure of these elements, studied by Broom (1900, 1915) and others early in this century, were considered to have far-reaching systematic value. Many of their generalizations are not supported by my investigations. Behind the lamina transversalis, the lower nasal duct opens into the funnel-shaped upper opening of the nasopalatine canal. The anterior wall of this canal is surrounded by a short extension of the lamina transversalis, the cartilago ductus nasopalatini. Both this cartilage and the nasopalatine canal pass through the foramen incisivum, which runs between the premaxillary and maxillary bone of the palate. The paired canals are separated by the papilla incisiva, a block of connective tissue containing an unpaired cartilago papillae palatini (not shown in Fig. 1). Most probably, the papilla and its cartilage are able to close the lumen of the slit-like nasopalatine canals when pressed from below by the tongue. The lower openings of these canals rostrally continue into the median raphe or philtrum, which, passing between the large medial incisors, forms a connection with the rhinarium. It seems most likely that pheromones are dissolved in the secretions of the eccrine sweat glands, which cover the moist rhinarium as a thin film. Probably this film is actively transported by the tongue to the lower openings of the nasopalatine canals, from where it is sniffed into the nasal cavity, thus passing by the entrance of the vomeronasal organ (Estes, 1972). In many mammals, the facial expressions, which are connected with the filling and emptying of the vomeronasal organ have been described as "Flehmen." To my knowledge, both in Tupaiiformes and in Primates this characteristic mimic feature has not been observed.

The microscopic anatomy of the vomeronasal organ in *Tupaia* has been described and discussed by Kolnberger (1971) and by Loo and Kanagasuntheram (1972). Therefore, we can confine our remarks to some topographic aspects (Fig. 2). The orifices of the vomeronasal organs, which form tubular pouches of about 10 mm in length, are situated right above the upper end of the nasopalatine canals. These orifices are bordered by small septal conchae, which are supported by the so-called fibula reuniens (= outer bar of Broom).

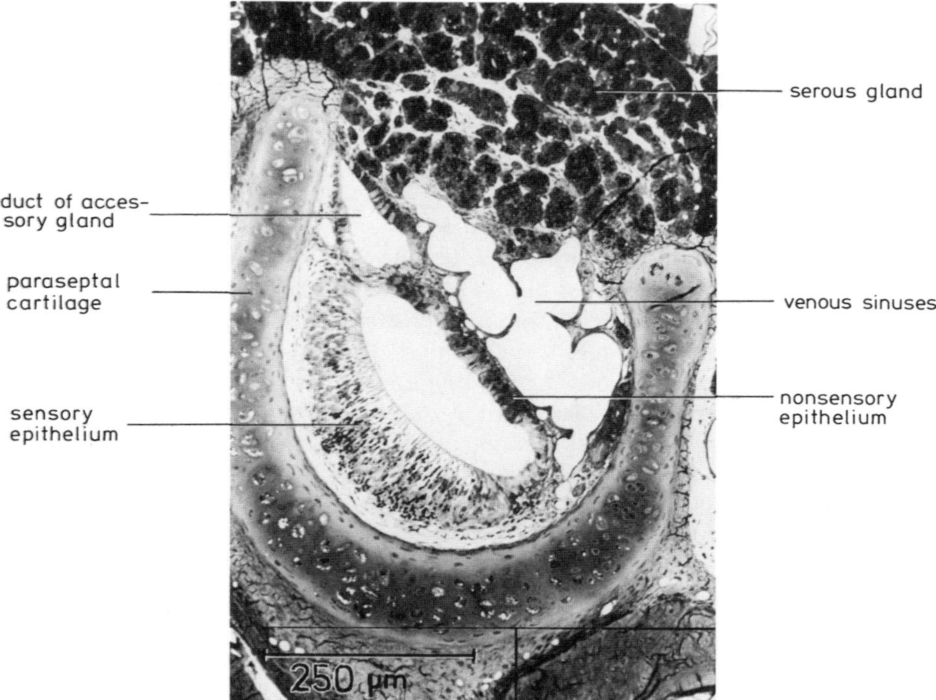

Fig. 2. Microscopic section through the distal part of a vomeronasal organ of *Tupaia glis.* Note the position of the organ within the gutterlike paraseptal cartilage, which provides an elastic counteracting frame. The secretions of the accessory glands and the pressure of the filled sinuses probably empty the organ; filling probably is caused by contraction of the vessels.

This fibula connects the lateral and the medial lamellae of the paraseptal cartilage, thus forming a narrow frame for the opening duct of the organ. The paraseptal cartilages are gutterlike structures which contain the vomeronasal organs with their accompanying vessels, nerves, and parts of the accessory serous glands. Mesially, these cartilages are backed by the anterior processes of the maxilla and by the vomer. The whole organs, with their accessory structures, bulge as rolls near the base of the anterior parts of the nasal septum.

The olfactory epithelium which lines the medial walls of the crescent-shaped organ tubes shows a number of differences in fine structure when compared with the proper olfactory epithelium (Andres, 1966). The neuritic processes of the receptor cells form the vomeronasal nerve, which ends at the accessory olfactory bulb (Stephan, 1965). At least in some rodents, this part of the olfactory bulb is connected with the posterior nucleus of the corpus amygdalae and from there with the nucleus ventromedialis of the hypothalamus, which is known to influence hormone metabolism. Hence, the vomeronasal organ probably is involved in intraspecific olfactory communication, such as the perception of sexual and territory-marking pheromones, which deeply influence social behavior (Engel, 1975; Epple, 1976).

Nasal Structures in Lower Primates

Having described *Tupaia* in some detail, it is possible to confine the following discussion of lower and higher primate nasal structures to a few comparative remarks. Lower primate nasal structures are relatively well known, but, again, mainly fetal stages were studied. Only Wen (1930) has provided some raw information on adult specimens. Frets (1914) and Henckel (1927) described chondrocranial features of *Nycticebus coucang,* Ramaswami (1957) studied earlier stages of *Loris tardigradus,* Eloff (1951) discussed nasal structures of *Galago senegalensis,* and Arnbäck-Christie-Linde (1914) mentioned some minor details of the nasal floor cartilages of *Galago* and *Cheirogaleus.* The chondrocranium of *Microcebus murinus* was studied by Bähler (1938) and Schilling (1970) provided a very thorough investigation of the vomeronasal organ of the same species. Frei (1938) described the skull of a fetus of *Avahi laniger* and Starck (1962) gave a detailed account of a late fetal stage of *Propithecus* sp. Finally, *Tarsius* is known from studies of Fischer (1905), Henckel (1927, 1928), and Wünsch (1975), the nasal region of this genus is presently being studied by Starck (in preparation).

I was offered the opportunity to describe a model of the nasal capsule of a fetus of *Galago demidovii,* reconstructed by Dr. W. B. Spatz and kindly put at my disposal for publication. Since this again is an immature specimen, my findings have been checked by examination of various sectional series (Figs. 3 and 4).

The general structure and many details of the external nose of *Galago* are quite similar to those of *Tupaia,* although the overall proportions are different. This does not necessarily mean, however, that "*Galago* comes closer to the Tupaiidae than to any other mammalian group, including even the higher primates" (Eloff, 1951, p. 653), since most similarities are best regarded as symplesiomorphies. Again, the nasal cavity of *Galago* is bordered anteriorly by relatively voluminous cupulae nasi. The nares open laterally and they are covered and surrounded by the naked skin of the rhinarium, which was briefly described by Pocock (1918). The upper margins of the nares roll inward to form a well-pronounced marginoturbinale, which internally continues as atrioturbinale and maxilloturbinale. The atrioturbinal part is supported by a lamina transversalis, which is much longer than in *Tupaia* because the rostrum of the Primates tends to become rather high instead of elongated. In all Galagidae, the lower end of the lamina transversalis does not become fused to the lower end of the nasal septum, but does abut against it. All examined specimens of *Galago* possess a processus alaris superior, but it is

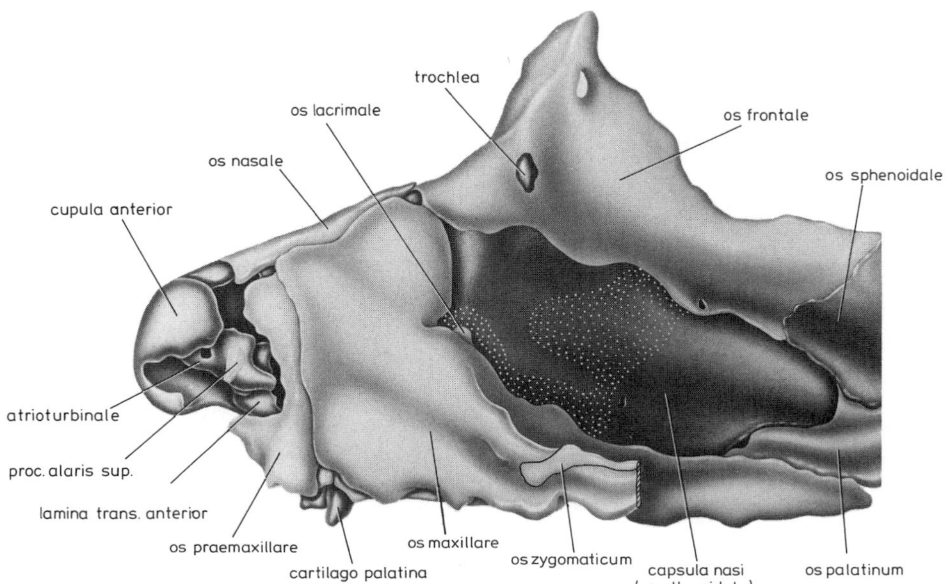

Fig. 3. Nasal capsule of a late fetus of *Galago demidovii* (lateral view). Note the protruding cartilages of the external nose, which support the rhinarium and the nares. The ethmoidal bone is only partially ossified. (Original drawing after a plate reconstruction of W. B. Spatz.)

relatively smaller than that seen in *Tupaia;* it is also fused to the lower margin of the atrioturbinale and it serves as an insertion for numerous bundles of facial muscles, which seem to be less complicated in their arrangement than in *Tupaia.* Immediately inside the fissure which separates the wing cartilage from the dorsal parts, a nasoturbinale projects from the dorsolateral wall of the atrium nasi; as in *Tupaia* its mass is largely formed by a serous nasal gland with a relatively wide duct.

Since the cartilages of the nasal floor are detached from the septum nasi throughout, the free upper end of the medial flange of the lamina transversalis projects from below into the lower nasal duct. In some cases this septal concha even covers this duct, and when the marginoturbinale is turned inward such as to lean against its upper border, there may be a strict separation of upper and lower air streams. As in *Tupaia* and most other primitive mammals, the nasolacrimal duct still opens into the atrium nasi below the atrioturbinale. The medial process of the lamina transversalis posteriorly continues into the paraseptal cartilage. Its ventral projection enters the foramen incisivum as the well-pronounced cartilago ductus nasopalatini. Laterally and posteriorly, the nasopalatine canal is accompanied by a large cartilago palatina, which fills most of the foramen incisivum. The upper opening of the nasopalatine canal is wider and deeper than in *Tupaia,* and the opening of the vomeronasal organ lies much further inside this funnel. The systematic impli-

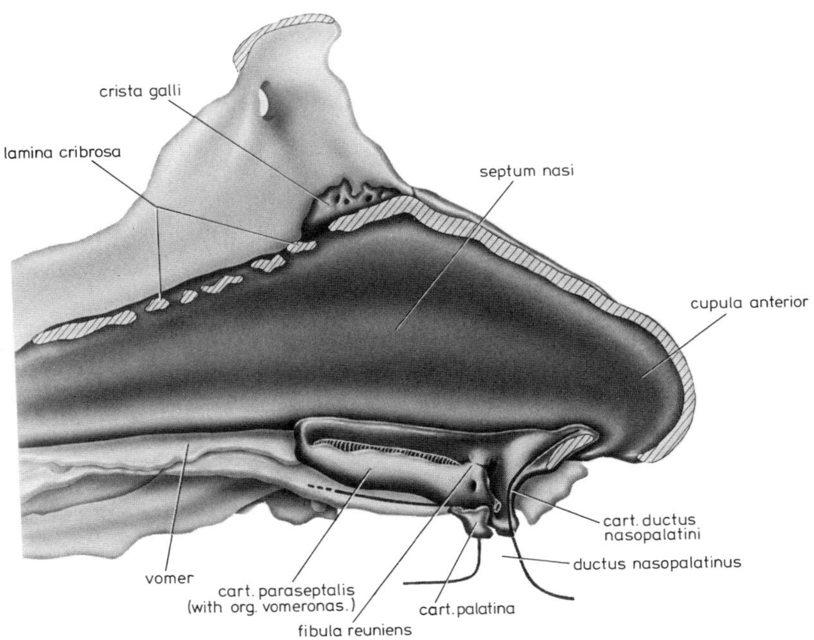

Fig. 4. Lateral parts of the nasal capsule of *Galago demidovii* are removed to show the position of the nasal floor cartilages, the vomeronasal organ, and the nasopalatine duct.

cations of this feature were greatly emphasized by Broom (1915); however, the functional meaning of this shifting should be much better understood before ascribing much significance to it. The opening of the Jacobson's organ is again covered by a fibula reuniens, which may be interrupted at times. A big venous vessel accompanies the vomeronasal duct dorsally through this cartilaginous passage.

In an exemplary manner, Schilling (1970) has provided detailed information on a lower primate vomeronasal organ (*Microcebus*). His findings seem to be referable more or less to all taxa of lower primates with the notable exception of *Tarsius* (Starck, 1975; in preparation). In *Microcebus,* as in all known lower primates, a highly differentiated organ of Jacobson is present, which seems distinctly more advanced than in some examined Insectivora (*Solenodon, Erinaceus, Setifer*). Whereas in *Tupaia* the maximum diameter of the epithelial duct of the organ measures about 0.7–0.9 mm, in *Microcebus* it is only 0.35 mm. This difference, however, is mainly due to a smaller internal lumen, whereas the thickness of the epithelia is actually very similar: sensory epithelium 90–100 μm and lateral nonsensory epithelium about 30 μm. The length of the organ is about 7–8 mm in *Microcebus.* As far as my material shows, the vomeronasal organ of *Galago demidovii* is slightly smaller than in *Microcebus,* and in *G. crassicaudatus* it seems to be still smaller. In contrast, *Loris tardigradus* possesses an organ with a very wide lumen (ca. 1.3 mm). It would certainly be worthwhile to undertake a comprehensive study of this structure in all the lower primates.

For *Microcebus,* Schilling (1970) calculated a total sensorial area of about 5 mm^2 with about 90,000 receptor cells per mm^2, and Stephan (1965) observed a well-developed bulbus olfactorius accessorius in all strepsirhine primates. Thus we may be allowed to conclude that the vomeronasal organ of Jacobson still plays an important role in the biology of lower primates (Epple, 1976). Many structures of the rostral part of the nose seem to be functionally connected with this organ, although it is not possible to fully understand the modes by which this organ is provided with pheromones. Broman (1920) suggested that the accompanying vessels form a kind of hydraulic system, enabling it to fill and empty the lumen of the organ (cf. Estes, 1972).

Nasal Structures in Platyrrhines

Although some reliable observations have been made on the vomeronasal organ of higher primates (Frets, 1913, 1914; Starck, 1960, 1975), its existence is sometimes ignored (Martin, 1973). Eisenberg (1977) attributes to the Callitrichidae a functional Jacobson's organ, whereas the "lack of a functional Jacobson's organ in the adult" is considered to be an apomorphic feature of the Cebidae (pp. 15–16). According to my observations all Platyrrhini possess a vomeronasal organ, which structurally should be able to function to some degree; however, its sensory epithelium and its accessory structures seem to be reduced in a way which is reminiscent of the condition in *Tarsius* (Starck, 1975). Surprisingly enough, the cartilages of the nasal floor, which are most functionally related to this organ, are highly developed in all platyrrhine taxa. Since there exist numerous morphological differences separating various platyrrhine taxa, most of the discussion presented here is based on *Saimiri,* which seems to be the most generalized form. The Callitrichidae appear greatly different in this respect, and *Callicebus* seems to be quite aberrant. *Pithecia* resembles *Saimiri* with regard to nasal floor structures, but exhibits differently formed cupulae nasi (Figs. 5–10).

The general organization of the outer nasal cartilages of the Platyrrhini basically corresponds to that seen in the Insectivora and the lower primates. The more or less broadened cupulae nasi house a voluminous nasal atrium resulting in a wide internarial area, which is covered by hairy skin (= haplorhine platyrrhiny). The nares, which open laterally, are framed anteriorly by the incisura cupularis and posteriorly by a small isolated processus alaris superior; this processus was observed in *Saimiri, Ateles,* and *Saguinus,* but not in *Pithecia* and *Alouatta.* The rostral cartilages of older fetuses and even later stages are gradually separated into several elements, thus allowing considera-

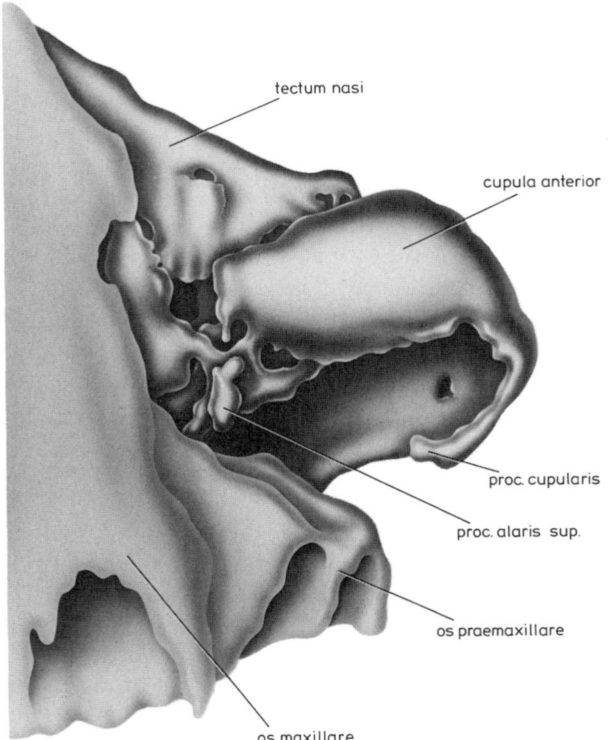

Fig. 5. Lateral view of the external nasal cartilages of an adult *Saimiri sciureus* (after a cardboard model). Although comparisons with *Galago* yield a number of structural similarities, the cartilages of *Saimiri* are actually less protruding in front of the enlarged premaxilla. The size of the wing cartilages (cupula anterior) exceeds that of the lower primates in size due to the exaggerated platyrrhiny of most South American monkeys.

ble movability. It is mainly the great wing cartilage (cartilago alaris) which becomes detached from the roof and wall of the nasal capsule. The narial border of this cartilage also becomes rolled inward to form the supporting skeleton of the marginoturbinale, which continues into the atrioturbinale and then into the maxilloturbinale. This gutter-shaped structure is separated by a broad suture from the remnants of the lateral wall cartilages, which become resorbed wherever they are covered by the exoskeletal membrane bones (here the nasal processes of the premaxillary). From below, the atrioturbinale is supported by a strong cartilage pillar, the lamina transversalis, which as a broad, platelike floor cartilage, rests on the dorsal surface of the premaxillary bone. This plate is the aboral continuation of the processus lateralis ventralis, whose medial edge becomes separated from the lower margin of the nasal septum shortly before changing into the H-shaped cartilage of Jacobson

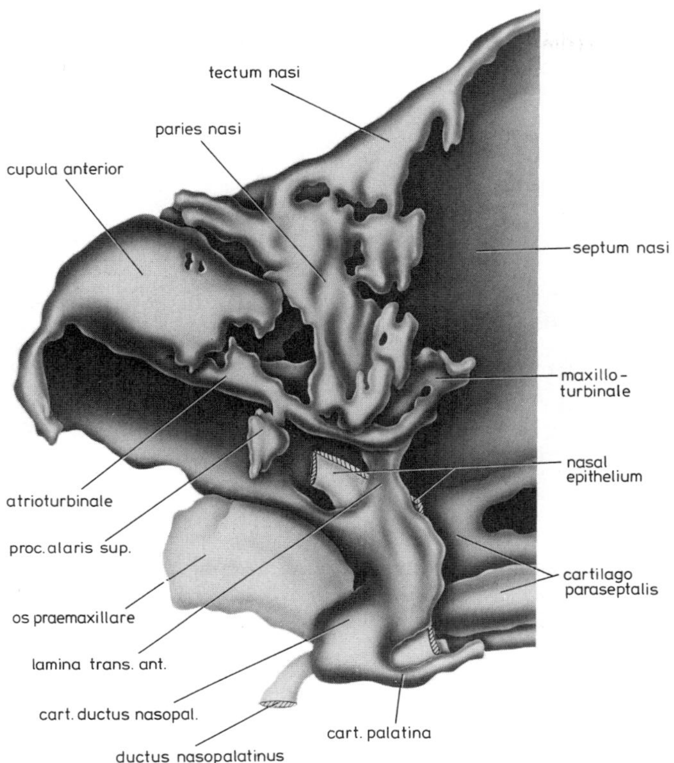

Fig. 6. Lateral view of the same model of *Saimiri sciureus* with the exoskeletal bones removed. The lateral walls of the original nasal capsule have been largely resorbed underneath the bones, whereas the floor cartilages are well developed. These cartilaginous structures surround the vomeronasal organ and the nasopalatine duct, and they are more differentiated morphologically than in lower primates.

(Broom, 1915). Posteriorly, this cartilage stock differentiates into the paraseptal and nasopalatine elements, which surround the vomeronasal organ and its ducts in a complicated manner. All of these cartilages are continuous and well-developed, more so than in most known lower primates, but resembling some Insectivora and Marsupialia (Broom, 1900). A papillary cartilage was not observed, however. Figure 10 shows the position of the vomeronasal organ in a detailed model of *Pithecia monacha,* which closely resembles *Saimiri* in this respect; a fibula reuniens is missing in *Saimiri,* for example.

As far as my observations go, *Saimiri* exhibits the most highly differentiated organon vomeronasale of all Platyrrhini, although it shows many signs of reduction in this species as well. The epithelial ducts have a total length of about 5 mm, its outer diameter being 0.6 mm at maximum. The epithelium has become relatively low (40–50 μm) and partially irregular (Fig.

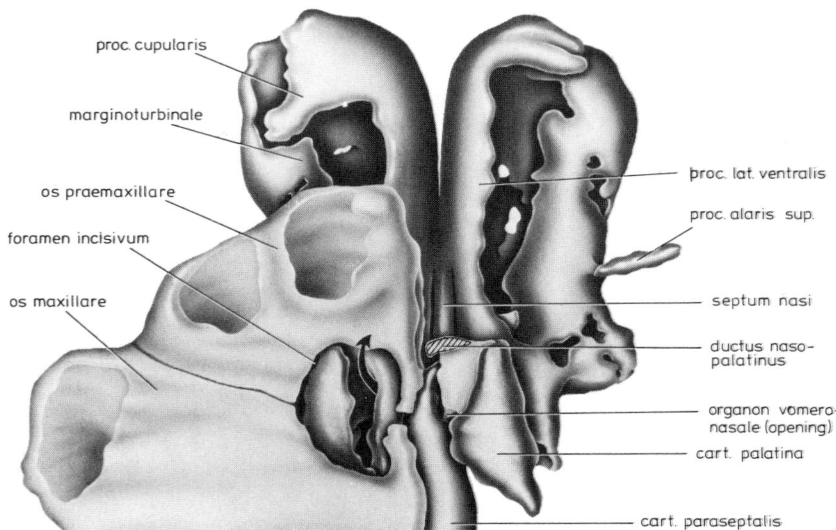

Fig. 7. Nasal region of *Saimiri sciureus* from the ventral aspect to demonstrate the relationship between the nasal floor cartilages and the maxillary bones. The incisal foramen is lined with cartilages, which support the nasopalatine canal. (The left cupula anterior is damaged postmortally.)

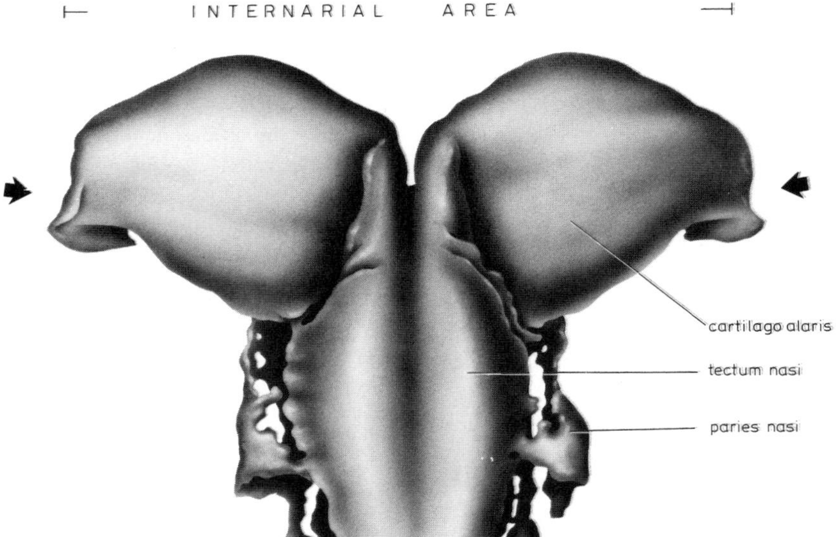

Fig. 8. Dorsal view of the nasal cartilages of *Pithecia monacha* (after a cardboard model of a subadult specimen). The cupulae nasi are extremely well developed in this species, resulting in a very broad internarial area and a very pronounced platyrrhiny. The nasal entrances are directed sidewards (see arrows).

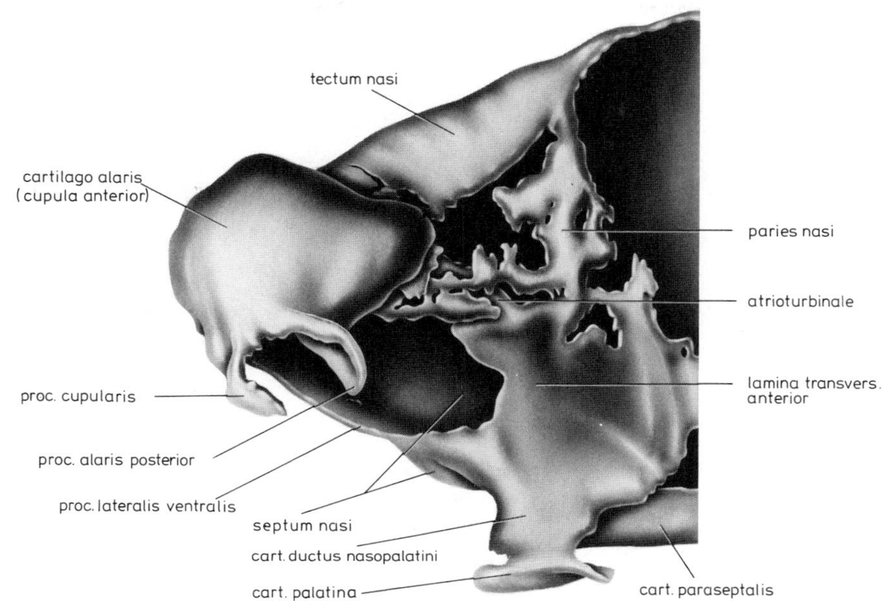

Fig. 9. Lateral view of the nasal cartilages of *Pithecia monacha* (subadult). A basic similarity in form exists between *Saimiri* and *Pithecia,* but all cartilages in the latter form appear to be exaggerated (note the very broad lamina transversalis and the fused nasal floor elements, for example). A processus alaris superior is missing in *Pithecia,* but instead a kind of processus alaris posterior supports the nares from above and behind.

11). There is no longer a clear epithelial distinction between a high sensorial wall on the medial and a low nonsensorial wall on the lateral side of the duct. Hence, the organs show oval or rounded cross sections and not the crescentlike shape of most primitive mammals. In many parts of the epithelium, no clear distinction between receptor and supporting cells is possible; in other parts layers of basal nuclei appear to exist, which are well rounded and largely devoid of heterochromatin, whereas the more superficial nuclei are more dense and elongate. The indisputable existence of receptors has not yet been demonstrated by electron microscopy, but the moderately well-developed vomeronasal nerve and the accessory olfactory bulb would seem to indicate their presence (Stephan, 1965). Many accessory serous-acinar glands and some smaller vessels still exist, but evidence of large sinuses is missing. In Callitrichidae, the vomeronasal epithelium seems even more degenerate, a fact which is hardly consistent with the hypothesis that this family represents the most primitive grade of higher primates (Hershkovitz, 1974). In quite an aberrant manner, the paraseptal cartilage mostly forms closed tubes, which even can ossify in some tamarins.

The persistence of a functioning vomeronasal organ in the Platyrrhini correlates well with the well-known marking behavior in this group of higher

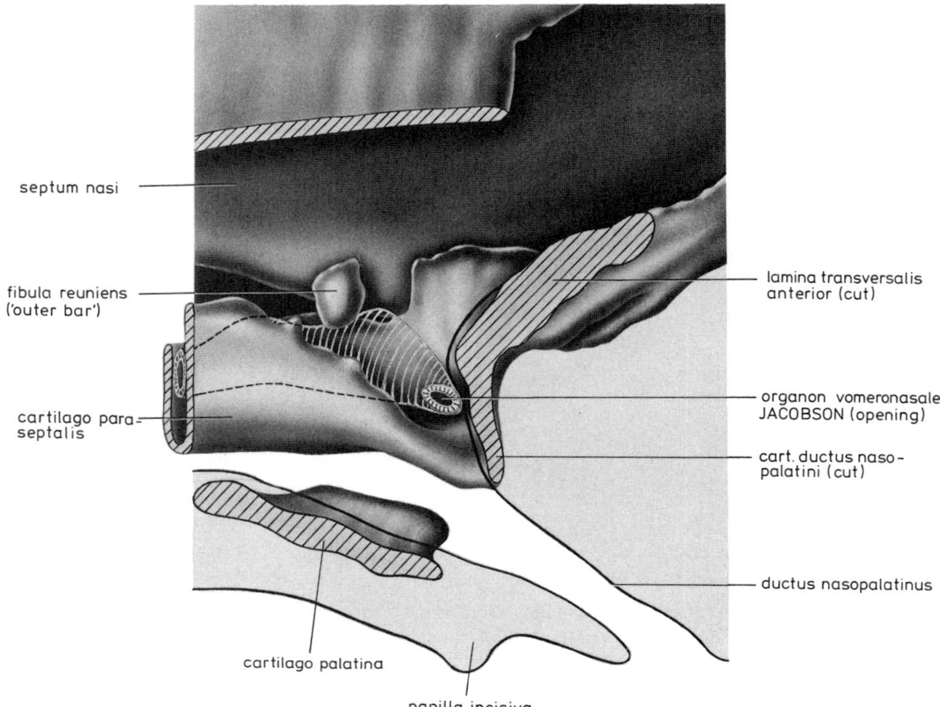

Fig. 10. Detailed lateral aspect of the model of *Pithecia monacha* showing the morphology of the vomeronasal organ and its accessory structures (the lamina transversalis and some other lateral cartilages are removed. Relicts of a fibula reuniens (outer bar) can be seen.

primates. Urine and the secretions of numerous glands are used in many behavioral contexts (Epple and Lorenz, 1967; Hershkovitz, 1977), but certainly visual communication and orientation predominates in these diurnal primates (*Aotus* most probably became nocturnal secondarily). Nevertheless, some features seem to link the Platyrrhini more closely with the lower primates: Hofer (1977) has pointed out that the loss of the rhinarium and the philtrum makes possible the development of a movable upper lip, and Martin (1973) has mentioned the interdependence between the philtrum and the interincisal diastema. In fact, the upper lip of many Platyrrhini, and especially of the Callitrichidae, is still partially fixed to the premaxillary by means of a thick upper frenulum, which possibly is to be interpreted as a relic of the primitive philtrum (cf. Fig. 12). In *Callicebus,* for example, the upper lip is fully movable, since the broad medial incisors can only become exposed through its retraction. On the other hand, in *Callicebus* the papilla incisiva between the two nasopalatine ducts is still more highly differentiated than in *Saguinus,* where these ducts open freely on the palate.

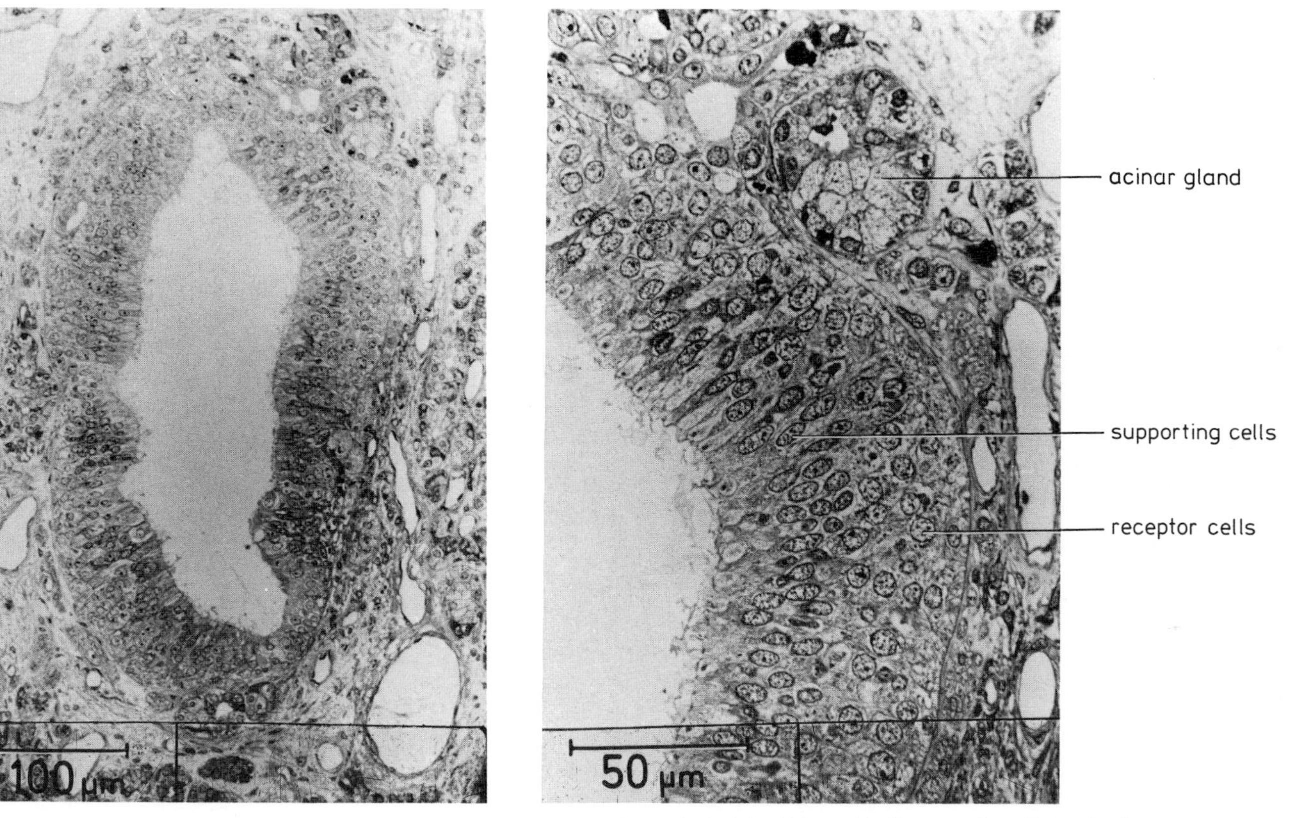

Fig. 11. Cross section of the vomeronasal organ of *Saimiri* (A). Note the irregular height of the epithelium; no clear distinction between sensory and nonsensory epithelium is evident. Detailed aspect of the same section (B). Different types of nuclei are visible: a superficial layer of dense, elongated nuclei belonging to supporting cells, and a basal stratum of large, rounded nuclei belonging to the receptor cells.

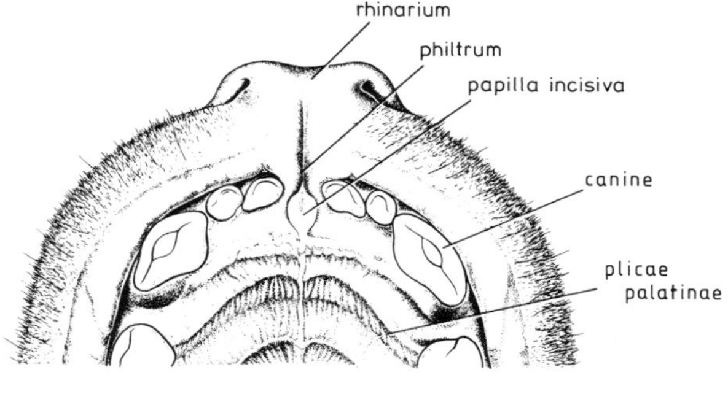

Nycticebus coucang

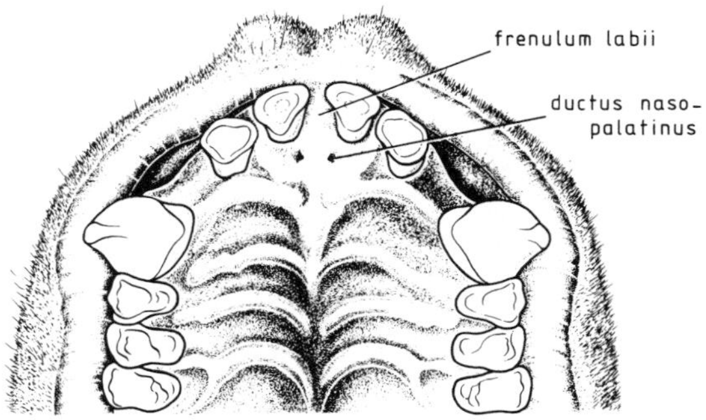

Saguinus midas

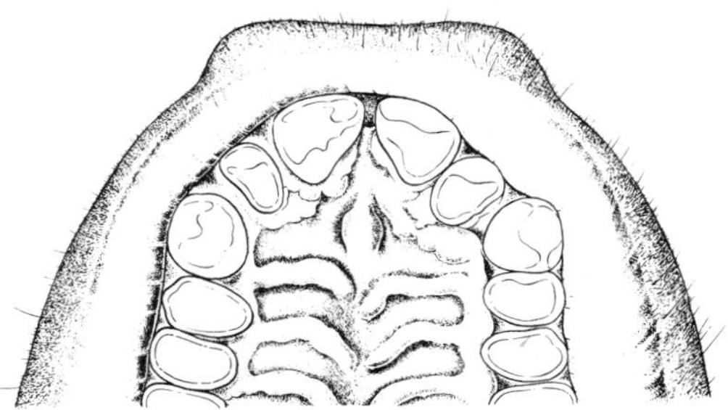

Callicebus moloch

Fig. 12. Palatinal views of the strepsirhine *Nycticebus coucang,* and the platyrrhines *Saguinus midas,* and *Callicebus moloch.* In *Saguinus* the broad upper frenulum labii could possibly be the remnant of the former philtrum; the upper lip is not very mobile in this form. *Callicebus,* however, shows a fully developed upper lip in connection with a well-differentiated papilla incisiva, which is much reduced in *Saguinus.*

Nasal Structures in Catarrhines

Catarrhines are often referred to as possessing a narrow "nasal septum." As mentioned above, it is not the septum nasi but the internarial area, which varies in different taxa; the breadth of the internarial area is largely dependent on the development of the cupulae nasi. Therefore, the difference between platyrrhines and catarrhines is not a matter of basically different construction of the nasal cartilages, but a mere difference in proportions (Figs. 13 and 14). In all the catarrhine primates, both Hominoidea and Cercopithecoidea, the wing cartilage, derived from the cupula nasi, becomes a narrow, sickle-shaped structure; its medial parts form only slender processes,

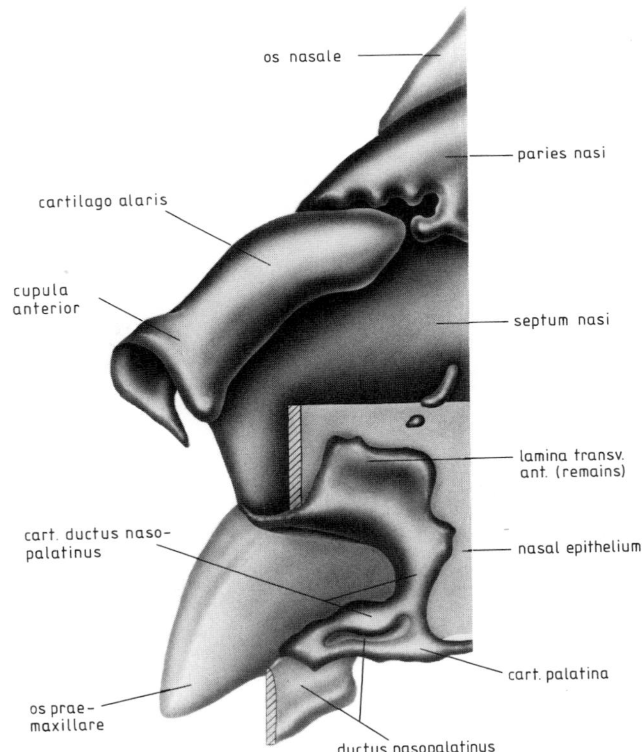

Fig. 13. Lateral view of the external nose of *Cercopithecus talapoin* (after a cardboard model of an adult specimen). Note the more slender and simplified structure of the wing cartilage and the partial reduction of the lateral wall cartilages; the lamina transversalis is broad in its lower part, but the connection with the atrioturbinale is lost. Note however, the well-developed floor cartilages and their relationship with the nasopalatine canal. These structures show much resemblance to the conditions in *Saimiri* and *Pithecia.* The platelike paraseptal cartilage is hidden behind the modeled lining of the nasal cavity. The premaxilla and the maxilla have been omitted in order to expose the floor cartilages.

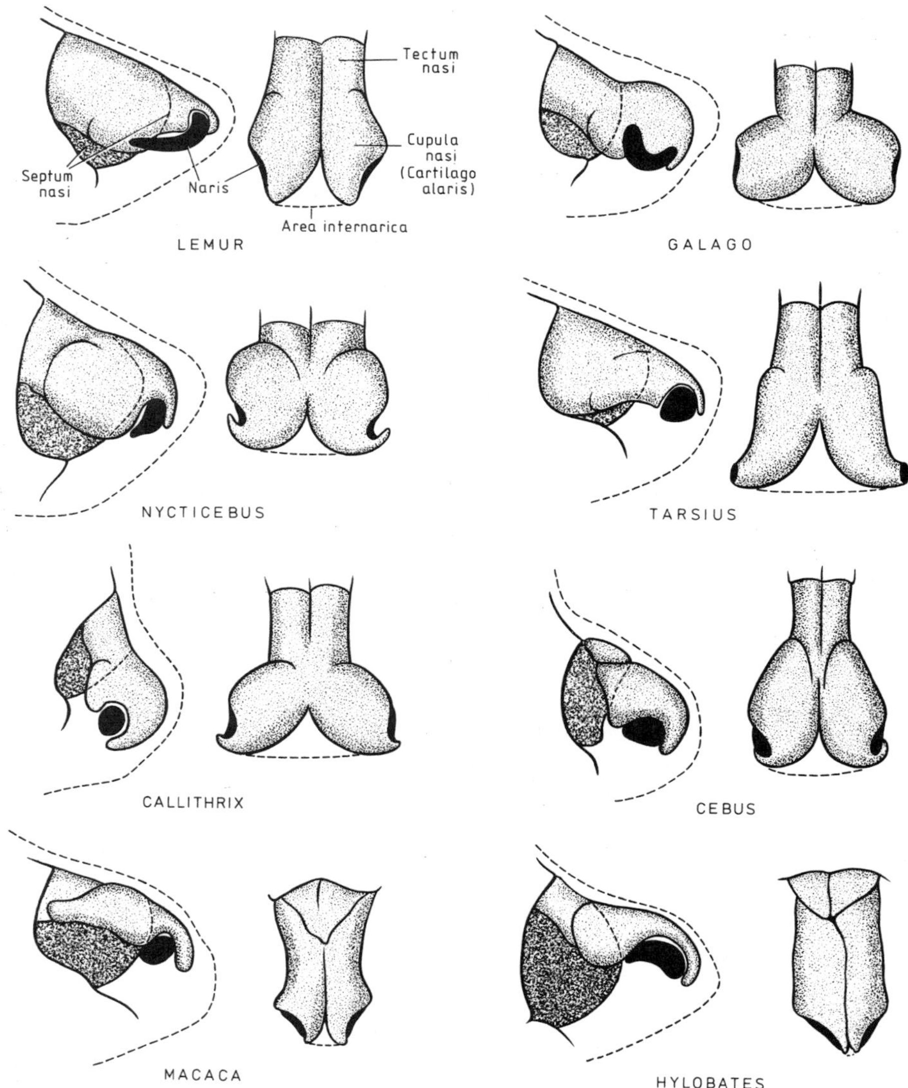

Fig. 14. Lateral and dorsal aspects of external nasal cartilages of various primate taxa (redrawn after Wen, 1930). In a slightly schematic way, these pictures show the main characteristics of the external nose in some Strepsirhini and Haplorhini; all taxa are represented by a lateral and a dorsal view. The lateral view shows the relationship between the nares and the wing cartilage. The external contours of the rostrum are shown by the broken line, the nasal cartilages are stippled, and the nares are black. Note the voluminous and wide wing cartilages of most Strepsirhini, of Tarsiiformes, and of Platyrrhini; due to a partial reduction of the cupula nasi, the internarial area of the catarrhine primates becomes very narrow. Some Strepsirhini (cf. *Lemur*) and Platyrrhini (*Alouatta, Brachyteles*) take an intermediate position. The true nasal septum never reaches the internarial area but ends somewhere in between the cupulae anteriora (broken line).

which cause the nares to point anteriorly instead of laterally. The anterior and superior margin of the nasal opening is still framed by cartilage, but there are no longer any well-pronounced margino- and atrioturbinalia. No processus alaris superior now exists, but the lateral wing process partly takes over its function; this process may be equivalent to the processus alaris posterior, previously described in *Tupaia* and *Pithecia.* The wing cartilage is not supported from below by a lamina transversalis, although relics of this structure may be observed in fetal stages. In the Catarrhini, the premaxilla and its teeth normally protrude over the nasal cartilages, thus failing to form a true rostrum. This change of proportions (and functions) may have caused the structural simplification of the external nose. It is now better protected from mechanical impacts, and its tactile functions probably are reduced, although not altogether absent. The forward orientation of the external nasal openings may further facilitate the passage of air through the small nasal cavities, and a small external nose may further open the visual field. Strong and protruding incisors in connection with highly mobile lips provide new possibilities of ingesting food.

Whereas all Platyrrhini still possess a vomeronasal organ, in the Catarrhini its presence has only been observed as an atavistic vestige in embryonic stages in man; an aberrant and doubtful occurrence in *Pan troglodytes* was also described by Starck (1960). To my knowledge, the organ of Jacobson has never been documented in the Cercopithecoidea, but this group has not been as completely studied as the Hominoidea. Therefore, it is quite surprising that the cartilages of the nasal floor seem to be fairly well-developed in the Catarrhini. They are especially well differentiated in the Cercopithecoidea, where they are mainly connected with the persisting nasopalatine canal, which normally becomes obliterated in the Hominoidea. In an adult specimen of *Cercopithecus talapoin* (Fig. 13), the floor cartilages show a remarkable similarity, for example, to *Saimiri,* which would seem difficult to be interpreted functionally. In both taxa, the lamina transversalis forms a broad plate, lying on the inner side of the premaxillary bone, although in *Cercopithecus* a direct connection with the maxilloturbinale no longer exists. The moderately wide nasopalatine duct, which runs obliquely through the incisal foramen, is partially surrounded dorsally and ventrolaterally by a cartilago ductus nasopalatini and a cartilago palatina which are fused at several points. Anteriorly, the connection with the cupula nasi is interrupted since a processus lateralis ventralis is largely missing.

Because a vomeronasal organ is totally absent in *Cercopithecus,* the cartilago paraseptalis is no longer a gutterlike structure, but a narrow vertical plate which becomes part of the compound septum nasi (Starck, 1967). Reinbach (1963) has described relatively simple floor cartilages in a younger human fetus, and Starck (1960) records a small paraseptal element in a chimpanzee fetus. An examined fetus of *Gorilla* shows a nearly obliterated nasopalatine duct, accompanied by some tiny relics of accessory cartilages.

Since the functional meaning of the nasopalatine canal in the Cercopithecoidea remains unclear, it is very difficult to evaluate the systematic significance of the nasal floor structures. Most probably, the complicated morphology of this region cannot be simply interpreted as a relic, whereas such an explanation would seem sufficient in the Hominoidea. Nevertheless, the morphological coincidences between *Saimiri* and *Pithecia* on the one hand, and *Cercopithecus* and *Macaca* on the other hand are quite striking. If a common African origin of both platyrrhine and catarrhine primates should be proven at a later date then we have to assume that this common ancestor of a higher primate grade probably possessed a vomeronasal organ similar to that of *Saimiri*. The organ would have undergone only moderate reduction or change in South American primates, but would have become entirely lost in all Old World Catarrhini. For some unknown reason, the Cercopithecoidea would have retained nearly the full set of the accessory structures of the organ of Jacobson and of the persisting nasopalatine canal, whereas they would have been largely reduced in the Hominoidea (the Hylobatidae are not yet studied, but adequate material is presently sectioned in our laboratory). Epple (1976) has recently pointed out the diminishing biological importance of olfactory communication in the Catarrhini, which would support our argument.

Conclusions

The foregoing presentation of old and new facts referring to the morphology of the anterior parts of the nasal capsule has revealed that the lower primates still show many resemblances to Insectivora such as the Tupaiidae. In the Primates, the nasal cartilages are proportioned differently, being less protruding and extending higher dorsoventrally. The active mobility of the skeletal structures around the external nares seems to become gradually reduced. The naked skin of the rhinarium is still important as a tactile field and an area for catching pheromones; both in the Tupaiidae and in lower primates it is linked with the palatinal opening of the nasopalatine canal by the grooved philtrum, which runs in between the mesial diastema of the upper incisors. In fossils, the existence of such a diastema is normally interpreted as indication of the presence of the organ of Jacobson; in the Soricidae, Solenodontidae, Daubentoniidae, Tarsiidae, and Platyrrhini the large upper incisors are closely approximated, but these taxa still possess a vomeronasal organ. The separation of the vomeronasal organ from the rhinarium at least in some cases seems to initiate the reduction of both structures; this holds especially true for both the Tarsiidae and the Platyrrhini, although other factors may be relevant. The transformation of the primitive strepsirhine condition into a haplorhine one may also have been caused by a general reduction of the olfactory system in favor of a more visual orientation. The transformation of the rhinarium and philtrum makes possible the differentiation of large upper incisors and of a mobile upper lip, which become increasingly important for ingesting a large variety of plant materials. There are indications in the South American Platyrrhini of a gradual disappearance of the philtrum. Their vomeronasal organ seems to be still functioning, but it displays many signs of structural reduction, becoming similar to the condition in *Tarsius.* It is surprising, however, that the accessory cartilages of the vomeronasal organ appear to be more differentiated than in the lower primates. The same is true for the cartilaginous structures of the external nose, whose broad and voluminous cupulae anterior provide a basis for a wide internarial area, a condition which is referred to as "platyrrhiny." Lower primates principally are "platyrrhine" as well. As far as the external nasal cartilages are concerned, the Catarrhini are apomorphic, but most probably an early simiiform forerunner would have possessed a moderate "platyrrhine" condition.

The great similarities between the nasal floor structures in Platyrrhini and Catarrhini could indeed indicate a monophyletic relationship between the two groups. This statement refers mainly to the Cercopithecoidea and some Cebidae, whereas the Hominoidea and the Callitrichidae are more aberrant. Obviously, all these findings are very difficult to interpret systematically, and therefore they can only shed a small amount of light on the problem of platyrrhine origins. Since no information about cartilaginous nasal structures is to be expected from the fossil record, we have to rely entirely on arguments of comparative anatomy of extant forms. But phylogenetic reconstruction always means the arrangement of form–function complexes according to evolutionary principles (Bock and von Wahlert, 1965; Gutmann and Peters, 1973).

A hypothetical common ancestor of the higher primates could have revealed nasal structures not unlike *Saimiri,* for example; this genus has also a fairly primitive dental morphology, which is quite similar to the Oligocene fossil *Branisella* (Maier, 1977). The Tarsiidae also show a morphology of the vomeronasal complex, which is not unlike that seen in the Platyrrhini, but convergence cannot be ruled out. This evidence would support the traditional haplorhine–tarsioid concept and not the Simio–Lemuriformes hypothesis of Gingerich (1976).

In summary, it is not in contradiction with this study of nasal morphology to assume that a primitive ceboid-like primate drifted from Africa to South America during the late Eocene, but quite significant changes subsequently occurred in the external nasal structures, which are now only partly understood in functional terms.

References

Andres, K. H., 1966, Der Feinbau der Regio olfactoria von Makrosmätikern, *Z. Zellforsch.* **69**:140–154.

Arnbäck-Cristie-Linde, A., 1914, On the cartilago palatina and the organ of Jacobson in some mammals, *Morphol. Jahrb.* **48**:343–364.

Bähler, H., 1938, Das Primordialcranium des Halbaffen *Microcebus murinus,* Medizinische Dissertation, University of Bern.

Bock, W. J., and von Wahlert, G., 1965, Adaptation and the form-function complex, *Evolution* **19**:269-299.

Broman, J., 1920, Das Organon vomeronasale Jacobson—ein Wassergeruchsorgan, *Anat. Hefte* **58**:143-191.

Broom, R., 1900, A contribution to the comparative anatomy of the mammalian organ of Jacobson, *Trans. R. Soc. Edinburgh* **39**:231-255.

Broom, R., 1915, On the organ of Jacobson and its relations in the Insectivora. Part I. *Tupaia* and *Gymnura, Proc. Zool. Soc. London* **1915**:157-162.

De Beer, G. R., 1929, The development of the skull of the shrew, *Phil. Trans. R. Soc. London, Ser. B* **217**:411-480.

Eisenberg, J. F., 1977, Comparative ecology and reproduction of New World monkeys, in: *The Biology and Conservation of the Callitrichidae* (D. G. Kleiman, ed.), pp. 13-22, Smithsonian Institution Press, Washington, D.C.

Eloff, F. C., 1951, On the organ of Jacobson and the nasal floor cartilages in the chondrocranium of *Galago senegalensis, Proc. Zool. Soc. London* **1951**:651-655.

Engel, K., 1975, Zur Kenntnis des Organon vomeronasale. Ethologische und elektrophysiologische Untersuchungen am Goldhamster (*Mesocricetus auratus,* Waterhouse 1839, Rodentia), Zoologische Dissertation, University of Hamburg.

Epple, G., 1976, Chemical communication and reproductive processes in nonhuman primates, in: *Mammalian Olfaction, Reproductive Processes, and Behavior* (R. L. Doty, ed.), pp. 257-282, Academic Press, New York.

Epple, G., and Lorenz, R., 1967, Vorkommen, Morphologie und Funktion der Sternaldrüse bei den Platyrrhini, *Folia Primatol.* **7**:98-126.

Estes, R. D., 1972, The role of the vomeronasal organ in mammalian reproduction, *Mammalia* **36**:315-341.

Fawcett, E., 1918, The primordial cranium of *Erinaceus europaeus, J. Anat.* **52**:210-250.

Fischer, E., 1905, On the primordial cranium of *Tarsius spectrum, Proc. Kon. Acad. Wetensch. Amsterdam* **8**:397-400.

Frei, H., 1938, Das Primordialcranium eines Fetus von *Avahis laniger,* Medizinische Dissertation, University of Bern.

Frets, G. P., 1913, Beiträge zur vergleichenden Anatomie und Embryologie der Nase der Primaten. II. Die Regio ethmoidalis des Primordialcraniums mit Deckknochen von einigen platyrrhinen Affen, *Morphol. Jahrb.* **45**:557-726.

Frets, G. P., 1914, Beiträge zur vergleichenden Anatomie und Embryologie der Nase der Primaten. III. Die Regio ethmoidalis des Primordialcraniums mit Deckknochen von einigen Catarrhinen, Prosimiae und dem Menschen, *Morphol. Jahrb.* **48**:239-279.

Gingerich, P. D., 1976, Cranial anatomy and evolution of early Tertiary Plesiadapidae (Mammalia, Primates), *Mus. Paleontol. Univ. Mich. Papers Paleontol.* **15**:1-140.

Gutmann, W. F., and Peters, D. S., 1973, Konstruktion und Selektion: Argumente gegen einen morphologisch verkürzten Selektionismus, *Acta Biotheor.* **22**:151-180.

Halata, Z., 1975, The mechanoreceptors of the mammalian skin. Ultrastructure and morphological classification, *Adv. Anat. Embryol. Cell Biol.* **50**:1-77.

Henckel, K. O., 1927, Zur Entwicklungsgeschichte des Halbaffenschädels, *Z. Morphol. Anthropol.* **26**:365-383.

Henckel, K. O., 1928, Studien über das Primordialcranium und die Stammesgeschichte der Primaten, *Morphol. Jahrb.* **59**:105-178.

Hershkovitz, P., 1974, A new genus of late Oligocene monkey (Cebidae, Platyrrhini) with notes on postorbital closure and platyrrhine evolution, *Folia Primatol.* **21**:1-35.

Hershkovitz, P., 1977, *Living New World Monkeys (Platyrrhini)* , Vol. I, University of Chicago Press, Chicago.

Hofer, H., 1976, Preliminary study of the comparative anatomy of the external nose of South American monkeys, *Folia Primatol.* **25**:193-214.

Hofer, H., 1977, The anatomical relations of the ductus vomeronasalis and the occurrence of taste buds in the papilla palatina of *Nycticebus coucang* (Primates, Prosimiae) with remarks on strepsirhinism, *Morpholog. Jahrb.* **123**:836–856.

Kay, R. F., and Cartmill, M., 1977, Cranial morphology and adaptations of *Palaechthon nacimienti* and other Paromomyidae (Plesiadapoidea, ?Primates), with a description of a new genus and species, *J. Hum. Evol.* **6**:19–53.

Klauer, G., 1976, Zum Bau und zur Innervation des Nasenspiegels von *Tupaia glis* (Diard 1820), Zoologische Thesis, University of Giessen.

Kolnberger, J., 1971, Vergleichende Untersuchungen am Riechepithel, insbesondere des Jacobsonschen Organs von Amphibien, Reptilien und Säugetieren, *Z. Zellforsch.* **122**:53–67.

Loo, S. K., and Kanagasuntheram, R., 1972, The vomeronasal organ in tree shrew and slow loris, *J. Anat.* **112**:165–172.

Maier, W., 1977, Die bilophodonten Molaren der Indriidae (Primates)—ein evolutionsmorphologischer Modellfall, *Z. Morphol. Anthropol.* **68**:307–344.

Maier, W., 1979, *Macrocranion tupaiodon,* an adapisoricid (?) Insectivore from the Eocene of "Grube Messel" (West Germany), *Paläontol. Z.* **53**:38–62.

Maier, W., Menzel, K. H., and Wiegand, M., Strukturen der rostralen Nasenregion bei verschiedenen Insectivoren (*Tupaia, Erinaceus, Setifer, Solenodon*) (in preparation).

Martin, R. D., 1973, Comparative anatomy and primate systematics, *Symp. Zool. Soc. London* **33**:301–337.

McKenna, M. C., 1966, Paleontology and the origin of the Primates, *Folia Primatol.* **4**:1–25.

Pocock, R. I., 1918, On the external characters of lemurs and of *Tarsius, Proc. Zool. Soc. London* **1918**:19–53.

Ramaswami, L. S., 1957, The development of the skull in the slender loris, *Loris tardigradus lydekkerianus* Cabr., *Acta Zool.* **38**:27–68.

Reinbach, W., 1963, Das Cranium eines menschlichen Feten von 93 mm Sch.-St.-Länge, *Z. Anat. Entwicklungsgesch.* **124**:1–50.

Roux, G. H., 1947, The cranial development of certain ethiopian "insectivores" and its bearing on the mutual affinities of the group, *Acta Zool.* **28**:1–233.

Schilling, A., 1970, L'organe de Jacobson du lémurien malgache *Microcebus murinus* (Miller, 1777), *Mém. Mus. Nat. Hist. Nat., N.S., A* **61**:203–280.

Schultz, A. H., 1935, The nasal cartilages in higher primates, *Am. J. Phys. Anthropol.* **20**:205–212.

Seiler, R., 1976, Die Gesichtsmuskeln, in: *Primatologia,* Vol. 4 (H. Hofer, A. H. Schultz, and D. Starck, eds.), pp. 1–252, S. Karger, White Plains, N.Y.

Spatz, W. B., 1964, Beitrag zur Ontogenese des Cranium von *Tupaia glis* (Diard, 1820), *Morpholog. Jahrb.* **106**:321–416.

Starck, D., 1960, Das Cranium eines Schimpansenfetus (*Pan troglodytes,* Blumenback 1799) von 71 mm SchStL., nebst Bemerkungen über die Körperform von Schimpansenfeten, *Morpholog. Jahrb.* **100**:559–647.

Starck, D., 1962, Das cranium von *Propithecus* spec. (Prosimiae, Lemuriformes, Indriidae), *Bibl. Primatol.* **1**:163–196.

Starck, D., 1967, Le crâne des Mammifères, in: *Traité de Zoologie,* Vol. 16 (P. P. Grassé, ed.), pp. 405–549.

Starck, D., 1975, The development of the chondrocranium in Primates, in: *Phylogeny of the Primates* (W. P. Luckett and F. S. Szalay, eds.), pp. 127–155, Plenum, New York.

Stephan, H., 1965, Der Bulbus olfactorius accessorius bei Insectivoren und Primaten, *Acta Anat.* **62**:215–253.

Szalay, F. S., 1975, Phylogeny of primate higher taxa. The basicranial evidence, in: *Phylogeny of the Primates* (W. P. Luckett and F. S. Szalay, eds.), pp. 91–125, Plenum, New York.

Van Valen, L., 1967, New Paleocene insectivorus and insectivore classification, *Bull. Am. Mus. Nat. Hist.* **135**:219–284.

Wen, I. C., 1930, Ontogeny and phylogeny of the nasal cartilages in primates, *Contrib. Embryol.* **130**:111–134.

Wünsch, D., 1975, Zur Kenntnis der Entwicklung des Craniums des Koboldmaki, *Tarsius Bancanus borneanus,* Horsefield, 1821, Zoologische Dissertation, University of Frankfurt.

Morphology, Function, and Evolution of the Anthropoid Postorbital Septum

12

M. CARTMILL

Introduction

Almost all the skeletal features that distinguish anthropoids from typical lower primates are features of the skull. To the untrained observer, perhaps the most obvious of these is the bony postorbital septum of anthropoids, which walls off the temporal fossa from the orbit proper and so converts the orbit into a shadowy eye socket. A somewhat less complete postorbital septum is found in *Tarsius*. As far as I know, no other vertebrate, living or fossil, has developed a bony partition between the temporal muscles and the periorbita. Because the postorbital septum is unique to anthropoids and *Tarsius*, systematists who lump tarsiers and anthropoids together as "Haplorhini" have generally regarded the septum as a sign of tarsiers' affinities to higher primates (Pocock, 1918; Jones, 1929; Hershkovitz, 1974; Cartmill and Kay, 1978; Luckett and Szalay, 1978). Conversely, those who believe that the closest allies of tarsiers are Eocene "tarsioids" (which lack the septum), or who think that lemurs are more closely related to anthropoids than tarsiers are, or who regard Anthropoidea as a polyphyletic taxon, have sought to prove that the

---

M. CARTMILL • Department of Anatomy, Duke University Medical Center, Durham, North Carolina 27710. Research made possible by a Research Career Development Award from the U.S. National Institutes of Health.

septum of *Tarsius* is not homologous with the septum of anthropoids (Simons and Russell, 1960; Gingerich, 1973; Cachel, 1976, 1979; Schwartz *et al.*, 1978).

Three general questions about the postorbital septum seem relevant to the problem of anthropoid origins: (1) What is the morphology of the septum in the various haplorhine groups? (2) What good does it do to have a septum? (3) How did the septum evolve? I do not have conclusive answers to the last two questions, but in what follows I hope to show that some of the answers that I and others have previously proposed are wrong, and to suggest some plausible alternatives.

Morphology of the Postorbital Septum

True primates ("Euprimates" of Hoffstetter, 1977) have a complete postorbital bar, formed by the junction of postorbital processes from the frontal and zygomatic bones. In primates that lack a postorbital septum, the bar appears roughly triangular in horizontal section (Fig. 1). The medial side of this triangle is in contact with the periorbita: the lateral side is subcutaneous; and the small posteriorly-directed base of the triangle contacts the anteriormost edge of the temporalis muscle.

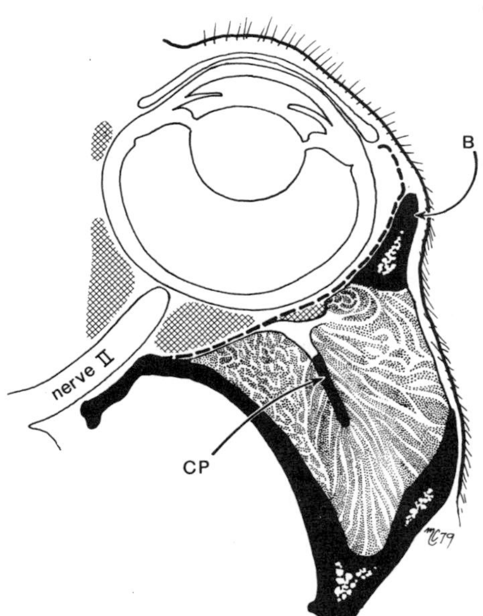

Fig. 1. Diagrammatic horizontal section through the right orbit and temporal fossa of *Lemur catta*. Stippling represents m. temporalis; hachure, fat; dashed line, periorbita. B, postorbital bar; CP, coronoid process of mandible. Based on a sectioned specimen from the Duke University Primate Center.

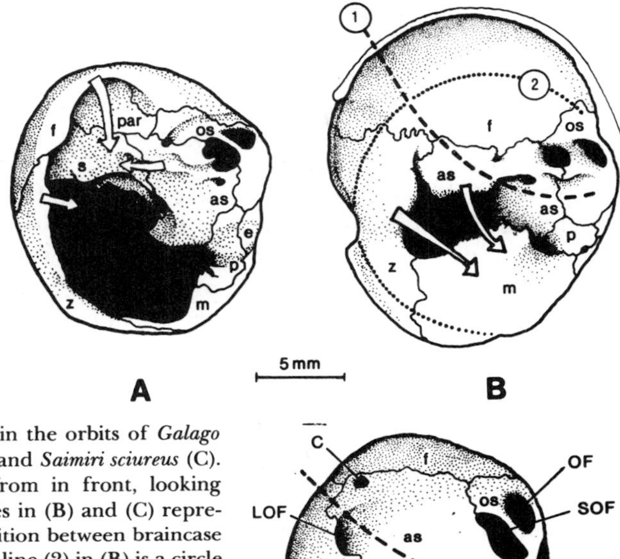

Fig. 2. Periorbital ossifications in the orbits of *Galago senegalensis* (A), *Tarsius* sp. (B), and *Saimiri sciureus* (C). All are right orbits, viewed from in front, looking slightly downward. Dashed lines in (B) and (C) represent approximate lines of transition between braincase and postorbital septum. Dotted line (2) in (B) is a circle of the same diameter as the *Galago* and *Saimiri* orbits. White arrows in the galago orbit indicate growth processes needed to produce tarsier morphology; those in the tarsier orbit show processes needed to produce *Saimiri*-like morphology. The drawing of *Tarsius* is a composite based on several specimens. Abbreviations: AOF, anterior orbital fissure; as, alisphenoid; C, cranio-orbital foramen (of sinus canal); e, ethmoid; f, frontal; FR, foramen rotundum (confluent with SOF in *Galago*); IOF, inferior orbital fissure; LOF, lateral orbital fissure; m, maxilla; OF, optic foramen; os, orbitosphenoid; p, palatine; par, parietal; s, squamous temporal; SOF, superior orbital fissure.

Each of the three edges of the bar (or, in section, the three apices of the triangle) provides anchorage for a sheet of connective tissue: the temporal fascia posterolaterally, the septum orbitale anteriorly, and the periorbita posteromedially. Bone may be deposited during development along any of these three edges. The postorbital septum is formed by bony processes that grow backward from the posteromedial edge of the bar (or forward from the braincase) along the cone-shaped sheet of free periorbita that demarcates the orbital contents from the anterior temporalis.

Figure 2 presents a pseudophylogenetic series illustrating successive structural phases in the development of the postorbital septum. In *Galago senegalensis* (Fig. 2A), as in many other strepsirhines, the posteromedial edge of the postorbital bar is prolonged posteriorly into a couple of small bony flanges—one from the frontal, one from the zygomatic—to which the periorbital cone is attached. The posterior (cranial) margin of the free periorbita is attached to the alisphenoid, where its attachment is marked by an indistinct ridge that ends medially in a blunt prong partially dividing the superior orbital fissure (inside the periorbita) from the foramen rotundum (which is outside). Expansion of these trivial periorbital processes of the frontal, zygomatic, and alisphenoid in *Galago* would yield a partial postorbital septum of the sort seen in *Tarsius* (Fig. 2B). In *Tarsius*, the septum is also supplemented by a delicate fringelike periorbital lamina of the maxilla, which extends 2–3 mm from the back of the molar alveoli up toward the septum proper but does not reach it (Figs. 2 and 6). Nothing much like this is seen in anthropoids, nor in strepsirhine primates (Fig. 3).

The postorbital septum of *Tarsius* does not extend down significantly below the level of the foramen rotundum. Although the primitive wide-open communication between orbit and temporal fossa is sufficiently constricted in *Tarsius* that it can realistically be called an inferior orbital fissure (IOF), the fissure is relatively larger than in any anthropoid, and much of it is visible above the zygomatic arch when the skull is viewed *in norma laterale*. In anthropoids, downward and forward growth of the periorbital processes of the alisphenoid and zygomatic (Fig. 2B,C) closes the uppermost end of the IOF and produces a long zygomatico-alisphenoid suture. In some anthropoids, the septal part of the zygomatic reaches the maxilla (Fig. 2C), obliterating the most anterior part of the IOF or partitioning it off as a separate aperture, which I will refer to as the *anterior orbital fissure* (AOF). A third aperture commonly persists high up on the septal zygomatico-alisphenoid suture of ceboids; Hershkovitz (1974) describes this aperture under the name of *lateral orbital fissure* (LOF).

The form and extent of the various orbital fissures differ considerably among anthropoids. The AOF is normally confluent with the IOF in *Aotus*

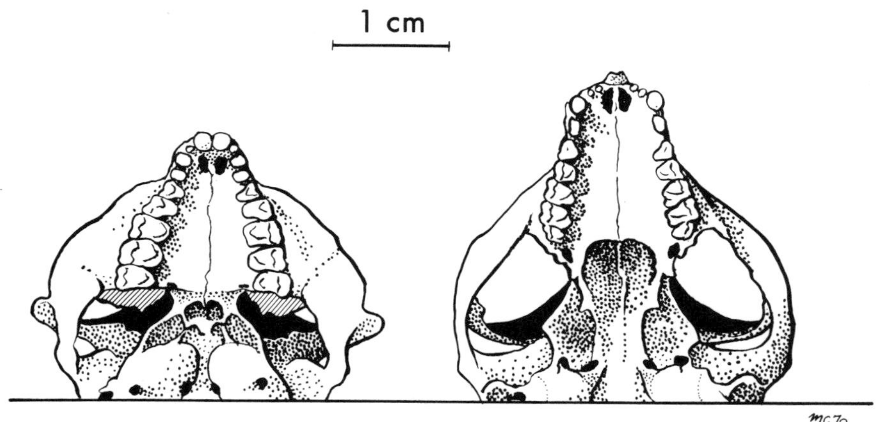

Fig. 3. Palatal regions of *Tarsius spectrum* (left) and *Galago senegalensis* (right) viewed *in norma basale*, showing periorbital processes of maxilla (diagonal hachure) in *Tarsius*. Semidiagrammatic.

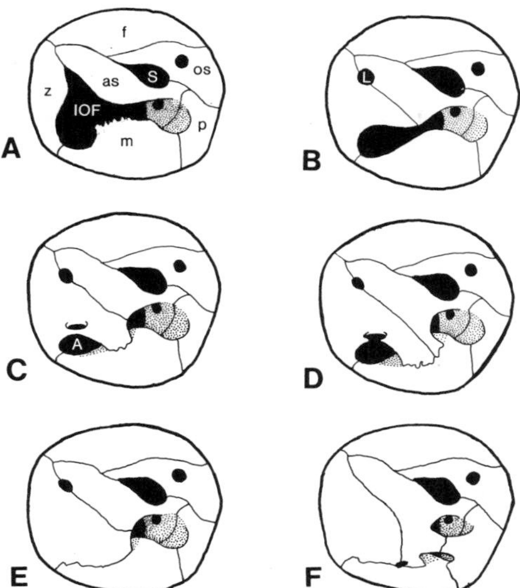

Fig. 4. Composition and form of the postorbital septum (diagrammatic); right orbits viewed from in front. (A) *Tarsius*; (B) *Aotus*; (C, D) variants seen in *Saimiri* and some other small platyrrhines; (E) *Ateles*; (F) *Alouatta*. Abbreviations: A, anterior orbital fissure; L, lateral orbital fissure; S, superior orbital fissure. Bones labelled as in Fig. 2.

(Fig. 4B) and frequently so in all other ceboids where an AOF occurs; but in *Callicebus, Alouatta,* the atelines (Fig. 4E,F), and all catarrhines with the possible exception of *Homo* (Fig. 5), the AOF is usually or invariably closed. (This closure cannot be taken as a synapomorphy of Rosenberger's (1979) family Atelidae, since a *Saimiri*-like AOF is found among the pithecines.) The LOF is almost invariably present in platyrrhines; rarely, in aged adults of *Alouatta* and *Pithecia*, it is vestigial or closed. A possibly homologous foramen occurs near the lower end of the zygomatico-alisphenoid suture in some species of *Cercopithecus* (Fig. 5C). In other adult catarrhines, the LOF is normally absent, though it may be counterfeited by senile decalcification of the postorbital septum in some hominoids (Duckworth, 1904; Schultz, 1944, 1973), and irregular ossification defects have been reported here in young human adults (Costa Ferreira, 1919).

The AOF and LOF can be regarded as vestiges of an extensive tarsierlike IOF. However, they are not functionless vestiges, but are correlated with a persistently primitive orbital vasculature. In strepsirhines, the internal jugular vein is diminutive, and most of the intracranial blood drains via the postglenoid foramen into the external jugular vein (Saben, 1963). In cheirogaleids, I find the external jugular vein also provides a major route of venous drainage from the orbit via a large vein which pierces the periorbital floor, runs downward on the medial surface of the masseter, and emerges superficially in front of the masseter to empty into the facial vein. A similar *reflex vein*

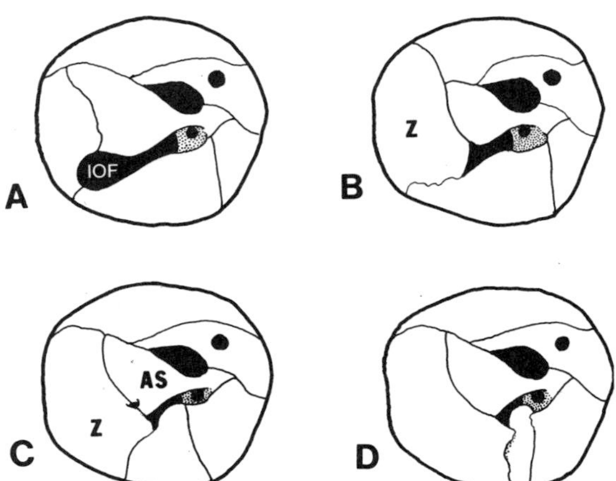

Fig. 5. Composition and form of the postorbital septum in some catarrhines (diagrammatic); right orbits viewed from in front. In *Homo* (A), the inferior orbital fissure (IOF) remains broadly open. Hylobatids (B) have a less open IOF and resemble ceboids in having a large orbital process of the zygomatic (Z; cf. Fig. 7). In *Cercopithecus* (C), the IOF is largely occluded by the maxilla (as in other cercopithecids) and a possible LOF occurs variably on the suture between zygomatic (Z) and alisphenoid (AS). The maxilla of macaques and baboons (D) occludes the IOF almost completely, and its orbital exposure is reduced to a small strip in the orbital floor.

(vena reflexa) or "deep facial vein" (probably not fully homologous with the vena facialis profunda of human anatomy) occurs in many nonprimate mammals (Preuss, 1954; Ruskell, 1964; Miller *et al.*, 1964; Dom *et al.*, 1970). It evidently represents the primitive therian route for venous drainage of the orbital floor. In two preserved specimens of *Saimiri sciureus* whose orbits I dissected, small veins in the fascia beneath the eyeball converge on the AOF and coalesce there to form a reflex vein that passes downward through the AOF. Inferiorly, this vein's relationships and connections are identical to those of the homologous vein in a dog, mouse lemur, or opossum. I conclude that the AOF in *Saimiri* and other platyrrhines may be taken as a sign of a persistently primitive pattern of orbital venous drainage. In adult catarrhines, there is ordinarily no AOF and hence no reflex vein in the strict sense; however, an apparent AOF occurs in some subadult ape skulls, and the anterior end of the exceptionally long IOF of *Homo* transmits small veins draining into the pterygoid venous plexus, which may represent vestigial homologs of the reflex vein. The deep facial vein of catarrhines probably represents a persistent remnant of the reflex vein's inferior extremity.

In the specimens of *Saimiri* that I dissected, the AOF also transmitted a nerve, which emerged from the infratemporal fossa and ran forward and upward to enter the so-called malar foramen. I interpret this as a branch of the maxillary nerve corresponding to the zygomaticofacial nerve of man. Oxnard (1957) found a similar nerve in *Callithrix* and *Saguinus*. Although the AOF and IOF were confluent in his specimens, the nerve in question ran beneath the medially-projecting extremity of the postorbital septum in the orbital floor, emerging beneath the periorbita at the anterior extremity of the large IOF (corresponding to the AOF: cf. Fig. 4B,C). The homologous nerve in the other ceboids he examined was intraorbital. We will consider the implications of this for the evolution of the postorbital septum later on.

Hershkovitz (1974), finding the LOF sometimes covered with a "membrane" in dried skulls of New World monkeys, inferred that it is not a real aperture in the living animal, but a functionless vestige of the primitively extensive IOF. In *Saimiri*, I find that it transmits a minute branch of the lacrimal artery that enters the temporal fossa. A similar vessel is found in *Microcebus*. Hill (1953) found a somewhat different arrangement in *Tarsius*, where a branch of a deep temporal artery (from the maxillary artery) passes forward through the LOF (represented in *Tarsius* by a notch) to supply the territory of the lacrimal artery. This is obviously a variant of the same pattern. While these observations contradict Hershkovitz's anatomical inferences, they strengthen his interpretation of the persistent LOF in ceboids as a primitive trait. It should be noted in passing that almost any aperture on a dried skull can become covered with a sheet of detached and desiccated connective tissue; on one *Cebuella* skull in the U.S. National Museum (U.S.N.M. No. 336323), not only the LOF, but also the superior orbital fissure and the carotid, jugular, and optic foramina are "occluded" in this way.

The bony orbital floor in strepsirhines, as in more primitive mammals, is largely unossified. The medial part of the floor is formed by the maxilla with a slight contribution from the palatine posteriorly and the zygomatic anteriorly. In most anthropoids, this primitive floor is augmented in the anterior part of the orbit by processes from the zygomatic and alisphenoid components of the postorbital septum, which extend down to or almost to the maxilla. *Aotus* and *Homo* are unique among extant anthropoids in retaining a broad and long IOF; in most adult specimens of other anthropoids, the front end of the IOF is either closed (catarrhines, some atelines, and *Alouatta*) or partitioned off more or less completely as a separate AOF (most platyrrhines). The septal contribution to the orbital floor may be so extensive that the maxilla is almost excluded from contact with the periorbita; this condition is seen in many cercopithecines, especially in baboons and macaques (Fig. 5D). In most *Alouatta* (Fig. 4F), and in some *Callicebus, Chiropotes,* and *Pithecia*, a process from the alisphenoid part of the postorbital septum spreads medially between periorbita and the infraorbital nerves and vessels, meeting a similar lateral process from the presphenopalatine lamina of the palatine bone and thus forming a bony roof over the posterior part of the inferior orbital fissure (Cartmill, 1978). In other anthropoids, the nerves and vessels running forward from the pterygopalatine fossa toward the infraorbital canal lie directly below the periorbita, though the maxillary groove in which they run may be quite deep. A caudad extension of the maxilla behind the dental arcade almost completely closes the IOF in most pongids and cercopithecids (Fig. 5C,D); in some cercopithecines, this maxillary "tail" reaches the alisphenoid in the infratemporal fossa, thus carrying the process of postorbital closure about as far as it can possibly be carried.

Functions of the Postorbital Septum

Some early authors treated the postorbital septum of anthropoids as a mere product of vaguely orthogenetic "trends" or "tendencies" supposedly characteristic of the primate order. Jones (1916) regarded postorbital ossifications as an essentially functionless byproduct of the recession of the snout. Le Gros Clark (1949) described the postorbital septum as the culmination of a trend toward "enclosure of the orbit," developed for presumably self-evident reasons in connection with "the increasing importance of vision and the consequent elaboration of the eye." Similarly vague accounts can be found in current textbooks. Standing (1908) proposed that the postorbital septum was characteristic of early primates but had been lost in Malagasy lemurs as a result of a degenerative reduction in brain size, which increased the distance between postorbital bar and neurocranium to the point where the alisphenoid was unable to bridge the gap. This bizarre account is refuted by the presence of a postorbital septum in *Tarsius*, where the orbital margin is proportionally about as far from the side walls of the braincase as it is in the larger Lemuriformes (Cartmill, 1970; Tattersall, 1973), as well as by the absence of the septum in *Necrolemur*, where the postorbital bar is quite close to the braincase (Simons and Russell, 1960).

Five different functions have been proposed for the postorbital septum: (1) support of the eye, (2) protection of the eye, (3) provision of increased area for temporalis origins, (4) transmission of masticatory stresses, and (5) insulation of the eye from temporalis movement. We will examine them in the order stated.

Support of the Eyeball

Simons and Russell (1960) suggested that the broad platelike margins of the tarsier orbit represent "circumorbital flanges" which are "concerned primarily with the support of the enormous eyes." Several later authors (Cartmill, 1970; Tattersall, 1973; Cachel, 1979) have invoked ophthalmic hypertrophy as an explanation for the appearance of a postorbital septum in *Tarsius*. It seems doubtful that the postorbital septum of *Tarsius* provides much support for the eyeball, and it is not clear why such support should be needed in any case. The eyeball of a tarsier is not likely to fall out, break loose from its moorings, or attempt to burrow backward into the temporal fossa. The orbit of tarsiers is enlarged because the periorbital cone and its contents have to cover the enlarged globe at least to its equator, where several adnexa (muscles, check ligaments, vorticose veins, etc.) are attached; but a simple postorbital bar would suffice to secure the periorbita around the equator of the eyeball. The eye of *Tarsius*, which is absolutely smaller than that of many other lower primates (Rohen, 1962), is adequately secured against gravity and the forces of locomotion by its extrinsic muscles and the surrounding periorbita and eyelids. No special adaptations are needed to preserve its shape or mechanical integrity.

Protection of the Eyeball

Prince (1953) suggested that the postorbital bar and septum of primates serve to shield the eyeball against attack from the side, which he regarded as a serious threat to animals with pronounced orbital convergence. "The primates," he noted, ". . . have moved the eyes right round to the front of the head, and if they had done this whilst leaving the orbits continuous with the temporal fossa, the eyes would have been extremely vulnerable, especially as visual awareness is at a minimum temporally to the rear." To defend the vulnerable lateral side of the eye, bony armor was developed: first a simple bar, and later a complete postorbital septum. Simons (1962) tentatively adopted a similar account of the septum's function.

The difficulty with this thesis is that many nonprimate mammals, especially carnivores, have forwardly-directed eyes; but few have postorbital bars, and none have postorbital septa. Prince tried to resolve this problem by suggesting that in carnivores "the orbital requirements may have remained subservient to the demands of jaw articulation required for wide gape and great power." But there is no reason to believe that postorbital ossifications limit gape. The Pliocene marsupial sabertooth *Thylacosmilus*, which must have had one of the largest gapes ever developed among mammals, is also one of the two marsupial genera ever to evolve a complete postorbital bar (Riggs, 1934). This example alone refutes Prince's argument.

If the postorbital septum is a protective device, it is hard to see why it should be restricted to haplorhine primates. Most mammals have a rather long lateral exposure of periorbita—not because their eyes have "moved right round to the front of the head," but because the anterior root of the zygomatic arch is primitively located far anterior to the optic foramen, as in didelphids or erinaceids. But no mammal other than *Tarsius* and anthropoids has developed periorbital ossifications to protect this supposedly vulnerable expanse of periorbita. Here again, we need to ask exactly what the threat is that the postorbital septum is supposed to be guarding against. Any injurious object that reaches the postorbital septum has already traversed the temporal or infratemporal fossa. I doubt that a mammal that has allowed a predator's teeth or beak to get that far into its head will find its chances of survival materially improved by the interposition of a thin bony sheet between jaw muscles and periorbita. If the side of the head needed to be armored, it would make more sense to develop a dermal head shield like that of primitive reptiles or the rodent *Lophiomys*, but no primate (nor, indeed, any mammal other than *Lophiomys*) has done so. The postorbital bar may help to protect the eye against accidental injury from twigs and the like (at least in anthropoids, where the eye does not protrude from the socket as far as it does in most lower primates), but it is highly unlikely that the postorbital septum serves any such purpose.

Provision of Increased Area for Temporalis Origins

Ehara (1969) suggested that the most important function of the postorbital septum was to augment the bony surface available for the origins of the anterior temporalis muscle. In a series of recent papers, Cachel (1976, 1977, 1979) has elaborated this notion into a theory of the origin of the septum and of the suborder Anthropoidea. Cachel's study of primate chewing muscles shows that the ratio of anterior temporalis weight to masseter weight (AT/M) is usually larger in frugivorous anthropoids than in folivorous ones, and usually larger in anthropoids than in a nonanthropoid group comprising *Didelphis, Tupaia, Microcebus, Cheirogaleus,* and *Tarsius*. Cachel infers from this that anthropoids evolved (diphyletically) from lower primates by becoming increasingly specialized for eating fruit. The postorbital septum, in Cachel's view, evolved to augment the area of origin available to the anterior temporalis. Cachel asserts that frugivores need a big anterior temporalis because their feeding habits involve powerful biting with the incisors, and electromyographic experiments show that the anterior temporalis is more active than the posterior temporalis during incisor biting. Since the AT/M ratio is not high in *Tarsius*, Cachel concludes that the postorbital septum of *Tarsius* has been independently evolved for different reasons, possibly to support the enlarged eyeball. The anthropoid specializations for frugivory were, Cachel proposes, originally selected for by the appearance and spread during the Oligocene of habitats having a marked seasonal change in climate, which "creates periods in which the availability of fruit is fairly predictable, and so may safely constitute a major dietary resource" (Cachel, 1979).

This argument, ingenious as it is, is not sound. Criticism of Cachel's paleoecological ideas is out of place here, but I cannot restrain myself from noting that animals are as a rule unlikely to specialize in eating something that is not available during most of their yearly cycle of activity. This is why all the specialized fruit-eaters among extant mammals inhabit tropical forests in which fruit is to be found during most of the year. The important flaws in the purely morphological aspects of Cachel's argument are the following.

1. According to Cachel's own figures, the AT/M ratio is higher (both absolutely and relative to body weight) in all "prosimians" than in some anthropoids, and higher in some "prosimians" than in some frugivorous anthropoids (e.g., *Microcebus* vs. *Saimiri*). An AT/M ratio in the range found among anthropoid fruit-eaters thus does not presuppose the existence of a postorbital septum, and a high ratio does not distinguish anthropoids from other primates. Some anthropoids have a lower AT/M ratio than any of Cachel's nonanthropoids has, including *Didelphis*.

2. Cachel's nonanthropoid "group" is also very heterogenous, and the five specimens included have nothing in common phyletically or ecologically. Her data lead us to expect that the supposedly significant differences she finds between anthropoids and nonanthropoids could be augmented or diminished at will; including more strepsirhines in her study would probably diminish the differences, whereas including more didelphids or some lipotyphlan Insectivora (to say nothing of rodents) would augment the differences. Her statistical tests comparing anthropoids to nonanthropoids accordingly have no biological significance.

3. The high AT/M values which Cachel finds for frugivorous anthropoids, as contrasted to folivorous anthropoids, might reflect any or all of the following apomorphies: (a) enlargement of the anterior temporalis in frugivores, (b) enlargement of the masseter in folivores, (c) reduction of the masseter in frugivores, or (d) reduction of the anterior temporalis in folivores. The data she presents do not allow us to determine which of these possibilities obtain, and therefore afford no grounds for believing that frugivory is associated with an "enlarging anterior segment of the temporalis muscle which requires additional bony areas of origin in the anterior temporal fossa" (Cachel, 1979). Similar criticisms apply to her comparisons involving nonanthropoids.

4. As Cachel herself notes, electromyographers find that the anterior temporalis is less active than the masseter during incisor biting, and is sometimes electrically silent. A frugivore requiring powerful incisor occlusion might therefore be expected to show specialized enlargement of the *masseter*, not of the anterior temporalis (cf. Hylander, 1978, 1979). The selection pressures proposed by Cachel thus have no intelligible connection with the effects they are supposed to have produced.

5. If Cachel's arguments were correct, one would expect the septal origin of anterior temporalis to be most extensive in frugivorous anthropoids. But *Ateles*, one of the most specialized of anthropoid frugivores, has a markedly reduced field of origin for anterior temporalis, extending only a short distance onto the postorbital septum (Starck, 1933).

6. Origin of temporalis fibers from the postorbital septum is not restricted to anthropoids but is also found in *Tarsius* (Fiedler, 1953). Thus, even if it is assumed that the septum originally evolved to provide temporalis with an expanded field of origin, there is no warrant for believing that the first appearance of the septum was associated with frugivory, or even with an anthropoid type of masticatory apparatus.

Though many of Cachel's conclusions about the function and evolution of the postorbital septum are either unwarranted or untenable, nothing that has been said so far contradicts Ehara's (1969) suggestion that the chief function of the septum in most modern anthropoids is to afford new areas for the origin of the temporalis. But even if this were true, it would not imply that this was the septum's *original* function; and two facts suggest that it was not. The first is that the temporalis of *Aotus* does not extend significantly onto the postorbital septum (Starck, 1933). The second is that the "zygomaticomandibularis" and "pars orbitalis" components of the temporalis, which originate from the septum in most platyrrhines, have a lorislike pattern of attachment in *Aotus* and *Saimiri* (Gaspard, 1972). These facts suggest that the first appearance of the postorbital septum antedates the spread of temporalis musculature onto the septum. However, these problems can be dealt with by asserting that the arrangement seen in *Aotus* is a specialization reflecting a secondary orbital enlargement (Starck, 1933), and by rejecting Gaspard's scheme of homologies for the various components of temporalis (Cachel, 1976). For the present, we will regard the temporalis-origins theory of the septum's function as a tenable hypothesis.

Transmission of Masticatory Stresses

Collins (1921) was the first to propose that the postorbital bar of primates evolved to ensure "steadiness of the movements of the eye in the interests of binocular vision." Earlier works of mine (Cartmill, 1970, 1972) elaborated Collin's thesis, concluding that the postorbital bar "replaces the postorbital ligament in order to lend increased rigidity to the orbital margin and prevent its deformation by . . . compression and tension induced by the action of the temporal musculature" (Cartmill, 1970, p. 384). Although I thought then that the postorbital septum of *Tarsius* had evolved to provide mechanical support for the eyeball, I suggested that the septum of *Tarsius* "may also lend support to the delicate, flaring postorbital bar" (Cartmill, 1970, p. 388) by resisting posteromedially-directed forces produced by tension in the temporalis fascia attached to the lateral orbital margin. Such tension is produced by contraction of the masseter (Eisenberg and Brodie, 1965), and probably by temporalis contractions as well. Although I still feel that the tarsier septum may help to brace the postorbital bar against this tension, such bracing is not found in strepsirhines; and I infer that it is needed, if at all, only in cases of tarsierlike optic hypertrophy.

Insulation of the Eye from Temporalis Movements

Cartmill (1968, 1970, 1972) and Ehara (1969) proposed that the postorbital septum of anthropoids serves to improve visual acuity, especially during mastication, by walling off the orbital contents from the temporal muscle and its contractions. There is no direct evidence for the hypothesis that these contractions are transmitted to the orbital contents in nonanthropoids, although the eyeball of a fresh *Lemur* cadaver moves perceptibly when severed anterior-temporalis fibers are pulled in the direction of fiber orientation (Cartmill, 1970, p. 387). The hypothesis is supported mainly by mechanical considerations (how could a spasming muscle *not* transmit vibrations to directly adjacent soft tissues?) and by the indirect evidence of retinal histology.

A retinal fovea is found in *Tarsius* and anthropoids, and in no other mammals (Walls, 1942; Rohen and Castenholz, 1967; Wolin and Massopust, 1970). Although Pariente (1975) claimed to have seen "zones fovéales vraies" during retinoscopy of *Lemur catta* and *Hapalemur griseus*, he noted that these areas were overlain by the endoretinal capillary network. They thus cannot be foveae, since the mammalian fovea represents an area where the endoretinal capillaries have been suppressed to allow light to reach photoreceptors directly without having to pass through blood vessels (Walls, 1942, p. 654; Weale, 1966). Mammals that have a fovea always have a postorbital septum, and vice versa—except for *Aotus*, which has lost its fovea (Jones, 1965; de Oliveira and Ripps, 1968), because it has become secondarily nocturnal. Because the fovea and septum can both be plausibly interpreted as devices for improving visual acuity, and because the presence of one implies the presence of the other, it makes sense to consider them as two aspects of a single adaptation, or else to regard the septum as a prerequisite for the appearance of a true fovea.

There are two principal difficulties with this theory. The first is that many nonmammals have a retinal fovea (some birds have two in each eye), but none has a postorbital septum (Walls, 1942). This is not a crippling objection. Reptiles and amphibians are ectothermic and eat less often than like-sized mammals; birds have relatively small jaw muscles; and nonmammals do not masticate food in any event. Temporalis contractions would thus be expected to interfere with vision in mammals more than in other terrestrial vertebrates. The second difficulty, which is more fundamental, is that *Tarsius* (like some other lower primates and more primitive eutherians) has a medial pterygoid muscle that arises from the medial orbital wall (Fiedler, 1953; Cartmill, 1970, 1978) (Fig. 6). Medial pterygoid contractions must therefore be transmitted to the orbital contents, and so the postorbital septum of *Tarsius* cannot insulate the orbital contents from masticatory movements. Yet *Tarsius* has a retinal fovea. This paradox led me, in earlier work, to propose that the septum of

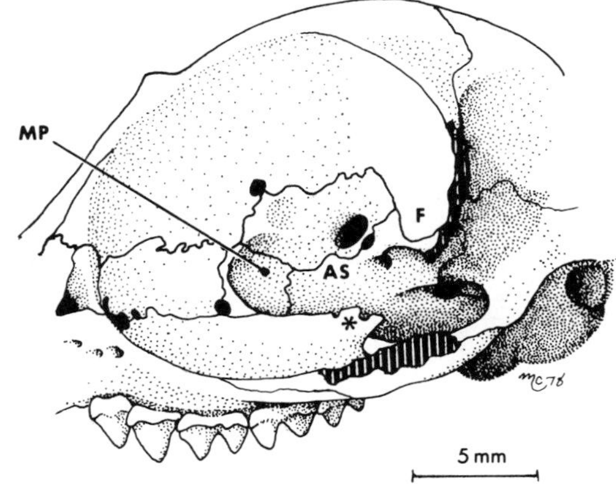

Fig. 6. Left orbit of a juvenile *Tarsius bancanus* (U.S. Natl. Mus. 300916), showing contributions of alisphenoid (AS) and frontal (F) to the postorbital septum. Vertical hachure indicates broken surfaces. The asterisk marks the periorbital process of the maxilla (cf. Fig. 3). MP, center of semicircular fossa of origin for the orbital head of the medial pterygoid muscle.

Tarsius evolved to provide support for the hypertrophied eyeball but was preadaptive for the evolution of a retina of anthropoid type (Cartmill, 1970, pp. 387–388). As shown above, this interpretation of the septum's function and evolution in *Tarsius* is implausible, and it does not account for the presence of a fovea in *Tarsius* itself.

No single function, then, can be attributed to the postorbital septum. In most anthropoids, it prevents the masticatory muscles from disturbing the delicate optical, neurological, and muscular coordination of the visual apparatus, but it does not effectively prevent this in *Tarsius*. In *Tarsius* and some anthropoids (Fiedler, 1953; Schwartz and Huelke, 1963; Gaspard, 1972), elements of the anterior temporalis originate from the temporal surface of the septum, but septal origins of the temporalis are negligible or nonexistent in other anthropoids. In *Tarsius,* the septum may brace the flaring postorbital bar against posteromedially-directed forces produced by tension in the temporal fascia, but the lower part of the septum, which is unique to anthropoids, cannot do this effectively, and the relative size and orientation of the anthropoid orbit probably renders such a brace unnecessary in any case.

The diverse functions of the septum might, of course, reflect parallel evolution of this feature in more than one lineage. If it could be shown that the septa of the various haplorhine groups are not homologous, it would be easier to develop a coherent set of hypotheses about their evolutionary histories. We will now examine the arguments bearing on this question of homology.

Homologies of the Postorbital Septa

The postorbital septa of extant catarrhines and platyrrhines are, as shown above, fundamentally similar in structure, although platyrrhines are generally more primitive in retaining patent lateral and anterior orbital fissures (and associated blood vessels) into adult life. The orbital floor (lateral to the maxilla) is incomplete in most platyrrhines, so that the AOF is confluent with the IOF. Those platyrrhines which have a more or less complete orbital floor (Fig. 2C) differ in the relationship between this floor and the zygomaticofacial nerve, which runs below the floor in *Saimiri* and callitrichids but is intraorbital in the other platyrrhines that have been studied (Oxnard, 1957). This strongly suggests that the primitive ceboid postorbital septum was broadly open inferiorly, as in *Aotus* (Fig. 4B), and that extensions of the septum downward to the maxilla have been developed in parallel in various ceboid lineages. The morphological diversity of such extensions and their absence in certain catarrhines (Figs. 4 and 5) also supports this conclusion. It follows that if the last common ancestor of the Anthropoidea had a postorbital septum, it was restricted to approximately the upper two-thirds of the lateral expanse of periorbita between braincase and maxilla.

Hershkovitz (1974) argues that postorbital closure in primitive platyrrhines was even less complete than that of *Tarsius*. His evidence for this assertion is the supposedly expansive IOF of the late Oligocene ceboid *Tremacebus harringtoni*. However, Rose and Fleagle (1980) conclude that parts of the postorbital septum of *Tremacebus* have been broken away on each side. Until we have some grounds for assessing the extent of this breakage, the size of the supposed IOF of *Tremacebus* cannot be taken as evidence for anything. Hershkovitz concludes that nothing can be said about the extent of the IOF in other fossil ceboids of similar antiquity, a judgment in which (after examining the relevant fossils) I concur.

Simons (1969) suggested that the postorbital septum of *Aegyptopithecus* was "not quite as complete as in most modern Anthropoidea"; he later modified this estimate and described it as having "about the extent seen typically in the platyrrhine monkeys" (Simons, 1972, p. 218). Photographs and casts of the original specimen (which I have not seen) suggest that the aperture connecting the orbit with the infratemporal fossa is asymmetrical and incorporates areas of breakage. The most complete of the smaller anthropoid frontal bones from the Egyptian Oligocene (Am. Mus. No. 14556) displays a clear frontal contribution to the postorbital septum, visible on the right side as a process projecting anterolaterally from the braincase; the extent of the septum is of course not determinable from the frontal alone, but there is no reason to conclude from the available material that it was less complete than that of modern anthropoids.

Simons (1972, pp. 84–85) suggests that postorbital closure may have been achieved independently in New and Old World anthropoids, because the zygomatic bone of platyrrhines extends further up on the postorbital septum and back onto the braincase than does that of catarrhines. If a cebid is compared with a hylobatid (Fig. 7), it is evident that there are indeed differences in the relationship of the zygomatic bone to its neighbors, but they are due more to differences in the size and shape of the temporal, frontal, and parietal than to any differences in the size and shape of the zygomatic contribution to the postorbital septum, which is about as extensive in hylobatids as in platyrrhines (Ashley-Montagu, 1933).

These differences in the relative contributions of various dermal bones to the formation of the braincases's side walls in different anthropoid groups have been extensively documented by Collins (1925) and Ashley-Montagu (1933). They surely do not imply that the last common ancestor of the anthropoids had a partly unossified braincase like that of reptiles, or that different anthropoid groups must be traced back to different premammalian ancestors. It is therefore equally unwarranted to infer from the fact that the frontal, alisphenoid, and zygomatic contributions to the postorbital septum differ in their relative sizes among haplorhines that the last common ancestor of the extant Haplorhini lacked a postorbital septum. Arguments to this effect have nevertheless been proposed. Simons and Russell (1960) suggest that the incomplete septum of *Tarsius* is formed largely by expansion of the frontal,

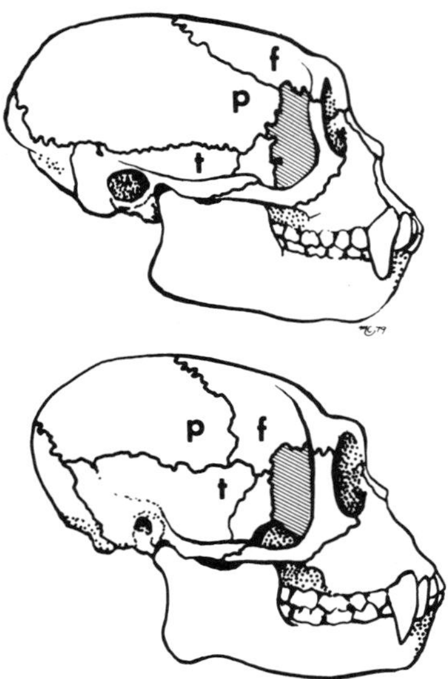

Fig. 7. The zygomatic component of the postorbital septum (diagonal hachure) in a ceboid (top) and a hylobatid (bottom). Abbreviations: f, frontal; p, parietal; t, temporal. The skulls are drawn semidiagrammatically; the sutures are based on specimens of *Cebus albifrons* (U.S. Natl. Mus. 398445) and *Hylobates lar* (Mus. Comp. Zool. 41429).

whereas "in higher Primates the greater part of the dorsolateral area of enclosure is contributed by the development of an orbital plate of the malar." They conclude that the postorbital septum of *Tarsius* is not homologous with that of anthropoids. Hershkovitz (1974) notes that this description omits mention of the alisphenoid component of the septum and suggests that Simons and Russell may have been misled by the premature fronto-alisphenoid fusion characteristic of *Tarsius*. The septum of *Tarsius* is actually formed chiefly by the zygomatic component (see below).

Schwartz *et al*. (1978) propose that the "expression of the bony contributions to postorbital closure in *Tarsius* is . . . distinct and markedly different from that in anthropoids." They conclude that the postorbital septa of platyrrhines and catarrhines are homologous but that of *Tarsius* was evolved independently. The only difference between anthropoids and *Tarsius* which they explicitly assert in support of this conclusion is that the postorbital part of the anthropoid "malar is broad inferiorly and superiorly narrows markedly," whereas the corresponding part of the tarsier zygomatic bone "is broad and flat superiorly" and narrows inferiorly. This observation is not precisely true; and if it were true, it would not be relevant. The superior part of the zygomatic bone's orbital plate is often broader than the inferior in many anthropoids, including *Homo*, and its exact shape is highly variable (Fig. 8). The reason the inferior part is normally broader in anthropoids is simply that anthropoids have a more complete postorbital septum than *Tarsius*; the broad, lower part of the anthropoid septum is not ossified in tarsiers (Fig. 2B,C). This has no bearing on the question of whether parts of the septum which do occur in both tarsiers and anthropoids represent an inheritance from their last common ancestor.

The alisphenoid contribution to the postorbital septum of *Tarsius* is restricted to a more or less narrow band (slightly expanded laterally) running along the lower edge of the frontal's postorbital lamina (Figs. 2 and 6). Schwartz *et al.* (1978, p. 111) imply (but do not explicitly claim) that the vertical narrowness of the alisphenoid and its relations to the other postorbital bones demonstrate that the septum of *Tarsius* is not homologous with that of anthropoids:

> In young individuals the alisphenoid is quite diminutive externally and internally is but a thin band running inferiorly for the breadth of the posterior orbital walls . . . In older *Tarsius*, the . . . alisphenoid, although expectedly larger than in juveniles, remains inferiorly a slender band which abuts the medial edge of the malar . . . An extraordinary feature of the orbital region of *Tarsius* is that the frontal not only contributes laterally but grows downward to overlap the malar posteriorly. We are unable to discern any external expression of the alisphenoid. The expression of the bony contributions to postorbital closure in *Tarsius* is thus distinct and markedly different from that in anthropoids.

Adult tarsiers that retain sutures in the postorbital septum do not conform to this description. One such skull is shown in Figure 9. The alisphenoid component of the septum is clearly visible posteriorly, and both it and the "malar" (zygomatic) overlap the posterior aspect of the frontal bone in this specimen, not the other way around.

Schwartz *et al.* (1978) are, however, correct in describing the periorbital lamina of the tarsier alisphenoid as a "thin band running inferiorly for the breadth of the posterior orbital wall" (Figs. 2 and 6). But this does not support their conclusion that the septum of *Tarsius* is not homologous with that of anthropoids. The slender alisphenoid process in the postorbital septum of *Tarsius* seems nearly redundant; whatever the septum's function may be, it could have been served at least as easily and directly by just extending the frontal part of the septum downward and laterally a few more millimeters. The fact that the alisphenoid contributes at all to the septum of *Tarsius* argues for its homology with that of anthropoids. The vertical shallowness of the alisphenoid part of the septum in *Tarsius*, like the shape of the zygomatic

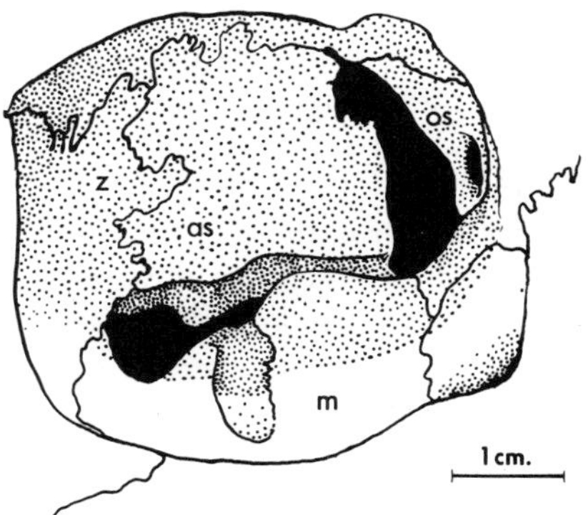

Fig. 8. Frontal view of right orbit of adult *Homo sapiens* (Duke University Anatomy Department collections). Abbreviations as in Fig. 2.

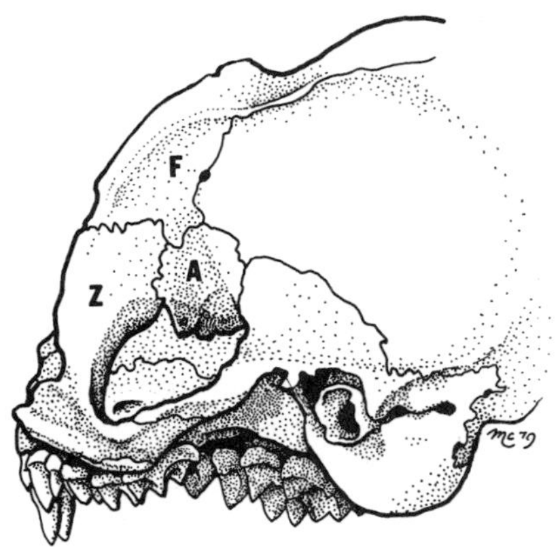

Fig. 9. Posterolateral view of left orbitotemporal region of an adult *Tarsius syrichta* with persistent sutures (Field Museum No. 56741), showing contributions of frontal (F), zygomatic (Z), and alisphenoid (A) to the postorbital septum.

component, is due simply to the fact that tarsiers have a much larger inferior orbital fissure than anthropoids, which is simply another way of saying that they have a less complete postorbital septum. Given this, it is not possible that the shapes of the bones comprising the septum *could* be identical in tarsiers and anthropoids. This does not imply that the septa were evolved independently from a simple postorbital bar, and none of the other arguments that have been advanced to prove such parallel evolution stand up to close scrutiny.

Many of the apparent differences between *Tarsius* and anthropoids are functions of the grotesque ocular hypertrophy which distinguishes *Tarsius* from all other mammals. Looking at the septal region of the tarsier orbit from the front (Fig. 2B), one is at once struck by how extensive the frontal is, and how diminutive the zygomatic and alisphenoid seem by comparison with anthropoids. But almost all of the intraorbital surface of the frontal in *Tarsius* contributes to the floor of the braincase's anterior fossa, not to the postorbital septum; this becomes clear when the boundary of the braincase is traced on the orbit's interior (line 1 in Fig. 2B). Increase in the height of the frontal here reflects the elevation of the braincase produced by the gross enlargement of the eyes (Spatz, 1968). This ocular enlargement has also resulted in expansion of the orbital margins to form the flaring "circumorbital flanges" noted by Simons and Russell (1960). When these peripheral additions are discounted (by disregarding those parts of the orbital margin which extend beyond a projection of the orbit of *Galago senegalensis* onto the tarsier skull: line 2 in Fig. 2B), the fundamental structural similarity between the septa of *Tarsius* and anthropoids becomes obvious.

To sum up, anatomical comparison indicates that the postorbital septa of extant haplorhines are homologous with each other, though the most inferior part of the septum was probably absent in the last common ancestor of Anthropoidea. Although the orbital lamina of the zygomatic is more extensive in platyrrhines than in most catarrhines, it is just as extensive in hylobatids, and the platyrrhine condition can accordingly be interpreted as a primitive anthropoid characteristic (Ashley-Montagu, 1933), not as a token of anthropoid diphyly. The differences between *Tarsius* and anthropoids with respect to the shape and relative size of the bones composing the septum appear to be wholly due to the fact that the septum is less complete in tarsiers. It is reasonable to conclude that a postorbital septum was present in the last common ancestor of the extant haplorhines. If so, then *Tarsius* and the Anthropoidea form a monophyletic clade which excludes the fossil "tarsioids" of the Eocene. This conclusion is still more strongly supported by basicranial morphology (Cartmill and Kay, 1978).

Origin and Evolution of the Postorbital Septum

Since anthropoids share many derived features of hard and soft anatomy not found in *Tarsius*, it is tempting to see *Tarsius* as a sort of living fossil, a little-changed survivor of the ancestral anthropoid stock, with several peculiar specializations of the eyes, teeth, and hind limbs superimposed on a fundamentally primitive ground plan. This picture is shattered, and efforts to reconstruct septum evolution greatly complicated, by a few lines of evidence which suggest that in some respects the last common ancestor of *Tarsius* and anthropoids was more like a small monkey than a modern tarsier, and that some of the seemingly primitive features of *Tarsius* may represent evolutionary reversals.

There are adequate grounds for believing that the last common ancestor of *Tarsius* and anthropoids was diurnal, and that the nocturnality of *Tarsius* is a recent specialization (Martin, 1973, 1978). *Tarsius* is the only nocturnal prosimian which has a fovea, and it is one of only three vertebrates known to have a foveate retina utterly lacking cones. The other two are the tuatara (*Sphenodon*) and a deep-sea fish (*Bathytroctes*), whose adaptations to an ill-lit environment are certainly secondary (Walls, 1942). Some predatory nocturnal birds possess a fovea, but again show clear signs of secondary specialization for nocturnality—for example, owls, which retain a raptorlike (but more poorly developed) temporal fovea for binocular vision but have lost the more primitive central fovea found in raptors and all other diurnal birds (Walls, 1942). Where foveae occur among vertebrates, they appear to have evolved as part of an adaptation involving acute vision in bright light, and this can be inferred for the last common ancestor of extant haplorhines as well.

A similar conclusion can be drawn from the distribution of the tapetum lucidum. This is a reflecting layer, either of shiny collagenous tissue (tapetum fibrosum) or of cells containing crystalline inclusions (tapetum cellulosum), which lies behind the retinal photoreceptors. Its function is to reflect incoming light back out through the retina again, so that quanta that are not absorbed by receptor cells on the way in have a second chance to be absorbed going out. In some strepsirhine tapeta cellulosa, the crystalline inclusions fluoresce yellow-green under illumination of shorter wavelengths, providing the retina with a sort of photomultiplier screen for radiation around the blue end of the visual spectrum (Hess, 1911; Pirie, 1959; Pariente, 1976). All nocturnal strepsirhines, and many diurnal or crepuscular ones, possess a choroidal tapetum (Rohen, 1962; Rohen and Castenholz, 1967; Wolin and Massopust, 1970; Pariente, 1976). Martin's (1972, 1973) inference, that the last common ancestor of the extant Strepsirhini was nocturnal, seems inescapable.

None of the extant Haplorhini, including the two nocturnal genera (*Tarsius* and *Aotus*), has a tapetum lucidum (Wolin and Massopust, 1970). The effective light-gathering powers of their eyes are correspondingly reduced. This is a problem for the nocturnal haplorhines, which must somehow compensate for the absence of the tapetum. The most direct way to do this is to increase the transverse diameter of the cornea and lens—a strategy equivalent to putting a faster lens on a camera. But to do so without increasing the anteroposterior diameter of the eyeball correspondingly would reduce the sharpness of the retinal image. It would thus be expected that both *Tarsius* and *Aotus* should have larger eyes, relative to body size, than nocturnal strepsirhines have. The available data (Fig. 10) confirm this expectation. So does a comparison of orbit diameter with skull length (Kay and Cartmill, 1977); the orbits of *Tarsius* and *Aotus*, like their eyes, are relatively larger than those of extant nocturnal strepsirhines.

Relative to body size, the lenses and corneas of *Tarsius* and *Aotus* are also larger than those of nocturnal strepsirhines, but the ratios of lens and cornea diameter to eyeball diameter are no larger in *Tarsius* and *Aotus* than in other nocturnal primates, though they are considerably larger than the corresponding ratios for diurnal anthropoids (Kolmer, 1930; Franz, 1934; Schultz, 1940;

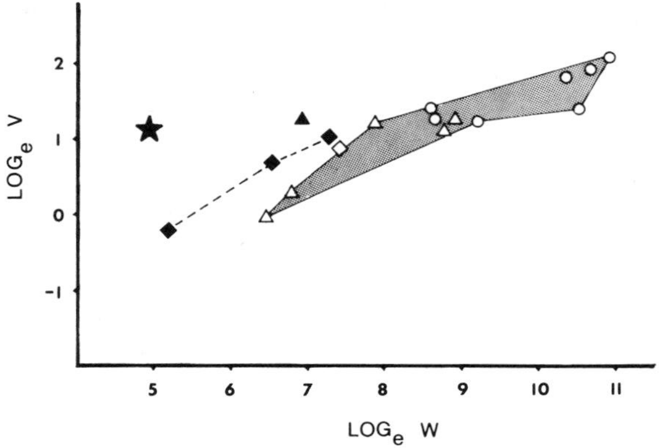

Fig. 10. Bivariate plot of the natural logarithm of body weight in grams (W) against that of eyeball volume in cubic centimeters (V) in nocturnal (black symbols) and diurnal (white symbols) primates. Black diamonds, nocturnal strepsirhines (*Galago, Nycticebus, Daubentonia*); white diamond, *Lemur*; black triangle, *Aotus*; white triangles, other platyrrhines (*Saguinus, Saimiri, Cebus, Alouatta, Ateles*); white circles, catarrhines (*Macaca, Papio, Pygathrix, Nasalis, Hylobates, Pan, Pongo, Homo*); black star, *Tarsius*. For a given body size, diurnal haplorhines (stippled area) have smaller eyes than nocturnal strepsirhines (dashed line), which in turn have smaller eyes than *Tarsius* and *Aotus*. The data are taken from Schultz (1940) and represent adults, all but three of them female (*Saimiri, Ateles,* and *Papio*). Nineteen genera yield 18 points because the data for *Macaca* and *Hylobates* are superimposed.

Table 1. Ontogenetic Changes in Relative Eye Size during Ontogeny in Catarrhines[a]

	Weight (kg)	Relative eye size[b]
Macaca mulatta		
Fetus	0.08	0.237
Newborn	0.46	0.233
Adult ♀	5.49	0.080
Pan troglodytes		
Fetus	0.57	0.227
Newborn	2.34	0.130
Infant	6.73	0.074
Juvenile	13.21	0.043
Adult ♀	44.05	0.016
Homo sapiens		
Fetus	2.00	0.136
Newborn	3.60	0.087
Adult ♀	54.89	0.015

[a] Data from Schultz (1940).
[b] Eyeball volume in cc, as a percentage of body weight in g.

Walls, 1942; Rohen, 1962). In other words, although *Tarsius* and *Aotus* have relatively much larger eyes than other extant haplorhines, their corneas and lenses are relatively larger still. During the evolution of *Aotus* from a diurnal ceboid ancestor, the cornea and lens must have enlarged faster than the eyeball itself, which expanded only to the size necessary to produce a typical nocturnal eye with a small retinal surface relative to the diameter of its cornea and lens (Walls, 1942). The optic proportions of *Tarsius* suggest a similar evolutionary history and a similar derivation from diurnal ancestors.

The marked secondary specializations of the tarsier's visual apparatus are also evident from ontogenetic data. In all vertebrates, the central nervous system is relatively large in early developmental stages and grows relatively smaller throughout ontogeny. The eye is derived from the brain, and in catarrhines (as in most vertebrates) it shows a similar growth curve, decreasing in relative size at every stage of development from fetus to adult (Table 1). But the eye of a late fetal tarsier is relatively smaller than that of the adult (Jones, 1951). *Aotus trivirgatus* seems to undergo a similar postnatal spurt of eye growth; the orbital diameter of a fetus measured by Hill (1960, p. 163) was 25.5% of skull length, whereas six adult skulls of *Aotus* that I measured had orbital diameters ranging from 31% to 33.2% of prosthion–inion length. If more extensive samples show that *Tarsius* and *Aotus* really differ from other primates in this respect, it will provide an additional bit of evidence that these two genera have gone through parallel processes of adapting a diurnal primate type of eye for a nocturnal primate mode of life.

The early Tertiary "tarsioids" (omomyids *sensu lato*) share certain specializations which suggest that they have closer phyletic affinities with the

Tarsius-anthropoid clade than with other primates, and they are commonly regarded as haplorhines (Szalay, 1975; Cartmill and Kay, 1978). Relative orbital diameter can be estimated on the skulls of three omomyids: *Tetonius, Necrolemur*, and *Rooneyia*. The orbits of *Tetonius* and *Necrolemur* are not proportionately as large as those of *Tarsius* or *Aotus*, nor as small as those of diurnal anthropoids; they fall into the nocturnal strepsirhine regression (Kay and Cartmill, 1977). If *Tetonius* and *Necrolemur* had lacked a tapetum, their orbits would be expected to be either larger or smaller than they are. We may infer that *Tetonius* and *Necrolemur* did have a tapetum lucidum, and that a tapetum was therefore present in the ancestral haplorhines. The absence of the tapetum appears to be a synapomorphy of the *Tarsius*-anthropoid clade, not of the Haplorhini as a whole. The Oligocene omomyid *Rooneyia viejaensis* had orbits of about the size that would be expected in sciurids or diurnal platyrrhines of similar skull length (Kay and Cartmill, 1977), suggesting diurnal habits and consequent reduction or loss of the tapetum; this, together with its markedly reduced stapedial artery (Szalay and Wilson, 1976), hints at phylogenetic affinities between *Rooneyia* and the *Tarsius*-anthropoid clade. However, *Rooneyia* lacks most of the derived traits of the orbit and ear region shared by *Tarsius* and anthropoids (Cartmill and Kay, 1978). *Rooneyia* may represent a phyletic sister of the *Tarsius*-anthropoid group, but it is almost certainly not descended from that group's last common ancestor.

The last common ancestor of *Tarsius* and the Anthropoidea was, then, probably a diurnal animal whose retina lacked a tapetum lucidum and had a fovea. If, as the available evidence indicates, the postorbital septum of *Tarsius* is homologous with the upper half or so of the septum found in anthropoids, it must represent an inheritance from that diurnal common ancestor. The most parsimonious conclusion from this is that the origin of the septum had nothing to do with optic hypertrophy of the sort seen in *Tarsius*, which resulted from a secondary shift to nocturnality. This conclusion immediately rules out the theory that the septum originated as a brace to resist masticatory stresses affecting the lateral margin of a grossly enlarged, tarsierlike orbit. Therefore, there are only only two viable theories of the septum's original function: either it provided additional area for temporalis origins, or it insulated a foveate retina from temporalis contractions.

Both theories have drawbacks. The notion that enlargement of the anterior temporalis would be selected for as part of an adaptation to a diet of fruit has already been rejected. This leaves no plausible account of how natural selection might have favored an expansion of temporalis origins in the anthropoid ancestry. A more fundamental question must be asked: why should increase in the size of the temporalis (or any other muscle) be expected to produce or necessitate an expansion of the muscle's origin? The data presented by Cachel (1979) indicate nothing about the size of the anterior temporalis relative to body weight, but they at any rate demonstrate that anterior temporalis size relative to *masseter size* in primates is not correlated with the presence or absence of the postorbital septum.

Another severe defect in the temporalis-origin theory of the septum is that it cannot account for the lower part of the septum. In many ceboids, septal elements send flanges medially downward to meet the maxilla (Fig. 2C) or the medial orbital wall (Fig. 4F), closing off much of the inferior orbital fissure and producing a more complete orbital floor. Similar trends toward the closure of the IOF appear among catarrhines (Fig. 5), though different elements are involved. From these multiple parallelisms, it is reasonable to conclude that closure of the IOF is favored by natural selection in anthropoids. The temporalis-origin theory cannot account for this; temporalis origins do not extend forward under the orbit. It also cannot account for the periorbital flanges of the tarsier maxilla (Fig. 3), which provide no additional surface for temporalis origins.

The eyeball-insulation theory of the septum's origins accounts for the multiple parallel trends toward IOF closure very well: closure of the IOF walls off the orbit from contractions of the pterygoid muscles, just as the upper part of the septum walls it off from the more intimate and disruptive contact of the contracting temporalis. This theory also explains why *Aotus*, with its afoveate and virtually coneless retina (Hamasaki, 1967), has not been subjected to selection pressures favoring more complete IOF closure. The retinal fovea (unlike large AT/M ratios) is invariably associated with a postorbital septum. In all these respects, the eyeball-insulation theory is preferable to the temporalis-origin theory.

The most serious flaw in the eyeball-insulation theory, as noted above, is the intraorbital origin of the medial pterygoid muscle in *Tarsius*. This flaw can be patched over only by making a radical assumption: that the medial pterygoid of the last common ancestor of *Tarsius* and anthropoids did not enter the orbit, and that the seemingly primitive condition seen in *Tarsius* is in fact a specialized evolutionary reversal.

Two lines of argument support this assumption. The first is that the slender strip of alisphenoid that forms the lower edge of the tarsier's postorbital septum (Fig. 6) is most easily accounted for by assuming that the septum of *Tarsius* is vestigial, and is derived from a septum of anthropoid type in which the alisphenoid was more extensive. The second is that the extreme ocular hypertrophy that distinguishes *Tarsius* has directly or indirectly eliminated much of the cranial space available for masticatory musculature. The temporal fossa of *Tarsius* has been constricted by backward displacement of the lateral orbital wall, and the flexion of the cranial base produced by expansion of the eye (Spatz, 1968) has brought the back edge of the palate so close to the enlarged bullae (Fig. 3) that little space is left for pterygoid musculature. I suggest that *Tarsius* has compensated for this in two principal ways: by developing an exceptionally large masseter (Fiedler, 1953), and by allowing the medial pterygoid to expand into the orbit through a reenlarged IOF. By this analysis, the periorbital fringes of maxilla and alisphenoid that stretch toward each other around the lateral side of the medial pterygoid's orbital head (Fig. 6) are ragged vestiges of larger processes that approached each other more closely in the diurnal ancestors of *Tarsius*. The upper part of the tarsier's postorbital septum may not be simply vestigial: it may have been retained to help brace the margins of the enlarged orbit against tension in the temporal fascia, or because anterior temporalis fibers had already developed attachments to that part of the septum.

If the fovea had been developed as an adaptation to a strictly predatory regime like that of *Tarsius*, the septum would have been of little value; as Cachel (1979) correctly notes, predators do not chew their last victim while they stalk their next one. The septum and fovea must have evolved in a population with a diurnal way of life involving a need for acute vision *during* mastication. Some of the smaller ceboids, which feed principally on fruits and other low-fiber plant foods but depend on insects for protein, provide a model for such a way of life. Squirrel monkeys (*Saimiri*), for example, move restlessly through the forest in groups of 20–40 animals, seldom remaining in one spot for more than a few seconds or in one tree for more than a few minutes, finding one or two items to masticate every minute and snatching concealed insects whenever they spy them or flush them out in moving past them (Thorington, 1967, 1968; Klein and Klein, 1975). Baldwin and Baldwin (1972) often saw squirrel monkeys eating their "last-discovered food from one hand and simultaneously searching visually and with the other hand for more food." Obviously, this mode of life demands that acute vision be maintained during mastication, for detecting and capturing cryptic insects; and a foveate retina isolated behind a postorbital septum is correspondingly advantageous. Similar opportunistic or methodical "browsing" on cryptic insects has been reported for *Cebuella pygmaea, Callicebus torquatus, Pithecia monacha,* and *Cebus* spp. (Moynihan, 1976; Kinzey, 1977*a,b*; Ramirez *et al.*, 1977; Izawa, 1979). This behavior contrasts with the stalk-and-pounce predatory pattern seen in nocturnal strepsirhines (Cartmill, 1970; Charles-Dominique, 1975, 1977). It is suggestive that, although *Tarsius* normally hunts by stalking and pouncing, it sometimes descends to the ground and feeds by running its hands through the leaf litter to flush concealed insects (Fogden, 1974).

It is now possible to attempt a reconstruction of the postorbital septum's evolutionary history. Primitive "tarsioids" (Omomyidae *sensu lato*) were small animals with relatively large eyes and well-developed shearing features of the molars; they were probably nocturnal insect- and fruit-eaters whose adaptations resembled those of moderately insectivorous strepsirhines like *Galago alleni* (R. F. Kay, unpublished analyses). Judging from *Tetonius* and *Necrolemur*, primitive omomyids had an afoveate retina backed by a tapetum lucidum. Later omomyids developed various herbivorous specializations (Szalay, 1976), attended in some cases by a shift to daytime activity (cf. *Rooneyia*). The *Tarsius*-anthropoid clade originated from a diurnal and predominantly frugivorous omomyid with an unfused mandibular symphysis, reduced conule cristae and other shearing features of the molars, a cone-rich retina lacking a tapetum lucidum, and *Saimiri*-like foraging behavior which made it desirable to maintain acute vision during mastication. This selection pressure led to the coordinated development of a retinal fovea and a postorbital septum; one may have preceded the other, but their unvarying co-occurrence in extant haplorhines (other than *Aotus*) makes it impossible to say.

Diurnal adaptations are not common among arboreal mammals, probably because most of the available niches are thoroughly exploited by birds (Charles-Dominique, 1975). Diurnal omomyids, including the postulated ancestors of the *Tarsius*-anthropoid clade, were undoubtedly in competition with birds to varying extents and were correspondingly pressed to develop unique specializations that could reduce this competition. [In Simpson's (1953) terminology, most of them were tachytelic lineages in an unstable adaptive zone; and they are accordingly less likely to be represented in the fossil record.] The lineage leading to *Tarsius* solved this problem by giving up the struggle and reverting to nocturnality. The adaptations of the ancestral anthropoids are less easily discerned. Any way of life that we postulate for them must account for the fusion of the two halves of the mandible, which is the most conspicuous anthropoid synapomorphy of the feeding apparatus. Partial fusion of the mandibular symphysis occurs in various leaf-eating animals, and this might suggest that the ancestral anthropoids were exploiting folivorous niches (Beecher, 1977, 1979*a*) effectively closed to birds; but early fossil anthropoids appear to have been fruit eaters (Kay and Simons, 1980), and the teeth of anthropoids do not exhibit any traces of shared ancestral specializations for shearing up fibrous plant tissues (Kay, this volume). Beecher (1979*b*) has recently suggested that anthropoid symphyseal morphology reflects an ancestral adaptation for eating unripe fruit. This would have effectively reduced competition with birds and would also have placed a greater premium on visual acuity than a diet of ripe fruits (which are often conspicuously colored).

In the last common ancestor of *Tarsius* and anthropoids, the postorbital septum covered the area of contact between periorbita and m. temporalis. It was probably no more extensive than that of *Homo*; and since the back of the maxilla was not pneumatized as it is in *Homo*, the IOF was relatively larger than that of any extant anthropoid. After diverging from the anthropoid ancestry, the lineage leading to *Tarsius* evolved a periorbital extension of the maxilla which helped to close off the IOF and insulate the eye more completely from masticatory movements. Subsequently, the tarsier lineage underwent a shift to a wholly nocturnal and predatory mode of life, developing secondary dental specializations for eating insects. Lacking a tapetum lucidum, the ancestral tarsier populations were compelled to adapt to nocturnality by greatly enlarging the eye. They were able to compensate for the resulting reduction of the temporal and pterygoid fossae by allowing the medial pterygoid to spread into the orbit through a reenlarging IOF, because loss of retinal cones and the adoption of stalk-and-pounce hunting behavior had removed selection pressures against contact between periorbita and masticatory musculature. The upper part of the septum may have been retained for one or more of the functional reasons discussed above; the periorbital processes of the alisphenoid and maxilla, like the retinal fovea, persist as functionless vestigal reminders of the more monkeylike morphology found in the tarsier's remote ancestors.

The stem anthropoids did not develop periorbital processes of the maxilla. Expansion of the maxillary air sinus has helped to plug the IOF in almost all anthropoids, but the variability of this pneumatization implies that it is not an anthropoid synapomorphy (Hershkovitz, 1977); the primitive anthropoid IOF appears to have been at least as large as that of *Aotus*. Further closure of the IOF has been achieved by septal extensions (especially in platyrrhines) and by pneumatization or elongation of the back end of the maxilla (especially in catarrhines), in different ways in different lineages. However, these differences provide no support for the notion that Anthropoidea is a diphyletic taxon. The catarrhine morphotype is derived relative to the platyrrhine morphotype in a few minor features of orbital morphology (loss of the lateral orbital fissure and associated blood vessels, reduction of the presphenopalatine lamina of the palatine, etc.), but a relatively primitive ceboid orbit like that of *Saguinus* could be easily changed into a primitive hominoid orbit like that of a gibbon. It is possible that the large IOF of *Aotus*, like the other peculiarities of its optic anatomy, represents a secondary reversal attendent upon ocular hypertrophy and loss of the photopic retina. The large IOF of *Homo* is difficult to account for; theorists who still believe that man is a neotenic chimpanzee or a specialized tarsioid are invited to add this feature to their armamentaria. Much light could be shed on this problem by a survey of the appropriate fossil hominoids.

The foregoing reconstruction is based upon several tenuous threads of speculative reasoning, and is highly vulnerable to refutation. Nevertheless, it is the sturdiest house of cards I can build with the available deck. I am not altogether happy with the notion that tarsiers are (so to speak) reprosimianized *Ur*-anthropoids. Much of what I believed at one time about anthropoid origins (Cartmill, 1970, 1972) must be rejected if this notion is accepted. Anthropoids can no longer be seen as descendants of an animal with tarsierlike ocular hypertrophy, and so the reduction of the olfactory apparatus and enlargement of the brain in anthropoids cannot be explained in those terms. The apical interorbital septum found in *Tarsius* and small anthropoids still seems like a satisfactory explanation of their extreme olfactory regression (Cave, 1948, 1967; Cartmill, 1972), but since a similar septum is present in some omomyids (Cartmill and Kay, 1978), it is probably a haplorhine synapomorphy. If so, no macrophthalmic phase in anthropoid evolution need be postulated.

We are left with no good explanation of brain enlargement in anthropoids. The conventional wisdom is that mammals living in trees need keen wits and large brains—except, of course, for those that are nocturnal, or nongregarious, or that live on Madagascar and have no predators. These qualifications, which are needed to exclude squirrels, sloths, lemurs, and so on from the scope of the explanation (Jerison, 1973, pp. 418–419), render the theory much less satisfactory; if neither arboreality nor gregariousness nor diurnality have any discernible effect on brain size singly, why should they jointly? The problem has no solution at present, and it may be irrelevant to the question of anthropoid origins. The only Oligocene anthropoid whose relative brain size can be guessed at is *Aegyptopithecus zeuxis* from the Fayum. As reconstructed by Radinsky (1973), its brain was slightly smaller, both absolutely and relative to cranial size, than that of *Indri*. Radinsky concluded that the brain of *Aegyptopithecus* was just the size that would be expected for an anthropoid of its body weight, which he estimated from foramen magnum area. As Jerison (1973) has shown, this is not a reliable way to estimate the denominator of brain/body weight ratios. The teeth, skull, and ulna of *Aegyptopithecus* all suggest that it weighed about as much as an *Alouatta* female (Fleagle *et al.*, 1975; Conroy, 1976; Kay and Simons, 1980), or somewhere in the general vicinity of 6 kg—which is just what *Indri* weighs (Kay, 1973). Gingerich (1977) and Kay and Simons (1980) have accordingly inferred that the relative brain size of *Aegyptopithecus* fell in the range seen among extant strepsirhines. Unless we wish to claim that *Aegyptopithecus* is a degenerate form or a product of "phyletic gigantism" [analogous to the "secondary dwarfing" posited by Stephan (1972) to explain why *Daubentonia's* brain is so big], we are forced to conclude that the ancestral anthropoid had a brain that was relatively no larger than that of *Indri*—which has a relatively smaller brain than most Malagasy lemurs (Bauchot and Stephan, 1969).

The question of anthropoid monophyly is not a cladistic question. There is universal agreement that catarrhines and platyrrhines are phyletic sisters, and that their last common ancestor could not have been ancestral to any other known primates. This point is not currently in doubt. The question of monophyly, so far as it is a live question, is simply one of grade: was the last common ancestor of platyrrhines and catarrhines enough like a monkey to call it an anthropoid? The evidence presented here suggests that the common ancestor had visual and masticatory apparatus of anthropoid grade, but a lemur-sized brain. Whether this description warrants calling it an anthropoid is not an objective question, and should be distinguished from the genuinely cladistic problem of determining whether the tarsiids, omomyids, adapids, or whatever represent the phyletic sister of the Anthropoidea. On this last question, as the present volume makes clear, studies of primate evolution are depressingly far from a consensus.

Conclusion

Tarsius and anthropoids differ from all other vertebrates in having a postorbital septum, and from all other extant mammals in having a retinal fovea. The septum of *Tarsius* appears to be homologous with that of anthropoids. Extant haplorhines also lack a tapetum lucidum. The tapetum was probably present in the ancestral strepsirhines. Judging from relative orbit size, early haplorhines had a tapetum too. Absence of the tapetum, like the presence of the postorbital septum and fovea, is thus a synapomorphy that links *Tarsius* more closely to anthropoids than to the early Tertiary "tarsioids." Since the last common ancestor of the *Tarsius*-anthropoid clade had a fovea and lacked a tapetum, it must have been diurnal. The postorbital septum of this common ancestor served to insulate the foveate retina from temporalis contractions, preserving visual acuity when the animal was simultaneously chewing and hunting for hard-to-find food items like insects. Nocturnal descendants of this ancestor, which have developed greatly enlarged eyes to compensate for the absence of the tapetum, have lost the cone-rich retina required for acute diurnal vision. Freed from selection pressures to insulate the orbital contents from the masticatory muscles, they retain larger inferior orbital fissures than other extant haplorhines. In *Tarsius*, the grotesque enlargement of the eyes has restricted the cranial space available for masticatory muscles. To compensate for this, the medial pterygoid has spread back into the orbit again through a secondarily reenlarged inferior orbital fissure. The postorbital septum of *Tarsius*, like its retinal fovea, is a vestige of a more anthropoid-like ancestral condition. Most diurnal anthropoids have tended to close off the inferior orbital fissure, presumably to prevent contractions of the medial pterygoid from affecting the orbital contents, but this closure has been accomplished differently in different groups of anthropoids. The last common ancestor of the extant Anthropoidea had an inferior orbital fissure at least as large as that of *Aotus*; a postorbital septum and orbital blood vessels of ceboid type; and characteristically anthropoid visual and masticatory apparatus. Its brain, however, was relatively no larger than a lemur's. Whether or not it should be called an anthropoid is a matter of definition.

ACKNOWLEDGMENTS

I am grateful to the staff and employees of the U.S. National Museum, the Field Museum of Natural History, the American Museum of Natural History, the Museum of Comparative Zoology, the Los Angeles County Museum of Natural History, and the Duke University Center for the Study of Primate Biology and History for their help and cooperation in the research on which this article is based. I thank K. Brown, W. L. Hylander, R. F. Kay, and J. Wible for their valuable comments and criticisms.

References

Ashley-Montagu, M. F., 1933, The anthropological significance of the pterion in the primates, *Am. J. Phys. Anthropol.* **18**:159–336.

Baldwin, J. D., and Baldwin, J., 1972, The ecology and behavior of squirrel monkeys (*Saimiri oerstedii*) in a natural forest in western Panama, *Folia Primatol.* **18**:161–184.

Bauchot, R., and Stephan, H., 1969, Encéphalisation et niveau évolutif chez les Simiens, *Mammalia* **33**:225–275.

Beecher, R. M., 1977, Function and fusion at the mandibular symphysis, *Am. J. Phys. Anthropol.* **47**:325–336.

Beecher, R. M., 1979*a*, Functional significance of the mandibular symphysis, *J. Morphol.* **159**:117–130.

Beecher, R. M., 1979*b*, Evolution of the mandibular symphysis in the Notharctinae: Implications for anthropoid origins, *Am. J. Phys. Anthropol.* **50**:418.

Cachel, S. M., 1976, The origins of the anthropoid grade, Ph.D. thesis, University of Chicago, Chicago.

Cachel, S. M., 1977, Function in primate masticatory musculature as demonstrated by muscle weights, *Am. J. Phys. Anthropol.* **47**:122.

Cachel, S. M., 1979, A functional analysis of the primate masticatory system and the origin of the anthropoid post-orbital septum, *Am. J. Phys. Anthropol.* **50**:1–18.

Cartmill, M., 1968, Morphology and orientation of the orbit in arboreal mammals, *Am. J. Phys. Anthropol.* **29**:131–132.

Cartmill, M., 1970, The orbits of arboreal mammals: A reassessment of the arboreal theory of primate evolution, Ph.D. thesis, University of Chicago, Chicago.

Cartmill, M., 1972, Arboreal adaptations and the origin of the order Primates, in: *The Functional and Evolutionary Biology of Primates* (R. H. Tuttle, ed.), pp. 97–122, Aldine-Atherton, Chicago.

Cartmill, M., 1978, The orbital mosaic in prosimians and the use of variable traits in systematics, *Folia Primatol.* **30**:89–114.

Cartmill, M., and Kay, R. F., 1978, Cranio-dental morphology, tarsier affinities, and primate sub-orders, in: *Recent Advances in Primatology* (D. J. Chivers and K. A. Joysey, eds.), Vol. 3, pp. 205–214, Academic Press, London.

Cave, A. J. E., 1948, The nasal fossa in the primates, *Proc. Annu. Meeting Br. Med. Assoc.* **1948**:363–366.

Cave, A. J. E., 1967, Observations on the platyrrhine nasal fossa, *Am. J. Phys. Anthropol.* **26**:277–288.

Charles-Dominique, P., 1975, Nocturnality and diurnality: An ecological interpretation of these two modes of life by an analysis of the higher vertebrate fauna in tropical forest ecosystems, in: *Phylogeny of the Primates* (W. P. Luckett and F. S. Szalay, eds.), pp. 69–88, Plenum Press, New York.

Charles-Dominique, P., 1977, *Ecology and Behaviour of Nocturnal Primates,* Columbia University Press, New York.

Collins, E. T., 1921, Changes in the visual organs correlated with the adoption of arboreal life and with the assumption of the erect posture, *Trans. Ophthalmol. Soc. U.K.* **41**:10–90.

Collins, H. B., Jr., 1925, The pterion in primates, *Am. J. Phys. Anthropol.* **8**:261–274.

Conroy, G.C., 1976, Primate postcranial remains from the Oligocene of Egypt, *Contrib. Primatol.* **8**:1–134.

Costa Ferreira, A. A. da, 1919, Sobre a formação da parede externa de órbita, *Arquiv. Anat. Antropol.* (Lisbon) **5**:289–294.

De Oliveira, L. F., and Ripps, H, 1968, The "area centralis" of the owl monkey (*Aotes trivirgatus*), *Vision Res.* **8**:223–228.

Dom, R., Fisher, B. L., and Martin, G. F., 1970, The venous sytem of the head and neck of the opossum (*Didelphis virginiana*), *J. Morphol.* **132**:487–496.

Duckworth, W. H. L., 1904, On irregularities in the conformation of the post-orbital wall in skulls of *Hylobates mulleri*, and of an aboriginal native of Australia, in: *Studies from the Anthropological Laboratory* (W. H. L. Duckworth, ed.), pp. 26–28, Cambridge University Press, Cambridge.

Ehara, A., 1969, Zur Phylogenese und Funktion des Orbitaseitenrandes der Primaten, *Z. Morphol. Anthropol.* **60**:263–271.

Eisenberg, N. A., and Brodie, A. G., 1965, Antagonism of temporal fascia to masseteric contraction, *Anat. Rec.* **152**:185–192.

Fiedler, W., 1953, Die Kaumuskulatur der Insectivora, *Acta Anat.* **18**:101-175.

Fleagle, J. G., Simons, E. L., and Conroy, G. C., 1975, Ape limb bone from the Oligocene of Egypt, *Science* **189**:135-137.

Fogden, M. P. L., 1974, A preliminary field study of the western tarsier, *Tarsius bancanus* Horsefield, in: *Prosimian Biology* (R. D. Martin, G. A. Doyle, and A. C. Walker, eds.), pp. 151-165, Duckworth, London.

Franz, V., 1934, Vergleichende Anatomie des Wirbeltierauges, in: *Handbuch der vergleichende Anatomie der Wirbeltiere* (L. Bolk, E. Goppert, E. Kallius, and W. Lubosch, eds.), pp. 989-1292, Urban and Schwarzenberg, Berlin.

Gaspard, M., 1972, *Les Muscles Masticateurs Superficiels des Singes à l'Homme. Anatomie Comparée et Anatomo-Physiologie*, Librairie Maloine S.A., Paris.

Gingerich, P. D., 1973, Anatomy of the temporal bone in the Oligocene anthropoid *Apidium* and the origin of Anthropoidea, *Folia Primatol.* **19**:329-337.

Gingerich, P. D., 1977, Correlation of tooth size and body size in living hominoid primates, with a note on relative brain size in *Aegyptopithecus* and *Proconsul, Am. J. Phys. Anthropol.* **47**:395-398.

Hamasaki, D. I., 1967, An anatomical and electrophysiological study of the retina of the owl monkey, *Aotes trivirgatus, J. Comp. Neurol.* **130**:163-169.

Hershkovitz, P., 1974, A new genus of Late Oligocene monkey (Cebidae, Platyrrhini) with notes on postorbital closure and platyrrhine evolution, *Folia Primatol.* **21**:1-35.

Hershkovitz, P., 1977, *Living New World Monkeys (Platyrrhini), With an Introduction to Primates*, Vol. 1, University of Chicago Press, Chicago.

Hess, C. von. 1911, Zur Kenntnis des Tapetum lucidum in Säugerauge, *Arch. Vergl. Ophthalmol.* **2**:3-11.

Hill, W. C. O., 1953, The blood-vascular system of *Tarsius, Proc. Zool. Soc. London* **123**:655-694.

Hill, W. C. O., 1960, *Primates: Comparative Anatomy and Taxonomy*. Vol. IV, *Platyrrhini, Cebidae,* Part A, University of Edinburgh Press, Edinburgh.

Hoffstetter, R., 1977, Phylogénie des Primates: Confrontation des résultats obtenus par les diverses voies d'approche du problème, *Bull. Mém. Soc. Anthropol. Paris* **4** (Sér. 13):327-346.

Hylander, W. L., 1978, Incisal bite force direction in humans and the functional significance of mammalian mandibular translation, *Am. J. Phys. Anthropol.* **48**:1-8.

Hylander, W. L., 1979, The functional significance of primate mandibular form, *J. Morphol.* **160**:223-240.

Izawa, K., 1979, Foods and feeding behavior of wild black-capped capuchin *(Cebus apella), Primates* **20**:57-76.

Jerison, H. J., 1973, *Evolution of the Brain and Intelligence,* Academic Press, New York.

Jones, A. E., 1965, The retinal structure of (*Aotes trivirgatus*) the owl monkey, *J. Comp. Neurol.* **125**:19-28.

Jones, F. Wood, 1916, *Arboreal Man,* E. Arnold, London.

Jones, F. Wood, 1929, *Man's Place Among the Mammals,* E. Arnold, London.

Jones, F. Wood, 1951, The external characters of a foetal tarsier, *Proc. Zool. Soc. London* **120**:723-730.

Kay, R. F., 1973, Mastication, molar tooth structure, and diet in primates, Ph.D. Thesis, Yale University, New Haven.

Kay, R. F., and Cartmill, M., 1977, Cranial morphology and adaptations of *Palaechthon nacimienti* and other Paromomyidae (Plesiadapoidea, ? Primates), with a description of a new genus and species, *J. Hum. Evol.* **6**:19-53.

Kay, R. F., and Simons, E. L., 1980, The ecology of Oligocene African Anthropoidea, *Int. J. Primatol.* **1**:21-37.

Kinzey, W. G., 1977*a*, Positional behavior and ecology in *Callicebus torquatus, Yearb. Phys. Anthropol.* **20**:468-480.

Kinzey, W. G., 1977*b*, Diet and feeding behavior of *Callicebus torquatus,* in: *Primate Ecology* (T. H. Clutton-Brook, ed.), pp. 127-151, Academic Press, London.

Klein, L. L., and Klein, D. J., 1975, Social and ecological contrasts between four taxa of neotropical primates, in: *Socioecology and Psychology of Primates* (R. H. Tuttle, ed.), pp. 59-86, Mouton, The Hague.

Kolmer, W., 1930, Zur Kenntnis des Auges der Primaten, *Z. Anat. Entwickl.* **93**:679–722.

Le Gros Clark, W. E., 1949, *History of the Primates: An Introduction to the Study of Fossil Man*, British Museum (Natural History), London.

Luckett, W. P., and Szalay, F. S., 1978, Clades versus grades in primate phylogeny, in: *Recent Advances in Primatology* (D. J. Chivers and K. A. Joysey, eds.), Vol. 3, pp. 227–235, Academic Press, New York.

Martin, R. D., 1972, Adaptive radiation and behavior of the Malagasy lemurs, *Phil. Trans. Roy. Soc. London, Ser. B* **264**:295–352.

Martin, R. D., 1973, Comparative anatomy and primate systematics, *Symp. Zool. Soc. London* **33**:301–337.

Martin, R. D., 1978, Major features of prosimian evolution: A discussion in the light of chromosomal evidence, in: *Recent Advances in Primatology* (D. J. Chivers and K. A. Joysey, eds.), Vol. 3, pp. 3–26, Academic Press, New York.

Miller, M. E., Christensen, G. C., and Evans, H. E., 1964, *Anatomy of the Dog,* Saunders, Philadelphia.

Moynihan, M., 1976, *The New World Primates,* Princeton University Press, Princeton.

Oxnard, C. E., 1957, The maxillary nerve in the Ceboidea, *Proc. Zool. Soc. London* **128**:113–117.

Pariente, G. F., 1975, Observation ophtalmologique de zones fovéales vraies chez *Lemur catta* et *Hapalemur griseus,* Primates de Madagascar, *Mammalia* **39**:487–497.

Pariente, G. F., 1976, Les différents aspects de la limite du tapetum lucidum chez les Prosimiens, *Vision Res.* **16**:387–391.

Pirie, A., 1959, Crystals of riboflavin making up the tapetum lucidum in the eye of a lemur, *Nature* **183**:985–986.

Pocock, R. I., 1918, On the external characters of the lemurs and of *Tarsius, Proc. Zool. Soc. London* **1918**:19–53.

Preuss, F., 1954, Gibt es eine V. reflexa?, *Tierarztl. Umschau* **9**:388–389.

Prince, J. H., 1953, Comparative anatomy of the orbit, *Br. J. Physiol. Optics, N.S.* **10**:144–154.

Radinsky, L., 1973, *Aegyptopithecus* endocasts: Oldest record of a pongid brain, *Am. J. Phys. Anthropol.* **39**:239–248.

Ramirez, M. F., Freese, C. H., and Revilla, C. J., 1977, Feeding ecology of the pygmy marmoset, *Cebuella pygmaea,* in northeastern Peru, *in: The Biology and Conservation of the Callitrichidae* (D. G. Kleiman, ed.), pp. 91–104, Smithsonian Institute Press, Washington, D.C.

Riggs, E. S., 1934, A new marsupial saber-tooth from the Pliocene of Argentina and its relationships to other South American predacious marsupials, *Trans. Am. Phil. Soc., N.S.* **24**:1–32.

Rohen, J. W., 1962, Sehorgane, *Primatologia* **2**[1(6)]:1–120.

Rohen, J. W., and Castenholz, A., 1967, Über die Zentralisation der Retina bei Primaten, *Folia Primatol.* **5**:92–147.

Rose, K. D., and Fleagle, J. G., 1980, The fossil history of nonhuman primates in the Americas, in: *Neotropical Primatology* (R. A. Mittermeier and A. F. Coimbra-Filho, eds.), Brazilian Academy of Sciences, Rio de Janeiro.

Rosenberger, A. L., 1979, Phyletic perspectives and platyrrhine classification, *Am. J. Phys. Anthropol.* **50**:476.

Ruskell, G. L., 1964, Blood vessels of the orbit and globe, in: *The Rabbit in Eye Research* (J. H. Prince, ed.), pp. 514–553, Charles C. Thomas, Springfield, Ill.

Saban, R., 1963, Contribution à l'étude de l'os temporal des Primates, *Mém. Mus. Natl. Hist. Nat., (N.S. 4, Ser. A)* **29**:1–378.

Schultz, A. H., 1940, The size of the orbit and of the eye in primates, *Am. J. Phys. Anthropol.* **26**:389–408.

Schultz, A. H., 1944, Age changes and variability in gibbons: A morphological study on a population sample of a man-like ape, *Am. J. Phys. Anthropol., N.S.* **2**:1–130.

Schultz, A. H., 1973, The skeleton of the Hylobatidae and other observations on their morphology, *Gibbon and Siamang* **2**:1–54.

Schwartz, D. J., and Huelke, D. F., 1963, Morphology of the head and neck of the macaque monkey: The muscles of mastication and the mandibular division of the trigeminal nerve, *J. Dent. Res.* **42**:1222–1233.

Schwartz, J. H., Tattersall, I., and Eldredge, N., 1978, Phylogeny and classification of the primates revisited, *Yearb. Phys. Anthropol.* **21**:95–133.

Simons, E. L., 1962, Fossil evidence relating to the early evolution of primate behavior, *Ann. N.Y. Acad. Sci.* **102**:282–294.

Simons, E. L., 1969, Recent advances in paleoanthropology, *Yearb. Phys. Anthropol.* **15**:14–23.

Simons, E. L., 1972, *Primate Evolution: An Introduction to Man's Place in Nature,* Macmillan, New York.

Simons, E. L., and Russell, D. E., 1960, Notes on the cranial anatomy of *Necrolemur, Breviora* **127**:1–14.

Simpson, G. G., 1953, *The Major Features of Evolution,* Columbia University Press, New York.

Spatz, W. B., 1968, Die Bedeutung der Augen für die sagittale Gestaltung des Schädels von *Tarsius* (Prosimiae, Tarsiiformes), *Folia Primatol.* **9**:22–40.

Standing, H. F., 1908, On recently discovered subfossil primates from Madagascar, *Trans. Zool. Soc. London* **18**:59–162.

Starck, D., 1933, Die Kaumuskulatur der Platyrrhinen, *Geg. Morph. Jb.* **72**:212–285.

Stephan, H., 1972, Evolution of primate brains: A comparative anatomical investigation, in: *The Functional and Evolutionary Biology of Primates* (R. H. Tuttle, ed.), pp. 155–174, Aldine, Chicago.

Szalay, F. S., 1975, Phylogeny of primate higher taxa: The basicranial evidence, in: *Phylogeny of the Primates* (W. P. Luckett and F. S. Szalay, eds.), pp. 91–126, Plenum Press, New York.

Szalay, F. S., 1976, Systematics of the Omomyidae (Tarsiiformes, Primates): Taxonomy, phylogeny, and adaptations, *Bull. Am. Mus. Nat. Hist.* **156**:157–450.

Szalay, F. S., and Wilson, J. A., 1976, Basicranial morphology of the early Tertiary tarsiiform *Rooneyia* from Texas, *Folia Primatol.* **25**:288–293.

Tattersall, I., 1973, Cranial anatomy of the Archaeolemurinae (Lemuroidea, Primates), *Anthropol. Papers Am. Mus. Nat. Hist.* **52**:1–110.

Thorington, R. W., Jr., 1967, Feeding and activity of *Cebus* and *Saimiri* in a Colombian forest, in: *Neue Ergebnisse der Primatologie* (D. Starck, R. Schneider, and H.-J. Kuhn, eds.), pp. 180–184, Fischer, Stuttgart.

Thornington, R. W., Jr., 1968, Observations of squirrel monkeys in a Colombian forest, in: *The Squirrel Monkey* (L. A. Rosenblum and R. W. Cooper, eds.), pp. 69–85, Academic Press, New York.

Walls, G. L., 1942, *The Vertebrate Eye and Its Adaptive Radiation* (1965 reprint), Hafner, New York.

Weale, R. A., 1966, Why does the human retina posses a fovea?, *Nature* **212**:255–256.

Wolin, L. R., and Massopust, L. C., 1970, Morphology of the primate retina, in: *The Primate Brain* (C. R. Noback and W. Montagna, eds.), pp. 1–28, Appleton-Century-Crofts, New York.

Comparative Study of the Endocranial Casts of New and Old World Monkeys

13

D. FALK

Introduction

The fossil record of brain evolution is poorly represented for monkeys. There is only one known fossil ceboid endocast and the described endocasts of fossil cercopithecoid monkeys can be counted on one hand (Radinsky, 1974). Thus, any effort to assess the evolutionary histories of Old and New World monkeys, based on neurological data, must rely heavily on comparisons of brains or endocasts from extant species.

Sulcal patterns of ceboid monkeys have previously been studied by Hershkovitz (1970) who prepared and analyzed endocasts from skulls, and Anthony (1946) who described real brains. Cortices of cercopithecoids have been described by Connolly (1950) who examined real brains, and Falk (1978*c*) who described and analyzed endocasts representing all living genera of cercopithecoids and determined which features of sulcal pattern seem relatively derived. In this report, 53 endocasts representing all extant ceboid genera are analyzed and details of sulcal patterns in Old and New World monkeys are compared and interpreted. An effort is made to identify homologous sulci in the two groups, based on criteria outlined by Simpson (1961) and elaborated for the central nervous system by Campbell (1976) and Campbell and Hodos (1970). The evolutionary history of Old and New World monkeys is discussed in light of neurological homologies, additional neurological data and recent

D. FALK • Department of Anatomy, University of Puerto Rico Medical Sciences Campus, San Juan, Puerto Rico 00936.

reinterpretations of potential biogeographic dispersal models (Hoffstetter, 1972).

Materials and Methods

Latex endocasts were prepared from 53 skulls, representing all 16 genera of extant ceboid monkeys (Napier and Napier, 1967), from collections of the Field Museum of Natural History, Chicago (Table 1). These endocasts, prepared according to the method of Radinsky (1968), reproduce details of external brain morphology, including sulcal patterns. Since sulci are present at birth (Falk, 1978*c*), endocasts were prepared from skulls of infants and juveniles as well as adult specimens. For each endocast, sulci were identified by comparison with brains and figures of brains and endocasts (Anthony, 1946; Connolly, 1950; Hershkovitz, 1970; Hill, 1960, 1962); see Fig. 1. Details of external brain morphology of New World monkeys were compared with those determined from my collection of 107 endocasts representing all 14 genera of extant cercopithecoids (Falk, 1978*c*), see Table 2.

The size of the paraflocculus of the cerebellum, located within the subarcuate fossa of the temporal bone, was estimated for those ceboid endocasts in my collection which (1) reproduce one or both paraflocculi and (2) for which I had determined cranial capacities. The volume of each paraflocculus was estimated by measuring its maximum width in centimeters on the endocast with a sliding caliper and substituting the obtained value (or the mean value for right and left sides if both paraflocculi were present) as the diameter (d) in the formula for the volume of a sphere ($v = 1/6\,\pi d^3$). These estimates must be

Table 1. Fifty-three Endocasts Prepared from Ceboid Skulls

Callitrichidae		Cebidae	
Genus	Number of endocasts	Genus	Number of endocasts
Callimico	2	*Alouatta*	4
Callithrix	4	*Aotus*	6
Cebuella	1	*Ateles*	3
Leontopithecus	3	*Brachyteles*	1
Saguinus	3	*Cacajao*	2
Total	13	*Callicebus*	4
		Cebus	5
		Chiropotes	3
		Lagothrix	4
		Pithecia	5
		Saimiri	3
		Total	40

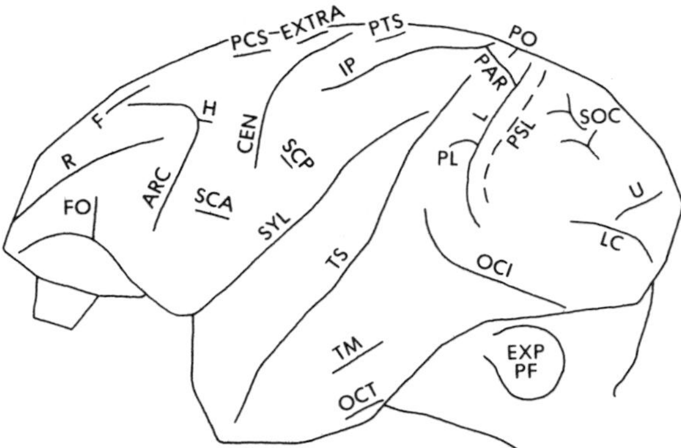

Fig. 1. Model of endocast showing sulci discussed in the text. Left lateral view, abbreviations as in Table 2.

viewed as rough approximations since paraflocculi are not perfectly spherical but rather tend to be egg-shaped.

An index of the relative size of the paraflocculus was devised by dividing its estimated volume by the cranial capacity of the specimen in question [Cranial capacities were determined by filling the braincase with millet seed; see Falk (1978*c*) for methodology].

For comparative purposes, estimates of relative paraflocculus size were determined in the above manner for eight cercopithecine and eight colobine endocasts. These are listed, along with the estimates for ceboids, in Table 3.

As noted in the figure legends, sulcal patterns of some of the largest-brained species are occasionally reproduced from figures in the literature, since the largest-brained species within primates (and other groups of mammals) fail to reproduce details of sulcal pattern on endocasts (Radinsky, 1972). Composite sulcal patterns shown in Figs. 2–5 exhibit all sulci reported present in this study and in the literature, for the species in question.

Observations

Callitrichidae

The most frequently seen sulcal pattern in the callitrichids consists simply of parallel Sylvian (S) and superior temporal (TS) sulci (Fig. 2). Hershkovitz (1970) figures the rectus sulcus (R) in endocasts of all callitrichids except *Cebuella*. None of my callitrichid endocasts, or callitrichid brains figured by Anthony (1946), show a distinct R; although occasionally a dimple is seen in its usual location (this study). The intraparietal sulcus (IP) has been figured in the literature for a few representatives of *Callithrix, Leontopithecus,* and *Saguinus*. Although Anthony does not show a central sulcus (C) for any callitrichids, Hershkovitz shows C for one *Callithrix* and one *Saguinus* specimen. My endocasts reproduce slight dimples or very short sulci in the position

Table 2. Comparative Data Determined from Endocasts of Old and New World Monkeys[a]

Superfamily		Cercopithecoidea		Ceboidea							
Family		Cercopithecidae		Callitrichidae	Cebidae						
Subfamily		Colobinae	Cercopithecinae		Aotinae		Alouattinae	Pitheciinae	Cebinae		Atelinae
Genus					*Aotus*	*Callicebus*			*Saimiri*	*Cebus*	
Feature	Numbers by Hershkovitz (1970)										
S	3	X	X	X	X	X	X	X	X	X	X
TS	4	X	X	X	X	X	X	X	X	X	X
IP	6	X	X	X	X	X	X	X	X	X	X
R	5	X	X	d	X	X	X	X	X	X	X
C	7	X	X	d	X	X	X	X	X	X	X
L	10	X	X	d	X	X	X	X	X	X	X
TM	11	X	X	?	X	X	X	X	X	X	X
OCI	13	X	X	?	X	X	X	X	X	X	X
SCA	—	X	X		X	X	X	X	X	X	X
LC	12	X	X		X	1	X	X	1	X	X
PAR	21	X				?	X	1	?	X	1
PO	15	X	X				1	X	X	X	X
PTS	17	X	X				X	X	1	1	X
PCS	16	X	X				X	X		X	X
OF	9	X	X					X		X	X
F	18,19	X	X					X		X	X
PL	—	?	X				1	X		X	1
ARC	8 + 25	X	X				1		?	X	1
H	26	X	X							X	X
SCP	24	X	X				1			X	1
S + TS	—	X	X				1		1	X	1
OCT	20		X								X
EXP PF	—							X			X
PSL	—									X	
SOC	14,23										X
EXTRA P	—						X				X
S + IP	—						X		1		X

[a] Feature abbreviations are listed below; numbers are those used by Hershkovitz (1970) to refer to sulci. Presence of X indicates that a feature has been observed in *some* of the endocasts (Table 1) representing the taxon listed at the top of Table 2; d means that the feature in question is represented by a dimple rather than a sulcus; 1 indicates that a feature has been figured at least once in the literature for a given taxon, although it is not reproduced on the endocasts reported on here, and a question mark (?) indicates that a given feature does not appear in the literature although it appears faintly on *one* of the endocasts studied in this report. Abbreviations for endocast features: (s. = sulcus) ARC, arcuate; C, central s., EXP PF, expanded paraflocculi of the cerebellum; EXTRA P, additional pre- or postcentral s. (associated with prehensile tails); F, mid- or superior-frontal s.; FO, fronto-orbital s.; H, caudal branch arcuate s.; IP, intraparietal s.; L, lunate s.; LC, lateral calcarine s.; OCI, inferior occipital s.; OCT, visibility of occipitotemporal s. in lateral view; PAR, paraoccipital s.; PCS, precentral superior s.; PL, prelunate s.; PSL, pseudolunate s.; PTS, postcentral superior s.; PO, parietooccipital s.; R, rectus s.; S, Sylvian s.; S + IP, confluence of S and IP; S + TS, merging and caudal meeting of S and TS in *most* representatives of the taxon; SCA, subcentral anterior s.; SCP, subcentral posterior s.; SOC, numerous secondary occipital sulci; TM, middle temporal s.; TS, superior temporal s.

Table 3. Relative Capacity of the Paraflocculus of the Cerebellum (CC)[a]

	n (endocasts)	$\bar{x}$	σ	Range
Callitrichidae	8	0.0016	0.0006	0.0009–0.0029
Aotinae	5	0.0027	0.0017	0.0013–0.0057
Alouattinae	1	0.0011		
Cebinae	6	0.0012	0.0004	0.0010–0.0017
Pitheciinae	6	0.0037	0.0011	0.0023–0.0056
Atelinae	4	0.0036	0.0012	0.0028–0.0053
Colobinae				
4 *Nasalis*				
4 *Simias*	8	0.0004	0.00008	0.0003–0.0005
Cercopithecinae				
4 *Macaca*				
4 *Papio*	8	0.0006	0.0003	0.0003–0.0010

[a] $\bar{x}$, mean volume of the paraflocculus divided by the cranial capacity. The volume of the paraflocculus is only a rough approximation; see text for discussion.

of central sulci in one specimen each of *Callithrix* and *Leontopithecus*. Other sulci which *infrequently* appear in figured specimens in the literature as dimples or very short sulci in callitrichids include: the middle temporal sulcus (TM), the lunate sulcus (L), and the inferior occipital sulcus (OCI; one *Leontopithecus* endocast, this study). Figure 2 illustrates typical callitrichid sulcal patterns as well as a composite pattern which exhibits each sulcus discussed above (see Table 2).

Cebidae

Sulcal patterns were determined for the Aotinae from four *Callicebus* and six *Aotus* endocasts (Table 2). The pattern shown for *Callicebus* (Fig. 2) is in agreement with corresponding figures in the literature. The sulcal pattern determined for *Aotus,* on the other hand, contradicts the findings of previous workers.

According to Anthony (1946), Hershkovitz (1970) and Hill (1960), S and IP are continuous in *Aotus.* However, the six endocasts of *Aotus* I studied suggest that S and IP are *not* continuous (Fig. 2). Previous workers may have labeled the caudal portion of an elongated S as IP. In one hemisphere of an *Aotus* endocast, IP appears to be confluent with R; in another hemisphere, there is but a slight break between the two longitudinally oriented sulci (Fig. 2). The central sulcus fails to appear in Anthony's (1946) four *Aotus* brains. Hershkovitz (1970), however, does show a short C based on his study of three *Aotus* endocast hemispheres (hemicasts). When C is reproduced on my endocasts, it appears as a slight sulcus that crosses IP (Fig. 2).

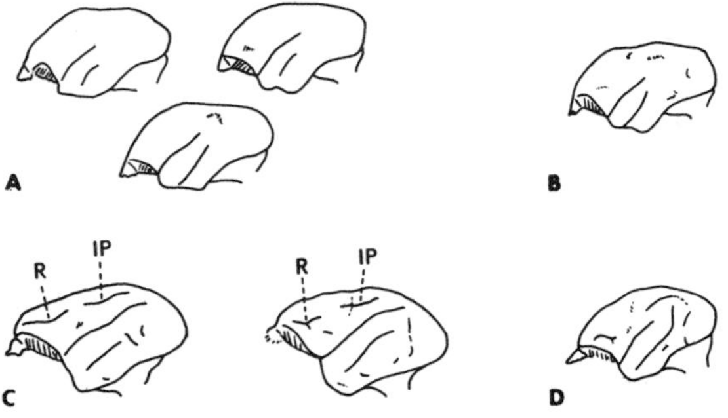

Fig. 2. (A) and (B) Callitrichid sulcal patterns: (A) typical callitrichid patterns; (B) composite pattern showing all sulci reported present for numerous individual endocasts. (C) and (D) Sulcal patterns of Aotinae: (C) patterns of two *Aotus* individuals (FM66430 and FM62074); (D) composite sulcal pattern based on four *Callicebus* endocasts. Abbreviations: R, rectus sulcus; IP, intraparietal sulcus—see text, Fig. 1, and footnote to Table 2 for identification of other sulci.

The configuration of R, IP, and C of *Aotus* is unique among both New and Old World monkeys. A lengthened S of *Aotus*, as well as a longitudinally oriented R and IP, appear similar to corresponding features in extant lower primates (Radinsky, 1970). However, the coronolateral sulcus of lower primates, which lack C, is a single sulcus rather than two sulci (i.e., R and IP) as is usually the case for *Aotus*. Electrophysiological mapping experiments are needed to confirm the identifications of the sulci of *Aotus*. Such studies would perhaps shed light on the functional significance of development of a transversely oriented, well-marked C in anthropoids in contrast to the longitudinally oriented, well-marked coronolateral sulcus of lower primates (Radinsky, 1975).

Composite sulcal patterns representing the three genera of Pitheciinae (*Pithecia, Chiropotes,* and *Cacajao*) are shown in Fig. 3. Because my endocasts do not reproduce much detail in the dorsal parietal/occipital region,

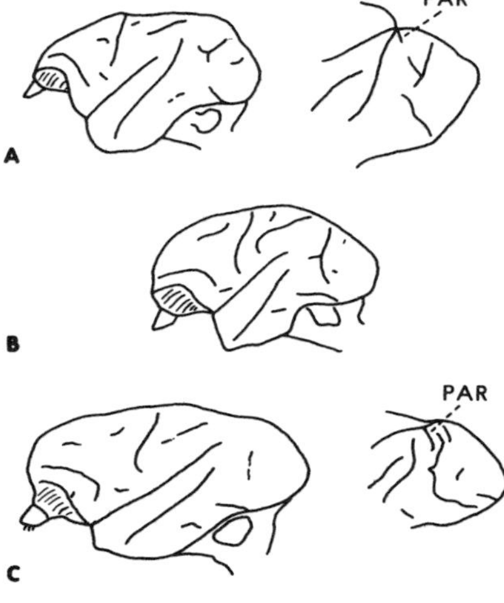

Fig. 3. Sulcal patterns of Pitheciinae. (A) Left, composite pattern based on five *Pithecia* endocasts; right, left occipital cortex of one *Pithecia* individual, after Hill (1960, p. 195). (B) Composite sulcal pattern based on three *Chiropotes* endocasts. (C) Left, composite pattern based on two *Cacajao* endocasts; right, left occipital cortex of one *Cacajao* brain, after Anthony (1946, p. 21). Abbreviations: PAR, paroccipital sulcus. See Fig. 1 and footnote to Table 2 for identification of other sulci.

supplementary illustrations of this region are included, based on figures from the literature. These figures show that, unlike previously discussed ceboids, pithecines exhibit a paroccipital sulcus (PAR). In addition to previously discussed sulci, pithecines may also have the following sulci: parieto-occipital (PO), frontal (F), prelunate (PL), extension of the fronto-orbital sulcus (FO) onto the dorsal surface of the frontal lobe, precentral superior (PCS) and postcentral superior (PTS); see Table 2.

The paraflocculus of the cerebellum, located within the subarcuate fossa of the petrous portion of the temporal bone (Smith, 1903) is reproduced on endocasts of monkeys (McClure and Daron, 1971; Straus, 1960). Both pithecines and atelines (see below) exhibit relatively larger paraflocculi than do other New or Old World monkeys (Table 3).

A composite sulcal pattern determined from four endocasts of *Alouatta*, the only genus of its subfamily, is shown in Fig. 4. The Sylvian and intraparietal sulci are confluent in *Alouatta* and an "extra" precentral sulcus appears medial to PCS. This additional sulcus is often mistaken for PCS as pointed out by Radinsky (1972). In one endocast, PCS and the extra sulcus appear to be continuous and thus give the impression of an extraordinarily elongated PCS.

Although previous workers (Anthony, 1946; Connolly, 1950; Hershkovitz, 1970; Hill, 1962) do not figure PAR for *Alouatta*, my endocasts exhibit PAR. [See Tilney (1928) for illustrations of a complex *Alouatta* sulcal pattern, complete with PAR.] The inferior occipital sulcus (OCI) of *Alouatta* courses in a relatively straight, rostral–caudal direction (versus a medial–lateral arch) and extends further caudally than is usual for other monkeys. This path of OCI is unique amongst ceboid species.

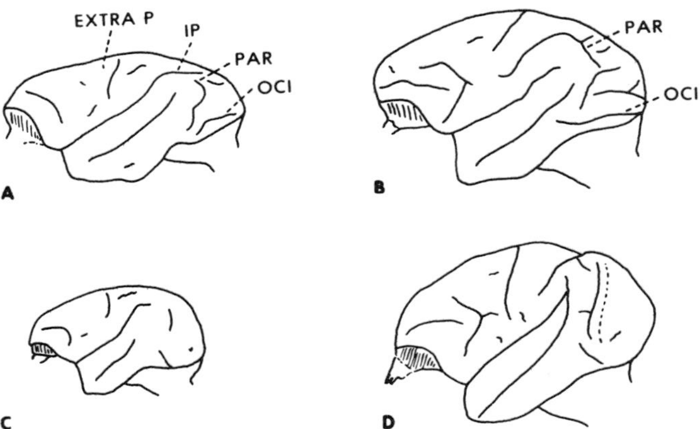

Fig. 4. (A) Composite sulcal pattern based on four *Alouatta* endocasts. (B) Pattern of one *Nasalis* endocast (MCZ37328). (C) Composite sulcal pattern based on three *Saimiri* endocasts. (D) composite pattern based on five *Cebus* endocasts. Abbreviations: EXTRA P, additional pre- or postcentral sulcus; IP, intraparietal sulcus; PAR, paroccipital sulcus; OCI, inferior occipital sulcus. See Fig. 1 and footnote to Table 2 for identification of other sulci.

The sulcal patterns of *Saimiri* and *Cebus* are quite different (Fig. 4), although Napier and Napier (1967) place both genera in the same subfamily (Cebinae). The sulcal pattern of *Cebus* is more complex than that of *Saimiri*. In addition to the occasional presence of PTS, PCS, and FO in *Cebus* endocasts (Table 2), other sulci are consistently present in *Cebus* but not *Saimiri*: prelunate (PL), a true arcuate sulcus (arc) and the caudal branch of *arc* (H). The paroccipital sulcus is sometimes exposed in *Cebus* and at other times is not (Anthony, 1946; this study). Four of the five endocasts of *Cebus* reported on here reproduce a long sulcus which courses parallel and caudal to L (Fig. 4). A gyrus-like ridge is present between this parallel sulcus and L. This sulcus is *not* the transverse occipital s. because it is not derived from a bifurcation of IP (Connolly, 1950, p. 84). Therefore, for lack of an obvious name, I have labeled this sulcus the pseudolunate (Figs. 1 and 4). The Sylvian and intraparietal sulci are discontinuous in *Cebus* endocasts; whereas S and IP have been shown as both continuous and separate for *Saimiri* in the literature (Anthony, 1946; Hershkovitz, 1970). It should be noted that, unlike the condition for other cebids, S and TS consistently merge caudally in *Cebus*.

The most complicated ceboid sulcal patterns are seen in the Atelinae (*Ateles, Brachyteles,* and *Lagothrix*); see Fig. 5 and Table 2. Ateline sulcal patterns exhibit extra pre- and postcentral sulci or, if extra sulci are not present, PCS and PTS may be relatively elongated; the caudal branch of ARC (H) appears elongated; and S and IP are usually (but not always) continuous. As was true for pithecines, endocasts of atelines indicate relatively large cerebel-

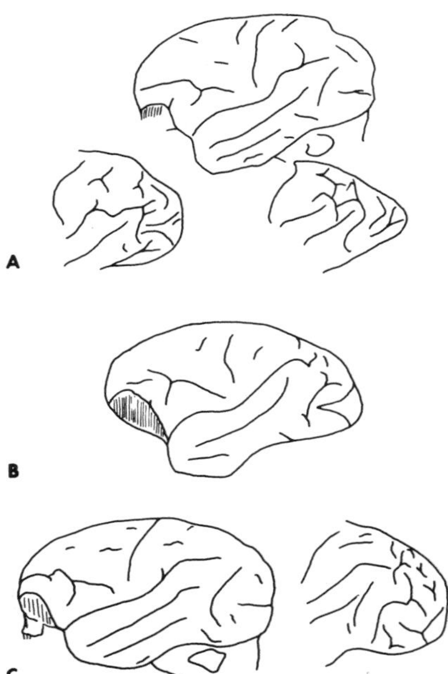

Fig. 5. Sulcal patterns of the prehensile-tailed cebids. (A) Composite sulcal pattern based on four *Lagothrix* endocasts; left, left occipital cortex of one *Lagothrix* brain after Anthony (1946, p. 29); right, left occipital cortex of one *Lagothrix* brain after Hill (1962, p. 227). (B) Pattern of one *Brachyteles* brain after Anthony (1946, p. 36). (C) Left, composite sulcal pattern based on three *Ateles* endocasts; right, left occipital cortex of one *Ateles* brain after Anthony (1946, p. 44). See Fig. 1 and footnote to Table 2 for identification of sulci.

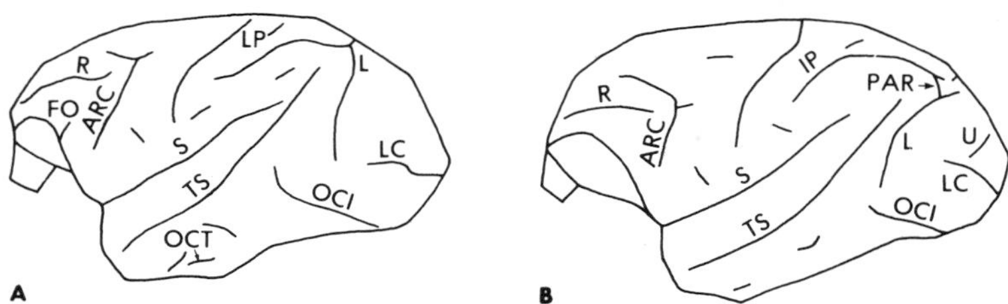

Fig. 6. Typical sulcal patterns of the two subfamilies of Old World monkeys, from Falk (1978*b*). See Fig. 1 and footnote to Table 2 for identification of sulci. (A) Cercopithecine. (B) Colobine.

lar paraflocculi (Table 3). A feature which seems to be unique to atelines amongst ceboids, is visibility of the occipitotemporal sulcus (OCT) in lateral views of some specimens. Finally, it should be noted that ateline occipital cortices exhibit numerous secondary sulci.

Cercopithecidae

I have recently reviewed sulcal patterns for all 14 genera of extant Old World monkeys (Falk, 1978*a*). Data on the presence of sulci for the two cercopithecoid subfamilies (colobines and cercopithecines) are summarized in Table 2, and typical colobine and cercopithecine sulcal patterns are shown in Fig. 6. There is considerably less diversity in sulcal patterns among cercopithecoids than among ceboids (Falk, 1978*a*).

Three characteristics distinquish colobine from cercopithecine sulcal patterns: (1) Colobine endocasts usually exhibit PAR; cercopithecine endocasts do so rarely if ever. (2) The Sylvian and superior temporal sulci converge and meet caudally in 75% of the cercopithecine hemispheres but do so in only 21% of the colobine hemispheres. (3) The occipitotemporal sulcus is visible in lateral view in 16% of the cercopithecine hemispheres but in none of the colobine hemispheres (Falk, 1978*a*). Examination of endocasts suggests that the inferior temporal gyrus of cercopithecines is expanded in a caudolateral direction—hence, visibility of OCT which delimits the ventral border of this gyrus. Other, more subtle distinctions between colobine and cercopithecine sulcal patterns are enumerated elsewhere (Falk, 1978*a,b,c*).

Tails

Both Anthony (1946, p. 89) and Connolly (1950, p. 64) attribute confluence of S and IP in certain genera of New World monkeys to lack of development of a gyrus ("supramarginal" of Connolly, "Leuret's third arcuate" of Anthony) situated in the insular region located within the Sylvian fissure. Radinsky (1972), on the other hand, summarizes evidence in favor of another explanation for confluence of S and IP which correlates with both anatomical and behavioral data. He suggests that IP converges with S as a result of expansion of medially located tail representations which causes a lateral shift of the remaining somatotopic cortex in monkeys with prehensile or nearly prehensile tails (see Fig. 7). Thus S and IP are confluent in the prehensile-tailed alouattines (Fig. 4) and atelines (Fig. 5).

Extra sulci bordering laterally expanded tail representations (see Fig. 7) appear medial to PCS and (sometimes) PTS in prehensile-tailed monkeys. In endocasts of prehensile-tailed monkeys which do not exhibit extra (separate) sulci, expansion of tail representations correlates with relatively elongated PCS and PTS. Atelines also exhibit an elongated caudal portion of ARC (H) which separates motor hand from face regions (Fig. 7). Thus, it seems that lateral expansion of cortical tail regions in fully prehensile-tailed monkeys correlates with elongation of the sulci that delimit leg, arm, and face areas of the sensory/motor cortex.

According to Napier and Napier (1967), *Cebus* is characterized by a slight amount of tail prehensility. Thus, less lateral expansion of cortical tail regions

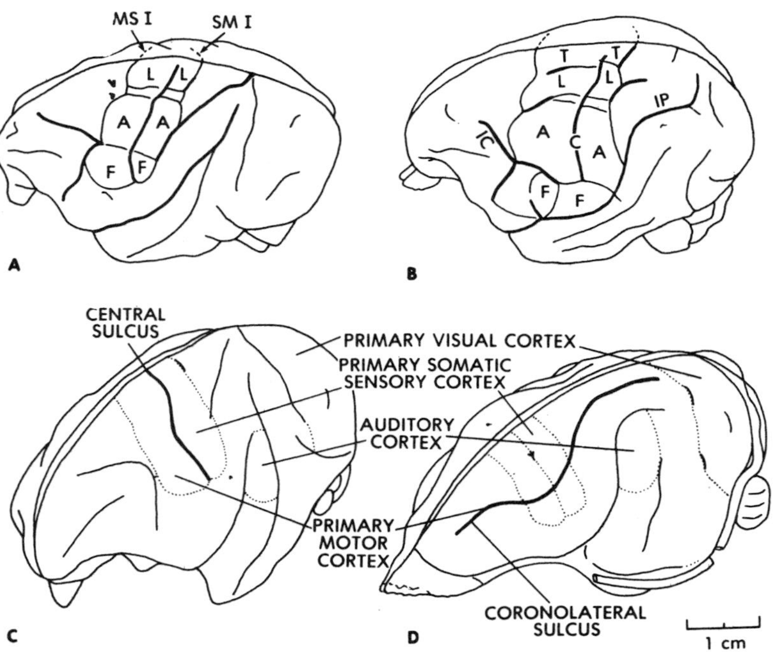

Fig. 7. (A) and (B) Cortical maps of *Macaca* (A) and *Ateles* (B) from Radinsky (1972). Abbreviations: T, L, A, F, cortical representations of tail, hindlimb, forelimb and face, respectively: SMI, primary somatosensory cortex; MSI, primary motor cortex. (C) and (D) Endocasts of *Cercopithecus* (C) and *Lemur* (D) from Radinsky (1975), which illustrate differences in lower primate and cercopithecoid sulcal patterns. Stippled areas outline major sensory and motor cortical areas. See text for discussion.

in *Cebus* than in the fully prehensile-tailed monkeys may account for discontinuity of S and IP. The Sylvian sulcus and IP are sometimes figured as confluent and at other times shown as separate for *Saimiri*, in the literature. Since cortical maps of *Saimiri* (Benjamin and Welker, 1957) show that tail regions are *not* expanded onto the lateral surface of the brain, occasional confluence of S and IP (assuming sulcal identifications in the literature are accurate) may be due to allometric effects in this small-brained species.

Radinsky (1972) notes that *Aotus* provides an exception to the anatomical/behavioral correlation discussed above—i.e., supposedly S and IP are confluent in *Aotus*, yet its tail is not at all prehensile. However, if my interpretation of six *Aotus* endocasts is correct, S and IP are *not* continuous in this species (contrary to the literature) and it therefore does not provide an exception to the rule. As noted above, electrophysiological experiments are needed to confirm the sulcal identifications of *Aotus*. Meanwhile, allometric effects (similar to those postulated above for *Saimiri*) should not be ruled out for small-brained *Aotus*.

As in the case with nonprehensile-tailed ceboids, S and IP of cercopithecoids (none of whom have prehensile tails) are discontinuous. Sulci that delimit face, arm, and leg regions in the sensory/motor cortex (PCS, PTS, and H) are relatively short in cercopithecoid species (Fig. 6).

Visual Areas

Presence of PAR distinguishes colobines from cercopithecines. Presence of PAR in colobines seems to be a primitive condition because in cercopithecines, PAR is opercularized by a rostrally expanded occipital cortex (the derived condition). The paroccipital sulcus is often present in larger brained cebids (Table 2). However, in general PAR is not as well defined or as large in cebid as it is in colobine endocasts, nor does it uniformly occur in most cebid endocasts as is the case for endocasts of colobines. The presence of PAR in pithecines and atelines suggests that their occipital cortices are no more rostrally expanded (i.e., derived) than are occipital cortices of Old World colobines. In fact, the caudal location and extreme concavity of L and the numerous secondary sulci which characterize atelines suggest that more occipital cortex of atelines may be folded *within* sulci (a primitive condition compared to rostral expansion of occipital cortex) than is the case for colobines.

The sulcal pattern of *Cebus*, on the other hand, appears to be more derived in the occipital region than is the case for other ceboids. The lunate sulcus appears to be relatively straight (or, if not straight, rostrally arched in its medial aspect) rather than arched with the concavity directed caudally as is the case for other cebids. The configuration and relatively rostral location of L in *Cebus* resembles the cercopithecine (derived) condition. Both my endocasts and brains of *Cebus* figured in the literature suggest that this species may be characterized by either presence or absence of PAR (with PAR opercularized by the occipital cortex in the latter case). In fact, in one of my endocasts and in two *Cebus* brains figured by Anthony (1946, p. 25), PAR is visible in one hemisphere and opercularized in the other. Thus, opercularization of PAR is less complete in *Cebus* than it is in cercopithecines.

Although *Cebus* lacks the numerous secondary sulci which are found in the occipital regions of the atelines and, to a lesser extent, the colobines, the occipital cortex of *Cebus* exhibits an extra sulcus which runs parallel to L (pseudolunate). Thus, with respect to numbers of secondary occipital sulci, *Cebus* may again be viewed as intermediary between colobines (primitive) and cercopithecines (derived).

Alouatta is the only cebid genus which seems to exhibit PAR as clearly as do the Old World colobines. The L/PAR configuration of *Alouatta* is similar to that which characterizes colobines. The course of OCI in *Alouatta* is unique among ceboids and is similar to the couse of OCI in *Nasalis* and, less frequently, other colobines (Falk, 1978*c*). Thus, the occipital cortices of *Alouatta* and *Nasalis* appear to be surprisingly similar (Fig. 4).

The occipitotemporal sulcus (OCT) is visible in lateral view in 16% of the cercopithecine hemispheres represented in my endocast collection; but is not visible in lateral view of any of the colobine hemispheres (Falk, 1978*a*). Visibility of OCT in lateral view seems to be due to expansion of the interior temporal gyrus in a caudolateral direction. The caudal portion of the inferotemporal convexity is apparently involved in visual pattern discriminations (Iwai and Mishkin, 1969) and the cercopithecine condition again seems to be relatively derived [see Falk (1978*a*) for further discussion]. Of ceboid monkeys, only the atelines occasionally show OCT in lateral view (Fig. 5). As with cercopithecines, this seems to be a derived condition which may have functional implications for visual processing. Thus, visual areas in atelines seem to be colobine-like (primitive) in the occipital cortex (above) but cercopithecine-like (derived) in the inferior temporal convexity.

Frontal Lobe

All genera of extant Old World monkeys exhibit a well-defined ARC which is usually separate from, and arches around the caudal end of, R (Fig. 6). Most ceboid genera, on the other hand, lack a discrete ARC and instead are characterized by an R which is either relatively straight or (more often) arches laterally in its caudal portion. I view the latter condition(s), which most closely resembles the simple, longitudinal orientation of the rostral portion of the lower primate coronolateral sulcus, as relatively primitive. The fossil record in the Old World lends credence to the suggestion that a discrete (separate) ARC represents the derived condition. Contrary to the usual condition in extant cercopithecoids, both the *Dryopithecus africanus* endocast and the Napak frontal from the African Miocene are characterized by lack of a discrete ARC

(Radinsky, 1974). The only New World monkey that consistently exhibits a well-defined ARC separate from, and arching around the caudal end of R, is *Cebus* (Fig. 4).

Other Features

In Old World monkeys (Falk, 1978*a*), S and TS converge and meet caudally in more cercopithecines (75%) than colobines (21%). Functional correlates of merging S and TS are unknown. The Sylvian and superior temporal sulci seldom meet caudally in ceboid monkeys except *Cebus*, in whom S and TS consistently merge caudally.

Unlike any cercopithecoid monkeys, pithecines and atelines manifest relatively enlarged cerebellar paraflocculi (Table 3). This condition may be relative to vestibular (equilibrium) functions.

Conclusions

It is widely believed that ceboids are descended from a lower primate group that originated in North America (Simons, 1972). Since the oldest known fossil primate in South America, *Branisella*, dates to the early Oligocene and since North and South America were separated prior to that during the Eocene, Simons (1972) suggests that the lower primate ancestors of the ceboids rafted (or island-hopped) to South America from Central America. Because of the relative proximity of Central and South America in the late Eocene and the fact that lower primates did inhabit North America during the early Cenozoic, Simon's suggestion seems plausible. However, recent reassessments of continental drift, especially in the South Atlantic Ocean (Tarling and Runcorn, 1973), has led some workers to argue that ceboid monkeys are directly descended from African ancestors who rafted to South America during the middle or late Eocene (Hoffstetter, 1972). Can a comparative study of sulcal patterns in Old and New World monkeys shed any light on this matter?

In order to use the comparative data in Table 2 to analyze the evolutionary histories of monkeys, one must determine which (if any) sulci are homologous in cercopithecoids and ceboids. Welker and Campos (1963) suggest that homologies in sulcal patterns should be determined on the basis of functional analyses. They stress the importance of similarities of thalamic projections to the cerebral cortex for determining sulcal homologies. However, the concept of homology based on structural correspondence is not particularly useful for determining evolutionary histories (Campbell, 1976) since, according to this concept, homologous sulci need not be traceable back to a common ancestor. For example, based on structural correspondence, the cruciate sulcus is homologous in living Carnivora, yet the fossil record shows that this sulcus has been evolved independently at least five times (Radinsky, 1971). Thus, for purposes of discussing the evolutionary history of monkeys, the classic (Simpson, 1961) concepts of homology and parallelism should be used (Campbell, 1976, p. 144):

> Structures and other entities are homologous when they can, *in principle*, be traced back through a genealogical series to a stipulated common ancestral precursor . . . [Parallelism is] the appearance of similar characters separately in two or more lineages of common ancestry and on the basis of, or channeled by, characteristics of that ancestry. [emphasis mine.]

Efforts to identify homologus sulci in Old and New World monkeys are complicated by allometric effects (Falk, 1979). With the exception of *Saimiri*, ceboid taxa are listed in Table 2 from left to right in order of increasing mean cranial capacities. Table 2 reveals the widely recognized correlation between number of sulci present and brain size. [See Gould (1975) for a discussion of allometry in primate brains.] Based on sulcal patterns, Hershkovitz (1970) placed *Saimiri* with the Aotinae and *Cebus* with the prehensile-tailed cebids. However, since the brain size of *Saimiri* is less than half the size of *Cebus*, differences in sulcal patterns may simply be due to allometric effects. Presence of a given sulcus in different species of large-brained monkeys (but lack of it in smaller-brained species) may be due to their size. If that sulcus is not present in a smaller-brained common ancestor of the larger-brained species, its absence may simply be due to allometry. [See Falk (1979) for further discussion of the implications of allometric effects for primate brain studies.]

Figure 7, after Radinsky (1972, 1975), illustrates differences in cortical folding in lower primates and cercopithecoids. As pointed out by Radinsky, the most dramatic difference between the two groups is reflected in the organization of the frontal/parietal lobes: cercopithecoids are characterized by a well-marked transverse C which separates primary sensory and motor cortices; the lower primate pattern, on the other hand, exhibits a longitudinally oriented coronolateral sulcus.

New and Old World monkeys share certain similarities of sulcal pattern which differ from the typical lower primate pattern [the reader is referred to Radinsky (1968, 1970) for thorough discussion of lower primate sulcal patterns]: ceboids, as well as cercopithecoids, are characterized by a well-marked, transverse C. Instead of exhibiting a lower primate-like coronolateral sulcus, monkeys from both worlds are characterized by discrete IP and R. The primary visual cortex of monkeys, but not of lower primates, is relatively expanded and consequently well delineated by sulci (L and OCI). Smaller sulci such as SCA and LC are present in Old and (most species of) New World monkeys but not in lower primates.

In addition to exhibiting the above similarities in sulcal patterns, cercopithecoids and ceboids are also very similar in degree and range of encephalization [for details see Bauchot and Stephan (1969) and Stephan

(1972)]. Unlike the evidence determined from sulcal patterns, encephalization indices reflect relative brain weight and are controlled for allometric effects. Monkeys are more encephalized than lower primates, although there is some overlap (Bauchot and Stephan, 1969; Stephan, 1972).

The above similarities of sulcal pattern in ceboids and cercopithecoids are not unique to monkeys, however. Ape and human brains are also characterized by transverse C, visual cortices demarcated by L (although this feature is variable in humans) and, as would be expected in bigger brained primates, numerous additional small sulci (some of which are listed in Table 2). Thus the similarities found in the sulcal patterns of ceboids and cercopithecoids are part of a basic anthropoid sulcal pattern.

Are any of the sulci that are present in both cercopithecoids and ceboids (Table 2) likely candidates for homologues that suggest a middle to late Eocene common ancestor? The fossil evidence suggests that most sulci common to Old and New World monkeys were not present in early lower primates (Radinsky, 1970). The lateral and/or Sylvian (or supra-Sylvian) are the only sulci reproduced on endocasts of the Eocene *Tetonius, Smilodectes, Adapis,* and *Necrolemur* and the early Oligocene *Rooneyia* reproduces only S.

The fossil record of anthropoid brains is more encouraging, however. Radinsky's (1973) analysis of three partial endocasts from the anthropoid *Aegyptopithecus* shows that well-developed C, L, and IP (all common to ceboids and cercopithecoids) had already appeared at least once by the Oligocene. The Oligocene fossil record of monkey brains is virtually blank, although caudal expansion of the occipital lobe in one fossil ceboid (*Dolichocebus*) suggests relative expansion of the visual cortex over the lower primate condition (Radinsky, 1973, 1974). It is interesting that *Aegyptopithecus* lacks sulci in its frontal lobe (Radinsky, 1973) since frontal lobe sulcal pattern (R/ARC) is one criterion that distinguishes ceboids from cercopithecoids (see below).

Although *Aegyptopithecus* is thought to be a basal pongid (Simons, 1972), its basic anthropoid sulcal pattern provides a reasonable model for the sulcal pattern of a hypothetical common ancestor of ceboids and cercopithecoids. If one explains the variation in numbers of sulci listed for different ceboid groups in Table 2 as the result of allometry [i.e., if callitrichids would have as many sulci as cebids if they were enlarged to a comparable size; Falk (1979)], then the data suggest that monkeys in both worlds exhibit the same *basic* sulcal pattern. This allometric hypothesis seems tenable to me (Falk, 1979), and since ceboids and cercopithecoids have many sulci in common (Table 2), a monophyletic origin seems probable.

Although ceboid and cercopithecoid endocasts exhibit the same basic sulcal pattern, they appear to be "overlain" with different specializations. Thus, certain features which are exhibited by cebid monkeys may be viewed as specializations which have been derived in the New World, regardless of the origins of ancestral ceboids—for example, features correlated with prehensile tails such as confluent IP and S and extra pre- and postcentral sulci. Other apparently derived features include an enlarged cerebellar paraflocculus in pithecines and atelines (perhaps related to vestibular functions), visibility of OCT in lateral view in some ateline endocasts (seemingly related to a laterally expanded inferior temporal lobe and independently evolved in some cercopithecines), and extensive proliferation of secondary sulci in occipital lobes of atelines. These specialized features, which are more numerous than the specializations found in cercopithecoid endocasts, may be useful for clarifying specific details of the adaptive radiation of New World monkeys (Falk, 1979).

Cercopithecoid brains exhibit one major specialization that distinguishes them from brains of all ceboid genera except *Cebus*. In the frontal lobe, ARC is consistently present and distinct from R. This condition is derived relative to the usual ceboid condition (lack of ARC, presence of R). *Cebus* also shows consistent merging of S and TS caudally, a characteristic that is also consistently found in the cercopithecine subfamily. Since Radinsky (1971) has shown that sulci may be developed independently in mammals, and since the number of features involved is small, these features may best be viewed as parallel developments in *Cebus* and cercopithecines. Nevertheless, one is struck by the fact that, except for the specialized pseudolunate sulcus (a feature derived in the New World), *Cebus* endocasts look like cercopithecoid endocasts.

In conclusion, because of the incompleteness of the fossil primate record and because of the complexities involved in interpreting comparative primate sulcal patterns (compounded by allometric effects), it is difficult to sort out the evolutionary history of monkeys based on a comparative study of endocranial casts. However, the above analysis suggests that a monophyletic origin seems reasonable for cercopithecoids and ceboids. The hypothetical common ancestor would have had a sulcal pattern similar to that of the Oligocene anthropoid *Aegyptopithecus* (i.e., already organized along basic anthropoid lines) and *may* also have had other sulci such as LC that are common to ceboids and cercopithecoids (Table 2). Specializations of sulcal pattern related to prehensile tails and vestibular functions would have occured in ceboids after dispersal to the New World. Similarly, ARC would have been selected for in cercopithecoids after separation of the platyrrhine stock. As discussed above, a number of other features such as PAR developed in parallel in various ceboid and cercopithecoid subgroups. Thus, a comparative study of endocasts of New and Old World monkeys reveals various unique specializations that were independently evolved in ceboids and cercopithecoids, some similar specializations that were independently evolved in subgroups of the two superfamilies, and many similarities that can best be explained (because of their relatively great number) as features that were retained from a common ancestor. These data are in keeping with the view that New and Old World monkeys were derived monophyletically and later became dispersed into two groups. Unfortunately, the limited fossil record of anthropoid brain evolution does little more than suggest that the dispersal of New World monkeys to South America had occurred by the time of the Oligocene.

ACKNOWLEDGMENTS

I thank authorities at the Field Museum of Natural History, Chicago, for permission to prepare endocasts and Dr. L. Radinsky for criticizing an earlier version of the manuscript.

References

Anthony, J., 1946, Morphologie externe du cerveau des singes Platyrrhines, *Ann. Sci. Nat. Zool. Biol. Annim.* **8**:1-150.

Bauchot, R., and Stephan, H., 1969, Encephalisation et niveau évolutif chex les simiens, *Mammalia* **33**:225-275.

Benjamin, R. M., and Welker, W. I., 1957, Somatic receiving areas of cerebral cortex of squirrel monkey (*Saimiri sciureus*), *J. Neurophysiol.* **20**:286-299.

Campbell, C. B. G., 1976, Morphological homology and the nervous system, in: *Evolution, Brain, and Behavior: Persistent Problems* (R. B. Masterton, W. Hodos, and H. Jerison, eds.), pp. 143-151, Wiley, New York.

Campbell, C. B. G., and Hodos, W., 1970, The concept of homology and the evolution of the nervous system, *Brain Behav. Evol.* **3**:353-367.

Connolly, C. J., 1950, *External Morphology of the Primate Brain,* Thomas, Springfield.

Falk, D., 1978*a*, Brain evolution in Old World monkeys, *Am. J. Phys. Anthropol.* **48**:315-320.

Falk, D., 1978*b,* Cerebral asymmetry in Old World monkeys, *Acta Anat.* **101**:334-339.

Falk, D., 1978*c,* External neuroanatomy of Old World monkeys (Cercopithecoidea), *Contrib. Primatol.* **15**:1-150.

Falk, D., 1979, Cladistic analysis of New World monkey sulcal patterns: Methodological implications for primate brain studies. *J. Hum. Evol.* **8**:637-645.

Gould, S. J., 1975, Allometry in primates, with emphases on scaling and the evolution of the brain, in: *Approaches to Primate Paleobiology* (F. Szalay, ed.), *Contrib. Primatol.* **5**:244-292.

Hershkovitz, P., 1970, Cerebral fissural patterns in platyrrhine monkeys, *Folia Primatol.* **13**: 213-240.

Hill, W. C. O., 1960, *Primates: Comparative Anatomy and Taxonomy.* Vol. IV, *Platyrrhini Cebidae,* Part A, Edinburgh University Press, Edinburgh.

Hill, W. C. O., 1962, *Primates: Comparative Anatomy and Taxonomy.* Vol. V, *Platyrrhini Cebidae,* Part B, Edinburgh University Press, Edinburgh.

Hoffstetter, R., 1972, Relationships, origins, and history of the ceboid monkeys and caviomorph rodents: A modern reinterpretation, in: *Evolutionary Biology,* Vol. 6 (Th. Dobzhansky, M. K. Hecht, and W. C. Steere, eds.), pp. 323-347, Appleton-Century-Crofts, New York.

Iwai, E., and Mishkin, M., 1969, Further evidence on the locus of the visual area in the temporal lobe of the monkey, *Exp. Neurol.* **25**:585-594.

McClure, T., and Daron, G., 1971, The relationship of the developing inner ear, subarcuate fossa and paraflocculus in the rat, *Am. J. Anat.* **130**:235-250.

Napier, J. R., and Napier, P. H., 1967, *A Handbook of Living Primates*, Academic Press, New York.

Radinsky, L. B., 1968, A new approach to mammalian cranial analysis, illustrated by examples of prosimian primates, *J. Morphol.* **124**:167-180.

Radinsky, L. B., 1970, The fossil evidence of prosimian brain evolution, in: *The Primate Brain, Advances in Primatology*, Vol. 1 (C. R. Noback and W. Montagna, eds.), pp. 209-224, Appleton-Century-Crofts, New York.

Radinsky, L. B., 1971, An example of parallelism in carnivore brain evolution, *Evolution* **25**:518-522.

Radinsky, L. B., 1972, Endocasts and studies of primate brain evolution, in: *The Functional and Evolutionary Biology of Primates* (R. Tuttle, ed.), pp. 175–184, Aldine, Chicago.

Radinsky, L. B., 1973, *Aegyptopithecus* endocasts: Oldest record of a pongid brain, *Am. J. Phys. Anthropol.* **39**:239–247.

Radinsky, L. B., 1974, The fossil evidence of anthropoid brain evolution, *Am. J. Phys. Anthropol.* **41**:15–28.

Radinsky, L. B., 1975, Primate brain evolution, *Am. Sci.* **63**:656–663.

Simons, E. L., 1972, *Primate Evolution: An Introduction to Man's Place in Nature,* MacMillan, New York.

Simpson, G. G., 1961, *Principles of Animal Taxonomy,* Columbia University Press, New York.

Smith, G. E., 1903, On the morphology of the brain in the Mammalia, with special reference to that of the lemurs, recent and extinct, *Trans. Linn. Soc. Lond.* **8**:319–432.

Stephan, H., 1972, Evolution of primate brains: A comparative anatomical investigation, in: *The Functional and Evolutionary Biology of Primates* (R. Tuttle, ed.), pp. 155–174, Aldine, Chicago.

Straus, W. L., 1960, The subarcuate fossa in primates, *Anat. Rec.* **138**:93–104.

Tarling, D., and Runcorn, S., 1973, *Implications of Continental Drift to the Earth Sciences,* Vol. 1, Academic Press, New York.

Tilney, F., 1928, *The Brain From Ape to Man,* Vol. 1, Hoeber, New York.

Welker, W. I., and Campos, G. B., 1963, Physiological significance of sulci in somatic sensory cerebral cortex in mammals of the family Procyonidae, *J. Comp. Neurol.* **120**:19–36.

Comparative Anatomical Study of the Carotid Circulation in New and Old World Primates

Implications for Their Evolutionary History

14

J. BUGGE

Introduction

A comprehensive comparative anatomical investigation of the carotid circulation in several mammalian orders, viz., insectivores, primates, rodents, lagomorphs, edentates, pangolins and carnivores (cf. Bugge, 1974*b*, 1975*a*, 1978), has shown the cephalic arterial pattern to be an important supplement to other morphological systems, especially tooth morphology and osteology, as a basis for a systematic classification of mammals, reflecting their probable evolutionary history and phylogenetic relationships.

In this respect it is important that in addition to odontological and osteological characters, which are distinct in fossil material, certain soft organ systems, for example the cephalic arterial system, can be followed back to the early placentals, since they have left traces in hard tissue (Bugge, 1972).

J. BUGGE • Department of Anatomy, Royal Dental College, DK-8000 Aarhus C Denmark. This research was supported by a grant from the Danish State Research Foundation.

During the Cenozoic era, changes in the relative position of the continental masses due to plate tectonics (continental drift) and, consequently, in the possible dispersal routes, have undoubtedly strongly influenced the evolution of mammals and must be taken into account in a discussion of their taxonomy and phylogenetic development. The results, for instance, of recent geophysical and oceanographic research on the paleogeographic history of the southern continents have revealed that the South Atlantic in the Eocene was probably not an impassable barrier for an immigration from Africa to South America, having a less formidable width at that time.

This revised paleogeographic situation is of special interest in the discussion of the possibility of deriving the ancestors of the New World hystricomorphs and monkeys from African immigrants which crossed the South Atlantic by natural rafting some time in the later Eocene. An investigation of the probable origin and phylogenetic relationships of the New World monkeys has been further actualized by the fact that an African ancestry for the New World hystricomorphs has been accepted on this basis as the most probable (cf. Bugge, 1974*a*). Considering the remarkable morphological similarity, including the cephalic arterial pattern, between the New and Old World hystricomorphs it becomes difficult to explain it only as pure parallelism.

The original research on the cephalic arterial pattern in insectivores and primates (Bugge, 1972) focused especially on the systematic position of the elephant shrews (Macroscelidoidea) and the tree shrews (Tupaioidea) and the probable insectivore–primate boundary. It is the intention here to discuss the

Table 1. Order: Primates Linnaeus, 1758 (s.l.)

Classification	Specimens
Grade O = Subprimates Lightoller, 1934 or separate order; cf. below	
Suborder: Tupaioidea Dobson, 1882 or order: Tupaioidea Straus, 1949	
Family: Tupaiidae	17 Common tree shrews (*Tupaia glis*)
Grade A = Strepsirhini Geoffroy, 1812	
Suborder: Lemuroidea Mivart, 1864	
Family: Lemuridae	2 Ring-tailed lemurs (*Lemur catta*)
Suborder: Lorisoidea Tate Regan, 1930	
Family: Lorisidae	8 Slow lorises (*Nycticebus coucang*)
Family: Galagidae	4 Lesser galagos (*Galago senegalensis*)
Grade B = Haplorhini Pocock, 1918	
Suborder: Tarsioidea Elliot Smith, 1907	
Family: Tarsiidae (0); cf. Hill (1953*a*)	
Suborder: Pithecoidea[a] Pocock, 1918	
Infraorder: Platyrrhini Geoffroy, 1812	
Superfamily: Ceboidea Simpson, 1931	
Family: Cebidae	2 Night monkeys (*Aotus trivirgatus*)
Family: Hapalidae[b]	2 Yellow-handed marmosets (*Saguinus midas*)
Infraorder: Catarrhini Geoffroy, 1812	
Superfamily: Cercopithecoidea Simpson, 1931	
Family: Cercopithecidae	4 Rhesus macaques (*Macaca mulatta*)
Superfamily: Hominoidea Simpson, 1931 (anthropoid apes + man) (0)	

[a] (= Simiiformes)
[b] (= Callitrichidae)

systematic classification of the New and Old World primates and their probable evolutionary history and phylogenetic relationship on the basis of the cephalic arterial pattern, adducing the new paleogeographic evidence.

Materials and Methods

The investigation is primarily based on the examination of corrosion casts of the cephalic arterial system, prepared by means of a special plastic injection and corrosion technique (Bugge, 1963, 1974*b*, 1975*b*). The seven species and 39 specimens used in this study are detailed in Table 1, the symbol (0) indicating that the group in question is not represented in the material. Besides the corrosion casts, the investigation comprised a few dissected animals.

The classification employed in the survey is nearly identical to that of Hill (1953*b*–1966) (cf. the discussion), apart from the fact that the tree shrews (Tupaioidea) are tentatively placed as subprimates by an expansion of the classical primate order, placing special emphasis on the pattern of cephalic arterial supply [cf. Bugge (1974*b*) and the discussion]. The placement of the taxonomically important tarsioids, however, which were not available for the present study, is based on the investigation of Hill (1953*a*) on the cephalic arterial system in *Tarsius syrichta*.

Results

Basic Pattern

The definitive pattern of cephalic arterial supply in primates is assumed to have been derived from a primary basic pattern, common to all mammals (Fig. 1A), comprising the internal carotid artery (ci), connected with the vertebral–basilar artery system by means of the posterior communicating artery (cp), the external carotid artery (ce), and the stapedial artery (st) with its three branches (rs, ri, rm).

Evidence from ontogenetic and paleontologic studies indicates that the basic pattern employed (Fig. 1A) is not only primitive in ontogenetic respects (cf. Grosser, 1901; Tandler, 1902; von Hofmann, 1914; Struthers, 1930; Padget, 1948), but must also be presumed to correspond fairly closely to the cephalic arterial pattern in the phylogenetically oldest placentals, apart from the fact that the internal carotid artery seems originally to have been divided into a lateral and a medial stem (cf. Matthew, 1909; McKenna, 1966; Van Valen, 1966). The lateral stem crossed the promontory branching off on its way into a distinct stapedial artery, while the medial stem followed the fissure between the petrous part and the basioccipital/basisphenoid.

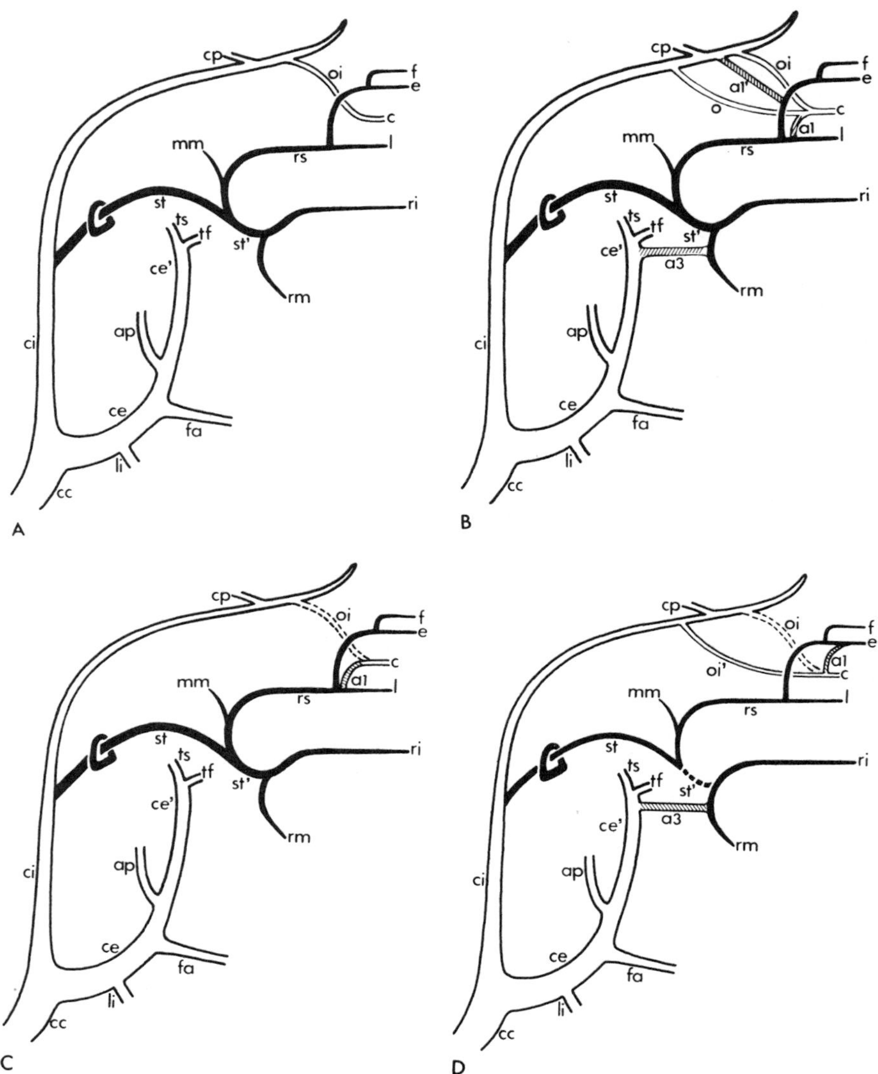

Fig. 1. Basic pattern and possible anastomoses (A,B) and the cephalic arterial pattern in primitive insectivores and primates, respectively (C,D). (A) Basic pattern; (B) basic pattern and three possible anastomoses (viz., a1, a1′ and a3); (C) primitive insectivores, viz., Cape golden mole (*Chrysochloris asiatica*); (D) tupaioids, viz., common tree shrew (*Tupaia glis*). White: internal–external carotid artery system; black: stapedial artery system; hatched: anastomoses; dashed: only temporarily developed. Abbreviations: ap, a. auricularis posterior; c, a. ciliaris; cc, a. carotis communis; ce, a. carotis externa (proximal part); ce′, a. carotis externa (distal part); ci, a. carotis interna; ci′, a. carotis anterior (apparent internal carotid artery); cp, a. communicans posterior; e, a. ethmoidalis; f, a. frontalis; fa, a. facialis; *l*, a. lacrimalis; li, a. lingualis; mm, a. meningea media; o, a. ophthalmica; oi, a. ophthalmica interna; oi′, a. ophthalmica "interna"; ri, r. infraorbitalis; rm, r. mandibularis; rs, r. supraorbitalis; st, a. stapedia; st′, distal part of the stapedial artery stem; tf, a. transversa faciei; ts, a. temporalis superficialis. (Reproduced from Bugge, 1972.)

Irrespective of whether this presumptive dichotomous branching of the primitive internal carotid artery into a lateral and a medial stem was a reality or not (it has never been found in recent material), it is a matter of fact that a typical lateral course of the internal carotid artery has been found in certain mammalian orders, such as insectivores and primates, and a medial course in others, like lagomorphs and those rodents and carnivores which have not entirely lost an internal carotid artery (see Bugge, 1974*b*). (As to the special course and homology of the "internal carotid artery" in lorisoids, see the discussion.) Therefore, as far as the phylogenetic development of the cephalic arterial system is concerned, the development of the internal carotid and stapedial arteries and the course of the internal carotid artery in relation to the tympanic cavity are of special importance in the systematic classification of mammals.

The primary area supplied by the internal carotid artery (ci) is the brain (with the assistance of the vertebral-basilar artery system) and the eyeball (oi-c), while the external carotid artery (ce) supplies the tongue (li) and face (fa, ts, tf). Of the three branches of the stapedial artery (st), the supraorbital branch (rs) supplies the dura (mm) and the extrabulbar part of the orbit (*l*, f, e), while the infraorbital branch (ri) supplies the upper jaw, and the mandibular branch (rm) the lower jaw.

Modifications of this basic pattern are presumed primarily to have occurred by obliteration of parts of the original system [i.e., the internal carotid artery (ci) and/or certain parts of the stapedial artery system (st, rs, ri, rm)] in connection with the presence of two anastomoses, designated a1 (or a1′) and a3 (see Fig. 1B), out of a total of six anastomoses, all of which occur as permanent or temporary anastomoses during human embryogenesis (see Padget, 1948; De la Torre and Netsky, 1960). Apart from the possible existence of anastomosis a1 (Fig. 1C), hardly any of these anastomoses can have been developed in the primitive placentals from the early Tertiary. [For further details concerning the system employed, see Bugge (1974*b*)].

Possible Anastomoses

The anastomosis a1 forms an important connection between the orbital arteries (*l*, f, e) from the supraorbital branch (rs) and the bulbar arteries (c) from the internal carotid artery (circulus arteriosus). It is shown in Fig. 1B as a connection between the ophthalmic (o)/internal ophthalmic (oi) and lacrimal (*l*) arteries. The internal ophthalmic artery (oi) from the circulus arteriosus is ordinarily attenuated or completely obliterated in nonprimates (Fig. 1C), anastomosis a1 forming the proximal part of the *ciliary artery* (c); in higher primates [Pithecoidea (= Simiiformes)], however, a1 becomes a part of the ophthalmic artery (Figs. 1B and 3C–D, o). The very specific anastomosis a1′ (Figs. 1B and 3B) connects in lorisoids the circulus arteriosus and the extrabulbar arteries (*l*, f, e) directly.

The anastomosis a3 between the distal end of the external carotid artery (Fig. 1B, ce′) and the proximal part of the mandibular branch (rm) is a permanent feature of all primates and many nonprimates, forming the first portion of the *maxillary artery*. By means of this anastomosis the external carotid artery system (ce) annexes an important part of the stapedial (st) area of supply, i.e., the upper and lower jaws (ri and rm) and, in higher primates (Fig. 3D), the dura (mm).

Primate Patterns (s.l.)

Grade O = Subprimates Lightoller, 1934 (or separate order)

Suborder: Tupaioidea Dobson, 1882 (or order, see Straus, 1949 and Table 1)
Tree shrews (Tupaiidae), see Table 1 and Figs. 1D, 2A,B, 5a, and 6.

In the tree shrews examined (see Table 1), both the internal carotid artery and most of the stapedial artery system are extant (Fig. 1D, ci and st). The internal carotid artery supplies the brain, assisted by the vertebral–basilar artery system (Fig. 2A, ci and v–ba; and Fig. 5a), while the stapedial area of supply (Fig. 1D, st) is confined to the dura and most of the orbit (Fig. 5a). Only the distal portion of the stapedial artery stem (Fig. 1D, st′) is obliterated, and two anastomoses are developed (a1 and a3).

The "internal" ophthalmic artery (Fig. 1D, oi′) from the distal part of the internal carotid artery (ci) is reduced (Fig. 2A,B), and the supraorbital branch (Figs. 1D and 2A, rs) from the stapedial artery system supplies the dura (mm) and the greater portion of the orbit (*l*, f, e, and c), assuming most of the supply of the eyeball (c) via anastomosis a1 (Fig. 2B). The supply of the upper and lower jaws (ri and rm), however, is annexed by the external carotid artery system (ce) by means of anastomosis a3, the distal part of the stapedial artery stem (st′) being obliterated.

Grade A = Strepsirhini Geoffroy, 1812

I. Suborder: Lemuroidea Mivart, 1864
Lemurs (Lemuridae), see Table 1 and Figs. 3A, 4A,B, 5b, and 6.

In the lemurs investigated (see Table 1), the internal carotid artery (Fig. 3A, ci) is obliterated just after the departure of the stapedial artery (Fig. 3A, st; cf. Fig. 4B) and the brain is supplied from the vertebral–basilar artery system alone (Fig. 4A, v–ba; and Fig. 5b). As in tupaioids, only the distal part of the stapedial artery stem (Fig. 3A, st′) is obliterated and two anastomoses are developed (a1 and a3).

In lemurids the "internal" ophthalmic artery (Fig. 3A, oi′) is a strong branch (Fig. 4A), which departs from the circulus arteriosus almost opposite the origin of the posterior communicating artery (cp). It is the main contributor to the supply of the eyeball (Fig. 3A, c), while the supraorbital branch

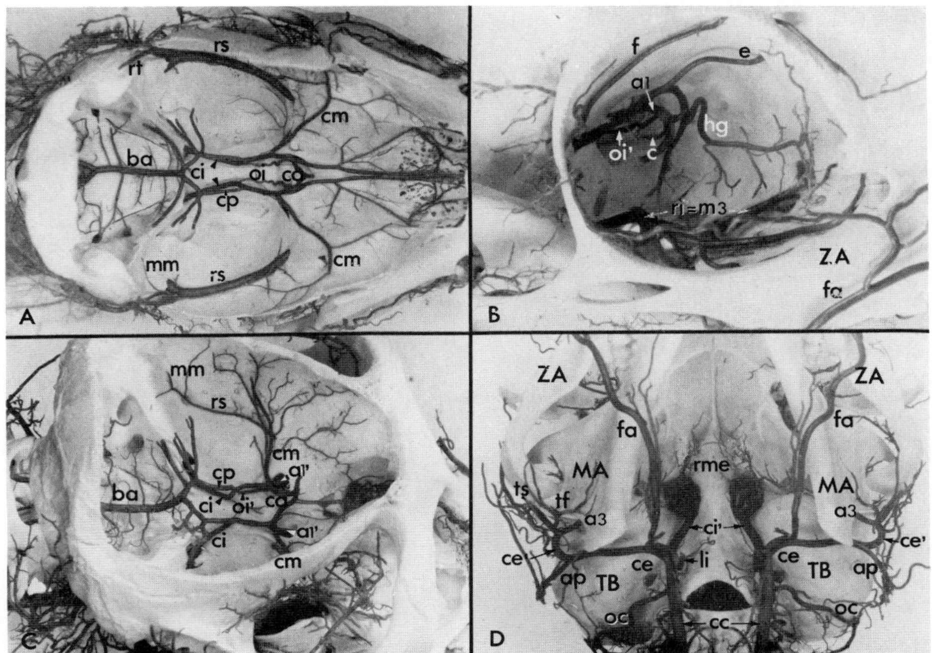

Fig. 2. Cephalic arterial supply in tupaioids (A,B) and certain strepsirhines, viz., lorisids (C) and galagids (D). Cranial cavity, dorsolateral view, rostral end to the right, cranial theca and brain removed (A,C); right orbit, dorsolateral view (B), and ventral view, rostral end upward (D). (A,B) Common tree shrew (*Tupaia glis*); (C) slow loris (*Nycticebus coucang*); (D) lesser galago (*Galago senegalensis*). ba, a. basilaris; ca, a. cerebri anterior; cm, a. cerebri media; hg, Harder gland artery; MA, mandible; m3, a. maxillaris III (ri); oc, a. occipitalis; rme, extracranial *rete mirabile*; rt, rr. temporales; TB, tympanic bulla; ZA, zygomatic arch. Other abbreviations as in Fig. 1. (Reproduced from Bugge, 1972, 1974*b*.)

(rs) from the stapedial artery system mainly supplies the dura (mm) and the extrabulbar portion of the orbit (*l*, f, e), the two supply systems being connected by anastomosis a1. The remainder of the stapedial area of supply, i.e., the upper and lower jaws (ri and rm), is, as in tupaiids, annexed by the external carotid artery system (ce) by means of anastomosis a3.

II. Suborder: Lorisoidea Tate Regan, 1930

Lorises (Lorisidae) and galagos (Galagidae); see Table 1 and Figs. 3B, 2C,D, 5c, and 6.

In the lorises and galagos studied (see Table 1), the ordinary internal carotid artery (Fig. 3B, ci) is supplanted by a more medial trunk. This atypically coursing "internal" carotid artery (Figs. 3B and 2D, ci′; see below) supplies the brain with assistance from the vertebral-basilar artery system (Fig. 2C, ba; and Fig. 5c), while the complete stapedial area of supply is assumed by the "internal"-external carotid artery system (Fig. 3B and 5e), by

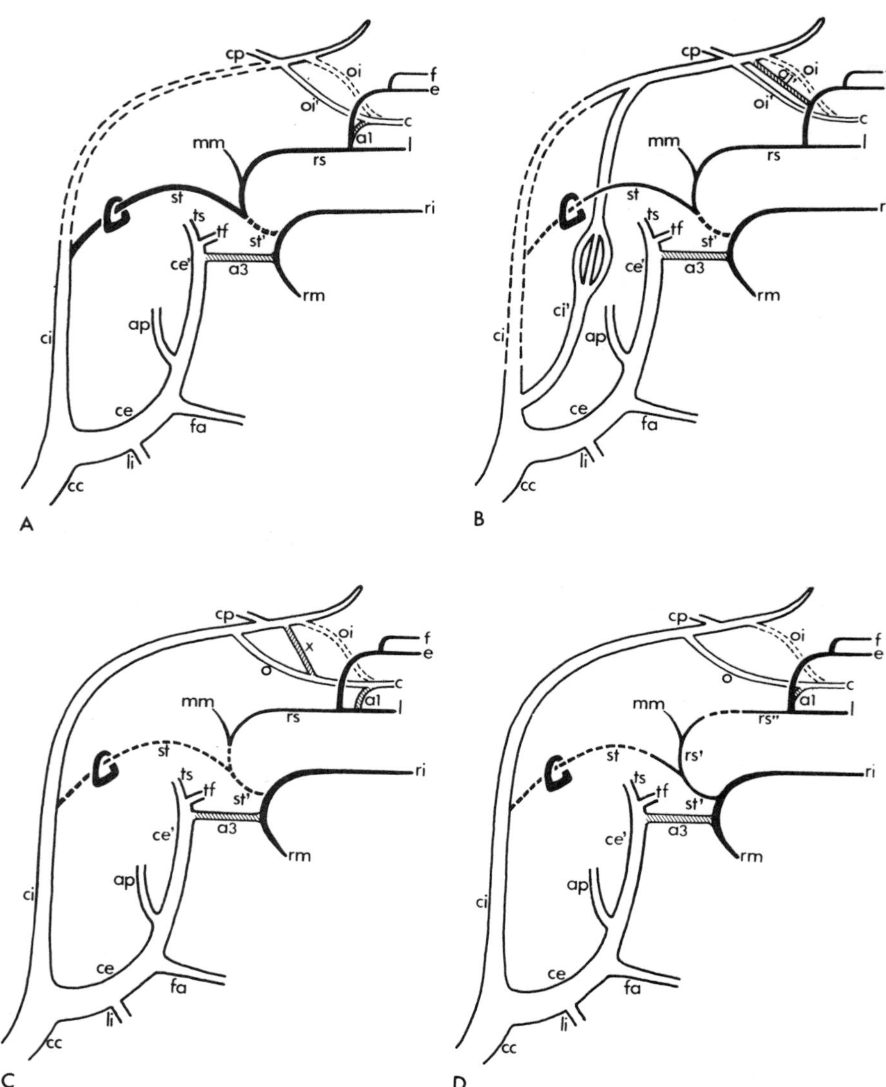

Fig. 3. Cephalic arterial pattern in Primates (s.s.), viz., lower primates (A,B) and simians (C,D). (A) Lemurids, viz., ring-tailed lemur (*Lemur catta*); (B) lorisids, viz., slow loris (*Nycticebus coucang*), and galagids, viz., lesser galago (*Galago senegalensis*); (C) certain platyrrhines, viz., night monkey (*Aotus trivirgatus*); (D) certain platyrrhines, viz., yellow-handed marmoset (*Saguinus midas*), and catarrhines, viz., rhesus macaque (*Macaca mulatta*). White: internal–external carotid artery system; black: stapedial artery system; hatched: anastomoses; dashed: only temporarily developed. See caption of Fig. 1 for explanation of abbreviations.

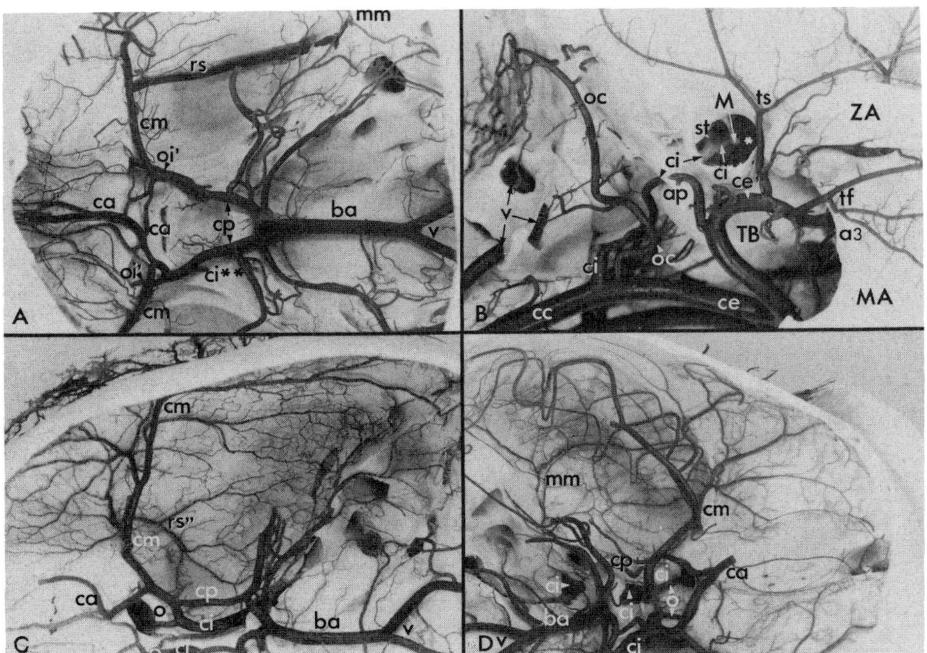

Fig. 4. Cephalic arterial supply in certain strepsirhines, viz., lemurids (A,B), and pithecoids, viz., advanced platyrrhines (C) and catarrhines (D). Cranial cavity, dorsolateral view, rostral end to the left, cranial theca and brain removed (A); right half, lateral view (B); cranial cavity, right half, medial view, brain removed (C), and cranial cavity, dorsolateral view, rostral end to the right, cranial theca and brain removed (D). (A,B) Ring-tailed lemur (*Lemur catta*); (C) yellow-handed marmoset (*Saguinus midas*); (D) rhesus macaque (*Macaca mulatta*). *, the point at which the obliteration of the internal carotid artery begins; ci**, the distal unobliterated part of the internal carotid artery; M, malleus; v, a. vertebralis. Other abbreviations as in Figs. 1 and 2. (Reproduced from Bugge, 1974*b*.)

means of anastomoses a1′ and a3, the stapedial artery stem (Fig. 3B, st–st′) aborting proximally and distally.

In the lorisoids the "internal" ophthalmic artery (Fig. 3B, oi′) is a weaker branch, originating from the circulus arteriosus almost opposite the posterior communicating artery (cp), and supplying only the eyeball (c). However, the stapedial artery stem (Fig. 3B, st–st′) is partly obliterated, and the "internal" carotid–vertebral artery system (circulus arteriosus) also supplies the remainder of the orbit (*l*, f, e) and the dura (mm) by means of anastomosis a1′, which connects the common trunk of the frontal and ethmoidal arteries (f + e) with the anterior portion of the circulus arteriosus (Figs. 2C and 3B) and not, as is usual (a1), with the "internal" ophthalmic artery (oi′). The external carotid artery system (ce) has, as in other primates, assumed the supply of the upper and lower jaws (ri and rm) via anastomosis a3.

Most of the ordinary internal carotid artery (Fig. 3B, ci), designated as

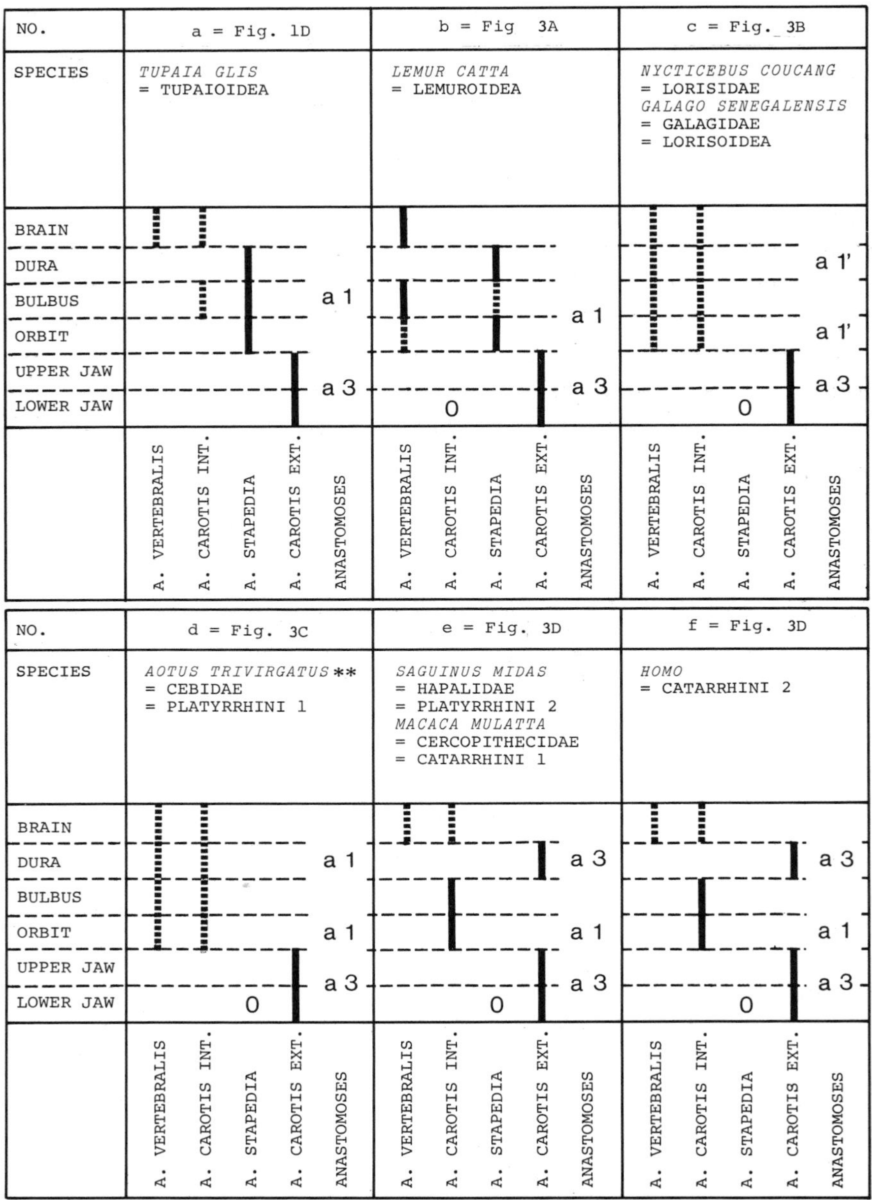

Fig. 5. Primates (s.l.): Cephalic pattern of supply. If two sources contribute unequally to the supply of the same area, the main source is shown with a solid line and the subsidiary with a broken one, while if two sources contribute approximately equally to the supply of the same area, they are both shown with broken lines. The anastomoses involved are placed opposite the areas mainly affected by their presence. (**) Cf. anastomosis X in Fig. 3C. (Reproduced from Bugge, 1974*b*.)

true internal carotid and entocarotide by Adams (1957) and Saban (1963), respectively, and corresponding to the promontory artery (Winge, 1895; cf. 1941), is obliterated and replaced by a more medially coursing arterial stem (Fig. 3B, ci′), described as apparent internal carotid (Adams, 1957) or carotide antérieure (Saban, 1963). This so-called internal carotid artery (ci′), which is presumably homologous with the ascending pharyngeal artery (see Bugge, 1972, 1974*b* and the discussion), follows the external cranial base along the fissure between the petrous part and the basioccipital/basisphenoid, forming an extracranial *rete mirabile* (Fig. 2D, rme; cf. Bugge, 1972, 1974*b* and Tandler, 1899; Ask-Upmark, 1953; Adams, 1957) before it enters the middle cranial fossa via the foramen lacerum (medium) to end in the circulus arteriosus.

Grade B = Haplorhini Pocock, 1918

I. Suborder: Tarsioidea Elliot Smith, 1907
Tarsiers (Tarsiidae), see Table 1 and Figs. 3C,D and 6.

In the Mindanao tarsier (*Tarsius syrichta*; see Hill, 1953*a*) the cephalic arterial pattern differs markedly from that in other classical lower primates (i.e., lemuroids and lorisoids) and corresponds fairly closely to that in the advanced pithecoids (i.e., *Saguinus midas* and *Macaca mulatta*), for example, in the presence of an internal carotid artery proper (cf. Fig. 3D, ci and the discussion), the only important difference being the more primitive supply pattern of the orbit (see below).

The brain is supplied, as in pithecoids, from the internal carotid/vertebral-basilar artery system, which also assumes the supply of the entire orbit (cf. Fig. 3D, *l*, f, e, and c) via anastomosis a1, while most of the remainder of the stapedial area of supply, i.e., the upper and lower jaws (ri and rm), is annexed by the external carotid artery system (ce) by means of anastomosis a3. The distal portion of the supraorbital branch (rs″) is partially obliterated and the atrophying stapedial artery (st) supplies only the dura (mm).

The most remarkable difference between the patterns of cephalic arterial supply in tarsiids and pithecoids is that the artery which supplies the orbit in the former does not originate from the distal portion of the internal carotid artery like the ophthalmic artery (Fig. 3C,D, o) in the pithecoids, but from the anterior part of the circulus arteriosus, like anastomosis X (Fig. 3C) in the most primitive pithecoids (see, Hill, 1953*a*, Fig. 8).

II. Suborder: Pithecoidea Pocock, 1918
1. Infraorder: Platyrrhini Geoffroy, 1812 (New World monkeys).
Cebid monkeys (Cebidae); see Table 1 and Figs. 3C, 5d, and 6.
Marmosets (Hapalidae = Callitrichidae); see Table 1 and Figs. 3D, 4C, and 5e.

In the ceboids examined (i.e., cebid monkeys and marmosets; see Table 1), the well-developed internal carotid artery (Figs. 3C–D and 4C, ci) supplies the brain together with the vertebral-basilar artery system (Fig. 4C, v–ba; and Fig. 5d,e), while the entire stapedial area of supply is taken over by the internal–external-carotid artery system alone in marmosets (i.e., *Saguinus midas*; see Fig. 5e), and assisted by the vertebral–basilar artery system in cebid monkeys (i.e., *Aotus trivirgatus*; see Fig. 5d). The complete stapedial artery stem (Fig. 3C, st–st′) is lacking in cebid monkeys, while only the tympanic portion of the stem (Fig. 3D, st) and the central part of the supraorbital branch (rs) are obliterated in marmosets, in connection with the development of the two ordinary anastomoses, a1 and a3, of which a1 becomes a permanent part of the definitive ophthalmic artery (Fig. 3C,D, o) and a3 forms the first portion of the maxillary artery.

In the cebid monkeys (*Aotus trivirgatus*), the bulbar part of the orbit (Fig. 3C, c) is supplied by the ophthalmic artery (o), which also annexes the supply of the extrabulbar portion of the orbit (*l*, f, e) and the dura (mm) via anastomosis a1, while the supply of the upper and lower jaws (ri and rm), as is usual, is taken over by the external carotid artery (ce) by means of anastomosis a3. However, although the ophthalmic artery (Fig. 3C, o) originates from the distal part of the internal carotid artery (ci), as in other pithecoids, the vertebral–basilar artery system also contributes to the orbital arterial supply in the cebids investigated (cf. Fig. 5d and the tarsiids), the anterior portion of the circulus arteriosus being connected with the ophthalmic artery (Fig. 3C, o) via anastomosis X.

In the marmosets (*Saguinus midas*), the ophthalmic artery (Fig. 3D, o) also supplies the entire orbit (*l*, f, e, and c), but without assistance from the circulus arteriosus, no anastomosis X is developed (Figs. 3D and 4C). The external carotid artery system (ce) takes over the supply of the upper and lower jaws (ri and rm) by means of anastomosis a3 and of most of the dura (mm) via the distal portion of the stapedial artery stem (st′) and the proximal part (rs′) of the supraorbital branch (see Fig. 5e).

2. Infraorder: Catarrhini Geoffroy, 1812 (Old World monkeys)

Guenons (Cercopithecidae); see Table 1 and Figs. 3D, 4D, 5e and 6.

In the cercopithecoids studied (see Table 1), the cephalic arterial system follows the “advanced” platyrrhine pattern, where the human pattern of supply has been attained (see Figs. 4C,D and 5e,f).

Discussion

Primate Concept and Taxonomy (see Table 1 and Fig. 6)

In the classification of Primates (s.s.) most interest has been concentrated on the phylogenetic relationships within the classical major divisions (i.e., lemuroids, lorisoids, tarsioids, and pithecoids; see Table 1) and the origin and relative systematic position of these main groups within the order. The validity of according the “prosimians” (= lower primates) and the simians (=

higher primates) the status of taxonomic entities within Primates has long called for special attention.

The discussion of primate concept and taxonomy, however, is intimately

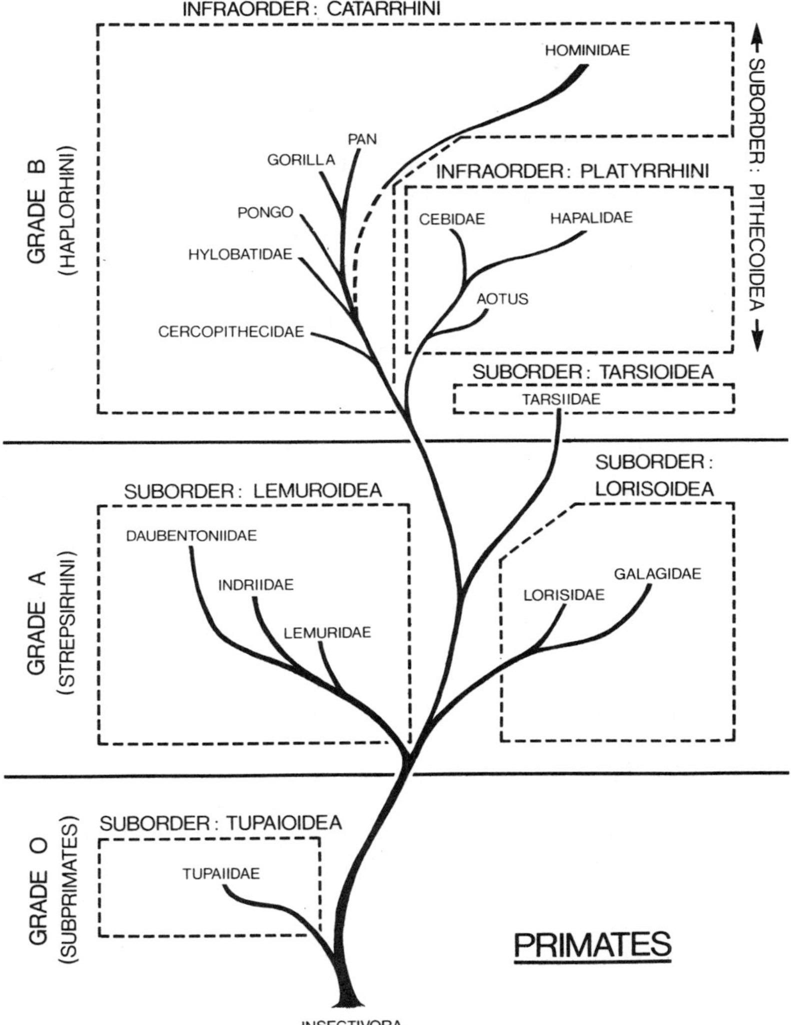

Fig. 6. Primates (s.l.). The presumptive phylogenetic development of the primates and the relation between grades O (Subprimates), A (Strepsirhini) and B (Haplorhini) and the suborders Tupaioidea, Lemuroidea, Lorisoidea, Tarsioidea, and Pithecoidea, the diagram showing the order in which the individual suborders are supposed to have differentiated from a common line of development. The diagram is based, not only on the cephalic pattern of supply, but especially on Remane (1956) and party on Fiedler (1956), Hill (1953*a*–1966), Le Gros Clark (1959), McKenna (1966), Romer (1966), and Starck (1956). (Reproduced from Bugge, 1974*b*.)

related to the problem of origin and systematic position of tree shrews (tupaioids). Especially Carlsson (1909, 1922) and Le Gros Clark (1925, 1926, 1934, 1959) advocated placing the tupaioids among the primates, in close relation to the lemuroids, in recent years followed by Simpson (1945), Fiedler (1956), and Remane (1956), who called attention to the close relationship between the tupaioids and the lower primates.

Fiedler (1956), however, included the tree shrews as a lower primate infraorder, Tupaiiformes, together with Lemuriformes, Lorisiformes, and Tarsiiformes, while Remane (1956) placed the primates including the tupaioids at three levels of development (see Fig. 6), which should not be confused with taxonomic entities. Of these three levels, the first one embraces the subprimates, of which the tupaioids are the only extant representatives, the others having died out in the early Tertiary. The second level comprises the classical lower primates, which Simpson (1945) considered a natural taxonomic unit. Within the lower primates, however, three relatively independent lines can be distinquished, one of which, Tarsiiformes, is regarded as being more closely related to the higher primates (cf. Haplorhini Pocock, 1918; see Hill, 1953*b*), while the two others, Lemuriformes and Lorisiformes, are considered to be more closely related to each other (cf. Strepsirhini Geoffroy, 1812; see Hill, 1953*b*). The third level includes only the two infraorders Platyrrhina Hemprich, 1820 [= Platyrrhini Geoffroy, 1812 (New World monkeys)] and Catarrhina Hemprich, 1820 [= Catarrhini Geoffroy, 1812 (Old World monkeys)].

The above classification, modified in Table 1 and Fig. 6 on the basis of the cephalic arterial pattern (see Bugge, 1974*b*), differs from Hill's (1953*b*–1966) among other things by the addition of the subprimates (= grade 0), comprising only the tupaioids. Although there was a lot to be said for placing the tree shrews among the primates, Hill found that this would require an extension of the primate concept, the boundaries of which were already none too clear.

Furthermore, Hill apparently conceives grades A and B (see Table 1 and Fig. 6) as taxonomic entities, comparable, for example, to suborders: "With the removal of tupaioids and tarsioids Simpson's concept of the Prosimii may be allowed to stand, and that is the attitude here taken" (Hill, 1953*b*). Hill's grade A thus corresponds to the classical primate suborder, "Prosimiae," excluding the tarsioids, which are transferred to grade B in the vicinity of the pithecoids in the group Haplorhini Pocock, 1918 (see Bugge, 1974*b*).

Grades O, A, and B, however, represent in the present classification primarily different degrees of development within the Primates (s.l.), expressing the order in which the groups concerned are supposed to have differentiated from a common line of development, each grade including one or two suborders of presumably monophyletic origin (see Fig. 6).

In the following sections some of the problems of primate concept and taxonomy will be discussed, primarily on the basis of the cephalic arterial system, but adducing the research of recent years in morphology, paleontology, and paleogeography. The discussion comprises the probable systematic position of the tupaioids in relation to the Primates (s.s.) and the contested validity of the grades A and B, representing a modification of the classical "prosimians" and simians, as truly taxonomic entities of monophyletic origin (Hennig, 1966), i.e., the probable phylogenetic relationship between the lemuroids and lorisoids and the systematic position and evolutionary history of tarsioids and pithecoids.

The placement of New World and Old World monkeys in the same suborder (Pithecoidea Pocock, 1918; see Table 1) has also been contested, because it indicates a monophyletic and not a paraphyletic origin. This discussion has been further actualized by the recent evidence regarding changes in the early Cenozoic of the relative position of the southern continents, revealing the South Atlantic as a possible migration route in the late Eocene, from Africa to South America.

The Systematic Position of Tupaioids

In the tupaioids (= Subprimates or grade O; see Table 1 and Fig. 6), the cephalic arterial pattern (Figs. 1D and 5a) is certainly more advanced than that generally observed in the primitive insectivores, where the original phylogenetic (= ancestral) system (Fig. 1C) still seems to be present. The orbital pattern of supply in tree shrews, however, also differs markedly from that in the most advanced insectivores, e.g., elephant shrews (Macroscelidoidea; see Bugge, 1972, 1974*b*), the "internal" ophthalmic artery (Fig. 1D, oi′) still being retained and like the ophthalmic artery in higher primates (Fig. 3C,D, o) departing as a branch from the distal end of the internal carotid artery (ci). Although the "internal" ophthalmic artery (Fig. 1D, oi′) in the tupaioids is weak, the incipient assumption by the internal carotid artery of the arterial supply to the orbit represents a distinct approach to the characteristic primate pattern.

In the classical literature the tree shrews (Tupaioidea) were as a rule placed together with the elephant shrews (Macroscelidoidea) under the insectivores, occasionally in a separate suborder, Menotyphla Haeckel, 1866, representing the highest insectivores. About the middle of the present century the tree shrews were placed as the lowest primates (see, for instance, Simpson, 1945; Fiedler, 1956), or as subprimates in an expanded primate order (Remane, 1956), and this placement was followed in a previous paper (see Bugge, 1974*b* and Fig. 6), taking into account the considerable difference between the cephalic arterial pattern of tupaioids and that of the most advanced insectivores (see above), and the weak but distinct approach to the primate cephalic supply pattern in tupaioids.

Van Valen (1965) and Campbell (1966), however, have together examined all the published similarities between tupaioids and primates without being able to accept any of these, alone or in combination, as indicating a special tupaioid–primate connection, and recently Haines and Swindler (1972) concluded, on the basis of comparative neuroanatomical studies, that the tree shrew central nervous system does not seem to suggest a tupaioid–primate or a tupaioid–insectivore affinity but tends to support a placement in a separate order, as suggested by Straus (1949) (see Table 1). The tupaioid orbital supply pattern represents without doubt an apomorphous (= derived) primate character which, however, may have been achieved by convergence, as a result of the same way of life, taking the evidence from studies of all other organ systems into account.

Nearly all taxonomists agree that the tupaioids separated from the ancestral group of the primates in the late Cretaceous or early Paleocene (Fig. 6), i.e., the tupaioids originated at the same time as most other mammalian orders. Following the modern principles of phylogenetic systematics, as advocated by Hennig (1966), the natural systematic position of the tupaioids would, therefore, probably be in a separate order of their own. (see Table 1).

Grade A: Strepsirhini

The Lemuroid–Lorisoid Relationship

In the lemuroids and lorisoids (= Strepsirhini or grade A; see Table 1 and Fig. 6), the cephalic arterial pattern (Figs. 3A,B and 5b,c) is generally more advanced than in tupaioids, especially with respect to the orbital and intracranial supply, but mutually very different. In the lemuroids (*Lemur catta,* Figs. 3A and 5b) the circulus arteriosus (i.e., the vertebral artery system) and the stapedial artery (via the supraorbital branch) contribute almost equally to the supply of the orbit, the eyeball (c), however, receiving most of its supply from the "internal" ophthalmic artery (oi′), while the stapedial artery system (st) supplies, besides the dura (mm), expecially the extrabulbar part of the orbit (*l,* f, e); cf. Fig. 5b. In the lorisoids (*Nycticebus coucang* and *Galago senegalensis,* Figs. 3B and 5c), however, the circulus arteriosus [i.e., the anterior carotid artery (= the apparent internal carotid artery)] plus the vertebral artery system) has assumed the complete supply of the orbit (and dura) via the "internal" ophthalmic artery (oi′) and the very specific anastomosis al′ (cf. Fig. 5c and below).

The lemuroids (*Lemur catta*) and lorisoids (*Nycticebus coucang* and *Galago senegalensis*) also differ markedly from each other and from all other primates with respect to the intracranial supply. In the lemuroids studied (*Lemur catta,* Fig. 3A), the promontory artery (i.e., the internal carotid artery after the departure of the stapedial artery) atrophies and the entire intracranial supply is assumed by the vertebral artery system (see Fig. 5b), while in the lorisoids (*Nycticebus coucang* and *Galago senegalensis*, Fig. 3B), the internal carotid artery proper (ci) is completely obliterated and replaced by an extracranial vessel, the anterior carotid artery (ci′). This, after forming a specific extracranial *rete mirabile* (Fig. 2D, rme) immediately before it enters the middle cranial fossa via the middle lacerate foramen (= the anterior carotid foramen), ends in the circulus arteriorus, supplying the brain together with the vertebral artery system (see Fig. 5c and below.).

The cephalic arterial patterns in the two strepsirhine suborders (Fig. 3A,B) differ markedly not only from each other but in several respects from the pattern in higher primates (Fig. 3C,D), and confirm the view of Remane (1956) that lemuroids and lorisoids probably represent two lines of development within the primate order, whose independent origin goes very far back (Fig. 6).

The cephalic arterial pattern in early placentals has undoubtedly strongly resembled the basic pattern used (Fig. 1A), apart from the possible bifurcation of the internal carotid artery (see above). Apart from anastomosis a1, hardly any of the anastomoses indicated in Fig. 1B could have been developed (Bugge, 1972). On the basis of fossil material, however, it is not possible to decide whether anastomosis a1 had been developed. Thus whether or not the stapedial artery system had already completely or partially annexed the supply of the eyeball in the early Tertiary mammals, as it appears to have done in most extant nonprimates, is not at present determinable. Most probably the cephalic arterial pattern at that time was very near that of primitive recent insectivores (Fig. 1C), the insectivores being a very ancient order whose members are as a rule regarded as the most direct descendants of primitive placentals.

This primitive pattern was probably still characteristic of the early Tertiary ancestor of lemuroids. Well-developed stapedial and promontory arteries are found, for example, in the Eocene adapoids (Saban, 1963), which probably represent a Paleocene offshoot from the lemuroid stem group (Hoffstetter, 1974), and the development in the direction of the recent lemuroid pattern has probably taken place since that time.

In Middle Eocene notharctids (*Notharctus* and *Smilodectes*), for instance, the inferior branch of the stapedial artery was already obliterated (Saban, 1963), the mandibular and infraorbital branches (rm and ri) probably being annexed by the external carotid artery system by means of anastomosis a3 as in all extant primates.

The increasing annexation by the intracranial circulation of the orbital supply at the expense of the stapedial artery system in the lemurids studied (Fig. 3A) compared with the supposed primitive condition (Fig. 1C) constitutes a distinct approach to the characteristic pattern of higher primates (Fig. 3C,D). The obliteration (Fig. 3A) or attenuation (see Bugge, 1974*b* and Saban, 1963) of the promontory artery (ci) in lemurids and indriids represents a trend opposite that of higher primates (Fig. 3C,D), where the internal carotid artery becomes enlarged (presumably with the increasing development of the brain) and moves forward in the tympanic cavity to run immediately in front of it in man.

The lorisoid cephalic arterial pattern, however, departs even more from the advanced primate pattern. The stapedial contribution to the orbital supply (Fig. 3B) is certainly annexed by the intracranial circulation as in higher primates (Fig. 3C,D), but in a highly unusual way by means of the very specific anastomosis a1′ between the circulus arteriosus and the extrabulbar arteries. This anastomosis is present in all lorisids and galagids investigated, but is not found in any other of the mammalian orders studied (see the Introduction). Also, the partial supply of the intracranial circulation by means of the anterior carotid artery (Fig. 3B, ci′) is a very specific feature of the two lorisoid families. In addition to being present in the lorisoids (and one of the two aberrant lemuroid groups, the Cheirogaleidae; see below), the anterior carotid artery is found only in certain fissipeds (cf. Bugge, 1978). From the investigation of the fissipeds it is obvious that the anterior carotid artery is homologous to the ascending pharyngeal artery and not to the medial entocarotid of primitive mammals, as tentatively proposed by Hoffstetter (1974). All intermediates are found in the fissipeds between a "normal" ascending pharyngeal artery and the condition in the herpestines, where the internal carotid artery, lodged in a bony tube within the medial wall of the bulla, and the anterior carotid artery contribute equally to the intracranial circulation.

Lorisids and galagids have undoubtedly inherited this very specific lorisoid pattern of cephalic arterial supply from a common pre-Miocene ancestor, and most probably the lemuroids (eventually minus the two aberrant groups, Daubentoniidae and Cheirogaleidae) and the lorisoids represent two natural taxonomic units whose age of origin (Hennig, 1966) corresponds to that of other suborders (see Table 1).

It is, however, still questionable whether the recent strepsirhines constitute a grade only (see Table 1) or a natural taxonomic entity of monophyletic origin, i.e., whether the lemuroids (s.s.) and the lorisoids are two sister-groups, stemming from a common Paleocene ancestor. In this respect, one of the two aberrant lemuroid groups, the Cheirogaleidae, is important. According to Saban (1963) the cheirogaleid cephalic arterial pattern is characterized by the presence of a typical anterior carotid artery, probably homologous to that of the lorisoids. Partly on this basis Szalay and Katz (1973) and Hoffstetter (1974) found it most probable that the cheirogaleids constitute an intermediate group between lemuroids (s.s.) and lorisoids. If future investigations reveal that the Cheirogaleidae also share the characteristic lorisoid orbital pattern of supply it would greatly substantiate their intermediate position between lemuroids and lorisoids.

The earlist known appearance of lorisoids and lemuroids is from the Lower Miocene of Africa and the Holocene of Madagascar, respectively, but irrespective of the lack of knowledge of pre-Miocene African Strepsirhini, Hoffstetter (1974) finds their existence in Africa at the beginning of the Tertiary most probable. In any event, the very peculiar cephalic arterial pattern of the lorisoids seems to presuppose a very early separation from a possible ancient African strepsirhine stock, probably not later than the Paleocene.

Grade B: Haplorhini

The Systematic Position and Relationship of the Tarsioids

The pattern of cephalic arterial supply of extant tarsioids (e.g., *Tarsius syrichta;* see Hill, 1953*a*) differs markedly from the very specific strepsirhine patterns, and corresponds fairly closely to that of the pithecoids. The most striking differences from the cephalic supply pattern of higher primates (Fig. 3C,D) concern two primitive features, viz., that the atrophying stapedial artery (st) still supplies the dura (mm) and that the artery supplying the orbit in the tarsioids, unlike the ophthalmic artery in the pithecoids (Fig. 3C,D, o), does not branch from the distal part of the internal carotid artery (ci), but instead is derived from the anterior part of the circulus arteriosus (see Hill, 1953*a*, Fig. 8); cf. anastomosis X in *Aotus trivirgatus* (Fig. 3C).

The presence of an acting internal carotid artery proper (ci), as in higher primates (Fig. 3C,D), is certainly a plesiomorphous character, inherited from an early primate ancestor, but the incipient enlargement of the internal carotid artery, correlated with a marked movement forward in relation to the tympanic cavity, is, on the other hand, a clearly apomorphous feature, representing a distinct approach to the characteristic pattern of higher primates.

On the basis of certain undoubted primitive features the tarsioids were regarded as lower primates in the classical taxonomy, and this arrangement has been followed by several recent taxonomists (cf., for instance, Simpson, 1945). Pocock (1918), however, emphasized the important apomorphous characters which the tarsioids shared with the higher primates and placed, on this basis, the tarsioids together with pithecoids in the grade of the Haplorhini, leaving the lemuroids and lorisoids in the grade of the Strepsirhini (cf. Table 1 and Fig. 6). Although the taxonomic value of some of these common apomorphous features is questionable (Remane, 1956), the classification of Pocock has been adopted, for instance, by Hill (1955), Remane (1956), and Hoffstetter (1972), and also by Bugge (1974*b*), the cephalic arterial pattern clearly substantiating this arrangement.

The tarsioids suddenly appeared in the Lower Eocene of Laurasia, but the cephalic arterial pattern of these early members was certainly more primitive than that of recent representatives. In the Middle Eocene tarsiid, *Necrolemur antiquus*, for instance, the internal carotid artery still followed the original course through the tympanic cavity from a posteriorly situated carotid foramen, just as the stapedial artery system was probably still rather complete (cf. Hürzeler, 1948; Saban, 1963). It is impossible, however, on this basis, to decide the probable time of separation of the tarsioids, the Eocene members being morphologically rather advanced in other respects (Hoffstetter, 1974); but the obvious pre-Eocene origin is alone sufficient to warrant a placement of the tarsioids in a separate suborder of their own (see Table 1).

The Relationship between New and Old World Monkeys (Platyrrhini and Catarrhini)

Among the pithecoids, constituting the most advanced group of Haplorhini (= grade B; see Table 1 and Fig. 6), New World monkeys (Platyrrhini) are generally more primitive than Old World monkeys (Catarrhini). The general line of development in the pithecoids with respect to the cephalic arterial system (Figs. 3C,D and 5d,e), however, is clearly in the direction of the hominoid pattern (i.e., anthropoid apes and man, Figs. 3D and 5f). This tendency is least marked in the most primitive New World monkey (*Aotus trivirgatus*, Figs. 3C and 5d), which most probably is close to the base of the platyrrhines (Romer, 1966), but already very distinct in the more advanced New World monkeys (e.g., *Saguinus midas*) and the most primitive Old World monkeys (e.g., *Macaca mulatta*).

The enlargement and forward movement of the internal carotid artery in relation to the tympanic cavity, correlated with an increasing obliteration of the stapedial artery system, are the most characteristic common features of higher primates, but also the complete annexation of the stapedial area of supply by the internal–external carotid artery system by means of anastomoses a1 and a3 (Figs. 3C,D and 5d,e), and, especially, the development of a true ophthalmic artery (o), departing from the distal part of the internal carotid artery and homologous with the human vessel, are characteristic higher primate features. [For details concerning the ontogenetic and probable phylogenetic development of the orbital supply pattern of primates, cf. the anastomosis X in primitive cebids (Fig. 3C); see Bugge, 1974*b*.]

New and Old World monkeys are known since the Lower Oligocene in South America and Africa (Fayum), respectively, but are completely unknown in Laurasian deposits before the connection between Africa and Eurasia was reestablished in the Miocene. The absence of Old World monkeys in the Oligocene of Laurasia, and the fact that the earliest representatives from the Fayum were highly differentiated together with the evidence that all the catarrhine families are or have been present in Africa some time since their first appearance, substantiate an African origin for the Old World monkeys, probably from a still unknown Eocene ancestor.

The two South American platyrrhine families Hapalidae (= Callitrichidae) and Cebidae first occur in the Upper Oligocene and Lower Miocene, respectively. Recently, however, Hoffstetter (1969) described a jaw fragment of a new ceboid from the Lower Oligocene of Bolivia, *Branisella boliviana,* possibly a primitive cebid, showing obvious monkeylike characteristics, associated with some archaic features (Hoffstetter, 1974). The presence of a primitive ceboid in Lower Oligocene deposits seems to confirm a Late Eocene platyrrhine invasion of the South American continent. From the dawn of the Tertiary until the Pliocene, South America was an island, well separated from North America and Africa, and early immigrants had to cross the sea by natural rafting regardless of where they came from.

Until recently most researchers accepted a North American origin for the early platyrrhine immigrants as the most probable, since the South Atlantic seemed to be an impassable barrier to a Late Eocene invasion from Africa. In this respect, however, the results of recent research on continental drift and rotation during the Tertiary are important, demonstrating that although the South Atlantic existed in the Eocene, it was not then of its present formidable width. The distance from Africa to South America then was probably no greater than that from North America, and the actual distance between the continents is, of course, important for the emergence of new dispersal routes across the sea by natural rafting. But the direction of the prevailing marine currents is probably more important, and the equatorial current from East to West markedly favored an invasion from Africa and presented difficulties for an invasion from North America.

The New World monkeys probably arose from a single stock, the two recent families sharing several complex morphological features—including the cephalic arterial pattern—which were probably inherited from a common, monkeylike ancestor. At the time in question, however, no North American primate had evolved beyond the tarsioid stage. Unfortunately, the Middle Eocene omomyids, the important North American tarsioids, most frequently referred to as a possible platyrrhine ancestor, are known mainly from jaw fragments, but the contemporary European tarsiid, *Necrolemur antiquus*, was certainly very primitive with respect to the cephalic arterial pattern (see above), and probably no Eocene tarsioid, including the North American omomyids, had reached the necessary level of development to be accepted as a probable platyrrhine ancestor. In contrast to the primitive tarsioids, even the most primitive catarrhine monkeys from the Lower Oligocene of Africa (e.g., the Parapithecidae) had attained the pattern of cephalic arterial supply typical of higher primates, characterized by the entire obliteration of the stapedial artery and the enlargement and movement forward of the internal carotid artery in relation to the tympanic cavity (see, for instance, Gingerich, 1973).

It is therefore most probable that a Late Eocene, monkeylike African ancestor of the Old World monkeys crossed the South Atlantic, giving rise in South America to the New World monkeys. Otherwise, an independent derivation of the New and Old World monkeys from primitive Laurasian tarsioids seems to presuppose a very complex set of convergences or parallelisms which, when taking their remarkable morphological similarity into account, including the cephalic arterial pattern, is hardly acceptable in the light of the probable paleogeographic situation at the time in question.

The characteristic cephalic arterial pattern of New and Old World monkeys therefore conforms with other common morphological features, with paleobiogeographic evidence, and with the recent revision of the relative position of the southern continents in the early Tertiary, in suggesting a probable Late Eocene common origin of the higher primates. The probable pre-Oligocene origin of the Platyrrhini and Catarrhini warrants a position as infraorders, while the probable pre-Eocene origin of the Pithecoidea (= Simiiformes) corresponds to that of the other suborders. However, it is still a question whether the recent haplorhines constitute a grade only (see Table 1) or a natural taxonomic entity of monophyletic origin, i.e., whether the tarsioids and the pithecoids are two sister-groups, early differentiated by geographical segregation on both sides of the Tethys, as proposed by Hoffstetter (1974). In all cases, the difference between the tarsioids and the pithecoids as far as the cephalic arterial pattern is concerned is much less than between the lemuroids and the lorisoids.

References

Adams, W. E., 1957, The extracranial carotid rete and carotid fork in *Nycticebus coucang, Ann. Zool.* **2**:21-37.

Ask-Upmark, E., 1953, On the entrance of the carotid artery into the cranial cavity in *Stenops gracilis* and *Otolicnus crassicaudatus, Acta anat.* **19**:101-103.

Bugge, J., 1963, A standardized plastic injection technique for anatomical purposes, *Acta anat.* **54**:177-192.

Bugge, J., 1972, The cephalic arterial system in the insectivores and the primates with special reference to the Macroscelidoidea and Tupaioidea and the insectivore-primate boundary, *Z. Anat. Entwick.* **135**:279-300.

Bugge, J., 1974*a*, The cephalic arteries of hystricomorph rodents, *Symp. Zool. Soc. Lond.* **34**:61-78.

Bugge, J., 1974*b*, The cephalic arterial system in insectivores, primates, rodents and lagomorphs, with special reference to the systematic classification, *Acta. Anat.* **87**(Suppl. **62**):1-160.

Bugge, J., 1975*a*, The cephalic arterial system in edentates (Xenarthra) and pangolins (Pholidota) with special reference to the systematic classification (abstract), *10th International Congress of Anatomists, Tokyo*, p. 100.

Bugge, J., 1975*b*, Corrosion casts of the cephalic arterial system, prepared by means of a standardized plastic injection and corrosion technique (abstract), *10th International Congress of Anatomists, Tokyo*, p. 528.

Bugge, J., 1978, The cephalic arterial system in carnivores, with special reference to the systematic classification, *Acta Anat.* **101**:45-61.

Campbell, C. B. G., 1966, The relationships of the tree shrews: The evidence of the nervous system, *Evolution* **20**:276-281.

Carlsson, A., 1909, Die *Macroscelididae* und ihre Beziehungen zu den übrigen Insectivoren, *Zool. Jahrb., Abt. Syst, Ökol. Geogr. Tiere* **28**:349-400.

Carlsson, A., 1922, Über die *Tupaiidae* und ihre Beziehungen zu den Insectivora und den Prosimiae, *Acta Zool., Stockholm* **3**:227-270.

De la Torre, E., and Netsky, M. G., 1960, Study of persistent primitive maxillary artery in the human fetus: Some homologies of cranial arteries in man and dog, *Am. J. Anat.* **106**:185-195.

Fiedler, W., 1956, Übersicht über das System der Primates, in: *Primatologia* (H. Hofer, A. H. Schultz, and D. Starck, eds.), Vol. 1, pp. 1-266, Karger, Basel.

Gingerich, P. D., 1973, Anatomy of the temporal bone in the Oligocene anthropoid *Apidium* and the origin of Anthropoidea, *Folia Primatol.* **19**:329–337.
Grosser, O., 1901, Zur Anatomie und Entwicklungsgeschichte des Gefässsystems der Chiropteren, *Arb. anat. Inst., Wiesbaden* **17**:203–424.
Haines, D. E., and Swindler, D. R., 1972, Comparative neuroanatomical evidence and the taxonomy of the tree shrews (*Tupaia*), *J. Hum. Evol.* **1**:407–420.
Hennig, W., 1966, *Phylogenetic Systematics,* University of Illinois Press, Urbana.
Hill, W. C. O., 1953*a,* The blood-vascular system of *Tarsius, Proc. Zool. Soc. London* **123**:655–694.
Hill, W. C. O., 1953*b, Primates: Comparative Anatomy and Taxonomy.* Vol. I, *Strepsirhini,* University of Edinburgh Press, Edinburgh.
Hill, W. C. O., 1955, *Primates: Comparative Anatomy and Taxonomy.* Vol. II, *Haplorhini: Tarsioidea,* University of Edinburgh Press, Edinburgh.
Hill, W. C. O., 1957, *Primates: Comparative Anatomy and Taxonomy.* Vol. III, *Pithecoidea: Platyrrhini, Hapalidae,* University of Edinburgh Press, Edinburgh.
Hill, W. C. O., 1960, *Primates: Comparative Anatomy and Taxonomy.* Vol. IV, *Platyrrhini, Cebidae,* Part A, University of Edinburgh Press, Edinburgh.
Hill, W. C. O., 1962, *Primates: Comparative Anatomy and Taxonomy.* Vol. V, *Platyrrhini, Cebidae,* Part B, University of Edinburgh Press, Edinburgh.
Hill, W. C. O., 1966, *Primates: Comparative Anatomy and Taxonomy.* Vol. VI, *Catarrhini: Cercopithecoidea-Cercopithecinae,* University of Edinburgh Press, Edinburgh.
Hoffstetter, R., 1969, Un primate de l'Oligocène inférieur sud-américain: *Branisella boliviana* gen. et sp. nov. *C.R. Acad. Sci., Paris, Sér. D* **269**:434–437.
Hoffstetter, R., 1972, Relationships, origins, and history of the ceboid monkeys and caviomorph rodents: A modern reinterpretation, in: *Evolutionary Biology,* Vol. 6 (Th. Dobzhansky, M. K. Hecht, and W. C. Steere, eds.), pp. 323–347, Appleton-Century-Crofts, New York.
Hoffstetter, R., 1974, Phylogeny and geographical deployment of the primates, *J. Hum. Evol.* **3**:327–350.
Hürzeler, J., 1948, Zur Stammesgeschichte der Necrolemuriden, *Schweiz. palaeontol. Abhandl.* **66**:1–46.
Le Gros Clark, W. E., 1925, On the skull of *Tupaia, Proc. Zool. Soc. London* **1925**:559–567.
Le Gros Clark, W. E., 1926, On the anatomy of the pen-tailed tree-shrew *(Ptilocercus lowii), Proc. Zool. Soc. London* **1926**:1179–1309.
Le Gros Clark, W. E., 1934, *Early Forerunners of Man. A Morphological Study of the Evolutionary Origin of the Primates*, pp. 222–252, Bailliere, Tindall, and Cox, London.
Le Gros Clark, W. E., 1959, *The Antecedents of Man. An Introduction to the Evolution of the Primates,* University of Edinburgh Press, Edinburgh.
Matthew, W. D., 1909, The Carnivora and Insectivora of the Bridger Basin, Middle Eocene, *Mem. Am. Mus. Nat. Hist.* **9**(Part 6):291–567.
McKenna, M. C., 1966, Paleontology and the origin of the primates, *Folia Primatol.* **4**:1–25.
Padget, D. H., 1948, The development of the cranial arteries in the human embryo, *Contrib. Embryol., Washington* **32**:205–261.
Pocock, R. I., 1918, On the external characters of the lemurs and of *Tarsius, Proc. Zool. Soc. Lond.* **1918**:19–53.
Remane, A., 1956, Paläontologie und Evolution der Primaten, in: *Primatologia* (H. Hofer, A. H. Schultz, and D. Starck, eds.), Vol. 1, pp. 267–378, Karger, Basel.
Romer, A. S., 1966, *Vertebrate Paleontology,* 3rd edn. University of Chicago Press, Chicago.
Saban, R., 1963, Contribution a l'étude de l'os temporal des primates. Description chez l'Homme et les prosimiens. Anatomie comparée et phylogénie, *Mém. Mus. Nat. Hist. Nat. N.S., Sér. A* **29**:1–378.
Simpson, G. G. 1945, The principles of classification and a classification of mammals, *Bull. Am. Mus. Nat. Hist.* **85**:1–350.
Starck, D., 1956, Primitiventwicklung und Plazentation der Primaten, in: *Primatologia* (H. Hofer, A. H. Schultz, and D. Starck, eds.), Vol. 1, pp. 723–886, Karger, Basel.

Straus, W. L., 1949, The riddle of man's ancestry, *Rev. Biol.* **24**:200–223.

Struthers, P. H., 1930, The aortic arches and their derivatives in the embryo porcupine *(Erethizon dorsatus)*, *J. Morphol.* **50**:361–392.

Szalay, F. S., and Katz, C. C., 1973, Phylogeny of lemurs, galagos and lorises, *Folia Primatol.* **19**:88–103.

Tandler, J., 1899, Zur vergleichenden Anatomie der Kopfarterien bei den Mammalia, *Denksch. Akad. Wissensch. Wien* **67**:677–784.

Tandler, J., 1902, Zur Entwicklungsgeschichte der Kopfarterien bei den Mammalia, *Morphol. Jahrb.* **30**:275–373.

van Valen, L., 1965, Treeshrews, primates and fossils, *Evolution* **19**:137–151.

van Valen, L., 1966, Deltatheridia, a new order of mammals, *Bull. Am. Mus. Nat. Hist.* **132**:1–126.

Von Hofmann, L., 1914, Die Entwicklung der Kopfarterien bei *Sus scrofa domesticus*, *Morphol. Jahrb.* **48**:645–671.

Winge, H., 1895, Jordfundne og nulevende Aber (Primates) fra Lagoa Santa, Minas Geraes, Brasilien, med Udsigt over Abernes indbyrdes Slægtskab (Fossil and recent monkeys (Primates) from Lagoa Santa, Minas Geraes, Brazil, with a review of the interrelationships of the monkeys), *E. Museo Lundii*, Vol. 2, Part 2, in: (H. Winge, ed., 1924) Pattedyr-Slægter II. *Rodentia, Carnivora, Primates,* pp. 250–313, Hagerup, Copenhagen. Also available in translation in: (A. S. Jensen, R. Spärck, and H. Volsøe, eds, 1941) *The Interrelationships of the Mammalian Genera,* Vol. 2. *Rodentia, Carnivora, Primates,* pp. 294–367, Reitzel, Copenhagen.

Evidence from Other Comparative Anatomy Studies

VI

Phylogenetic Relationships of the Platyrrhini

The Evidence of the Femur

15

S. M. FORD

Introduction

The major problem addressed in this volume concerns the origin of the New World monkeys. This problem is sometimes simplified to the question of whether they came from North America or from Africa. However, the biological problem is actually more complex and can be treated as four separate hypotheses: (1) descent of platyrrhines from an omomyid ancestor from North America, independent of catarrhines (Gazin, 1958; Orlosky and Swindler, 1975; Patterson and Pascual, 1972, Stirton, 1951; Stirton and Savage, 1950; Szalay, 1975; Wood, 1973, 1977); (2) descent from adapids, again independent of catarrhines (Gingerich, 1973); (3) origin of a monophyletic Anthropoidea, from either omomyids or adapids. Anthropoidea may have evolved in Holarctica, with later independent southern migrations of catarrhines into Africa and platyrrhines into South America (Gingerich, 1975, 1977; Gingerich and Schoeninger, 1977; Simons, 1969, 1976; Simpson, 1978). Alternately, anthropoids may have evolved in Africa, with the subsequent introduction of platyrrhines into South America from Africa, most likely via rafting (Genet-Varcin, 1974; Hoffstetter, 1972, 1974; Lavocat, 1974;

S. M. FORD • Department of Anthropology, Southern Illinois University, Carbondale, Illinois 62901.

see also Szalay, 1976, for the possibility of a South American origin with subsequent rafting of catarrhines to Africa). In either case, the closest relatives of platyrrhines would be the Catarrhini. (4) The fourth possibility is that platyrrhines have a Cretaceous, pre-rift, African origin and may even be polyphyletic (Hershkovitz, 1972, 1974, 1977). Should this be the case, platyrrhines would not be expected to show any close, unified relationship to other primate groups of later origin, e.g., omomyids, adapids, or catarrhines.

In cladistic terms, the problem becomes a search for the sibling-group of the platyrrhines. This is accomplished by finding derived character states that are present in both platyrrhines and one or more of the other primate taxa in question. Once shared character states are recognized, it must be demonstrated that these represent derived states that are homologous, as opposed to convergent or retained primitive character states. It is imperative that the primitive or ancestral condition of each group is compared, rather than various specialized or derived conditions that arose subsequent to the origin of that group. I have attempted, for each character or character complex, to determine the primitive condition for primates and the primitive condition for each pertinent taxon within primates (i.e., Platyrrhini, Catarrhini, Omomyidae, and Strepsirhini, or, more specifically, Adapidae).

There is much evidence that the investigation of postcranial characters can yield valuable phylogenetic information (e.g., Calhoun, 1977; Cartmill and Milton, 1977; Ciochon and Corruccini, 1975; Conroy, 1976; Decker and Szalay, 1974; Delson and Andrews, 1975; Godfrey, 1976; Lewis, 1974; MacFadden, 1976; Novacek, 1980; Robinette, 1968; Robinette and Stains, 1970; Romankowowa, 1963*a,b*; Romer, 1922; Stains, 1962; Szalay, 1975, 1976, 1977*a,b*; Szalay and Decker, 1974; Szalay *et al.*, 1975). Locomotor adaptations have been very significant in the evolution of primates, and postcranial elements may contain a number of synapomorphies which can elucidate phylogenetic relationships. Also, the inclusion of postcranial data increases the number of independent characters and provides a test of hypotheses derived from studies of the skull and teeth. In this study, qualitative and quantitative data on a number of femoral features have been collected and analyzed in an attempt to clarify the phylogenetic affinities of the New World monkeys (see Fig. 1).

Assuming one has chosen "good" characters (i.e., definable, recognizable, and relatively invariable within the smallest taxonomic unit under consideration), the key problem is the determination of the polarity of the morphocline for each character (which character state is primitive). Criteria most often used to determine polarity include: (1) frequency of occurrence within the group under study, (2) comparison with closely related groups, (3) character covariance in a "form–function complex," (4) ontogeny of characters, and (5) occurrence of various character states in fossils of progressively earlier age (see, e.g., Hecht, 1976; Hecht and Edwards, 1977; Kavenaugh, 1972; Kluge and Farris, 1969; Novacek, 1980). Various authors favor different ones

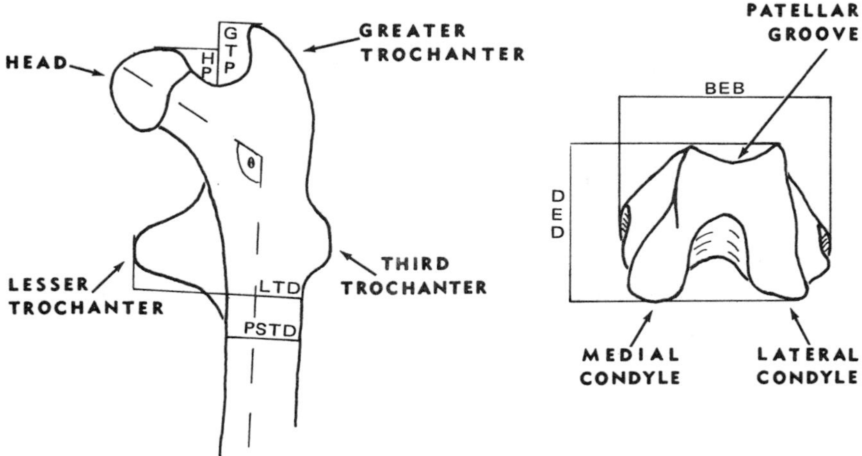

Fig. 1. Generalized primate left femur, illustrating some of the measurements used in this study. Proximal and distal ends not to scale. (A) proximal femur, anterior view; (B) distal femur, inferior view. Abbreviations: HP, head projection, measured parallel to long axis of bone; GTP, greater trochanter projection, measured parallel to long axis of bone; LTD, diameter at lesser trochanter, measured parallel to direction of projection of lesser trochanter and perpendicular to long axis of bone, and not including projection of third trochanter (if present); PSTD, proximal shaft transverse diameter, measured parallel to head–greater trochanter axis and taken just distal to lesser trochanter; θ, angle of neck to long axis of bone; BEB, biepicondylar breadth, maximum; DED, distal epiphysis depth, three-point measurement taken across both condyles to maximum anterior extension of patellar groove margins.

among these criteria, but clearly, the more that can be applied to a given morphocline and result in the same polarity determination, the more confident one can be that the polarity is correct.

Two of the criteria were not applied in this analysis. The paucity of relevant ontogenetic data rules out that criterion. Statistical correlation tests show that none of the characters used in this study covary, and therefore polarity cannot easily be determined by studying a unified "form–function complex." Napier and Walker (1967) hypothesized that the primitive primate locomotor pattern was vertical clinging and leaping. This has been largely discredited by Decker and Szalay (1974). I found that two additional characters cited by Napier and Walker (1967) as characteristic of vertical clingers and leapers, a cylindrical head of the femur and a large third trochanter, were not uniformly present in strepsirhine vertical clingers and leapers, nor were they present in *Cebuella* and *Callithrix*, which have been described as vertical clingers (Kinzey *et al.,* 1975; Moynihan, 1976). In fact, Howell (1932) found the third trochanter was absent in *Pedetes* and jerboas, which are saltatorial rodents. Based on this and other studies, Howell concluded that this feature "has considerable phylogenetic, but little or no adaptive, significance, except possibly in the broadest sense" (1944, p. 172). Therefore, a complex of characters associated with vertical clinging and leaping has not been identified, and even if one were, those associated characters could not automatically be considered primitive primate characteristics.

I did rely heavily on in-group distribution of characters with one state present in the vast majority of the group. Out-group comparisons were used particularly for those characters whose distribution was not overwhelmingly clear-cut [what one scholar refers to as the 80% rule (T. Erwin, personal communication)]. If the majority of extinct taxa shared the hypothesized primitive state, or if it was present in the earliest known member of the group, this was considered as evidence supporting the accuracy of the inferred polarity.

Materials and Methods

The taxa studied included all extant platyrrhine genera, both extinct platyrrhines for which femora are known [*Homunculus* and those referred to *Cebupithecia*, although the postcranium assigned to the latter may not belong to the same taxon as the type cranial material (A. Rosenberger, personal communication)], selected generalized catarrhines (specifically *Cercopithecus, Macaca, Nasalis, Presbytis, Rhinopithecus,* and *Pan*), the adapids *Notharctus* and *Smilodectes* as well as some extant strepsirhines [including *Lichanotus (Avahi), Daubentonia, Cheirogaleus, Galago, Perodicticus, Loris, Arctocebus, Nycticebus,* and three species of *Lemur*], and the omomyid *Hemiacodon*. *Pseudoloris* and *Necrolemur* (as described in Simpson, 1940) and *Tarsius* were also included, to increase the tarsiiform [although not strictly omomyid; but see also Krishtalka and Schwartz (1978)] sample, and *Plesiadapis,* as described in Simpson (1935), Szalay *et al.* (1975), and Novacek (1977, 1980). For out-group comparisons in the determination of primitive character states for the order Primates, I relied on the distributions given by Novacek (1977, 1980) for Dermoptera, Macroscelididae, Leptictidae, Ptilocercinae, Tupaiinae, Erinaceoidea, Soricidae, Talpidae, Tenrecidae, Chrysochloridae, and Chiroptera. Also included were data taken from published descriptions of taeniodonts and condylarths (Matthew, 1937; Radinsky, 1966), marsupials (Gregory, 1910), monotremes (Gregory, 1910; Jenkins and Parrington, 1976), the Triassic mammals *Erythrotherium, Eozostrodon,* and *Megazostrodon* (Jenkins and Parrington, 1976), and cynodont (Gregory, 1910; Jenkins, 1971; Parrington, 1961) and therocephalian (Kemp, 1978) therapsids.

From sixty original characters, the study was narrowed to 23 for which the primitive condition in primates as a whole can be identified with reasonable confidence. However, even if the polarities are correct, there is still the possibility that the derived character states in different taxa are not homologous.

Results

The characters fall into three categories. The first group is those for which the distributions of derived states appear to be random, falling into no discernible phylogenetic, functional, or allometric patterns. Some of these are the morphology of the trochanteric fossa, the relative length of the lesser trochanter, and some aspects of the morphology of the patellar groove.

The second group include characters which appear to unite some or all members of one group, but which fail to align them conclusively with any other single group. Most of the characters fall into this category, and several of them have interesting, if enigmatic, distributions. For example, based on the distribution in the eutherians mentioned earlier, it appears that having the greater trochanter extend farther superiorly than the head is primitive. The derived condition of the head extending farther superiorly is exhibited by most of the cebids. It also occurs in some species of *Saguinus*, the fossil ceboid *Homunculus, Pan, Loris, Nycticebus, Galago,* and *Tarsius.* This is particularly intriguing because Howell (1944) found that a high greater trochanter, denoting powerful, efficient deep gluteal muscles, is characteristic of leapers and runners, although also occurring in other mammals. Yet many of the classic primate vertical clingers and leapers show the opposite condition. An intermediate condition of a fairly equal head and greater trochanter occurs in the rest of the cebids, some species of *Saguinus, Cebupithecia, Cebuella, Rhinopithecus* (but none of the other cercopithecids), *Arctocebus, Daubentonia,* and *Notharctus.* Another character of this group is the angle that the neck makes with the shaft of the femur. The neck forms a very low angle (varying around 130°) in all of the Tarsiiformes, but also in *Loris, Galago,* and some *Ateles.* The relative distance between the greater and lesser trochanters also falls into this category. This distance is particularly large in the callitrichids and in *Homunculus*, but also in *Cheirogaleus* and *Perodicticus*. As an alternate derived condition, it is especially small in the tarsiiforms, but also in *Loris, Presbytis,* the pithecines, and *Ateles* (although not in *Lagothrix, Brachyteles,* or *Alouatta*). In none of these cases do the distributions correlate with size or locomotor specializations of the animals.

The last category is those derived character states which may link the platyrrhines phylogenetically with one or more other primate groups. Only a few characters fall into this category. The size or extension of the lesser trochanter is the first of these characters. The lesser trochanter is large relative to proximal shaft diameter, extending almost as far or as far as the head medially, in plesiadapiforms, tarsiiforms (*contra* Novacek, 1980), and some strepsirhines, including *Cheirogaleus*, the lorisines, and the adapids *Notharctus* and *Smilodectes* (see Fig. 2). It is also large in ptilocercines, tupaiines, soricids, talpids, and chiropterans (Novacek, 1977, 1980), as well as taeniodonts (Matthew, 1937). This led Novacek (1977, 1980) to conclude that a large lesser trochanter is primitive for eutherians. However, it is reduced in some strepsirhines, platyrrhines, catarrhines, dermopterans, macroscelidids, leptictids,

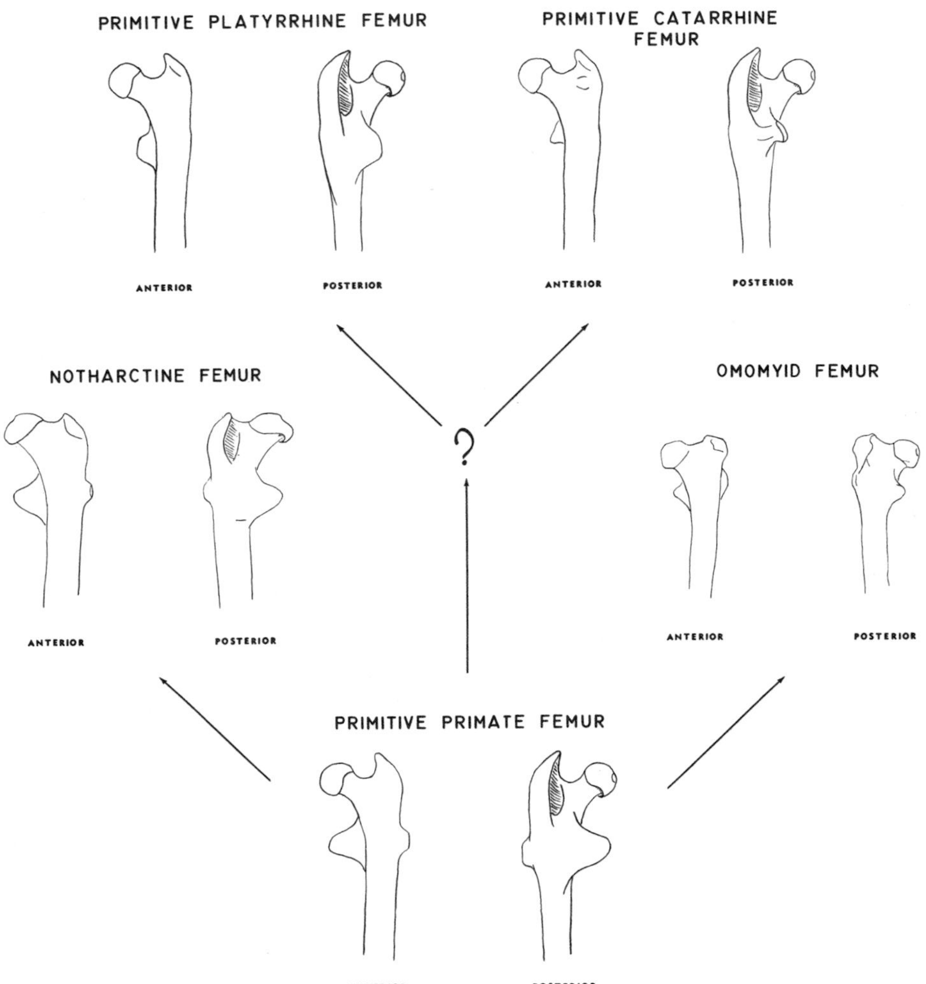

Fig. 2. Evolutionary change of the proximal femur within primates. Question mark indicates uncertainty as to the monophyly of Anthropoidea. See text for discussion.

erinaceids, and tenrecs, an almost equally large number of groups. It is fairly large in monotremes and Triassic mammals (Jenkins and Parrington, 1976), although the positions of the greater and lesser trochanters and the head of the femur differ from that in eutherians. For this character, distribution alone is insufficient for polarity determination. However, a large number of scholars, including Butler (1977), Kielan-Jaworowska (1977), Matthew (1937), Szalay and Decker (1974), and Van Valen and Sloan (1965), have suggested that condylarths, in particular *Protungulatum*, may be among the most primitive of eutherians. Matthew (1937) stated that the skeleton and most of the cranial and dental features of *Loxolophus*, an arctocyonid condylarth, probably "represent the closest known actual approach to the common primitive type from which the placental mammals, or most of them, are derived" (1937, p. 51). Most of the condylarths have prominent lesser trochanters, although *Tetraclaenodon,* a primitive phenacodontid, does not (Matthew, 1937; Radinsky, 1966). This distribution, combined with the fact that plesiadapiforms have a large lesser trochanter, leads to agreement with Novacek that a large lesser trochanter is primitive. If this polarity is correct, then the derived condition of a reduced lesser trochanter (presumably associated with reduced importance of the iliopsoas muscle, a flexor of the thigh) is characteristic of *Lichanotus (Avahi), Daubentonia, Galago, Lemur,* all catarrhines, and all platyrrhines except *Cebuella*. This distribution can be interpreted in a number of ways. Ignoring the split distribution in extant strepsirhines, the most pertinent interpretations are: (1) Platyrrhines and catarrhines share a common ancestor with a reduced lesser trochanter, with the condition in *Cebuella* being a secondary specialization. This interpretation would support a monophyletic Anthropoidea, and perhaps suggest an African origin of the platyrrhines, although not ruling out two independent migrations of anthropoids into the southern continents. (2) Platyrrhines and catarrhines have independently reduced the lesser trochanter. Again, *Cebuella* has secondarily enlarged this process. This would argue against an African origin of the platyrrhines. (3) *Cebuella* retains the primitive condition, and catarrhines and all other platyrrhines have independently reduced the lesser trochanter. This would also argue against an African origin of the platyrrhines.

The second character which may link the platyrrhines with another group is the size of the third trochanter or gluteal tuberosity, the location of insertion of m. gluteus superficialis. As stated earlier, Howell (1944) found that the presence or absence and size of the third trochanter was of unknown functional significance and of primarily phylogenetic significance. I find that a large third trochanter is primitive for primates (see Fig. 2). Novacek (1977, 1980) concluded that a weak or absent third trochanter was primitive, based on the occurrence of this character state in anthropoids, dermopterans, ptilocercines, tupaiines, erinaceoids, tenrecids, and chiropterans. However, the third trochanter is present and even prominent in marsupials (Gregory, 1910), macroscelidids, leptictids, soricids, talpids (Novacek, 1977, 1980), lagomorphs, tubulidentates, hyracoids, terrestrial edentates, a variety of rodents, perissodactyls (Howell, 1944), condylarths (Matthew, 1937; Radinsky, 1966), as well as plesiadapids, tarsiiforms, and most strepsirhines, including adapids. Stern (1971) pointed out that the gluteus superficialis insertion in cercopithecoids is unique within primates. This suggests that the lack of a third trochanter is derived at least for Old World monkeys. The overwhelming evidence seems to indicate that a prominent third trochanter is primitive for eutherians and for primates, a conclusion also reached by Gregory (1910). If so, then the derived condition of no third trochanter, or of only a slight

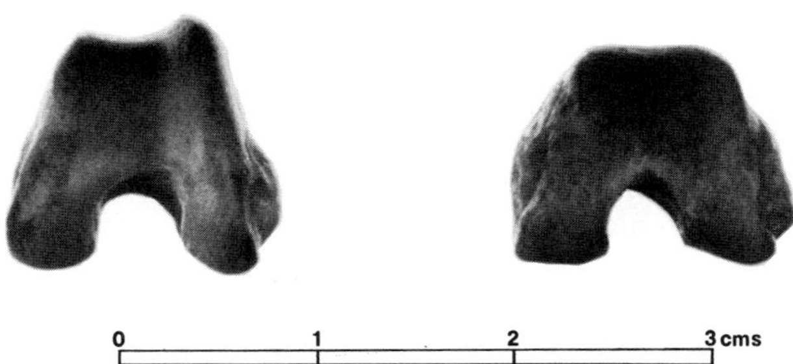

Fig. 3. Photograph illustrating the two alternate character states in the morphology of the distal femur. Left: *Lemur mongoz* (USNM 337946), exhibiting the primitive condition of a squared distal epiphysis. Right: *Callicebus pallescens* (USNM 269827), exhibiting the derived condition of a compressed anteroposteriorly (or elongated mediolaterally) distal epiphysis. Both specimens are inferior views of left femora.

rugosity which may be slightly raised but no clear trochanter, is characteristic of all catarrhines and all platyrrhines except *Aotus*. The most parsimonious explanation of this distribution is that the catarrhines and platyrrhines shared a common ancestor, and that *Aotus* has independently reacquired a (small) third trochanter. Of course, it is also possible that the third trochanter has been lost independently in these groups, as it apparently has been in the lorisines and possibly in *Pseudoloris* (Simpson, 1940). Certainly, the derived muscle insertion of the cercopithecoids could lead to this conclusion, in which case the third trochanter would also have been lost independently in the hominoids.

A last feature which may fall into this category is the shape of the distal epiphysis (see Fig. 3). In *Tarsius, Hemiacodon, Galago, Lichanotus (Avahi), Lemur, Smilodectes, Notharctus,* and *Plesiadapis,* and to a lesser degree in *Daubentonia* and *Arctocebus*, the distal epiphysis is squared, the ratio of anteroposterior depth to mediolateral width approaching one. This distribution suggests that this morphology is primitive for primates. All New World monkeys and all catarrhines exhibit the derived condition of a distal femur which is compressed anteroposteriorly (or elongated mediolaterally). The most parsimonious interpretation of this distribution is that platyrrhines and catarrhines form a monophyletic group, as is also suggested by the distribution of derived states of the two characters discussed previously. However, it must be noted that the distal epiphysis is compressed anteroposteriorly in all four lorisine genera as well. This is almost certainly a convergent adaptation, perhaps due to the lorisine locomotor specialization of slow-climbing. This may be supported by the fact that the distal epiphysis is also greatly compressed anteroposteriorly (or elongated mediolaterally) in the sloths, *Bradypus* and *Choloepus*, but squared (the hypothesized primitive condition) in armadillos (*Dasypus*) and the giant anteater, *Myrmecophaga*. The fossil indriid *Palaeopropithecus* exhibits the greatest degree of mediolateral elongation (or anteroposterior compression) of the distal epiphysis. If this morphology is associated with a slow-climbing adaptation in lorisines and sloths, it would support Jungers' (1979) suggestion of a sloth model for *Palaeopropithecus.* It is intriguing to note, however, that the derived condition is also present in *Tamandua*, the collared anteater, which is not a slow-climber (Enders, 1930, 1935) but which is prehensile-tailed (Walker, 1975) and spends more than half of its time in arboreal environments, unlike the almost exclusively terrestrial giant anteater (Taylor, 1978), and in *Potos,* the kinkajou, which is also prehensile-tailed and moves rapidly in trees (Enders, 1935). It is clear that this derived morphology of the distal femur can evolve convergently in mammals, and in fact has done so at least twice in primates (lorisines and anthropoids), although probably for different reasons. Therefore, it is not impossible, although less likely, that it evolved separately in platyrrhines and in catarrhines.

Conclusion

The distributions of most femoral traits do not shed any light at present on the problem of platyrrhine origins. However, several interesting observa-

Table 1. Number of Primitive and Derived Character States in Platyrrhine Genera[a]

Genus	Number of primitive states	Number of derived states
Callimico	11	12
Callithrix	10	13
Cebuella	13	10
Leontopithecus	13	10
Saguinus	12	11
Cebupithecia[b]	14	9
Homunculus[b]	13	10
Alouatta	13	10
Aotus	14	9
Ateles	9	14
Brachyteles	9	14
Cacajao	11	12
Callicebus	11	12
Cebus	14	9
Chiropotes	12	11
Lagothrix	12	11
Pithecia	13	10
Saimiri	11	12

[a] Genera with the highest number of primitive character states are not intended to be interpreted as the most primitive platyrrhines nor as representative of the primitive New World monkey morphotype in all features.

[b] Extinct platyrrhines.

tions can be made. Neither of the two fossil platyrrhine femora exhibited any greater number of primitive character traits than did any other platyrrhine genus. Also none of the callitrichids were more primitive than any of the cebids (see Table 1). Omomyids, as represented by *Hemiacodon*, and tarsioids in general do not share any derived features with platyrrhines, nor do adapids, as represented by *Notharctus* and *Smilodectes*. Catarrhines and platyrrhines share at least three derived features. The most parsimonious (although not necessarily the correct) explanation is that these three features are homologous in catarrhines and platyrrhines, and that catarrhines and platyrrhines share a more recent common ancestor than does either taxon with any other group of primates. Therefore, distributions do not rule out the possibility of a monophyletic Anthropoidea, or of either an African or a North American origin of platyrrhines. However, they do not support either an omomyid or a notharctine (adapid) ancestry. The primary difficulties in this analysis have been the determination of the primitive character states within Platyrrhini, a very diverse group morphologically and behaviorally, and the scarcity of fossil postcranial remains for all groups. The discovery of additional early fossil postcranial material from all continents should aid greatly in elucidating the relationships.

Acknowledgments

I would like to thank the following persons for allowing me to examine specimens in their care: Drs. R. Emry, C. L. Gazin, and R. W. Thorington, Jr. of the Smithsonian Institution, P. Freeman of the Field Museum of Natural History, S. Anderson and M. McKenna of the American Museum of Natural History, J. Biegert and P. Schmid of the Anthropologisches Institut, Zurich, P. Napier of the British Museum (Natural History), and J. Fleagle of State University of New York at Stony Brook; W. Jungers kindly provided the data on *Palaeopropithecus*. This work benefited from discussions with Mr. G. Morgan; earlier drafts were read by Drs. L. Krishtalka and D. Domning. I am particularly grateful for discussions, criticism, support, and critical readings of the manuscript by Mr. S. M. Peters, and Dr. R. W. Thorington, Jr., and for expert advice in the preparation of illustrations from Mr. G. Venable. This work was completed while under the support of a Smithsonian Institution Predoctoral Fellowship, an Andrew W. Mellon Fellowship, and a Sigma Xi Grant-in-Aid of Research.

References

Butler, P. M. 1977, Evolutionary radiation of the cheek teeth of Cretaceous placentals, *Acta Palaeontol. Polon.* **22**:240–269.

Calhoun, T. P., 1977, Sesamoid structures in primate hands, *Yearb. Phys. Anthropol.* **20**:525–537.
Cartmill, M., and Milton, K., 1977, The lorisiform wrist joint and the evolution of "brachiating" adaptations in the Hominoidea, *Am. J. Phys. Anthropol.* **47**:249–272.
Ciochon, R. L., and Corruccini, R. S., 1975, Morphometric analysis of platyrrhine femora with taxonomic implications and notes on two fossil forms, *J. Hum. Evol.* **4**:193–217.
Conroy, G. C., 1976, Primate postcranial remains from the Oligocene of Egypt, *Contrib. Primatol.* **8**:1–134.
Decker, R. L., and Szalay, F. S., 1974, Origins and function of the pes in the Eocene Adapidae (Lemuriformes, Primates), in: *Primate Locomotion* (F. A. Jenkins, ed.), pp. 261–291, Academic Press, New York.
Delson, E., and Andrews, P., 1975, Evolution and interrelationships of the catarrhine primates, in: *Phylogeny of the Primates* (W. P. Luckett and F. S. Szalay, eds.), pp. 405–446, Plenum Press, New York.
Enders, R. K., 1930, Notes on some mammals from Barro Colorado Island, Canal Zone, *J. Mammal.* **11**:280–292.
Enders, R. K., 1935, Mammalian life histories from Barro Colorado Island, Panama, *Bull. Mus. Comp. Zool.* **78**:385–502.
Gazin, C. L., 1958, A review of the Middle and Upper Eocene primates of North America, *Smithsonian Misc. Coll.* **136**(1):1–112.
Genet-Varcin, E., 1974, Platyrrhine contribution to the phylogeny of the primates, *J. Hum. Evol.* **3**:259–263.
Gingerich, P. D., 1973, Anatomy of the temporal bone in the Oligocene anthropoid *Apidium* and the origin of Anthropoidea, *Folia Primatol.* **19**:329–337.
Gingerich, P. D., 1975, A new genus of Adapidae (Mammalia, Primates) from the Late Eocene of Southern France and its significance for the origin of higher primates, *Contrib. Mus. Paleontol. Univ. Mich.* **24**:163–170.
Gingerich, P. D., 1977, Radiation of Eocene Adapidae in Europe, *Geobios, Mém. Spéc.* **1**:165–185.
Gingerich, P. D., and Schoeninger, M., 1977, The fossil record and primate phylogeny, *J. Hum. Evol.* **6**:463–505.
Godfrey, L., 1976, Postcranial osteology and the positional and postural behavior of the genus *Lepilemur;* implications for lemur phylogeny (abstract), *Am. J. Phys. Anthropol.* **44**:180–181.
Gregory, W. K., 1910, The orders of mammals, *Bull. Am. Mus. Nat. Hist.* **27**:1–524.
Hecht, M. K., 1976, Phylogenetic inference and methodology as applied to the vertebrate record, *Evol. Biol.* **9**:335–363.
Hecht, M. K., and Edwards, J. L., 1977, The methodology of phylogenetic inference above the species level, in: *Major Patterns in Vertebrate Evolution* (M. K. Hecht, P. C. Goody, and B. M. Hecht, eds.), pp. 3–51, Plenum Press, New York.
Hershkovitz, P., 1972, The recent mammals of the Neotropical region: A zoogeographic and ecological review, in: *Evolution, Mammals and Southern Continents* (A. Keast, F. C. Erk, and B. Glass, eds.), pp. 311–431. State University of New York Press, Albany.
Hershkovitz, P., 1974, A new genus of Late Oligocene monkey (Cebidae, Platyrrhini) with notes on postorbital closure and platyrrhine evolution. *Folia Primatol.* **21**:1–35.
Hershkovitz, P., 1977, *Living New World Monkeys (Platyrrhini) with an Introduction to the Primates,* Vol. 1, University of Chicago Press, Chicago.
Hoffstetter, R., 1972, Relationships, origins, and history of the ceboid monkeys and caviomorph rodents: A modern reinterpretation, in: *Evolutionary Biology,* Vol. 6 (Th. Dobzhansky, M. K. Hecht, and W. C. Steere, eds.), pp. 323–347, Appleton-Century-Crofts, New York.
Hoffstetter, R., 1974, Phylogeny and geographical deployment of the primates, *J. Hum. Evol.* **3**:327–350.
Howell, A. B., 1932, The saltatorial rodent *Dipodomys*: The functional and comparative anatomy of its muscular and osseous systems. *Proc. Am. Acad. Arts. Sci.* **67**:377–536.
Howell, A. B., 1944, *Speed in Animals* (1965 facsimile of 1944 edition), Hafner, New York.
Jenkins, F. A., Jr., 1971, The post-cranial skeleton of African cynodonts, *Bull. Peabody Mus. Nat. Hist.* **36**:1–216.

Jenkins, F. A., Jr., and Parrington, F. R., 1976, The postcranial skeletons of the Triassic mammals *Eozostrodon, Megazostrodon* and *Erythrotherium, Phil. Trans. Roy. Soc. Lond., Ser. B* **273**:387–431.

Jungers, W. L., 1979, Adaptive diversity in subfossil Malagasy prosimians, *Abstracts of the VIIth International Congress of Primatology,* Bangalore, India.

Kavenaugh, D. H., 1972, Hennig's principles and methods of phylogenetic systematics, *Biologist* **54**:115–127.

Kemp, T. S., 1978, Stance and gait in the hindlimb of a therocephalian mammal-like reptile, *J. Zool., London* **186**:143–161.

Kielan-Jaworowska, Z., 1977, Evolution of the therian mammals in the Late Cretaceous of Asia. Part II: Postcranial skeleton in *Kennalestes* and *Asioryctes*, *Paleontol. Polon.* **37**:65–83.

Kinzey, W. G., Rosenberger, A. L., and Ramirez, M., 1975, Vertical clinging and leaping in a Neotropical anthropoid, *Nature* **255**:327–328.

Kluge, A. G., and Farris, J. S., 1969, Quantitative phyletics and the evolution of anurans, *Syst. Zool.* **18**:1–32.

Krishtalka, L., and Schwartz, J. H., 1978, Phylogenetic relationships of plesiadapiform-tarsiiform primates, *Ann. Carnegie Mus.* **47**:515–540.

Lavocat, R., 1974, The interrelationships between the African and South American rodents and their bearing on the problems of the origin of South American monkeys, *J. Hum. Evol.* **3**:323–326.

Lewis, O. J., 1974. The wrist articulations of the Anthropoidea, in: *Primate Locomotion* (F. A. Jenkins, Jr., ed.), pp. 143–169, Academic Press, New York.

MacFadden, B. J., 1976, Cladistic analysis of primitive equids, with notes on other perissodactyls, *Syst. Zool.* **25**:1–14.

Matthew, N. D., 1937, Paleocene faunas of the San Juan Basin, New Mexico, *Trans. Am. Phil. Soc., N.S.* **30**:1–510.

Moynihan, M., 1976, *The New World Primates*, Princeton University Press, Princeton.

Napier, J. R., and Walker, A. C., 1967, Vertical clinging and leaping in living and fossil primates, in: *Progress in Primatology* (D. Starck, R. Schneider, and H. J. Kuhn, eds.), pp. 66–69, Gustav Fischer Verlag, Stuttgart.

Novacek, M. J., 1977, Evolution and relationships of the Leptictidae (Eutheria: Mammalia), Ph.D. thesis, University of California, Berkeley.

Novacek, M. J., 1980, Cranioskeletal features in tupaiids and selected Eutheria as phylogenetic evidence, in: *Comparative Biology and Evolutionary Relationships of Tree Shrews* (W. P. Luckett, ed.), *Advances in Primatology,* Vol. 4, pp. 35–93, Plenum Press, New York.

Orlosky, F. J., and Swindler, D. R., 1975, Origins of New World monkeys, *J. Hum. Evol.* **4**:77–83.

Parrington, F. R., 1961, The evolution of the mammalian femur, *Proc. Zool. Soc. Lond.* **137**:285–298.

Patterson, B., and Pascual, R., 1972, The fossil mammal fauna of South America, in: *Evolution, Mammals, and Southern Continents* (A. Keast, F. C. Erk, and B. Glass, eds.), pp. 247–310, State University of New York Press, Albany.

Radinsky, L. B., 1966, The adaptive radiation of the phenacodontid condylarths and the origin of the Perissodactyla, *Evolution* **20**:408–417.

Robinette, H. R., 1968, Comparative study of the calcanea of some members of the Pinnipedia, M.S. thesis, Department of Zoology, South Illinois University, Carbondale.

Robinette, H. R., and Stains, H. J., 1970, Comparative study of the calcanea of the Pinnipedia, *J. Mammal.* **51**:527–541.

Romankowowa, A., 1963*a*, Comparative study of the skeleton of the hyoid apparatus in some bat species, *Acta Theriol.* **7**:15–23.

Romankowowa, A., 1963*b*, Comparative study of the structure of the *os calcaneum* in insectivores and rodents, *Acta Theriol.* **7**:91–126.

Romer, A. S., 1922, The locomotor apparatus of certain primitive and mammal-like reptiles, *Bull. Am. Mus. Nat. Hist.* **46**:517–606.

Simon, E. L., 1969, The origin and radiation of the Primates, *Ann. N. Y. Acad. Sci.* **167**:319–331.

Simons, E. L., 1976, The fossil record of primate phylogeny, in: *Molecular Anthropology* (M. Goodman and R. E. Tashian, eds.), pp. 35–62, Plenum Press, New York.

Simpson, G. G., 1935, The Tiffany fauna, Upper Paleocene: II. Structure and relationships of *Plesiadapis, Am. Mus. Nov.,* No. 816.

Simpson, G. G., 1940, Studies on the earliest primates, *Bull. Am. Mus. Nat. Hist.* **77**:185–212.

Simpson, G. G., 1978, Early mammals in South America, *Proc. Am. Phil. Soc.* **122**:318–328.

Stains, H. J., 1962, Osteological data used in mammal classification, *Syst. Zool.* **11**:127–130.

Stern, J. T., Jr., 1971, Functional myology of the hip and thigh of cebid monkeys and its implications for the evolution of erect posture, *Bib. Primatol.* No. 14.

Stirton, R. A., 1951, Ceboid monkeys from the Miocene of Colombia, *Bull. Univ. Calif. Publ. Geol. Sci.* **28**:315–356.

Stirton, R. A., and Savage, D. E., 1950, A new monkey from the La Venta Miocene of Colombia, *Compil. Estud. Geol. Ofic. Colombia, Serv. Geol. Nac., Bogotá* **8**:345–356.

Szalay, F. S., 1975, Phylogeny, adaptations, and dispersal of tarsiiform primates, in: *Phylogeny of the Primates* (W. P. Luckett and F. S. Szalay, eds.), pp. 357–404, Plenum Press, New York.

Szalay, F. S., 1976, Systematics of the Omomyidae (Tarsiiformes, Primates): Taxonomy phylogeny, and adaptations, *Bull. Am. Mus. Nat. Hist.* **156**:157–450.

Szalay, F. S., 1977*a*, Phylogenetic relationships and a classification of the eutherian Mammalia, in: *Major Patterns in Vertebrate Evolution* (M. K. Hecht, P. C. Goody, and B. M. Hecht, eds.), pp. 315–374, Plenum Press, New York.

Szalay, F. S., 1977*b*, Constructing primate phylogenies: A search for testable hypotheses with maximum empirical content, *J. Hum. Evol.* **6**:3–18.

Szalay, F. S., and Decker, R. L., 1974, Origins, evolution, and function of the tarsus in Late Cretaceous Eutheria and Paleocene primates, in: *Primate Locomotion* (F. A. Jenkins, Jr., ed.), pp. 223–259, Academic Press, New York.

Szalay, F. S., Tattersall, I., and Decker, R. L., 1975, Phylogenetic relationships of *Plesiadapis*—postcranial evidence, *Contrib. Primatol.* **5**:136–166.

Taylor, B. K., 1978, The anatomy of the forelimb in the anteater (*Tamandua*) and its functional implications, *J. Morphol.* **157**:347–368.

Van Valen, L., and Sloan, R. E., 1965, The earliest primates, *Science* **150**:743–745.

Walker, E. P., 1975, *Mammals of the World,* 3rd Ed., Vol. I, Johns Hopkins University Press, Baltimore.

Wood, A. E., 1973, Eocene rodents, Pruett Formation, Southwest Texas: Their pertinence to the origin of the South American Caviomorpha, *Texas Mem. Mus., Pearce-Sellards Ser.* **20**:1–41.

Wood, A. E., 1977, The Rodentia as clues to Cenozoic migrations between the Americas and Europe and Africa, in: *Paleontology and Plate Tectonics* (R. M. West, ed.), *Milwaukee Public Mus. Spec. Publ. Biol. Geol.* No. 2, pp. 95–109.

The Phylogenetic Significance of the Skin of Primates

16

Implications for the Origin of New World Monkeys

E. M. PERKINS and W. C. MEYER

Introduction

A wealth of information concerning the skin of subhuman primates has been generated over the past two decades. Encompassing more than 60 articles that address the specific cutaneous attributes of more than 50 primate species, these details have contributed substantially to a baseline understanding of the primate integumentary organ system. In each study, histological and histochemical data established cutaneous profiles unique to the species examined. As more studies accumulated, species-specific nuances became even more apparent, as did certain basic, underlying trends. For example, patterns in categories such as numbers and distribution of melanocytes (Machida and Perkins, 1967) and arrangement of hair follicles (Perkins *et al.*, 1969) began to emerge, i.e., the synthesis and interpretation of this previously fragmentary data revealed cutaneous signatures which, in effect, recorded primate evolutionary affinities.

E. M. PERKINS • Department of Biological Sciences, University of Southern California, Los Angeles, California 90007. W. C. MEYER • Department of Earth Sciences, Pierce College, Woodland Hills, California 91371.

Two recent studies (Perkins, 1975; Grant and Hoff, 1975), more comprehensive in scope, further substantiated the employment of skin characteristics as a useful tool in primate taxonomy. The former dealt primarily with the systematics of callitrichid monkeys, whereas the latter subjected 84 parameters of primate skin to cluster analysis in order to test the hypothesis of nonspecificity (Sokal and Sneath, 1963). Conclusions reached in both of these papers generally tended to support more orthodox taxonomic procedures.

Until now, however, no single study has attempted to assimilate this mass of data—in conjunction with the realities of continental drift—in order to gain insight into both the evolution and distribution of extant primates. The intent of this study, therefore, is to condense and integrate salient integumentary features into a workable, systematic overview that will contribute to our understanding of the ancestry and dispersal of modern anthropoid species.

Materials and Methods

All species included in this study are listed in Table 1. Their taxonomic ranking (left column) largely incorporates the classification scheme proposed by Hershkovitz (1977). Taxonomic clarifications and corrections are indicated by footnotes. Works specifically devoted to the integument of a single species are cited by author (right column). Additional histological studies, which also make reference to a given species, are indicated by numbers in parentheses.

The particular skin characteristics employed by these numerous investigators fall into three basic categories: gross, microscopic, and histochemical. Gross characters include specialized entities such as axillary organs, ischial callosities, penile spines, prehensile tails, pubic cushions, sex skin, sinus hairs, sternal glands, tactile papillae, and ulnar vibrissae. Microscopic characters pertain to various histological traits associated with the composition and configuration of integument, i.e., skin (epidermis and dermis) and its derivatives (hairs, sebaceous glands, eccrine sweat glands, and apocrine glands occurring over the general body surface). Histochemical criteria record general degrees of enzymatic activity exhibited by various integumental components. A succinct tabulation of all of these characteristics by Grant and Hoff (1975) precludes their arduous repetition here.

Results

When 20 of the more readily discernible characteristics of primate integument are appraised relative to representative members of the seven monophyletic taxa, distinctive patterns emerge. These characters, tabulated in Table 2, were chosen because they optimally lend themselves to accurate and consistent interpretation.

Table 1. Systematic Arrangement of the Order Primates[a]

Taxon	Reference
Order Primates[b]	
Suborder Strepsirhini	
Infraorder Lemuriformes	
Superfamily Lemuroidea	
Family Lemuridae	
Lemur catta	Montagna and Yun (1962*a*) (1,2,4)[j]
Lemur macaco[c]	Montagna *et al.* (1961*b*) (1,2,4)
Lemur mongoz	Montagna and Yun (1963*b*) (1,2,3,4)
Infraorder Lorisiformes	
Superfamily Lorisoidea	
Family Galagidae	
Galago senegalensis	Yasuda *et al.* (1961) (1,2,3,4)
Galago crassicaudatus	Montagna and Yun (1962*c*) (1,2,3,4)
Galago (= *Gallagoides*) *demidovii*	Machida *et al.* (1966) (1,2,3,4)
Family Lorisidae	
Loris tardigradus	Montagna and Ellis (1960) (1,2,3,4)
Nycticebus coucang	Montagna *et al.* (1961*a*) (1,2,4)
Arctocebus calabarensis	Montagna *et al.* (1966*b*) (1,2,3,4)
Perodicticus potto	Montagna and Ellis (1959); Montagna and Yun (1962*d*) (1,2,3,4)
Suborder Haplorhini	
Infraorder Tarsiiformes (= Tarsii)	
Superfamily Tarsioidea	
Family Tarsiidae	
Tarsius syrichta	Montagna and Machida (1966); Arao and Perkins (1969) (1,2,3,4)
Infraorder Platyrrhini	
Superfamily Ceboidea	
Family Callitrichidae	
Subfamily Callitrichinae	
Callithrix (= *Mico*) *argentata*	Perkins (1969*c*) (3,5)
Callithrix (= *Hapale*) *aurita*	(3)
Callithrix (= *Hapale*) *humeralifer*	(4)
Callithrix (= *Cebuella*) *pygmaea*	Perkins (1968) (4,5)
Subfamily Saguininae	
Saguinus (= *Tamarinus*) *fuscicollis*[d]	Perkins (1966) (1,2,4,5)
Saguinus (= *Tamarin*) *midas*	(2)
Saguinus (= *Oedipomidas*) *oedipus*	Perkins (1969*a*) (2,3,4,5)
Saguinus (= *Oedipomidas*) *geoffroyi*[e]	(2,4)
Family Callimiconidae	
Callimico goeldii	Perkins (1969*b*) (5)
Family Cebidae	
Subfamily Saimiriinae	
Saimiri sciureus	Machida *et al.* (1967) (1,2,3,4,5)
Subfamily Aotinae	
Aotus trivirgatus	Hanson and Montagna (1962) (2,4,5)
Subfamily Callicebinae	
Callicebus moloch	(5)
Callicebus torquatus	(5)
Subfamily Alouattinae	
Alouatta caraya	Machida and Giacometti (1968) (1,2,3,4)

(*continued*)

Table 1. (*Continued*)

Subfamily Pitheciinae	
Pithecia monachus	Perkins and Ford (1975) (2,4,5)
Cacajao rubicundus	Perkins *et al.* (1968*b*) (2,3,4,5)
Subfamily Cebinae	
Cebus capucinus	(2)
Cebus albifrons	Perkins and Ford (1969) (3,5)
Cebus apella	(2)
Subfamily Atelinae	
Ateles geoffroyi	Perkins and Machida (1967) (2,4,5)
Lagothrix lagothricha	Machida and Perkins (1966) (1,2,4,5)
Infraorder Catarrhini	
Superfamily Cercopithecoidea	
Family Cercopithecidae	
Subfamily Cercopithecinae	
Cercopithecus aethiops	Machida *et al.* (1964) (1,2,3,4)
Cercopithecus mitis	Machida *et al.* (1964) (1,2,3,4)
Cercopithecus mona	(2)
Cercopithecus neglectus	(2,4)
Erythrocebus patas	(2,4)
Cercocebus atys[f]	Machida *et al.* (1965) (1,2,4)
Macaca silenus	(2)
Macaca nemestrina	Perkins *et al.* (1968*a*) (1,2,3,4)
Macaca mulatta	Montagna *et al.* (1964) (1,2,3,4)
Macaca speciosa	Montagna *et al.* (1966*a*) (1,2,4)
Macaca fascicularis (= *irus*)	(2,3,4)
Macaca fuscata	(2,4)
Macaca (= *Cynopithecus*) *niger*	(1,2,4)
Papio anubis[g]	Montagna and Yun (1962*b*) (1,2,4)
Papio cynocephalus	(1,2)
Papio papio	(4)
Mandrillus sphinx	(2)
Subfamily Colobinae	
Presbytis cristatus[h]	Machida and Montagna (1964) (1,2,4)
Presbytis entellus	(2,4)
Superfamily Hominoidea	
Family Hylobatidae	
Hylobates hoolock	Parakkal *et al.* (1962) (2,4)
Family Pongidae	
Pongo pygmaeus	(2,3)
Chimpanzee (= *Pan*) *troglodytes*[i]	Montagna and Yun (1963*a*); Ford and Perkins (1970) (2,3,4)
Gorilla gorilla	Ellis and Montagna (1962) (1,2,3,4)

[a] Species cited are accompanied by references to their major cutaneous investigators.
[b] Older systems of classification divided the Primates into two large suborders: the Prosimii (which included tree shrews, all Lemuriformes, Lorisiformes, and Tarsiiformes) and the Anthropoidea (which encompassed all Platyrrhini and Catarrhini).
[c] *L. fulvus*, described in Machida and Giacometti (1967), is a subspecies of *L. macaco*.
[d] Subspecies *illigeri*, misidentified as *S. nigricollis* in Perkins (1966) and (1,2,4).
[e] This species was designated *spixi* in publications (2,4).
[f] Subspecies *lunulatus*, termed *C. fuliginosus* in (4).
[g] Erroneously termed *P. doguera* in publications ("doguera" is, in fact, one of several common names).
[h] Authors described this species as *P. pyrrhus* (*pyrrhus* is a subspecies of *P. cristatus*).
[i] Reported as *Pan satyrus*.
[j] Numbers in parentheses in this column indicate references that deal with the respective taxa, but not as the *main* topic of the reference: (1) Machida and Giacometti (1967); (2) Machida and Perkins (1967); (3) Arao and Perkins (1968); (4) Perkins *et al.* (1969); (5) Perkins (1975).

In general terms, commonality of integumental traits reduces the seven monophyletic taxa (whose ordinal and familial rankings are cited for the reader's convenience) to four major groups: the lower primates (the Lemuriformes, Lorisiformes, and Tarsiiformes), Platyrrhini, Cercopithecoidea, and Hominoidea. Although the lower primate taxa share more characters in common than they do with higher ranking anthropoid taxa, each of the three is nevertheless distinct in its own right. The Platyrrhini, by contrast, are a highly variable taxon, compared to the Cercopithecoidea. Like lower primates, both hominoid families also have unique integumentary profiles; together, however, they share more in common than either does with other anthropoid taxa.

Other patterns also become apparent. In Table 2, note that the monophyletic taxa (whose common names are provided) are arranged from left to right, in ascending taxonomic order. Inspection of a given trait often reveals progressive trends (e.g., epidermal thickness, amount of epidermal melanin, and degree of development of the dermal papillary body), series of distinct structural configurations (e.g., hair groupings), reduction or loss of structures (e.g., sinus hairs and ulnar vibrissae), amplitudes in the frequency of other entities (e.g., arrector pili muscles, sebaceous glands, and ratios of eccrine to apocrine glands), and modal variations common to certain combinations of these taxa (e.g., the site where apocrine excretory ducts open).

These observations are so consistent that it is often possible to very closely approximate the identity of a primate, based solely on properties of its integument. The following three examples suffice to illustrate this point. The designation CCS—meaning body hairs arranged as circular clustered sets of 20 or less—applies exclusively to the Lorisiformes, i.e., every species of *Loris, Nycticebus, Arctocebus, Perodicticus,* and *Galago* falls into this category. Similarly, only the apocrine excretory ducts of Lemuriformes, Lorisiformes, and Tarsiiformes open at the skin surface (SS); this trait has not been discovered in any other primate. Man is singular among all primates in that his sebaceous glands and eccrine sweat glands are best developed and most plentiful. He is also unique in that he is the only primate devoid of sinus hairs.

A number of comparative studies of primate skin, most recently those of Perkins (1975) and Grant and Hoff (1975), confirm that cutaneous traits shared by the Lemuroidea, Lorisoidea, Tarsioidea, Cercopithecoidea, and Hominoidea are relatively similar and consistent within and between the familial rankings of each of these major taxa. The Ceboidea or New World monkeys, however, are the major exception to this general rule. Although collectively their integumental characteristics set them apart from other monophyletic taxa (Table 2), careful examination at the familial level reveals a considerable lack of internal homogeneity.

Table 3 lists major cutaneous characteristics possessed by the Platyrrhini. Note that traits shared by the Callitrichidae and Callimiconidae agree quite well. Within the Cebidae, however, a wide spectrum of variation occurs. Each of the five categories of cebid subfamilies is distinctive. This phenomenon can

Table 2. Basic Integumental Traits of Seven Monophyletic Taxa[a]

Suborder:	Strepsirhini		Haplorhini				
Infraorder:	Lemuriformes	Lorisiformes	Tarsiiformes	Platyrrhini	Catarrhini		
Superfamily:	Lemuroidea	Lorisoidea	Tarsiioidea	Ceboidea	Cercopithecoidea	Hominoidea	
Families:	Lemuridae	Galagidae Lorisidae	Tarsiidae	Callitrichidae Callimiconidae Cebidae	Cercopithecidae	Hylobatidae Pongidae	Hominidae
Common names:	Lemurs	Galagos Lorises	Tarsiers	New World monkeys	Old World monkeys	Great apes	Man
Glabrous specializations[b]	–	–	–	PT	IC	–	–
Epidermal thickness	–	–	–	+	++	+++	+++
Alkaline phosphatase cells	–	+	–	±	–	–	–
Epidermal melanin	++	–	–	±±±	–	+++	+++
Adnexal blood supply	–	–	–	±	±	+	+++
Defined papillary body	–	–	–	±	+	++	+++
Abundant elastic fibers	–	–	–	±	±	+	++
Hair groupings[c]	CCI	CCS	LPS	~	LPI	LI	~

Number of hairs per group	<24	<20	<9	~	<7	<5	~
Sinus hairs	+	+	+	+	+	+	−
Ulnar vibrissae	+	−	+	+	−	−	−
Arrector pili muscles	−	±	−	+	+	+	+
Admixed suprapubic glands	−	±	−	+	−	−	−
Sebaceous gland size	++	−	−	±	+	+	+++
Numbers of apocrine glands	++	++	++	+	±	−	−
Apocrine glands open at[d]	SS	SS	SS	PO	PO	PO	PC
Sternal apocrine glands	−	−	+	+	−	±	−
Axillary organ	−	−	−	−	±	+	++
Eccrine sweat glands							
Confined to volar surfaces	+	+	+	±	−	−	−
In hairy skin as well	−	−	−	±	+	++	+++
Ratio to apocrine glands	−	−	−	E<A	E≤A	E≥A	E>A
Differentiated secretory cells	−	+	−	±	+	+	+
Phosphorylase and glycogen	+	±	+	±	+	+	+

[a] General symbols correspond to the following: (~) highly variable; (−) absent or poorly developed; (+) present or few in number; (++) moderate; (+++) plentiful; (± to ±±±) varies with body region or species.

[b] PT, prehensile tails; IC, ischial callosities.

[c] CCI, independent circular clusters; CCS, circular clustered sets; LPS, linear perfect sets; LPI, independent perfect lines; LI, imperfect lines.

[d] PC, pilary canal; SS, skin surface; PO, pilary orifice.

Table 3. Integumentary Characteristics of the Platyrrhini

Suborder:	Haplorhini							
Infraorder:	Platyrrhini							
Superfamily:	Ceboidea							
Family:	Callitrichidae		Callimiconidae	Cebidae				
Subfamilies:	Callitrichinae	Saguininae		Aotinae Callicebinae	Pitheciinae	Cebinae Saimiriinae	Atelinae	Alouattinae
Genera:	*Callithrix*	*Saguinus*	*Callimico*	*Aotus* *Callicebus*	*Pithecia* *Cacajao*	*Cebus* *Saimiri*	*Ateles* *Lagothrix*	*Alouatta*
Prehensile tail[b]	−	−	−	−	−	H	G	G
Epidermal thickness	−	−	−	−	−	−	+	+
Epidermal melanin	++	+++	+++	−	+	−	+++	+++
Epidermal MAO activity[c]	+	+	+	+	++	++	+++	+++
Well-developed dermis	−	−	−	−	−	−	+	+
Dermal melanin	+	+	+	+	++	++	−	−
Tactile papillae	±	±	±	−	−	−	−	−
Hair groupings[d]	LPS	LPS	LPS	CCI/CE	LPI/LPS	LI/LPI	LI/CCI	LPI
Hairs grouped on cheek	+	+	+	−	±	−	−	−
Ulnar vibrissae	+	+	+	−	−	−	−	−
Suprapubic gland	+	+	+	−	−	−	−	−
Large sternal gland	+	+	+	+	−	−	−	−
Apocrine glands, dorsally	−	+	+	±	+	+	+++	+++
Eccrine sweat glands								
Confined to volar surfaces	+	±	+	+	+	+	−	−
In hairy skin as well	−	±	−	−	−	−	+	+
Phosphorylase activity	++	−	++	+++	++	++	+++	+++
Glycogen in clear cells	−	−	−	+	±	+	−	+
Glycogen in dark cells	+	−	−	+	±	+	−	+

[a] General symbols correspond to the following: (−) absent or poorly developed; (+) present or few in number; (++) moderate; (+++) plentiful; (±) varies with body region or species.

[b] H, haired; G, glabrous.

[c] MAO, monoamine oxidase.

[d] LPS, linear perfect sets; CCI, independent circular clusters; CE, elongated clusters; LPI, independent perfect lines; LI, imperfect lines.

not be attributed to a fortuitous selection of cutaneous characters (cf. Grant and Hoff, 1975), nor can it be ascribed to inadequate sampling based on a single species or genus. How then can it be accounted for? More specifically, could these distinct yet varied sets of data exhibited by each subfamilial grouping be clues to the origin or origins of New World monkeys?

Discussion

In order to address these questions, it will first be necessary to focus on several basic trends that pervade the integument of all primates. Next, it will be helpful to identify those characteristics which various taxa within the Platyrrhini share with other members of the order. Finally, we shall attempt to ascertain whether or not such affinities are concordant with accepted views of plate tectonics and the various hypotheses concerning platyrrhine derivation.

In Tables 2 and 3, it is apparent that certain trends are directly proportional to the degree of a primate's phylogenetic ascension. Examples include epidermal thickness, epidermal monoamine oxidase activity, the degree to which the dermis is developed, numbers of dermal elastic fibers, adnexal blood supply, numbers of eccrine sweat glands, and the degree of their secretory cell differentiation. Within the Platyrrhini and Catarrhini (Anthropoidea), the following traits tend to be inversely proportional to phylogenetic ascension: presence of alkaline-phosphatase-positive cells, ulnar vibrissae, admixed suprapubic glands, numbers of apocrine glands, and the degree to which eccrine sweat glands are confined to volar friction surfaces.

Excluding the tree shrews, whose cutaneous traits resist relegation to either of the primate suborders, all other taxa may be appraised relative to their particular lower primate and/or anthropoid similarities. New World and Old World monkeys, for example, are separable on the basis of ulnar vibrissae, suprapubic glands, and (to a lesser extent) abundance of apocrine glands. Among all anthropoids, only New World monkeys possess these particular traits, which they share with lower primate taxa cited in Table 2.

In similar fashion, it is also possible to assess the standing of platyrrhine subfamilies relative to one another. Inspection of Table 3 reveals the subfamilies Aotinae and Callicebinae—and to a lesser extent, Pitheciinae—to be the most cutaneously primitive among cebids; conversely, the Alouattinae—and to a lesser extent, Atelinae—are the most advanced.

Given the preceding, it is tempting to fashion the following evolutionary scenario: lower primate ancestors invaded South America, giving rise to the Aotinae (Fig. 1). From the Aotinae arose the Callitrichidae and Pitheciinae. The family Callitrichidae, which closely resembled the *Callithrix* grade of organization, progressed to the Saguininae and Callimiconidae; the subfamily Pitheciinae served as the common ancestral stem for the remaining cebids, i.e., the Cebinae, Saimiriinae, Atelinae, and Alouattinae.

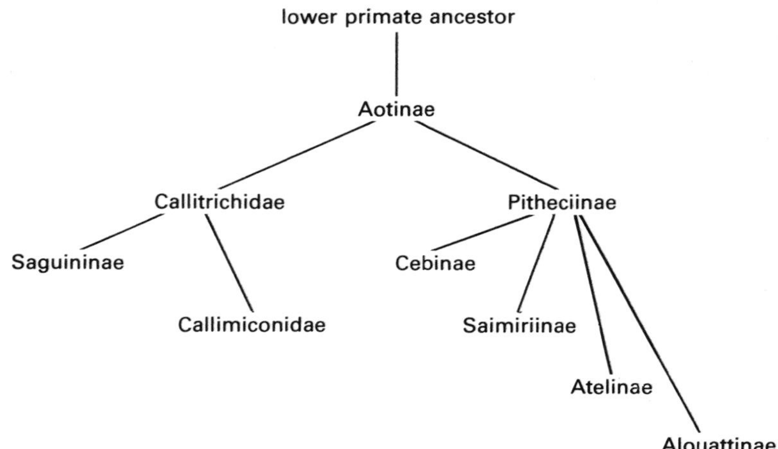

Fig. 1. Proposed phylogeny of platyrrhines, based on general integumentary traits.

A second possibility (Fig. 2) is that the Pitheciinae, having arisen from the Aotinae, produced two divergent lineages: on the one hand, the Callitrichidae and Callimiconidae—their archetype being *Pithecia,* and, on the other, the Cebinae, Saimiriinae, Atelinae and, Alouattinae—their archetype being

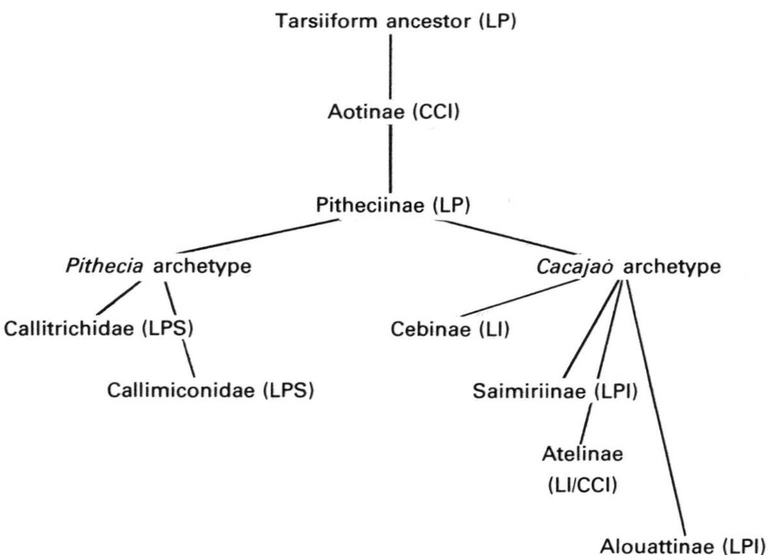

Fig. 2. Proposed phylogeny of platyrrhines, based on hair follicle goupings. LP, linear perfect; CCI, independent circular clusters; CE elongated clusters; LPS, linear perfect sets; LPI, independent perfect lines; LI, imperfect lines.

Cacajao. Superficially, this scheme appears to be particularly attractive in light of what may well be the single most reliable and constant cutaneous characteristic, namely, the configuration of hair follicle groupings (Perkins *et al.*, 1969). Details regarding this criterion appear in Fig. 3.

Modern lower primates (Lemuriformes and Lorisiformes) are all characterized by clustered hair follicle groupings (Table 2). So too are the Aotinae and Callicebinae (Table 3), which are generally believed to be descended from lower primate ancestors. The subfamily Pitheciinae is comprised of three

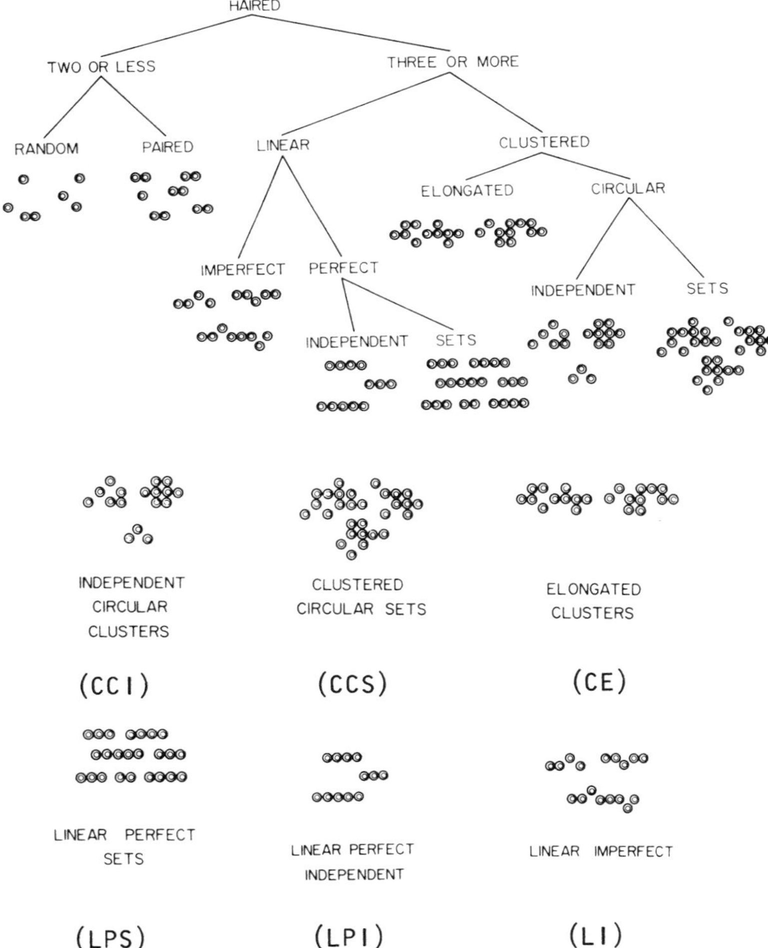

Fig. 3. The configuration and classification of hair follicle groupings. (Above) Diagram showing the various categories of hair follicle configuration, and their relationships to one another. (Below) Examples of those classifications referred to in this paper; each small circle represents a single hair in cross-section. From *A Study of Hair Groupings in Primates,* Perkins *et al.*, (69).

extant genera: *Pithecia, Chiropotes* (for which no information is available), and *Cacajao. Pithecia,* the proposed progenitor of the Callitrichidae—like the marmosets, tamarins, and pinchés, possesses linear perfect sets of hair follicles (Fig. 2); *Cacajao,* the proposed progenitor of the Cebinae, Saimiriinae, Atelinae and, Alouattinae—like the squirrel monkeys and howler monkeys, possesses hair follicles arranged in independent perfect lines (Fig. 2). Note, however, that the Cebinae and Atelinae do not adhere to this classification scheme: the genus *Cebus,* like *Ateles,* possesses imperfect lines of hairs, whereas *Lagothrix* is distinguished by circular clusters of follicles. Although the former occurrences might be mutations of the type that are independently acquired but more successful within the Pongidae (Table 2), the latter obviously contradicts the principle of irreversibility (Dollo's law).

According to this principle, structures lost in the course of evolution do not reappear in the same line. Assuming that the lower primate ancestor of New World monkeys was an omomyid tarsiiform, it follows that its hairs were most likely grouped in linear fashion, as are those of its modern-day relatives—the infraorder Tarsiiformes (Table 2). Because this is not compatible with the more primitive, clustered arrangement of hairs possessed by the Aotinae (Fig. 2), the scheme depicted in Fig. 2 is unacceptable—even if *Lagothrix* were dismissed as a violation of Dollo's principle (which it probably is).

This raises an interesting question about the Aotinae, whose integumental characteristics are remarkably primitive among the New World monkeys. Hanson and Montagna (1962), while examining the skin of *Aotus,* noted such a striking similarity to the skin of lemurs and lorises that they dubbed it a

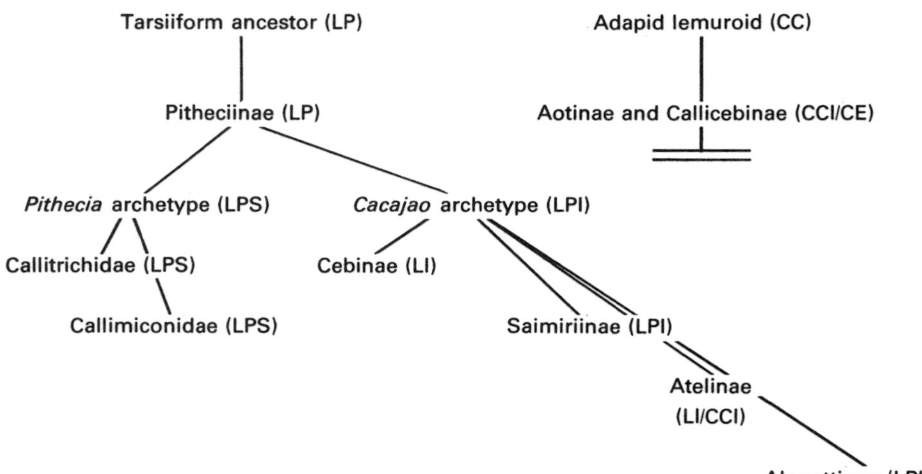

Fig. 4. Proposed phylogeny of platyrrhines, based on hair follicle groupings and degree of cellular differentiation in eccrine sweat glands. LP, linear perfect; CC, circular clusters; CCI, independent circular clusters; CE, elongated clusters; LPS, linear perfect sets; LIP, independent perfect lines; LI, imperfect lines.

"New World prosimian." The senior author's observations of the skin of both *Aotus* and *Callicebus* (unpublished) conform with Hanson and Montagna's initial impression. Statistical analyses by Grant and Hoff (1975) failed to place *Aotus* into any New World monkey plot; instead, it fell in with the lower primates in two plots and with Old World monkeys in all other plots.

In light of these observations, the scheme of New World monkey lineages (Fig. 2) was modified slightly, in order to accommodate the enigmatic Aotinae, as well as the Callicebinae. In Fig. 4, these subfamilies are represented as evolutionary dead ends, deriving from an adapid lemuroid stock. A clustered grouping of hair follicles, characteristic of both the Aotinae/Callicebinae and modern lemuroids, is consistent with this arrangement. Lorisoids, although they also have clustered hair groupings, were ruled out because unlike lemuroids (whose eccrine secretory cells are undifferentiated) and aotines (whose cells are poorly differentiated at best), those of the lorisoids are well differentiated—so well so, in fact, that this is a hallmark among all lower primate taxa (Table 2).

Conclusions

If the sequence of events outlined in Fig. 4 proves to be correct, then the New World monkeys derived from at least two ancestral stocks, not one. This somewhat novel interpretation of platyrrhine origins tends to agree with two currently existing models. Before elaborating upon these, however, it should be pointed out that even though fossil remains of early primates are sparse, the limitation of available evidence should not prevent interpretations of what happened. It should, however, emphasize a need for caution. Questionable inferences should not be extrapolated dogmatically to support pet theories, and no theory based on an insecure data base should be considered more than tentative. In the absence of hard evidence, one must scrupulously adhere to "Occam's razor," accepting the simplest explanation as the most probable.

The first of these current models holds that a family of tarsiiform primates (the Omomyidae) forms the ancestral stock of both the platyrrhines and catarrhines. These omomyids were present throughout the interconnected continents of North America and Eurasia during the early Tertiary. Once the Atlantic land bridge between North America and Eurasia was severed in the Eocene, they became separated. Under conditions of geographic isolation, these two groups then evolved in parallel fashion into ancestral platyrrhines and catarrhines, respectively.

Considering that the Holarctic realm began to cool at the time gene flow between these two populations was severed, it is reasonable to suggest that contracting isotherms helped promote their spread into the lower regions of South America and Africa as they followed the southward migration of warm tropical climates.

Although South and North America did not abut at this time, it would be erroneous to assume that they were not tenuously connected by island arcs associated with the Bolivar Trench. The activities of this trench, which separated these two continents during the early Paleogene, are ongoing, even today. Hence, these island arcs would have provided a suitable, island-hopping dispersal route which allowed the ancestral platyrrhines to reach South America by the middle or late Eocene (Simons, 1961). The discovery of *Branisella* from the Oligocene is consistent with this interpretation.

The second model, proposed by Gingerich (1977) proposes that early anthropoids were derived from lemuriform—not tarsiiform—primates. According to this hypothesis, Gingerich proposes that an adapid stock from Asia migrated across the Bering land bridge to North America, from whence it arrived in South America by the Eocene. Considering the meager state of knowledge about arctic areas, it is perhaps a little premature to base this hypothesis on the existence of a land bridge that is at present only speculative.

The suggestion does, however, raise an important issue: namely, is it not possible that several dispersal routes existed? Also, would it not have been possible for Asian adapids to migrate westward, reaching North America via the Atlantic land bridge? During the Paleocene and Eocene, the Eurasian continents were farther south than they are today, and tropical climate zones were greatly expanded (Berggren and Phillips, 1971). Hence, it is likely that Siberia may have served as a trans-Eurasian dispersal route.

Based upon those cutaneous traits which were earlier presented, we therefore suggest that an adapid lemuroid stock invaded South America from North America, giving rise to the Aotinae and Callicebinae. Similarly, an omomyid tarsiiform stock (also present in the pre-Eocene fossil record of North America) migrated into South America, where it evolved into the Pitheciinae, from which all other extant platyrrhines arose. Yet to be resolved, however, is how either of these early stocks arrived in North America in the first place.

References

Arao, T., and Perkins, E., 1968, The nictitating membranes of primates, *Anat. Rec.* **162:**53–70.

Arao, T., and Perkins, E., 1969, The skin of primates. XLIII. Further observations on the Philippine tarsier (*Tarsius syrichta*), *Am. J. Phys. Anthropol.* **31:**93–96.

Berggren, W. A., and Phillips, J. D., 1971, Influence of continental drift on the distribution of the Tertiary benthonic Foraminifera in the Caribbean and Mediterranean regions, in: *Symposium on the Geology of Libya, Woods Hole Contrib.* No. 2376, pp. 263–299.

Ellis, R. A., and Montagna, W., 1962, The skin of primates. VI. The skin of the gorilla (*Gorilla gorilla*), *Am. J. Phys. Anthropol.* **20:**79–94.

Ford, D. M., and Perkins, E. M., 1970, The skin of the chimpanzee, in: *The Chimpanzee,* Vol. 3 (G. Bourne, ed.), pp. 82–119, Karger, Basel.

Gingerich, P. D., 1977, Radiation of Eocene Adapidae in Europe, *Géobios, Mém. Spéc.* **1:**165–185.

Grant, P. G., and Hoff, C. J., 1975, The skin of primates. XLIV. Numerical taxonomy of primate skin, *Am. J. Phys. Anthropol.* **42:**151–166.

Hanson, G., and Montagna, W., 1962, The skin of primates. XII. The skin of the owl monkey (*Aotus trivirgatus*), *Am. J. Phys. Anthropol.* **20:**421–430.

Hershkovitz, P., 1977, *Living New World Monkeys (Platyrrhini), with an Introduction to the Primates,* Vol. 1, University of Chicago Press, Chicago.

Machida, H., and Giacometti, L., 1967, The anatomical and histochemical properties of the skin of the external genitalia of primates, *Folia Primatol.* **6:**48–69.

Machida, H., and Giacometti, L., 1968, The skin. X, in: *Biology of the Howler Monkey (Alouatta caraya)* (R. M. Malinow, ed.) *Bibl. Primatol.* No. 7, pp. 126–140, Karger, Basel.

Machida, H., and Montagna, W., 1964, The skin of primates. XXII. The skin of the lutong (*Presbytis pyrrus*), *Am. J. Phys. Anthropol.* **22:**443–452.

Machida, H., and Perkins, E., 1966, The skin of primates. XXX. The skin of the woolly monkey (*Lagothrix lagothricha*), *Am. J. Phys. Anthropol.* **24:**309–320.

Machida, H., and Perkins, E., 1967, The distribution of melanotic melanocytes in the skin of subhuman primates, in: *Advances in the Biology of Skin.* Vol. 8, *The Pigmentary System* (W. Montagna and F. Hu, eds.), pp. 41–58, Pergamon Press, Oxford.

Machida, H., Perkins, E., and Montagna, W., 1964, The skin of primates. XXIII. A comparative study of the skin of the green monkey (*Cercopithecus aethiops*) and the Sykes monkey (*Cercopithecus mitis*), *Am. J. Phys. Anthropol.* **22:**453–466.

Machida, H., Perkins, E., Montagna, W., and Giacometti, L., 1965, The skin of primates. XXVII. The skin of the white-crowned mangabey (*Cercocebus atys*), *Am. J. Phys. Anthropol.* **23:**165–180.

Machida, H., Perkins, E., and Giacometti, L., 1966, The skin of primates. XXIX. The skin of the pigmy bushbaby (*Galago demidovii*), *Am. J. Phys. Anthropol.* **24:**199–204.

Machida, H., Perkins, E., and Hu, F., 1967, The skin of primates. XXXV. The skin of the squirrel monkey (*Saimiri sciureus*), *Am. J. Phys. Anthropol.* **26:**45–54.

Montagna, W., and Ellis, R. A., 1959, The skin of primates. I. The skin of the potto (*Perodicticus potto*), *Am. J. Phys. Anthropol.* **17:**137–161.

Montagna, W., and Ellis, R. A., 1960, The skin of primates. II. The skin of the slender loris (*Loris tardigradus*), *Am. J. Phys. Anthropol.* **18:**19–44.

Montagna, W., and Machida, H., 1966, The skin of primates. XXXII. The skin of the Philippine tarsier (*Tarsius syrichta*), *Am. J. Phys. Anthropol.* **25:**71–84.

Montagna, W., and Yun, J. S., 1962*a,* The skin of primates. X. The skin of the ring-tailed lemur (*Lemur catta*), *Am. J. Phys. Anthropol.* **20:**95–118.

Montagna, W., and Yun, J. S., 1962*b,* The skin of primates. VIII. The skin of the Anubis baboon (*Papio doguera*), *Am. J. Phys. Anthropol.* **20:**131–142.

Montagna, W., and Yun, J. S., 1962*c,* The skin of primates. VII. The skin of the great bushbaby (*Galago crassicaudatus*), *Am. J. Phys. Anthropol.* **20:**149–165.

Montagna, W., and Yun, J. S., 1962*d,* The skin of primates. XIV. Further observations on *Perodicticus potto, Am. J. Phys. Anthropol.* **20:**441–450.

Montagna, W., and Yun, J. S., 1963*a,* The skin of primates. XV. The skin of the chimpanzee (*Pan satyrus*), *Am. J. Phys. Anthropol.* **21:**189–204.

Montagna, W., and Yun, J. S., 1963*b,* The skin of primates. XVI. The skin of *Lemus mongoz, Am. J. Phys. Anthropol.* **21:**371–382.

Montagna, W., Yasuda, K., and Ellis, R. A., 1961*a,* The skin of primates. III. The skin of the slow loris (*Nycticebus coucang*), *Am. J. Phys. Anthropol.* **19:**1–22.

Montagna, W., Yasuda, K., and Ellis, R. A., 1961*b,* The skin of primates. V. The skin of the black lemur (*Lemur macaco*), *Am. J. Phys. Anthropol.* **19:**115–130.

Montagna, W., Yun, J. S., and Machida, H., 1964, The skin of primates. XVIII. The skin of the rhesus monkey (*Macaca mulatta*), *Am. J. Phys. Anthropol.* **22:**307–320.

Montagna, W., Machida, H., and Perkins, E., 1966*a,* The skin of primates. XXVIII. The skin of the stump-tail macaque (*Macaca speciosa*), *Am. J. Phys. Anthropol.* **24:**71–86.

Montagna, W., Machida, H., and Perkins, E., 1966*b,* The skin of primates. XXXIII. The skin of the angwantibo (*Arctocebus calabarensis*), *Am. J. Phys. Anthropol.* **25:**277–290.

Parakkal, P., Montagna, W., and Ellis, R. A., 1962, The skin of primates. XI. The skin of the white-browed gibbon (*Hylobates hoolock*), *Anat. Rec.* **143:**169–178.

Perkins, E., 1966, The skin of primates. XXXI. The skin of the black-collared tamarin (*Tamarinus nigricollis*), *Am. J. Phys. Anthropol.* **25:**41–70.

Perkins, E., 1968, The skin of primates. XXXVI. The skin of the pigmy marmoset (*Callithrix* [= *Cebuella*] *pygmaea*), *Am. J. Phys. Anthropol.* **29**:349–364.

Perkins, E., 1969*a*, The skin of primates. XL. The skin of the cottontop pinché (*Saguinus* [= *Oedipomidas*] *oedipus*), *Am. J. Phys. Anthropol.* **30**:13–28.

Perkins, E., 1969*b*, The skin of primates. XXIV. The skin of Goeldi's marmoset (*Callimico goeldii*), *Am. J. Phys. Anthropol.* **30:**231–250.

Perkins, E., 1969*c*, The skin of primates. XLI. The skin of the silver marmoset (*Callithrix* [= *Mico*] *argentata*), *Am. J. Phys. Anthropol.* **30:**361–388.

Perkins, E., 1975, Phylogenetic significance of the skin of New World monkeys (Order Primates, Infraorder Platyrrhini), *Am. J. Phys. Anthropol.* **42:**395–423.

Perkins, E., and Ford, D. M., 1969, The skin of primates. XXXIX. The skin of the white-browed capuchin (*Cebus albifrons*), *Am. J. Phys. Anthropol.* **30:**1–12.

Perkins, E., and Ford, D. M., 1975, The skin of primates. XLII. The skin of the silvered sakiwinki (*Pithecia monachus*), *Am. J. Phys. Anthropol.* **42:**383–394.

Perkins, E., and Machida, H., 1967, The skin of primates. XXXIV. The skin of the golden spider monkey (*Ateles geoffroyi*), *Am. J. Phys. Anthropol.* **26:**35–44.

Perkins, E., Arao, T., and Dolnick, E. H., 1968*a*, The skin of primates. XXXVII. The skin of the pigtail macaque (*Macaca nemestrina*), *Am. J. Phys. Anthropol.* **28:**75–84.

Perkins, E., Arao, T., and Uno, H., 1968*b*, The skin of primates. XXXVIII. The skin of the red uacari (*Cacajao rubicundus*), *Am. J. Phys. Anthropol.* **29:**57–80.

Perkins, E., Smith, A. A., and Ford, D. M., 1969, A study of hair groupings in primates, in: *Advances in the Biology of Skin. IX. Hair Growth* (W. Montagna and R. L. Dobson, eds.), pp. 357–367, Pergamon Press, Oxford.

Simons, E. L., 1961, The dentition of Ourayia: Its bearing on relationships of omomyid prosimians, *Postilla* **54:**1–120.

Sokal, R. R., and Sneath, P. H. A., 1963, *The Principles of Numerical Taxonomy,* Freeman, San Francisco.

Yasuda, K., Aoki, T., and Montagna, W., 1961, The skin of primates. IV. The skin of the lesser bushbaby (*Galago senegalensis*), *Am. J. Phys. Anthropol.* **19:**23–34.

Monophyletic or Diphyletic Origins of Anthropoidea and Hystricognathi

17

Evidence of the Fetal Membranes

W. P. LUCKETT

Introduction

In recent years, much of the discussion on phylogenetic relationships between South American platyrrhine primates and Afro-Asian catarrhines has focused on the possible effects of continental drift and transoceanic rafting on the Oligocene–Recent distributional patterns of these mammals in the southern continents (Hoffstetter, 1972, 1974; Keast, 1972; Lavocat, 1974*a*; Cracraft, 1974). The Oligocene–Recent distribution of hystricognathous rodents in the southern continents is strikingly similar to the primate pattern, and most students of mammalian evolution and paleozoogeography are in agreement that the mode of dispersal was probably the same for both taxa. There is considerable disagreement, however, concerning the monophyletic or diphyletic relationships of the South American and African primates and

W. P. LUCKETT • Department of Anatomy, School of Medicine, Creighton University, Omaha, Nebraska 68178.

hystricognathous rodents, as well as contrasting hypotheses on the area of origin of these taxa (cf. Hoffstetter, 1972, 1974; Patterson and Pascual, 1972; Lavocat, 1969; 1974*a,b,*; Wood, 1972, 1974; Wood and Patterson, 1970; Hershkovitz, 1972). Unfortunately, speculations on zoogeographic aspects of primate evolution have not been accompanied by careful phylogenetic analyses of characters which might corroborate or refute the hypothesis of anthropoidean monophyly. Most students of primate comparative anatomy and paleontology tend to specialize on catarrhines, platyrrhines, or strepsirhines, and, as a result, few have evaluated the phylogenetic significance of morphological features across a broad spectrum of primates. Similar limitations are encountered in studies on rodent phylogeny.

If both Anthropoidea and Hystricognathi are diphyletic taxa, then there is no reason to invoke transatlantic rafting to explain the similarities between South American and African forms; shared similarities would be the result of extreme parallelisms, as suggested by Simpson (1945, 1961) for Anthropoidea, and by Wood (1950, 1965, 1974) for Hystricognathi. However, these authors have utilized the disjunct distribution of Oligocene–Recent primates and rodents in the southern continents as evidence that the shared similarities within each group are due to parallelisms, and thus an element of circularity has been introduced into their evolutionary analyses. Evolutionary relationships among taxa should be evaluated initially on the basis of the biological evidence from extant and fossil organisms, whereas the zoogeographic and geological events which may have led to their historical and present distribution should be inferred only after there is a sound basis for their phylogeny.

Preliminary evaluation of fetal membrane and placental features in primate higher taxa (Luckett, 1975) demonstrated several shared and derived homologous character states (synapomorphies) which support the monophyly of Anthropoidea, and corroboration of this hypothesis was provided by a multidisciplinary analysis of soft anatomical and cranial characters (Luckett and Szalay, 1978). The present study was undertaken to evaluate developmental, reproductive, and other anatomical features in primates and rodents in order to test hypotheses of monophyly or diphyly of anthropoids and hystricognaths in South America and Africa. Phylogenetic analysis of characters in these groups is considered as a necessary prerequisite for assessment of possible historical pathways for deployment of primates and rodents to the southern continents.

Materials and Methods

Most of the primate and rodent embryos examined during the present study are housed at the Hubrecht Laboratory, Utrecht, the Netherlands. Additional specimens were examined at the Carnegie Laboratories of Embryology, Davis, California, and in the collection of Dr. H. W. Mossman, Department of Anatomy, University of Wisconsin.

The rationale and methodology for using developmental data in assessments of mammalian phylogeny have been discussed in detail elsewhere (Luckett, 1975, 1977) and are only summarized here. The mammalian fetal membranes and placenta comprise a genetically complex and interrelated organ system which is essential for the normal prenatal development of the embryo. Developmental patterns of the fetal membrane complex may vary considerably between different suborders and orders, whereas these features are relatively conservative at the generic and familial levels. This combination of complexity, variability, and conservatism, coupled with the availability of their entire ontogenetic pattern for study, enhance the use of fetal membrane data for assessing phylogenetic relationships among higher taxonomic categories of mammals. The advantages for phylogenetic analysis of such a complex organ system serve to offset the absence of these features in the fossil record. Moreover, ontogenetic studies facilitate the recognition of convergent or parallel evolution in individual fetal membrane traits.

Cladistic or sister-group relationships of phylogeny are defined in terms of relative recency of common ancestry. Two taxa are more closely related cladistically to each other than to any other taxa, and comprise a monophyletic group, when it can be demonstrated that they share derived, homologous character states (synapomorphies) which were inherited from a relatively recent common ancestor not shared with other taxa (Hennig, 1966). Further corroboration for hypotheses of monophyly (or *holophyly*) is provided when some of the synapomorphies shared by sister-groups can be shown to be uniquely derived (*autapomorphous*) character states which do not occur in other more distantly related taxa.

Evolutionary analyses of characters are used to test hypotheses of phylogenetic relationships among taxa (Bock, 1977); thus, emphasis should be placed on the methodology of character analysis. Essential tasks in the phylogenetic analysis of characters include: (1) determination of the homologous or homoplastic nature of shared similarities; (2) identification of all character states of homologous traits, and the arrangement of these character states in a transformation series or morphocline; and (3) assessment of the relative primitive or derived nature of character states (determination of morphocline polarity). The development of hypotheses concerning the pattern and direction of evolutionary change in character states of a morphocline is the most challenging and crucial aspect of character analysis, and the resulting hypotheses of character phylogeny comprise the best test of phylogenetic hypotheses about taxa (Bock, 1977).

No single criterion for analysis of morphocline polarity is infallible. In the present study, consideration was given to ontogenetic character precedence, distribution pattern of character states among a wide range of taxa (the "commonality" principle), and the form–function interrelationships between individual traits and among the entire complex of features during evaluation of morphocline polarity of fetal membrane and reproductive characters. Form–function analysis often proves invaluable when there is incongruence between polarities suggested by ontogenetic and commonality data.

Character Analysis of Developmental and Reproductive Traits in Primates

No *a priori* assumptions were made concerning the possible sister-group relationships of Anthropoidea to other Primates, and character states in all primate higher taxa have been evaluated in order to test hypotheses of anthropoidean monoplyly or diphyly. Many of the developmental features of the primate fetal membranes and placenta have been analyzed previously for phylogenetic inference (Luckett, 1975). Anthropoids share several apomorphous fetal membrane features with *Tarsius,* including invasive implantation and subsequent differentiation of a hemochorial placenta, development of a primordial amniotic cavity, precocious differentiation of the primitive streak correlated with development of a mesodermal body stalk, rudimentary nature of the allantoic diverticulum, and the absence of a transitory choriovitelline placenta (Fig. 1). This suite of synapomorphous traits supports the hypothesis of a monophyletic sister-group relationship between Anthropoidea and Tarsiiformes (as the suborder Haplorhini). This hypothesis is strongly corroborated by phylogenetic analysis of a variety of cranioskeletal and soft anatomical attributes in both fossil (where applicable) and extant primates (Luckett and Szalay, 1978). Anthropoid monophyly is corroborated by synapomorphous features of implantation, amniogenesis, yolk sac differentiation, double discoidal placentation, and the occurrence of a simplex uterus.

Further assessment of primate placentation reveals the occurrence of two autapomorphous traits in cebids and callitrichids which corroborate the monophyly of extant Platyrrhini. Endothelial-lined maternal vascular channels which communicate with endometrial arterioles are retained within the definitive placental discs of both families (Hill, 1932; Luckett, 1974). Persistence of these intraplacental vessels is probably a consequence of the relatively slow pattern of trophoblastic invasion into the endometrium during early pregnancy.

Another unique feature of platyrrhine placentation is the development of hematopoietic foci within sinusoidal vessels of the placental trabeculae (Hill, 1932; Wislocki, 1943; Jollie et al., 1975). This phenomenon is most pronounced during midgestation, and it is unclear whether the placental foci originate *in situ* or are derived from hematopoietic cells of the yolk sac splanchnopleure. In contrast to the usual mammalian developmental pattern, there is little evidence of hematopoiesis in the embryonic liver or spleen during the first half of pregnancy in marmosets (and presumably in other platyrrhines),

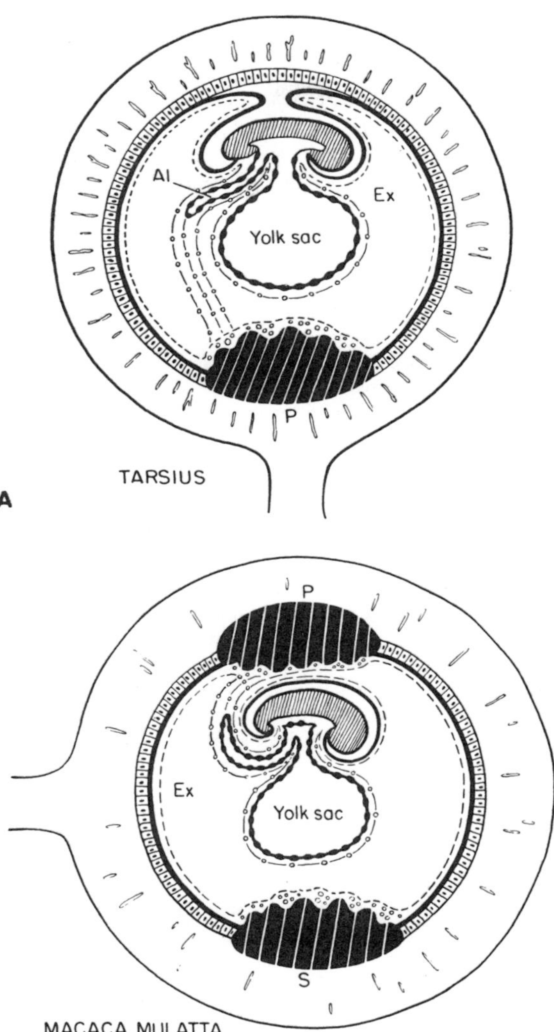

Fig. 1. Diagram of fetal membranes and placenta in haplorhine primates. (A) *Tarsius spectrum in utero.* Note the tubular allantoic diverticulum (Al) projecting into the body stalk, the reduced yolk sac, amniogenesis occurring by folding, and the mesometrially situated single placental disc (P). Ex = Exocoelom. (B) *Macaca mulatta in utero.* This developmental pattern approximates the reconstructed morphotype of Anthropoidea. The secondary placental disc (S) of anthropoids is homologous with the single disc of the tarsier. Note the similar relationships of the allantois and yolk sac in both haplorhine groups, whereas amniogenesis occurs by cavitation in Anthropoidea. P = Primary placental disc.

and placental hematopoiesis appears to compensate for this absence of embryonic hematopoiesis during early pregnancy (Wislocki, 1943). Such a unique phenomenon has not been clearly identified in any other mammalian taxon.

The development of dizygotic twins is a normal occurrence in all genera of callitrichids, whereas it is a relatively rare event in cebids, *Callimico,* and other genera of anthropoids. Consideration of nipple count, uterine morphology, and the rarity of twinning in primates suggests that the ancestral anthropoid and platyrrhine conditions were characterized by the develop-

Table 1. Primitive and Derived Character States of Selected Primate Traits

Primitive	Derived[a]
1. Naked rhinarium; unfused nasal processes	1. Haired rhinarium; fused nasal processes
2. Transitory choriovitelline placenta	2. No choriovitelline placenta
3. Large vesicular allantois; no body stalk	3. Rudimentary allantois; well-developed body stalk
4. Ovarian bursa well developed	4. Ovarian bursa reduced or absent
5. Retina lacking area centralis or fovea	5. Retina with area centralis and fovea
6. No primordial amniotic cavity; amniogenesis by folding	6. Primordial amniotic cavity present; amniogenesis by cavitation (1)
7. Noninvasive attachment; diffuse epitheliochorial placenta	7. Invasive attachment; bidiscoidal hemochorial placenta (2)
8. Blastocyst attachment by paraembryonic pole	8. Blastocyst attachment by embryonic pole
9. Primary yolk sac only	9. Primary and secondary yolk sacs
10. Uterus bicornuate	10. Uterus simplex
11. Head-head agglutination of sperm	11. No head-head agglutination of sperm
12. Sublingua present	12. Sublingua absent
13. No cytotrophoblastic shell	13. Cytotrophoblastic shell present (3)
14. Placental disc labyrinthine	14. Placental disc villous (4)
15. Implantation superficial; no decidua capsularis	15. Implantation interstitial; decidua capsularis present
16. Intraplacental maternal vessels absent	16. Intraplacental maternal vessels present
17. No placental hematopoiesis	17. Placental hematopoiesis present
18. Little or moderate development of ovarian interstitial gland tissue	18. Abundant ovarian interstitial gland tissue
19. Twinning rare	19. Twinning normal
20. Clitoris relatively small	20. Clitoris hypertrophied, peniform (5)

[a] The numbers in parentheses refer to the following intermediately derived character states: (1) primordial amniotic cavity transitory; definitive amniogenesis by folding; (2) invasive attachment; monodiscoidal hemochorial placenta; (3) rudimentary development of cytotrophoblastic anchoring villi; (4) placental disc trabecular; (5) clitoris hypertrophied, grooved ventrally.

ment of a single young. Leutenegger (1973) has presented compelling evidence that the production of twins (or triplets) in callitrichids is a derived trait which results from strong selective pressure to guarantee successful delivery of relatively large neonates in species that have undergone an evolutionary decrease in adult body size. Further support for the hypothesis that twinning is a derived attribute in callitrichids is the fact that it is associated with the unique condition of hematopoietic chimerism which is facilitated by anastamoses of the fetal placental circulations (Benirschke and Layton, 1969). Hershkovitz's (1977, p. 441) rejection of twinning as a derived condition in callitrichids does not appear to be based on a careful consideration of the evidence; it was doubtlessly influenced by his belief that small-bodied, insectivorous callitrichids approximate the ancestral ceboid condition.

Platyrrhine monophyly is also corroborated by the autapomorphous occurrence of abundant interstitial gland tissue and/or accessory corpora luteal

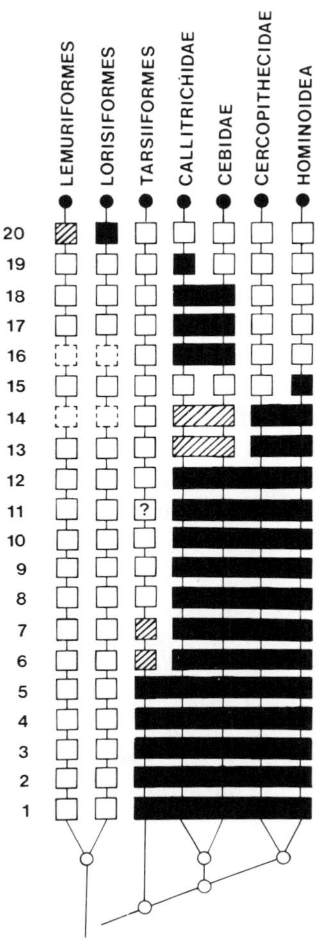

Fig. 2. Character phylogeny of fetal membrane, reproductive, and other primate anatomical features evaluated during the present study. Monophyly of the higher taxa Haplorhini, Anthropoidea, Platyrrhini, and Catarrhini is corroborated by the demonstration of synapomorphies. Primitive (plesiomorphous) character states of primates are indicated by an open square, and derived (apomorphous) character states are indicated by black squares. Shared, derived homologous character states are represented by black rectangles. Intermediately derived character states are indicated by diagonal lines. See Table 1 for the list of characters evaluated. Characters 14 and 16 are not applicable to lemuriforms and lorisiforms.

tissue in the ovaries of cebids and callitrichids (for a review, see Koering, 1974). This tissue becomes so extensive that it is impossible to distinguish the corpus luteum from the interstitial gland tissue during the second half of pregnancy. Ovarian interstitial gland tissue is only moderately developed in other primate families, and the corpus luteum remains distinct.

The male reproductive system of primates has received less attention in comparative and phylogenetic studies; this is related in part to its greater evolutionary conservatism. Preliminary investigation of spermatozoa and analysis of their surface biochemistry (Bedford, 1974) revealed several features which appear to be of value in assessing primate phylogeny. Of particular relevance to the present study was Bedford's demonstration of a marked difference in the biochemical nature of the sperm surface in Strepsirhini and Anthropoidea. These differences are reflected in the distribution of negative surface charge on the plasma membrane and the ability of sperm to exhibit head-head agglutination in physiological solutions or serum. Spermatozoa of strepsirhines exhibit surface features, including head-head agglutination, comparable to those of most other eutherians examined, and these are believed to approximate the primitive eutherian and primate condition. In contrast, anthropoid spermatozoa bear a heavy negative surface charge and lack head-head agglutination, and Bedford suggested that this relatively rare pattern is derived. Subtle differences in charge distribution also serve to distinguish between cercopithecoids and hominoids. Unfortunately, no data are available as yet on surface features of tarsier sperm. The functional significance of these surface charge differences is unclear, but they may have implications for the specificity of sperm maturation and their fertilizing ability (Bedford, 1974).

The relatively primitive (plesiomorphous) and derived (apomorphous) character states of developmental and reproductive traits analyzed during the present study and in a previous report (Luckett, 1975) are listed in Table 1, and the distribution of these character states in primate higher taxa is illustrated in Fig. 2. Several additional soft anatomical characters, including the sublingua (see Hofer, 1977), rhinarium, and retinal fovea, were also incorporated into the present analysis. Synapomorphous features corroborate hypotheses of monophyly for the higher taxa Haplorhini, Anthropoidea, Platyrrhini, and Catarrhini. Developmental and reproductive characters generally have provided little evidence for assessing the affinities of Lemuriformes and Lorisiformes; this is due to their retention of numerous primitive primate and eutherian attributes. An exception is the occurrence of a hypertrophied clitoris in all extant strepsirhines. This character state is rare in other primates (with the notable exception of some cebid genera), and its infrequent development in other eutherians suggests that it is a derived character state. Martin (1975) has suggested that the hypertrophied clitoris of strepsirhines may be an adaptation for increased control over urine marking on fine branches, and that this plays an important role in olfactory communication between the sexes. To date, no developmental, reproductive, or other soft anatomical synapomorphies have been identified which would corroborate a hypothesis of sister-group relationship between Strepsirhini and Anthropoidea (*contra* Gingerich, 1976; Schwartz *et al.*, 1978).

Character Analysis of Developmental and Reproductive Traits in Rodents

Developmental features of the fetal membranes and placenta in most rodent higher taxa have been summarized by several authors (Mossman, 1937; Fischer and Mossman, 1969; Luckett, 1971), and Roberts and Perry (1974) have presented comparative data on fetal membrane development in several families of caviomorphs. These investigators also discussed the relatively primitive and derived nature of fetal membrane traits, although their data were used mainly to evaluate patristic rather than cladistic relationships. A number of reproductive attributes common to caviomorph rodents and at least some African hystricognaths have been enumerated by Weir (1974); however, the morphocline polarity of these traits in comparison to those in other rodents was not considered. The following discussion summarizes developmental and reproductive attributes in the major rodent higher taxa, with emphasis placed on assessing the relatively primitive and derived character states of these features.

Implantation

The development of preimplantation blastocysts in rodents is similar to that in other eutherians, and differences are evident only in the degree of blastocyst expansion and the fate of the polar trophoblast. All rodents are characterized by mesometrial orientation of the embryonic disc at the time of implantation, and initial attachment is effected by localized invasion of the abembryonic trophoblast into the antimesometrial endometrium. Superficial implantation of a moderately expanded blastocyst in an eccentric, antimesometrial implantation chamber occurs in sciurids and aplodontids and probably approximates the primitive rodent condition. A similar method of implantation is evident in castorids, geomyoids, dipodoids, pedetids, ctenodactylids, and, probably, anomalurids. The blastocyst of muroids is relatively smaller at the time of implantation, and the antimesometrial implantation chamber becomes isolated from the remainder of the uterine lumen. Obliteration of the uterine lumen distal to the blastocyst results in a relationship which can be described as secondary interstitial implantation. The most derived mechanism of implantation occurs in caviomorphs. The abembryonic trophoblast penetrates the uterine epithelium, and the blastocyst comes to lie completely within the uterine stroma. Examination of more mature implanted blastocysts in the African hystricognathous families Bathyergidae and Hystricidae (Mossman and Luckett, manuscript) indicates that a similar unique pattern of *primary* insterstitial implantation also characterizes these taxa.

Amniogenesis

Differing patterns of amniogenesis in rodents reflect in part the fate of the polar trophoblast which overlies the inner cell mass, as well as the rapidity with which the polar trophoblast makes contact with the uterine endometrium. The polar trophoblast is lost by the time of early implantation in sciurids, aplodontids, castorids, and geomyids, and uterine epithelium at the mesometrial pole of the uterus does not come into intimate contact with the exposed embryonic disc at this stage. In these families, amniogenesis occurs by somatopleuric folding (Fig. 3a).

The polar trophoblast persists during implantation in muroid rodents, and clasping of the blastocyst in the antimesometrial implantation chamber brings the polar trophoblast and uterine epithelium into intimate apposition. The polar trophoblast subsequently thickens to form the ectoplacental cone (preplacental trophoblast), and, concomitant with this, the embryonic cell

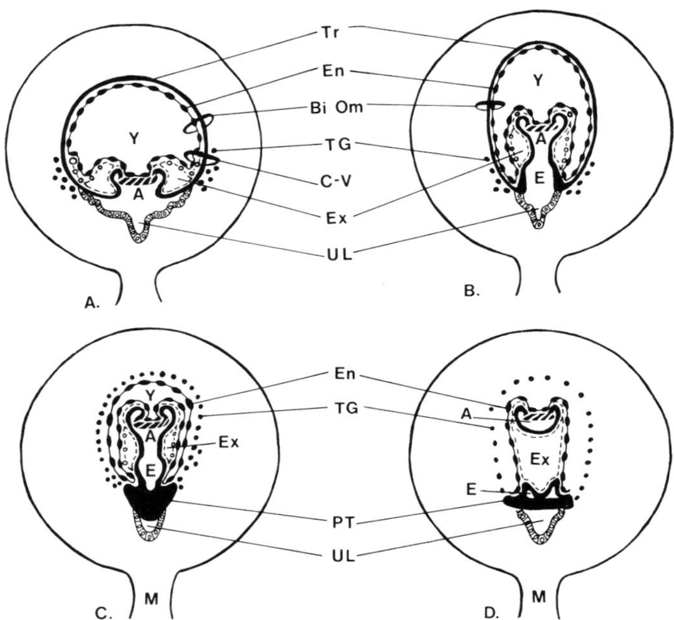

Fig. 3. Diagram illustrating four major patterns of early fetal membrane development in rodent higher taxa. (A) Sciurid and aplodontid pattern, approximating the rodent morphotype. Amniogenesis occurs by folding, and the embryo is not invaginated into the large yolk sac. A transitory choriovitelline placenta occurs paraembryonically, and the extensive bilaminar omphalopleure persists abembryonically. (B) Geomyoid and dipodoid pattern. In contrast to the sciurid condition, the embryonic region is inverted into the yolk sac, and this results in the formation of an elongate epamniotic cavity in continuity with the developing amniotic cavity. A choriovitelline placenta is absent, probably correlated with inversion of the embryonic disc. (C) Muroid pattern. Embryonic inversion has occurred relatively early, and the amniotic and epamniotic cavities have become secondarily continuous. Note that the roof of the epamniotic cavity is formed by the cone-shaped thickening of the preplacental trophoblast, in contrast to the geomyoid pattern. The trophoblastic layer of the bilaminar omphalopleure is disrupted and represented primarily by an incomplete layer of trophoblastic giant cells. (D) Hystricognath pattern. Amniotic and epamniotic cavities form by cavitation and do not become continuous. The distal layer of yolk sac endoderm fails to develop, and the trophoblastic layer consists only of scattered giant cells. Abbreviations: A, amnion; Bi Om, bilaminar omphalopleure; C-V, choriovitelline placenta; E, epamniotic cavity; En, yolk sac endoderm; Ex, exocoelom; M, mesometrium; PT, preplacental trophoblast; Tr, trophoblast; TG, trophoblastic giant cells; UL, uterine lumen; Y, yolk sac.

mass elongates and becomes inverted into the developing yolk sac (Fig. 3c). Separate cavities develop within the ectoplacental cone (epamniotic cavity) and the embryonic cell mass (primordial amniotic cavity), although these cavities become secondarily confluent (Fig. 3c). Following differentiation of the primitive streak and exocoelom, somatopleuric amniotic folds fuse with each other to separate a definitive amniotic cavity from the epamniotic cavity. An intermediate condition between the primitive sciurid pattern of amniogenesis and the highly derived muroid pattern is found in geomyoids (Mossman, 1937). An ectoplacental cone does not develop in this group, because the polar trophoblast has disappeared in early implantation stages. Nevertheless, the embryonic mass becomes inverted into the yolk sac (Fig. 3b), and, following development of somatopleuric folds, the definitive amniotic cavity becomes separated from an epamniotic cavity. The trophoblastic roof of the epamniotic cavity remains incomplete, however, because of the absence of the polar trophoblast. Differentiation of the amniotic and epamniotic cavities in Dipodoidea is essentially identical to that of Geomyoidea (King and Mossman, 1974).

The most derived pattern of amniogenesis occurs in the Caviomorpha. In all families for which data are available, the primordial amniotic cavity develops by cavitation and persists to form the definitive amniotic cavity. An epamniotic cavity also originates by cavitation within the moderately developed preplacental trophoblast, but it remains separated from the primordial amniotic cavity by a transitory proexocoelomic cavity. Following differentiation of the primitive streak, the proexocoelom is converted into an exocoelom by the spread of extraembryonic mesoderm (Fig. 3d). Although the initial stages of amniogenesis have not been observed in bathyergids and hystricids, evidence from slightly later stages suggests that development of the amniotic, epamniotic, and proexocoelomic cavities occurs in a manner identical to that of caviomorphs. Amniogenesis in pedetids (Fischer and Mossman, 1969) and ctenodactylids (Luckett, unpublished) is characterized by development of a primordial amniotic cavity, and by the subsequent rupture of its roof to form a temporary, slitlike trophoepiblastic cavity. In contrast to the condition in geomyoids, dipodoids, muroids, and hystricognaths, there is no inversion of the embryonic mass into the yolk sac, and the definitive amnion is formed by somatopleuric folding.

Yolk Sac and Choriovitelline Placentation

In most rodent families, endoderm spreads peripherally from the embryonic disc to form an endodermal lining for the yolk sac (Fig. 3a,b,c). All rodents are characterized by the failure of extraembryonic mesoderm to extend into the abembryonic half of the blastocyst; consequently, only the proximal (embryonic) half of the yolk sac becomes vascularized during pregnancy. In sciurids, aplodontids, pedetids, and anomalurids, the original bilaminar omphalopleure (trophoblast plus endoderm) of the distal half of the blastocyst persists throughout gestation (Figs. 3a and 4a), and this apparently represents the primitive rodent condition. In other families, there is a trend toward loss of both the trophoblastic and endodermal layers of the bilaminar omphalopleure as gestation progresses; as a result, the vascular splanchnopleure of the proximal yolk sac becomes inverted against the uterine endometrium (Fig. 4b,c,d). In castorids, geomyoids, and dipodoids, the bilaminar omphalopleure is lost relatively late during gestation, whereas it disappears relatively earlier in muroids.

The most uniquely derived pattern of yolk sac development in rodents occurs in caviomorphs, hystricids, and bathyergids. In these groups, extraembryonic endoderm never spreads to the abembryonic (distal) half of the early blastocyst, and there is a precocious disruption and degeneration of the abembryonic trophoblast in preprimitive streak embryos (Fig. 3d). The distal endoderm of the yolk sac also fails to develop in ctenodactylids, but the attenuated layer of abembryonic trophoblast persists until at least midpregnancy (Luckett, unpublished). The inverted yolk sac splanchnopleure of muroids and caviomorphs has been shown to play an important functional role in the absorption of immunoglobulins and other proteins from the endometrium (for a recent summary, see King, 1977).

A transitory choriovitelline placenta develops in sciurids, aplodontids, castorids, anomalurids, pedetids, and ctenodactylids following the vascularization of paraembryonic mesoderm (Fig. 3a). This relationship is undoubtedly primitive for rodents as it is for eutherian mammals in general. In those families which exhibit early inversion of the embryonic mass into the yolk sac (geomyoids, dipodoids, muroids, caviomorphs, bathyergids, and hystricids), the vascular mesoderm of the yolk sac does not contact the chorion that is apposed to the endometrium, and, consequently, there is no differentiation of a choriovitelline placenta (Fig. 3b,c,d).

Allantoic Vesicle

An endodermal allantoic vesicle develops in continuity with the hindgut and is evident before differentiation of the limb buds in sciurids, aplodontids, castorids, anomalurids, pedetids, and ctenodactylids. This vesicle and its associated vascular mesoderm have fused with the mesometrial chorion or preplacenta by early limb bud stages to initiate the differentiation of the chorioallantoic placenta. The small allantoic vesicle persists at the fetal surface of the placenta throughout the remainder of gestation in these families (Fig. 4a). In geomyoids, dipodoids, muroids, and hystricognaths, a bud of allantoic mesoderm differentiates precociously during late presomite–early somite stages, in the absence of an endodermal allantoic diverticulum. This mesodermal bud becomes vascularized and grows across the exocoelom to fuse with the preplacenta, initiating the differentiation of a vascular chorioallantoic placenta. The vascular mesodermal body stalk remains devoid of an endodermal component throughout gestation in these taxa (Fig. 4b,c,d).

Chorioallantoic Placenta

All rodents are characterized by the development of a discoidal, hemochorial placenta at the mesometrial pole of the uterus (Fig. 4). Following the development of the definitive chorioallantoic placenta, a paraplacental region of somatopleuric chorion persists throughout pregnancy in sciurids, aplodontids, and castorids (Fig. 4a). In all other rodent families the true chorion is a temporary structure which is obliterated by the development of the definitive chorioallantoic placenta.

Two uniquely derived (autapomorphous) features of the definitive placenta have been identified in all hystricognaths examined. One of these, the subplacenta, was previously considered to be a unique attribute of caviomorphs (Perrotta, 1959; Roberts and Perry, 1974). However, an identical pattern of subplacental development also occurs in bathyergids and hystricids (Mossman and Luckett, manuscript), and in thryonomyids (Oduor-Okelo, personal communication). The subplacenta appears to be a growth zone for the proliferation of the trophoblast in the lobulated placental disc, and it has also been suggested as a possible site of placental endocrine activity (Davies *et al.*, 1961).

Another unique feature of caviomorphs, first identified by Perrotta (1959), is a fibrovascular ring which surrounds the site of attachment of the yolk sac splanchnopleure to the mesodermal body stalk at the fetal surface of the placenta (Fig. 4d). We have also observed an identical fibrobascular ring in hystricids and bathyergids. The functional significance of this structure remains unclear at present.

Reproductive Cycles and Anatomy

Weir (1974) has recently reviewed a number of reproductive attributes which characterize caviomorph rodents, and she emphasized that these features *in toto* serve to distinguish caviomorphs from other rodent taxa. These include a long estrous cycle and gestation period, a vaginal closure membrane which is open only during estrus and parturition, lateral position of the nipples, and a sacculus urethralis in the penis. Character states for some of these features are poorly known for some other rodents, and this limits their usefulness in phylogenetic analysis. However, both long estrous cycles and gestation periods appear to be derived attributes in rodents, and they also characterize African hystricognaths. The sacculus urethralis is a uniquely derived feature shared by caviomorphs and African hystricognaths, although a possible rudimentary homologue has been described in ctenodactylids (Tullberg,

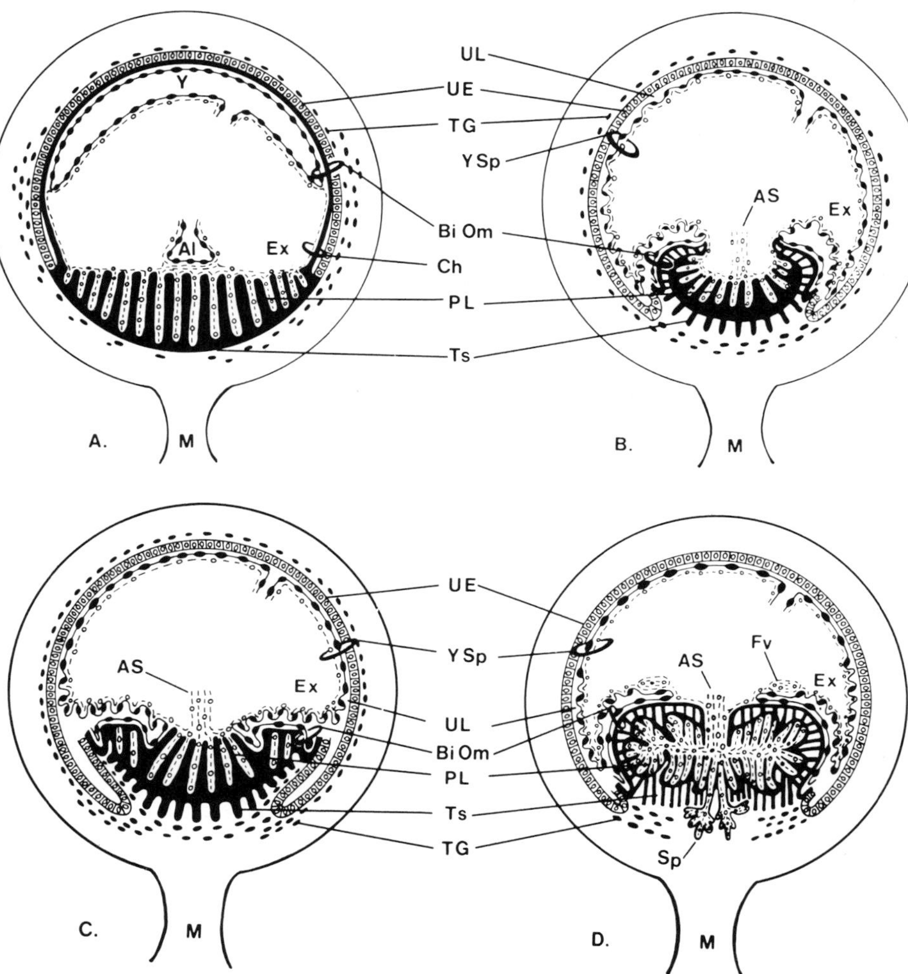

Fig. 4. Diagram representing definitive arrangement of fetal membranes and placenta in rodents. (A) Sciurid and aplodontid pattern. Note persistence of the bilaminar omphalopleure, paraplacental chorion, and small allantoic vesicle. (B) Geomyoid and dipodoid pattern. Note the absence of the bilaminar omphalopleure (except on the surface of the placental disc), paraplacental chorion, and allantoic vesicle. The cup-shaped placental disc is characteristic of geomyoids. (C) Muroid pattern. Note the general resemblance to the geomyoid pattern. (D) Hystricognath pattern. The highly lobulated nature of the placental disc, and the presence of a subplacenta and fibrovascular ring are characteristic features of all hystricognaths. Abbreviations: AS, allantoic mesodermal stalk; Fv, fibrovascular ring; PL, placental labyrinth; Sp, subplacenta; Ts, trophospongium; UE, uterine epithelium; Y Sp, yolk sac splanchnopleure. Other abbreviations as in Fig. 3.

1899; Wood, 1974). Preliminary analysis of reproductive parameters in female ctenodactylids (George, 1978) reveals several derived traits which are shared with hystricognaths, including a relatively long estrous cycle and gestation period, small litter size, relatively large neonates, and a vaginal closure membrane.

Phylogenetic Analysis of Developmental and Anatomical Traits in Rodents

The relatively primitive and derived character states of developmental and reproductive traits of rodents which were discussed during the present study are listed in Table 2, and the distribution of these character states in rodent higher taxa is illustrated in Fig. 5. The superfamily Muroidea is represented only by the families Cricetidae and Muridae in this analysis; little or no data are available for the families Rhizomyidae, Spalacidae, and Gliridae. Fetal membrane and reproductive attributes are known in some detail for the caviomorph families Octodontidae, Echimyidae, Erethizontidae, Caviidae, Dasyproctidae, Chinchillidae, and Capromyidae (for a summary, see Roberts and Perry, 1974; Weir, 1974).

Several musculoskeletal features which have been utilized in previous assessments of rodent relationships (Tullberg, 1899; Landry, 1957; Wahlert, 1968; Woods, 1972; Lavocat, 1974*b*; Wood, 1974) are also incorporated into the present analysis. These include the nature of the pterygoid fossa, the relationship between malleus and incus, the pattern of incisor enamel, morphology of the mandible, and the degree of differentiation of the masseter muscle complex. The morphocline polarity of character states for some of these features has not been adequately discussed, although most investigators agree that character states in hystricognaths are relatively "specialized" (= derived). For certain traits, such as incisor enamel patterns, both the multiserial pattern characteristic of hystricognaths and the uniserial pattern characteristic of extant geomyoids, muroids, and sciurids appear to be derived character states which were derived independently from an ancestral pauciserial condition (Wahlert, 1968). Likewise, the sciuromorphous, myomorphous, and hystricomorphous patterns of the masseter muscle complex are each derived character states, although it is possible that the myomorphous pattern is structurally and phyletically derived from an intermediate sciuromorphous condition, rather than independently from the ancestral rodent protrogomorphous condition.

Phylogenetic analysis of developmental, reproductive, and musculoskeletal traits in all rodent higher taxa corroborates the hypothesis of monophyly for Afro-Asian and South American hystricognathous rodents, based on their common possession of numerous synapomorphies (Fig. 5). Identification of these synapomorphies does not complete the analysis of phylogenetic relationships, however. Two essential tasks in such an analysis are to consider the probability of whether shared, derived similarities are the result of homology or convergence, and to identify homologous synapomorphies which are uniquely derived (= autapomorphies). Autapomorphous

Table 2. Primitive and Derived Character States of Selected Rodent Traits

Primtive	Derived[a]
1. Mandible sciurognathous	1. Mandible hystricognathous
2. Subplacenta absent	2. Subplacenta present
3. Pars reflexa of superficial masseter muscle small or indistinct	3. Pars reflexa of superficial masseter well developed
4. Pars posterior, deep division of masseter lateralis profundus indistinct	4. Pars posterior, deep division of masseter lateralis profundus well developed
5. Fibrovascular ring absent	5. Fibrovascular ring present
6. Implantation superficial, eccentric	6. Implantation primary interstitial (1)
7. Transitory proexocoelomic cavity absent	7. Transitory proexocoelomic cavity present
8. Pterygoid fossa does not open anteriorly	8. Pterygoid fossa opens anteriorly into orbit
9. Complete endodermal lining develops in early blastocyst	9. Parietal endodermal layer fails to develop
10. Gestation period relatively short	10. Gestation period relatively long
11. Malleus and incus unfused	11. Malleus and incus fused (2)
12. Sacculus urethralis absent	12. Sacculus urethralis present (3)
13. Abembryonic trophoblast persists throughout gestation	13. Abembryonic trophoblast disappears precociously in preprimitive streak stage (4)
14. No primordial amniotic cavity; amniogenesis by folding	14. Primordial amniotic cavity develops by cavitation; persists to form definitive amniotic cavity (5)
15. True chorion permanent and paraplacental	15. True chorion temporary and placental
16. Incisor enamel pauciserial	16. Incisor enamel multiserial or uniserial
17. Masseter muscle complex protrogomorphous	17. Masseter muscle complex myomorphous or hystricomorphous (6)
18. Allantoic vesicle small, permanent	18. Allantoic vesicle absent
19. Transitory choriovitelline placenta present	19. Choriovitelline placenta absent
20. Epamniotic cavity absent	20. Epamniotic cavity develops by cavitation, remains distinct from amniotic cavity (7)

[a] The numbers in parentheses refer to the following intermediately derived character states: (1) implantation secondary interstitial; (2) malleus and incus closely appressed; (3) sacculus urethralis rudimentary; (4) abembryonic trophoblast disappears late (as in geomyoids, dipodoids, and castorids); or, relatively earlier during gestation (as in muroids); (5) primordial amniotic cavity transitory; definitive amniogenesis by folding (as in pedetids and ctenodactlyids); or, primordial amniotic cavity transitory, then becomes continuous secondarily with epamniotic cavity; definitive amniogenesis by folding (as in muroids); (6) the sciuromorphous masseter pattern is interpreted here as intermediate between the protrogomorphous and myomorphous conditions; (7) open epamniotic cavity develops in continuity with amniotic cavity (as in geomyoids and dipodoids); or, in a more derived condition, the epamniotic and amniotic cavities arise independently by cavitation but become continuous secondarily (as in muroids).

traits provide the strongest corroboration for monophyly of a taxon. Much of the disagreement concerning the monophyletic or diphyletic origin of hystricognathous rodents results from insufficient evaluation of these factors.

As emphasized by Bock (1977), there are not absolute criteria for recognizing homologous similarities between two taxa, and detailed phenetic resemblances (including similar ontogenetic patterns) provide the best evidence for identifying shared homologies. As an example, a hemochorial placenta characterizes rodents, haplorhine primates, lagomorphs, erinaceids, and a variety of other eutherian mammals, and some authors have suggested that this widespread feature may represent the primitive eutherian condition. However, ontogenetic analysis of the hemochorial placenta in all these eutherians reveals that this character state is attained by different developmental pathways in most of these taxa. Such analysis provides compelling evidence that this shared similarity is the result of convergence rather than homology. This hypothesis is further corroborated by the "reciprocal illumination" obtained from phylogenetic analysis of other fetal membrane features (Luckett, 1977).

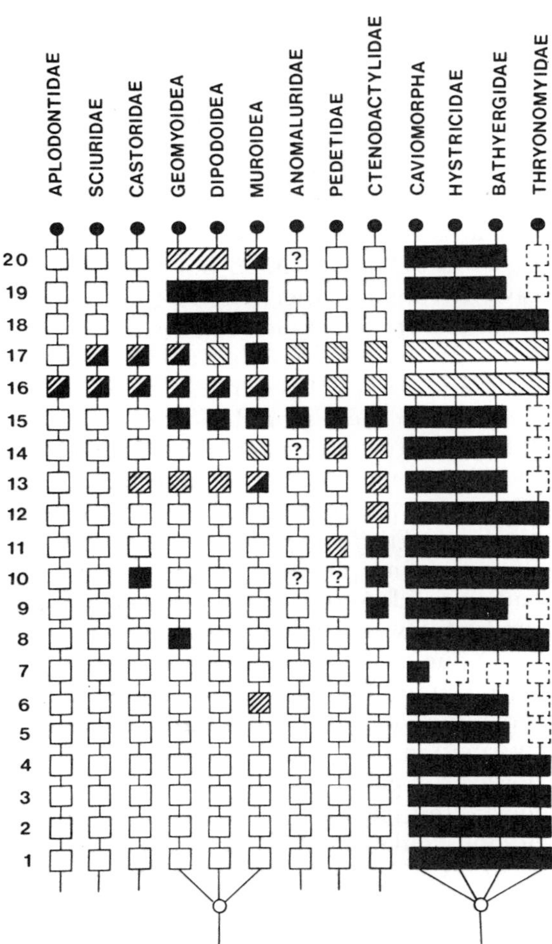

Fig. 5. Character phylogeny of fetal membrane, reproductive, and other rodent anatomical features evaluated during the present study. Monophyly of the suborder Hystricognathi is strongly corroborated by the demonstration of numerous synapomorphies and autapomorphies. Character state symbols as in Fig. 2. Broken-lined squares represent unknown character states. See Table 2 for the list of characters evaluated. Note: Both character states illustrated for character 16 are equally derived, as are the myomorphous and hystricomorphous conditions for character 17.

Ontogenetic evidence evaluated during the present study indicates that the individual features and the entire developmental complex of the fetal membranes in caviomorphs are the most derived patterns which occur in rodents (and, indeed, in eutherian mammals). Available data on fetal membrane and placental development in the African families Bathyergidae and Hystricidae reveal a homologous pattern which is identical in every detail to that of caviomorphs (Mossman and Luckett, 1968, manuscript). Furthermore, the developmental pattern of the fetal membranes and placenta in the African family Thryonomyidae is also identical to that of caviomorphs (Oduor-Okelo, personal communication), although details of this study are not yet published. Both individual fetal membrane features (characters 2, 5–7, 13, 14, and 20 in Table 2 and Fig. 5) and the entire developmental complex of the fetal membranes and placenta comprise autapomorphous traits which strongly corroborate the monophyly of Old and New World hystricognathous rodents, and it seems probable that this derived development pattern also characterized the last common ancestor of these taxa.

Other autapomorphous traits of hystricognathous rodents include the hystricognathous mandible and associated modifications of the masseter muscle (masseter superficialis, pars reflexa, and masseter lateralis profundus, pars posterior, deep division), and the sacculus urethralis of the penis (characters 1, 3, 4, and 12 in Table 2 and Fig. 5). Hystricognathy has also been identified in mandibular fragments from three genera of Eocene North American rodents (see Wood, 1974), but the possible affinities of these genera with caviomorphs or other hystricognaths are unclear. A number of additional synapomorphous traits which characterize all hystricognath families (characters 8, 10, 11, and 15–19 in Table 2 and Fig. 5) occur also in some other rodent taxa, and Wood (1974) has cited this as evidence for the possible parallel evolution of these shared similarities in Old and New World hystricognaths. While it is true that each of these apomorphous character states may be found in at least one nonhystricognath family, only caviomorphs and African hystricognaths are characterized by the common possession of *all* these derived attributes (Fig. 5). If parallelism is used as an argument to explain this suite of shared apomorphies in Old and New World hystricognaths, then it would be equally valid to conclude that parallelism also accounts for these same shared similarities within Caviomorpha.

None of the characters evaluated in the present study provide evidence for distinguishing between Old and New World hystricognaths. Taxonomic separation of these groups has been based primarily on their zoogeographic distribution and differences in dental morphology. There is little agreement, however, concerning the primitive vs. derived character states of differing molar crest patterns, and uncertainty also surrounds the suggested homology of lophs and crests of molar teeth in these groups (cf. Lavocat, 1974*b*; Wood, 1974). Variations in dental morphology are common at the familial and generic level in rodents, and, in the absence of adequate morphotype reconstruction of these features in higher taxonomic categories, undue weight has been given to these traits in evaluating hystricognath phylogeny.

Many of the controversies encountered in rodent systematics are related to a common failure to distinguish between phylogenetic reconstruction and classification. Most classifications of rodents have been typological, with major emphasis placed on either the: (1) masseter muscle complex and associated modification of the infraorbital foramen, (2) structure of the mandible, or (3) dental morphology. Little attention has been devoted, however, to phylogenetic analysis of the distribution of these traits. Most students of rodent systematics have enumerated patristic resemblances among taxa as the basis for classification, without distinguishing between shared primitive and derived resemblances. Shared primitive (plesiomorphous) features neither corroborate nor falsify hypotheses of phylogenetic relationships, because these traits may have been inherited unchanged from a more distant ancestor (Hennig, 1966). As an example, many fetal membrane features of pedetids, ctenodactylids, and anomalurids are similar to each other and to those of sciurids and aplodontids (Fischer and Mossman, 1969; Luckett, 1971) and these resemblances were considered to support a hypothesis of relationships among these taxa. However, phylogenetic analysis indicates that all these shared resemblances are the result of retention of primitive rodent attributes (Fig. 5); as such, they provide no evidence of possible affinities among the three families.

Phylogenetic analysis and the search for synapomorphies should be an essential prerequisite for the construction or modification of rodent classifications. The numerous synapomorphies identified in the present study provide strong corroboration for the monophyletic relationship of Caviomorpha, Hystricidae, Bathyergidae, and Thryonomyidae, and for the classification of these taxa in Tullberg's (1899) suborder Hystricognathi. Hystricognathy is always accompanied by hystricomorphy and multiserial enamel in extant hystricognaths and their unquestioned fossil relatives, but the occurrence of hystricomorphy and multiserial enamel is not always accompanied by hystricognathy. Consequently, it is logical to hypothesize that hystricognathy evolved in an ancestral stock which was already hystricomorphous and possessed multiserial enamel. This working hypothesis facilitates the search for the possible sister group of the monophyletic taxon Hystricognathi. Both pedetids and ctenodactylids are hystricomorphous and possess multiserial enamel, and ctenodactylids exhibit several additional derived or intermediately derived character states which are shared with hystricognaths (characters 9–12 in Table 2 and Fig. 5). Retention of the primitive rodent condition of sciurognathy in ctenodactylids does not falsify a hypothesis of sister-group relationship between ctenodactylids and hystricognaths, contrary to the implications of Lavocat (1974*b*) and Wood (1974), nor do any of the other attributes evaluated in Fig. 5.

The Oligocene rodent *Tsaganomys* from Mongolia exhibits hystricognathy, a bathyergid-like infraorbital foramen, multiserial enamel, and a perforated pterygoid fossa (see Lavocat, 1974*b*: Hoffstetter, 1975), and this combination of apomorphous traits increases the probability that this genus belongs in the Hystricognathi. Ctenodactylids are also abundant in the same fauna (Mellett, 1968), and the possibility of a common (Eocene?) Asian ancestry for these two taxa should be evaluated by further character analysis. A similar hypothesis suggesting a common ancestry for Hystricognathi and Ctenodactylidae was also proposed by Hussain *et al.* (1978) on the basis of some dental remains of hystricomorphous, sciurognathous rodents with multiserial incisor enamel from the Middle Eocene of Pakistan.

Conclusions

Phylogenetic analysis of developmental, reproductive, and other anatomical features in primates and rodents corroborates the hypothesis of monophyly for Anthropoidea and Hystricognathi. This is based on the identification of autapomorphous and other synapomorphous traits which characterize each group. The possibility that parallelism could account for numerous synapomorphies in several unrelated and complex organ systems appears unlikely. Morphotype reconstruction of these features in Anthropoidea and Hystricognathi and the search for sister-groups of each taxon provide an essential step for identification of the possible ancestral stock for these higher taxa. In contrast, shared primitive traits are of little or no value in assessing phylogenetic relationships.

Although the monophyly of South American and Afro-Asian representatives of the Anthropoidea and Hystricognathi is corroborated by the present analysis, no direct evidence is presented which might account for the paleodistributional pattern of these taxa in the southern continents. The known distribution of fossil and extant anthropoids and hystricognaths does suggest, however, that a similar sweepstakes route was probably involved in their southern deployment. The reconstruction of the South Atlantic during the Eocene-Oligocene presented by Tarling (this volume) suggests that the most likely location of this route may have been from West Africa to eastern South America.

References

Bedford, J. M., 1974, Biology of primate spermatozoa, *Contrib. Primatol.* **3:**97–139.

Benirschke, K., and Layton, W., 1969, An early twin blastocyst of the golden lion marmoset, *Leontocebus rosalia* L., *Folia Primatol.* **10:**131–138.

Bock, W. J., 1977, Foundations and methods of evolutionary classification, in: *Major Patterns in Vertebrate Evolution* (M. K. Hecht, P. C. Goody, and B. M. Hecht, eds.), pp. 851–895, Plenum Press, New York.

Cracraft, J., 1974, Continental drift and vertebrate distribution, *Annu. Rev. Ecol. Syst.* **5**:215–261.

Davies, J., Dempsey, E. W., and Amoroso, E. C., 1961, The subplacenta of the guinea pig: Development, histology and histochemistry, *J. Anat.* **95**:457–473.

Fischer, T. V., and Mossman, H. W., 1969, The fetal membranes of *Pedetes capensis,* and their taxonomic significance, *Am. J. Anat.* **124:**89–116.

George, W., 1978, Reproduction in female gundis (Rodentia: Ctenodactylidae), *J. Zool.* **185:**57-71.
Gingerich, P. D., 1976, Cranial anatomy and evolution of early Tertiary Plesiadapidae (Mammalia, Primates), *Univ. Mich. Mus. Paleontol. Pap. Paleontol.* **15:**1-140.
Hennig, W., 1966, *Phylogenetic Systematics,* University of Illinois Press, Urbana.
Hershkovitz, P., 1972, The recent mammals of the Neotropical Region: A zoogeographic and ecological review, in: *Evolution, Mammals, and Southern Continents* (A. Keast, F. C. Erk, and B. Glass, eds.), pp. 311-431, State University of New York Press, Albany.
Hershkovitz, P., 1977, *Living New World Monkeys (Platyrrhini),* Vol. 1, University of Chicago Press, Chicago.
Hill, J. P., 1932, The developmental history of the Primates, *Phil. Trans. Roy. Soc.* **221:**45-178.
Hofer, H. O., 1977, On the sublingual structures of *Tarsius* (Prosimiae, Tarsiiformes) and some platyrrhine monkeys (Platyrrhina, Simiae, Primates) with casual remarks on the histology of the tongue, *Folia Primatol.* **27:**297-314.
Hoffstetter, R., 1972, Relationships, origins, and history of the ceboid monkeys and caviomorph rodents: A modern reinterpretation, in: *Evolutionary Biology,* Vol. 6 (Th. Dobzhansky, M. K. Hecht, and W. C. Steere, eds.), pp. 323-347, Appleton-Century-Crofts, New York.
Hoffstetter, R., 1974, Phylogeny and geographical deployment of the primates, *J. Hum. Evol.* **3:**327-350.
Hoffstetter, R., 1975, El origen de los Caviomorpha y el problema de los Hystricognathi (Rodentia), in: *Actas del Primer Congreso Argentino de Paleontologia y Bioestratigrafia,* pp. 505-528, Tucumán.
Hussain, S. T., de Bruijn, H., and Leinders, J. M., 1978, Middle Eocene rodents from the Kala Chitta Range (Punjab, Pakistan), *Proc. Kon. Ned. Akad. Wetenschap., Amsterdam, Ser. B.* **81:**74-112.
Jollie, W. P., Haar, J. L., and Craig, S. S., 1975, Fine structural observations on hemopoiesis in the chorioallantoic placenta of the marmoset, *Am. J. Anat.* **144:**9-38.
Keast, A., 1972, Continental drift and the evolution of the biota on southern continents, in: *Evolution, Mammals, and Southern Continents* (A. Keast, F. C. Erk, and B. Glass, eds.), pp. 23-87, State University of New York Press, Albany.
King, B. F., 1977, An electron microscopic study of absorption of peroxidase-conjugated immunoglobulin G by guinea pig visceral yolk sac *in vitro, Am. J. Anat.* **148:**447-456.
King, B. F., and Mossman, H. W., 1974, The fetal membranes and unusual giant cell placenta of the jerboa (*Jaculus*) and jumping mouse (*Zapus*), *Am. J. Anat.* **140:**405-432.
Koering, M. J., 1974, Comparative morphology of the primate ovary, *Contrib. Primatol.* **3:**38-81.
Landry, S. O., 1957, The interrelationships of the New and Old World hystricomorph rodents, *Univ. Calif. Publ. Zool.* **56:**1-118.
Lavocat, R., 1969, La systématique des Rongeurs hystricomorphes et la dérive des continents, *C. R. Acad. Sci. Paris, Ser. D* **269:**1496-1497.
Lavocat, R., 1974*a*, The interrelationships between the African and South American rodents and their bearing on the problem of the origin of South American monkeys, *J. Hum. Evol.* **3:**323-326.
Lavocat, R., 1974*b*, What is an hystricomorph? in: *The Biology of Hystricomorph Rodents* (I. W. Rowlands and B. J. Weir, eds.), pp. 7-20, Academic Press, London.
Leutenegger, W., 1973, Maternal-fetal weight relationships in primates, *Folia Primatol.* **20:**280-293.
Luckett, W. P., 1971, The development of the chorio-allantoic placenta of the African scaly-tailed squirrels (family Anomaluridae), *Am. J. Anat.* **130:**159-178.
Luckett, W. P., 1974, Comparative development and evolution of the placenta in Primates, *Contrib. Primatol.* **3:**142-234.
Luckett, W. P., 1975, Ontogeny of the fetal membranes and placenta: Their bearing on primate phylogeny, in: *Phylogeny of the Primates* (W. P. Luckett, and F. S Szalay, eds.), pp. 157-182, Plenum Press, New York.
Luckett, W. P., 1977, Ontogeny of amniote fetal membranes and their application to phylogeny,

in: *Major Patterns in Vertebrate Evolution* (M. K. Hecht, P. C. Goody, and B. M. Hecht, eds.), pp. 439–516, Plenum Press, New York.

Luckett, W. P., and Szalay, F. S., 1978, Clades versus grades in primate phylogeny, in: *Recent Advances in Primatology,* Vol. 3 (D. J. Chivers and K. A. Joysey, eds.), pp. 227–237, Academic Press, London.

Martin, R. D., 1975, The bearing of reproductive behavior and ontogeny on strepsirhine phylogeny, in: *Phylogeny of the Primates* (W. P. Luckett and F. S. Szalay, eds.), pp. 265–297, Plenum Press, New York.

Mellett, J. S., 1968, The Oligocene Hsanda Gol formation, Mongolia: A revised faunal list, *Am. Mus. Nov.,* **No. 2318:**1–16.

Mossman, H. W., 1937, Comparative morphogenesis of the fetal membranes and accessory uterine structures, *Contrib. Embryol. Carnegie Inst.* **26**:129–246.

Mossman, H. W., and Luckett, W. P., 1968, Phylogenetic relationship of the African mole rat, *Bathyergus janetta,* as indicated by the fetal membranes, *Am. Zool.* **8:**806.

Patterson, B., and Pascual, R., 1972, The fossil mammal fauna of South America, in: *Evolution, Mammals, and Southern Continents* (A. Keast, F. C. Erk, and B. Glass, eds.), pp. 247–309, State University of New York Press, Albany.

Perrotta, C. A., 1959, Fetal membranes of the Canadian porcupine, *Erethizon dorsatum, Am. J. Anat.* **104:**35–59.

Roberts, C. M., and Perry, J. S., 1974, Hystricomorph embryology, in: *The Biology of the Hystricomorph Rodents* (I. W. Rowlands and B. J. Weir, eds.), *Symp. Zool. Soc. Lond.,* No. 34, pp. 333–360, Academic Press, London.

Schwartz, J. H., Tattersall, I., and Eldredge, N., 1978, Phylogeny and classification of the Primates revisited, *Yearb. Phys. Anthropol.* **21:**95–133.

Simpson, G. G., 1945, The principles of classification and a classification of mammals, *Bull. Am. Mus. Nat. Hist.* **85:**1–350.

Simpson, G. G., 1961, *Principles of Animal Taxonomy,* Columbia University Press, New York.

Tullberg, T., 1899, Ueber das System der Nagethiere, eine phylogenetische Studie, *Nova Acta Reg. Soc. Sci. Upsala* **18:**1–514.

Wahlert, J. H., 1968, Variability of rodent incisor enamel as viewed in thin section, and the microstructure of the enamel in fossil and recent rodent groups, *Breviora* **369:**1–18.

Weir, B. J., 1974, Reproductive characteristics of hystricomorph rodents, in: *The Biology of Hystricomorph Rodents* (I. W. Rowlands and B. J. Weir, eds.), *Symp. Zool. Soc. Lond.,* No. 34, pp. 265–301, Academic Press, London.

Wislocki, G. B., 1943, Hemopoiesis in the chorionic villi of the placenta of platyrrhine monkeys, *Anat. Rec.* **85:**349–364.

Wood, A. E., 1950, Porcupines, paleogeography, and parallelism, *Evolution* **4:**87–98.

Wood, A. E., 1965, Grades and clades among rodents, *Evolution* **19:**115–130.

Wood, A. E., 1972, An Eocene hystricognathous rodent from Texas: Its significance in interpretations of continental drift, *Science* **175:**1250–1251.

Wood, A. E., 1974, The evolution of the Old World and New World hystricomorphs, in: *The Biology of Hystricomorph Rodents* (I. W. Rowlands and B. J. Weir, eds.), *Symp. Zool. Soc. Lond.,* No. 34, pp. 21–60, Academic Press, London.

Wood, A. E., and Patterson, B., 1970, Relationships among hystricognathous and hystricomorphous rodents, *Mammalia* **34:**628–639.

Woods, C. A., 1972, Comparative myology of jaw, hyoid, and pectoral appendicular regions of New and Old World hystricomorph rodents, *Bull. Am. Mus. Nat. Hist.* **147:**115–198.

Comparative Study of the Sperm Morphology of South American Primates and Those of the Old World

18

D. E. MARTIN and K. G. GOULD

Introduction

Primate species exist today in four regions of the world: in South America, Africa, Madagascar and Southeast Asia, all separated by water expanses of differing magnitudes. Various theories have been proposed as to how these species, if derived from a common ancestor, could have populated such widely separated areas of the world. Recent resurgence of interest in Wegener's theory of continental drift (Hallam, 1973) has revived an interest in global migration of various species during the late Mesozoic and Cenozoic Eras. It is the purpose of this chapter to use some parameters of comparative sperm morphology to seek evidence of the means by which catarrhine (Old World) and platyrrhine (New World) primates achieved their present biogeographic distribution. To this end this study has been specifically di-

D. E. MARTIN • Yerkes Regional Primate Center, Emory University, Atlanta, Georgia 30322 and Georgia State University, Atlanta, Georgia 30303. K. G. GOULD • Yerkes Regional Primate Center, Emory University, Atlanta, Georgia 30322. This research was supported by NIH Grant #RR00165; Ford Foundation grant #690-0645A, and a grant from the McCandless Fund of Emory University.

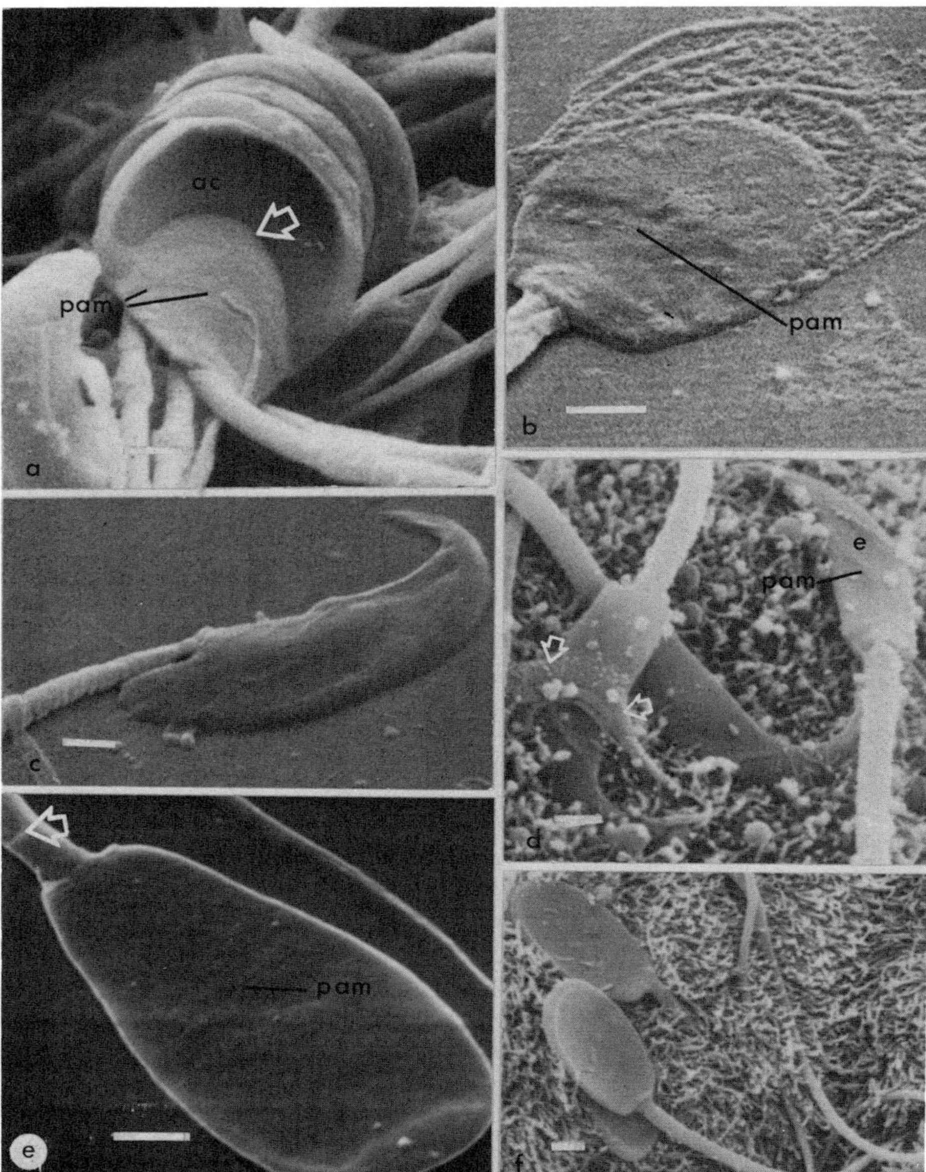

Fig. 1. (a) Guinea pig (*Cavia porcellus*). When obtained from the cauda epididymis as shown here, sperm are frequently stacked together in "rouleaux" fashion. The nuclear region is clearly visible, as is the acrosome (ac), reflected back at the anterior nuclear margin (arrow) and continuing onto the nuclear surface as far as its posterior acrosomal margin (pam). (b) Guinea pig (*Cavia porcellus*). Cauda epididymal sperm, air dried. Disruption of the acrosome in this specimen demonstrates the true size of the nucleus, which has a paddle-shaped appearance reminiscent of primate spermatozoa. (c) Mouse (*Mus musculus*). Cauda epididymal sperm, air dried. The head

rected toward an investigation of the phylogenetic relationships of cercopithecoid (catarrhine) and ceboid (platyrrhine) species based on analysis of their sperm morphology.

Among mammals, there appears to have been a marked centripetal selective force acting with regard to sperm morphology. This is reflected in the almost universal maintenance of small, motile sperm with a very similar basic structure. This structure involves packaging of genetic material within a cytoplasm-poor head, an energy-producing midpiece section wrapped with mitochondria, and a motile tail. The basic structure of the tail is the classic nine-plus-two represented in ciliated structures of diverse phyla, and that basic structure is here reinforced with a second ring of nine fibers. Superimposed upon this basic conservative structure, however, is a wide variety of more minor characteristics. Thus, one observes (Fig. 1), in species from widely divergent mammalian families, a marked variation in the structure and overall appearance of the sperm head (Gould *et al.*, 1975). Among the primates, however, there is much more interspecies uniformity (Martin and Gould, 1975), and bizarre and grossly asymmetric forms of the sperm head are not found. There is considerable variation in overall size of the cell itself, its acrosome, and in the ratio of dimensions of head, midpiece, and tail (Bedford, 1974; Martin *et al.*, 1975).

The variation in head size is largely accounted for by variations in size and shape of the acrosome, a membranous structure carried on the anterior portion of the head. Described by Allison and Hartree (1970) as a "modified lysosome" containing various enzymes associated with the fertilization pro-

---

is considerably larger than that of rat or guinea pig sperm. The mitochondria are easily visible, arranged in a fine spiral along the midpiece. Air-drying permits better viewing of the mitochondrial gyres due to a tighter apposition of the plasma membrane to underlying structures. The sperm nucleus does not extend the full length of the head, but the relationship of the acrosome, perforatorium and nucleus is better demonstrated using TEM (Bedford, 1974). (Reprinted from Gould, 1973; with permission of the Williams and Wilkins Co.). (d) Golden (Syrian) hamster (*Mesocricetus auratus*). The individual sperm shown here, on the microvillous surface of the vitellus (ovum), clearly demonstrate the hook-shaped nature of the sperm head. The extent of the acrosome is delineated by a distinct posterior acrosomal margin (pam), and, in this species, the equatorial segment is also clear (e). During preparation of the sperm for fertilization (capacitation), the acrosome is lost, detaching at the anterior margin of this segment. Such a released acrosome is visible (arrows). In common with other rodents the mitochondria of the midpiece are arranged in a fine, multiturn spiral, and the midpiece is long relative to the total length of the sperm cell. (e) Bull (*Bos taurus*). The sperm head is flat, paddle-shaped and very large. The posterior margin of the acrosome (pam) is barely visible behind the center of the sperm head. An anterior acrosomal swelling can be seen as an apical ridge, beneath which may be stored acrosomal enzymes. A defect in the plasma membrane overlying the midpiece is visible (arrow). (f) Rabbit (*Oryctolagus cuniculus*). The apical ridge of the acrosome is clearly observed here. Although not visible on these spermatozoa, the posterior acrosomal margin of rabbit sperm is commonly demarcated as a distinct serrated line about one third the distance of the head from its base (Gould, 1973; Koehler and Kinsey, 1977). These sperm are within the vas deferens, and the microvillous lining of that organ is evident. In overall form, rabbit sperm resemble those of the bull, but are smaller. Notable is their lack of similarity with those of rodents.

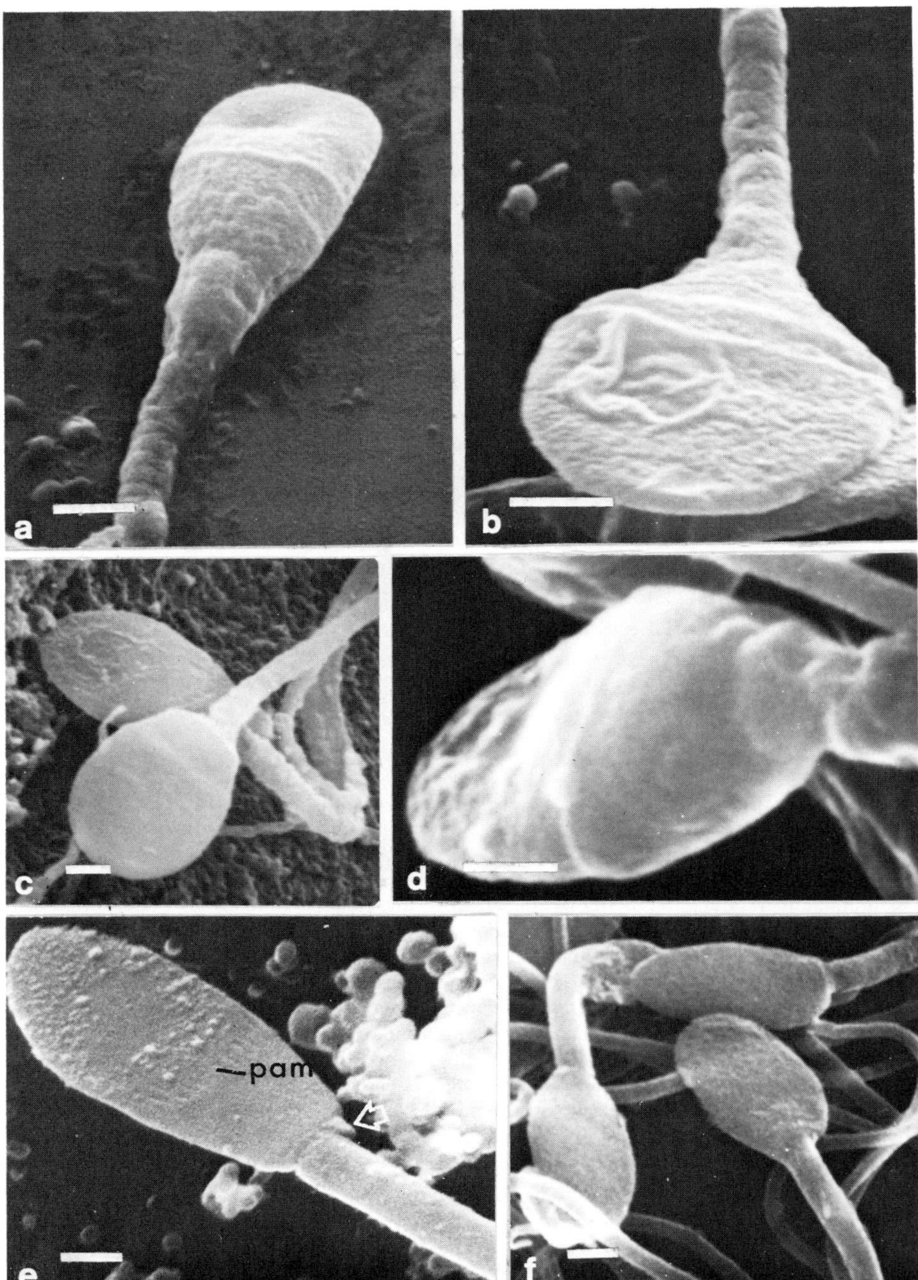

Fig. 2. (a) Man (*Homo sapiens*). Although great size pleiomorphism exists in this species, the form depicted here is commonly seen. Note the characteristically rounded posterior head region, obvious posterior acrosomal margin (pam) and anterior flattening with cavitation. The insertion

cess, these enzymes and their activities are themselves subject to centripetal selection forces and are remarkably uniform from species to species, both within and outside the primate orders. The nuclear genetic content in terms of DNA weight varies among species, but is usually constant within species. However, hominoid species show a greater variation in this regard than do nonhominoid species (Seuanez *et al.*, 1977).

The surface anatomy of the sperm midpiece is also largely characteristic of the individual species observed, especially regarding the size of the mitochondria around the midpiece and the number of turns (gyres) of the mitochondrial spiral. By simply examining head shape and midpiece features, one can readily identify many primate species (Fig. 2) (Bedford, 1974). Such conservatism in maintenance of basic structures is of value in an investigation of evolutionary changes. It permits one to make inferences concerning the relationship of groups from observation of relatively minor structural variations of no universal functional significance and, presumably, free of centripetal selection pressure. The existence of these variations in more than one group of individuals is more likely to have resulted from divergent than

---

of the midpiece into the head is obscured, probably by redundant cytoplasm. Mitochondrial gyres are rather large, and coiled tightly. Human sperm are among the smallest of the primate sperm thus far studied. (b) Man (*Homo sapiens*). A common morphological variation is the flattening of the entire head, with no cavitation, as depicted here, making it almost indistinguishable from sperm from other anthropoid primate families. Again, the posterior acrosomal region is easily visible, as well as the large mitochondrial gyres. (Reprinted from Martin *et al.*, 1975; with permission of Academic Press). (c) Orangutan (*Pongo pygmaeus*). Sperm from this ape, as well as those from all anthropoid primates examined thus far except for man and the gorilla, show extreme morphological similarity among individuals *within* each species, in addition to having many similar features *among* the species. These general features, as shown here, include a flattened, paddle-shaped head, and a visible demarcation of the acrosome from the remainder of the head. Depending upon species, the head may be rounded (in the orangutan this roundedness is developed to the maximum), the midpiece may insert eccentrically into the head (it does not in the orangutan), and mitochondrial gyres may be loosely or tightly (as shown here) apposed about the axial filament complex. (d) Gorilla (*Gorilla gorilla*). Displaying as much size pleiomorphism as human sperm, with a typical head shape also having a thick posterior portion, obvious acrosomal demarcation, and anterior narrowing, it would appear difficult to distinguish sperm from the two species. Cavitation seems much less frequent, however, and gorilla sperm are a bit larger, with an almost universal incidence of an annular ring at the posterior margin of the midpiece. (e) Ruffed lemur [*Varecia variegatus (= Lemur variegatus)*]. The sperm head is flattened, but somewhat elliptical rather than rounded; the posterior acrosomal margin (pam) indicates that this structure covers the anterior two-thirds of the head. Notice the eccentric insertion of the midpiece into the head, and the presence of a small knoblike structure adjacent to the neck (arrow). This structure, composed of redundant membranous material, is not invariably present in this species. The definite difference in morphology between *V. variegatus* and other lemurs examined (see Fig. 5) supports the suggestion, made on the basis of other morphological characteristics, that this species be classified in the separate genus *Varecia* (Petter, 1962; Seligsohn and Szalay, 1974) as followed in this study. (f) Rhesus macaque (*Macaca mulatta*). Sperm heads are flattened, but not as rounded, and there is a symmetrical insertion of the midpiece into the head. Mitochondria are fairly small, and regularly arranged. Spermatozoan morphology is remarkably uniform among the macaques, and micrographs of several species have been published elsewhere (Martin *et al.*, 1975; Hafez and Kanagawa, 1973).

convergent or parallel evolution. With these underlying facts in mind, the surface morphology of South American and African primates was compared.

Methodology

The scanning electron microscope permits detailed analysis of surface morphology while retaining a high degree of sampling power. Thus a large number of sperm from a given specimen can be examined in great detail to enhance accurate morphologic interpretation.

Unless otherwise stated in the discussion concerning a specific species, sperm for this study were recovered by rectal probe electroejaculation using techniques developed and now routinely utilized at the Yerkes Regional Primate Research Center (Warner *et al.*, 1974; Gould *et al.*, 1978). Sperm in the fluid portion of the ejaculate were centrifuged from the seminal plasma, washed in isotonic buffered saline, fixed with 2.5% glutaraldehyde, dehydrated through an ascending series of alcohols in a manner similar to that used for conventional histology, and critical-point dried (Gould, 1973). Prior to the critical-point drying procedure (Cohen, 1977), specimens were mounted on glass slides. Subsequent to drying they were coated with a thin layer of gold–palladium using a diode sputter coater. Observations were made using an ISI Super III scanning electron microscope with photographic recording via Polaroid Type-55 or Kodak Plus-X 35 mm film.

In conjunction with scanning electron microscopy (SEM), light microscopy was utilized to provide a direct point of contact with previous studies and to permit comparison of results with those existing in the literature on the species we have examined. It is interesting that SEM has permitted the identification of sperm structures which can be recognized as existing when seen in the light microscope, but not reported or observed in studies using light microscopy alone. Wet semen smears were stained using either eosin blue-nigrosin or eosin blue-aniline blue and air dried (Martin and Davidson, 1976).

Results: Introduction to Figures 1–5

Specific details regarding magnification and individual structural features seen in each photograph are provided in the respective figure legends. In each case the horizontal bar represents one micron. All specimens were critical-point dried except those in Fig. 1b,c, which were air dried from absolute alcohol. Some general comments concerning organization of the plates is appropriate.

Figure 1 provides comparison photomicrographs of sperm from a variety of mammalian orders. Representative species from the orders Rodentia (a,b,c,d), Lagomorpha (f), and Artiodactyla (e), in addition to primates shown in subsequent figures, all show the basic conservatism in structure. There is, however, a considerable difference in sperm size and shape in nonprimate families, even between species within a given family. This is demonstrated by sperm from the guinea pig (a), mouse (c), and hamster (d). Sperm of the bull (e) and rabbit (f) approach the characteristics of the primitive mammalian sperm cell more closely than those of any other nonprimate family.

The presence of an acrosomal sac is shown by the swollen anterior portion of the sperm head. The magnitude of this structure varies; it is very well developed in the guinea pig (a), as seen by the much reduced size of the sperm head when destruction or dispersion of the acrosome (b) leaves only the nucleus remaining. The mouse and rat possess sperm with the most eccentrically positioned midpiece/head junction reported for any mammalian species.

Figure 2 illustrates several representative primate species, and depicts the considerable conservatism in sperm surface morphology. With the partial exception of the gorilla and man, in which there is considerable size pleiomorphism (a,b,c), primate sperm have basically a paddle-shaped head. It now appears likely that the "characteristic" or "typical" human sperm head is also paddle-shaped, and that the cavitated head (a) is the less frequently observed form. Such forms, however, do retain fertilizing ability, since fertile individuals may produce semen samples in which this type comprises almost the entire sperm population in the ejaculate.

Figure 3 includes examples of sperm from Old World primate species. Two groups are represented: the baboons (a,b,c) representing the higher primate Cercopithecidae, and the bushbabies (d,e,f) representing the lower primate Lorisidae (Gould and Martin, 1978). Again, even in these evolutionarily diverse groups, the basic conservatism in sperm morphology is preserved. Two unusual features are notable, however. In the baboons, the midpiece inserts or attaches eccentrically to the head, rather than symmetrically. Although baboon sperm have been described by other workers, using SEM (Flechon *et al.*, 1976), this eccentricity has not been described except in earlier work by the present authors (Martin *et al.*, 1975). In the bushbabies, there is an asymmetry in the anterior portion of the acrosome found in mature or nearly mature cells. There is an accumulation of material within it (d,e) of unknown functional significance. It has been described by other workers (Bedford, 1974).

Figures 4 and 5 contain examples of present-day South American primate species. On this continent, only two families are represented: the higher primate *Cebidae* and *Callitrichidae*. Again, the adherence to a basically conservative structure is notable.

Discussion

The details of sperm morphology which have been both described and illustrated provide strong evidence for centripetal selection in the evolution of

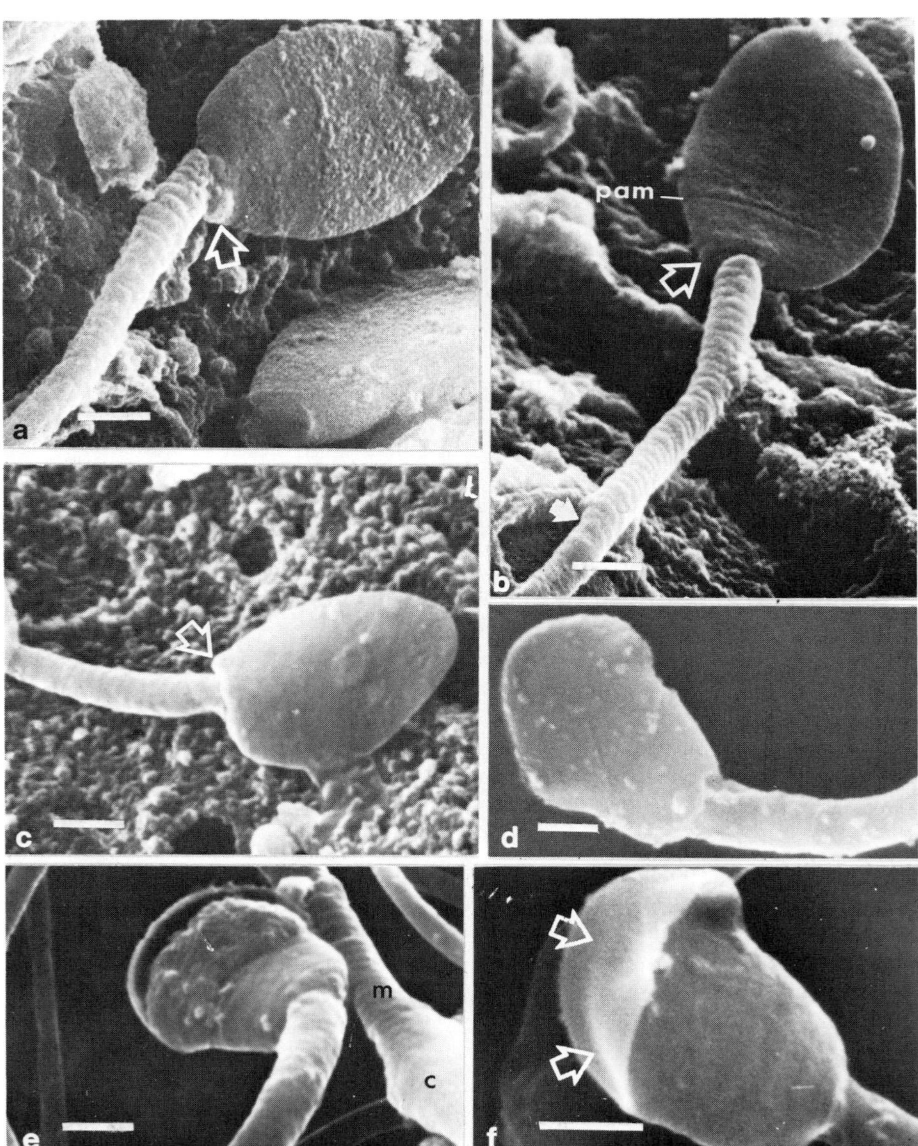

Fig. 3. (a) Gelada "baboon" (*Theropithecus gelada*). In this individual's sperm the acrosome region is delineated from the posterior nuclear region more by a generalized membrane wrinkling than by a specific marginal line. This wrinkled appearance may, however, merely represent the accretion of superfluous material on the sperm surface. Note the eccentric midpiece insertion, seen among baboons generally but apparently not among the other cercopithecids. Note also the knoblike structure (arrow), always located on the inner insertion surface. Mitochondrial gyres are large and wrapped rather loosely around the axial filament complex. The shape of the sperm head is characteristic of all the cercopithecids. (b) Mandrill "baboon" (*Mandrillus sphinx*). This

sperm structure. Bearing in mind the dual requirements of motility and production of gametes in vast number, such a simple, yet metabolically active "no frills" anatomical package would seem evolutionarily attractive. The paddle-shaped head is eminently suited for dense packaging within the male genital tract. Moreover, it is intuitively logical to reduce the acrosome, as a means of transporting essential enzymes, to a minimum size.

For our purpose here, however, it is the minor structural features, and their similarity or dissimilarity between species and families, which are of potential value in assessing an evolutionary link between Old and New World primates. These derived features are less likely to have arisen and been maintained spontaneously in different evolutionary lines than they are to have persisted along one or more branches of a diverging evolutionary tree. The existence of similarities among such derived characteristics may thus suggest a common ancestry of the populations possessing them.

An immediate problem arising in the application of this approach to determine relationships between phyla lies in the identification of those characteristics of sperm morphology which represent the primitive state and those which represent the derived state. Among eutherian mammals the following characteristics are currently regarded as primitive: the head is of relatively large dimension compared to the overall length of the spermatozoa, is flattened dorsoventrally, and has a symmetrical outline; the acrosome is well developed, with either a large apical ridge or a symmetric formation of the acrosomal membrane over the anterior portion of the nucleus; the ratio of the

---

species, though very different phenotypically from the other two baboon species depicted here (a and c), nevertheless has sperm which are almost indistinguishable from the two. The sperm shown here has a distinct posterior acrosomal margin (pam), eccentric midpiece insertion into the head, and large mitochondrial gyres. Note the junction where the midpiece terminates, leaving the axial filament complex of the tail covered only by the plasma membrane (solid arrow). The accessory knob noted in (a) above (arrow) is present but reduced in size (arrow). (c) Yellow baboon (*Papio cynocephalus*). Although the demarcation of acrosome from postacrosomal region is indistinct, the characteristic paddle-shaped head of the cercopithecids in general, and the eccentric midpiece insertion found specifically among the baboon genera of this group, are both easily seen. The accessory knob is clearly visible (arrow). (Reprinted from Martin *et al.*, 1975; with permission of Academic Press). (d) Thick-tailed bushbaby (*Galago crassicaudatus argentatus*). This sperm is of similar maturity to the one in (f), but is pictured from the opposite side to reveal the convex surface of the acrosome, betraying a slight ballooning of this structure anterior to the nucleus. The midpiece appears relatively thicker than in most other primate species due to the presence of redundant membranous material. (e) Thick-tailed bushbaby (*Galago crassicaudatus argentatus*). This nearly mature sperm, with a cytoplasmic droplet (c), demonstrates the presence and amount of extraneous material carried on the concave acrosomal surface, above. This specimen also demonstrates the fine coiling of mitochondria with small cross-sectional area around the midpiece (m). (f) Thick-tailed bushbaby (*Galago crassicaudatus argentatus*). This African lower primate, a member of the *Lorisidae* has spermatozoa with a unique concave acrosomal structure. The cuplike shape shown here is characteristic of an almost-mature sperm. In more immature sperm a prominent anterior ridge is present in addition to the lateral fold, and is lost during maturation; it is barely present in this specimen (arrows). At the same time there is an accumulation of electron-dense material under this acrosomal expansion, the function of which is unknown.

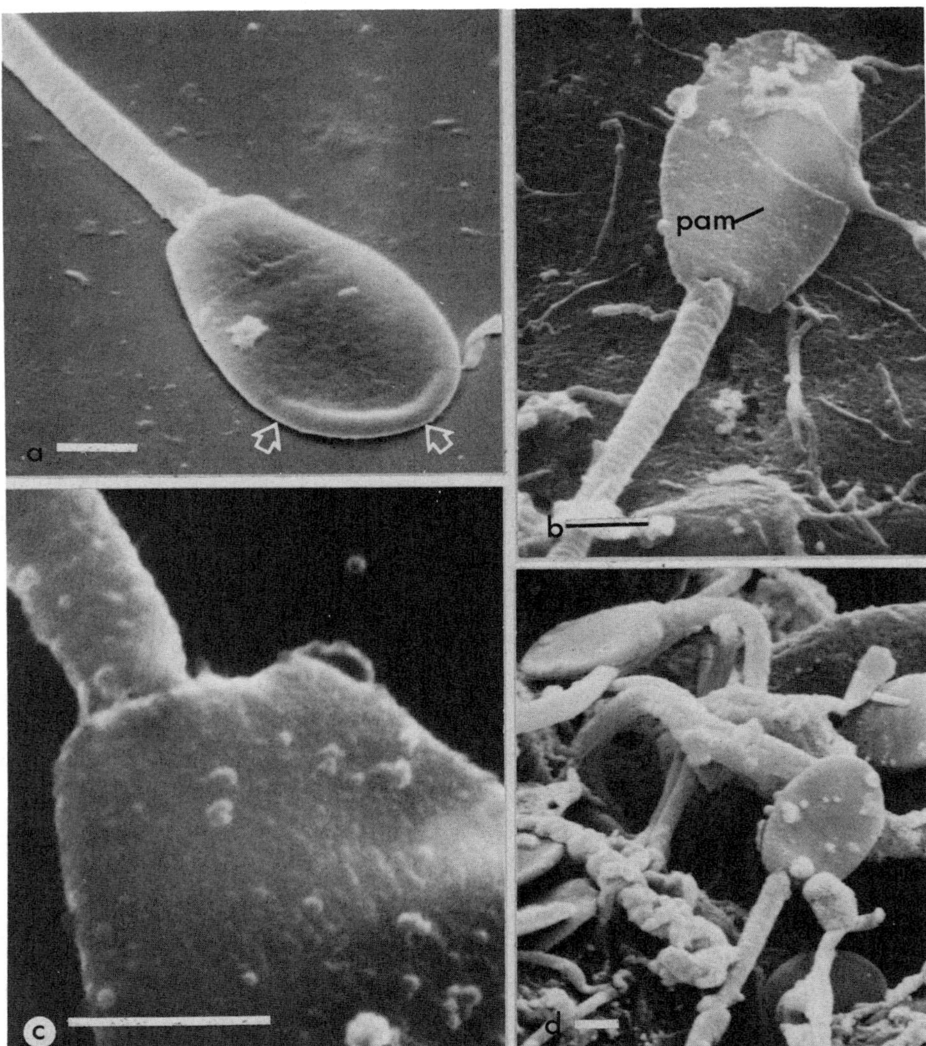

Fig. 4. This plate contains examples of two cebid genera, *Cebus* and *Saimiri*. There is some intergeneric sperm variation within the Cebidae as there is within the Cercopithecidae. (a) The spermatozoa of *Cebus apella* possess the typical primate paddle-shaped head, the acrosome is somewhat reduced and is demonstrated by the anterior swelling (arrows). The posterior acrosomal margin is distinct, the mitochondrial gyres numerous and regular. This midpiece often is inserted eccentrically, but this is barely detectable in this micrograph. (b–d) Squirrel monkey (*Saimiri sciureus*). Spermatozoa of this genus uniformly demonstrate an eccentric insertion of the midpiece. This is most dramatically seen in (c). There is no evidence in these specimens, however, of the accessory knoblike structure visible in the baboons (Fig. 3). (b) A sperm on the surface of ovarian follicle cells in *in vitro* culture. By virtue of the absence here of the anterior ridge otherwise visible in this genus (as in *Cebus*), it appears that the acrosome has been lost, a change in the texture of the surface of the head betraying the position of the postacrosomal margin (pam).

area of the head (nucleus plus acrosome) to the area of the nucleus alone is in excess of 1.2:1. The overall size of the spermatozoa is larger than that seen in more developed genera. Mitochondria, arranged in the midpiece section of the sperm, are numerous, relatively small in individual size and arranged in a regular spiral. The spermatozoa exhibit a uniformity of structure within a given ejaculate and between ejaculates from a given species. These features were utilized in development of the cladogram (Fig. 6) and are defined in Table 1. It must be remembered that some distinctions are made on an arbitrary basis, for example, the nucleus area associated with the primitive state, with guidance from the observed distribution in genera studied. Also, the classification of intermediately derived characters does not imply a uniformity within the group. For example, the presence of asymmetry in head shape is considered a derived character, but not all sperm which are asymmetrical show the same asymmetry, and it would be wrong to equate the two characters. This consideration, together with consideration of results from other phylogenetic analyses precludes the linking of certain genera in Fig. 6. Variations from these primitive characteristics are regarded as being derived. When a number of such derived characters are shared among various genera, this provides evidence for a common generic ancestry. Sharing of primitive characters among genera is of much less importance in this regard, representing, as it does, maintenance of the primitive condition throughout the period of evolution under study. A diagram of several of the features utilized in the current evaluation is provided (Fig. 7).

The structure and number of mitochondria surrounding the axial filament complex (together forming the midpiece) is one such feature. Organization in this region has tended toward one of two extremes. On the one hand, the apes and man all possess spermatozoa with a relatively few large, irregularly spaced mitochondria, frequently terminated with an obvious annular ring. On the other hand, the cercopithecoid and ceboid species often possess many mitochondria with reduced cross-sectional dimension, and arranged in a regular, finely spaced array. Measurement of the number of mitochondrial gyres permits partial identification between genera, and apparently is a stable characteristic within a genus. Such counting of gyres is more easily done using transmission electron microscopy (TEM) (Bedford, 1974), since some forms of tissue preparation for SEM, especially critical-point drying, leave the plasma membrane loosely arranged about the mitochondria.

What inferences can be drawn from the morphological data presented here? Unfortunately the amount of data presently available for New World species is inadequate to permit the creation of a cladistic table sufficiently

---

(d) Spermatozoa in the seminal coagulum, showing a smooth intact sperm surface, an anterior bulge indicating presence of the acrosome, and an eccentric midpiece insertion. As with *Cebus* (a) and *Alouatta* (Fig. 4d) the mitochondrial gyres are fine, numerous, and regular. At this time it is regrettable that so few members of the Ceboidea have been examined using SEM with a view of increasing our knowledge of the intergeneric similarities and differences that exist.

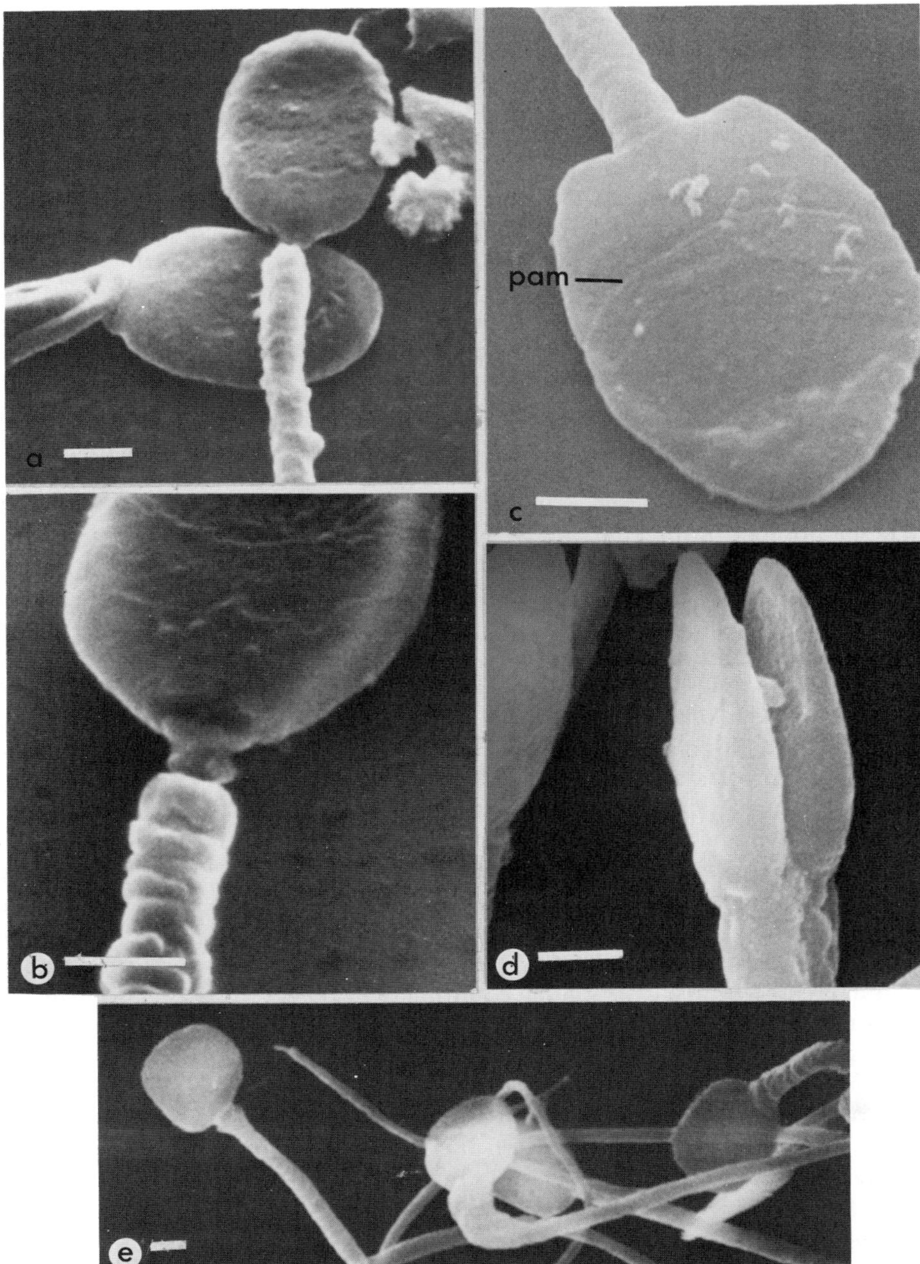

Fig. 5. (a,b) Spermatozoa of the common marmoset (*Callithrix jacchus*). These spermatozoa demonstrate typical features which have been observed in other cebid families. The head is flat and paddle-shaped, and the insertion of the midpiece is definitely central. The mitochondria are

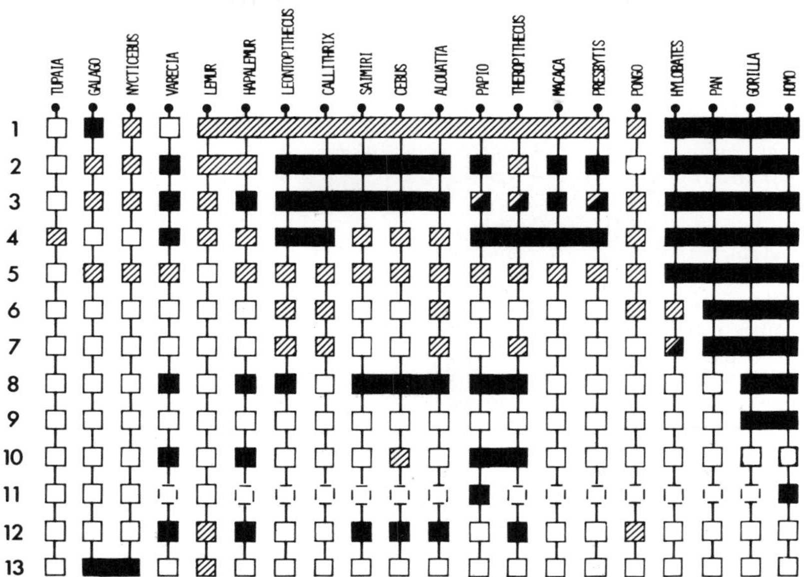

Fig. 6. Cladogram of various primate genera derived from a cladistic analysis of 13 sperm characters listed in Table 1.

complete to allow definite statements of relationships between platyrrhine and catarrhine species. Such evidence as we have clearly shown demonstrates the affinities between *Pan, Gorilla,* and *Homo* and provides evidence of a closer affinity between *Gorilla* and *Homo* than between either of these species and *Pan*. The affinity of *Gorilla* and *Homo* is especially striking with regard to character 9, pleiomorphism, which is only seen in these genera among the eutherian mammals, let alone the primates. *Hylobates* shows affinity with this group, and appears closer to *Pan* than *Homo* and *Gorilla*. As had been sus-

---

more irregular in shape, and occupy a smaller number of gyres than is the case for *Saimiri* and *Cebus*. (c) Black lemur (*Lemur macaco*). This photomicrograph should be compared with that of *Varecia variegatus* (Fig. 2e) from which it differs markedly with respect to sperm head structure. The insertion of the midpiece is central, the overall shape of the head is larger, asymmetrical, and the acrosomal margin is further anterior (pam). The acrosome itself is symmetrically applied to the head, as evidenced by the anterior swelling, in contradistinction to the situation in the Lorisidae. (d) Sperm from *Alouatta* appear to share more characters with *Callithrix* than with *Saimiri*. The midpiece is relatively broad, implying the presence of larger mitochondria than in *Saimiri*. The posterior aspect of the head is somewhat thickened. We do not yet know, as a result of restricted sample size, if this is a uniform characteristic within the genus. (e) Golden marmoset (*Leontopithecus rosalia*). Sperm from this species, also a member of the Callitrichidae, illustrate midpiece and mitochondrial morphology similar to *Callithrix jacchus* (a,b). The midpiece is fairly wide, with a central insertion, and the mitochondrial sheath relatively short. The paddle-shaped head is asymmetrical in outline, a feature reminiscent of certain of the common lemurs (cf. Fig. 5c). More specimens need to be examined from this genus, however, before categoric statements can be made concerning the uniformity of this characteristic.

Table 1. Definitions of Potentially Useful Primate Sperm Characters Utilized in This Study

Character	Primitive	Derived
1. Length (μm)	Longer	Shorter
2. Head area (μm^2)	Larger	Smaller
3. Nucleus area (μm^2)	Larger	Smaller
4. Acrosome area (μm^2)	Larger	Smaller
5. Head thickness (uniform/not uniform)	Thin (uniform)	Thick (not uniform)
6. Mitochondria, number of gyres	Many	Few
7. Mitochondria (organization)	Organized	Not organized
8. Lateral head symmetry	Symmetrical	Asymmetrical
9. Pleiomorphism	Uniform (monomorphic)	Nonuniform (pleiomorphic)
10. Tail insertion	Central	Asymmetrical
11. Agglutination	Positive	Negative
12. Anteroposterior head symmetry	Symmetrical	Asymmetrical
13. Acrosome: symmetry of rostral projection	Symmetrical	Asymmetrical
14. Ratio head/tail	Low	High
15. Ratio midpiece/head	< 2:1	> 2:1
16. Decondensation	Readily decondensed	Not readily decondensed
17. Surface charge	Uniform—head and midpiece	Not uniform

[a] Only the first 13 of these characters have been surveyed widely enough to be of use in the construction of the cladogram appearing in Figure 6.

pected, *Pongo* is removed by some distance from the other apes. The data show a tendency to group the Old World monkeys together as they do the New World monkeys; however, several derived characters can be found which link *Papio* with *Saimiri* and to some extent with *Cebus*. The data tend to suggest a closer affinity between *Saimiri* and *Cebus* than between *Cebus* and *Callithrix* or *Alouatta*. Unfortunate in this regard is a present lack of data concerning *Ateles* and *Lagothrix*. Not surprisingly, *Callithrix* is extremely similar to *Leontopithecus* with the sharing of several derived characteristics. Among characters utilized in this preliminary study, it is intriguing to note a high coincidence of shared characters between *Varecia, Hapalemur,* and *Saimiri*.

The uniformity of sperm morphology among primate species is somewhat surprising in light of research which has demonstrated that sperm morphology in the mouse is a relatively labile characteristic, rapidly alterable by selective breeding. This potential lability in the presence of uniform sperm structure, as yet not demonstrated in primates, implies that selective forces have been directed to maintenance of the basic form, with little pressure for diversification, the genetic coding for altered sperm form being very recessive. This interpretation attributes greater significance to the development and maintenance of minor alterations in sperm morphology when they are observed. Among the African primates considered here, similarity of derived characters between lower primate and higher primate species is observed in regard to *Hapalemur, Varecia,* and *Papio.* Both groups demonstrate eccentric insertion of the midpiece and presence to a greater or lesser degree of the lateral accessory structure adjacent to the tail insertion.

With regard to the New World primates, it is intriguing to note that the cebids, to a greater or lesser extent, also demonstrate the eccentric tail insertion observed in *Papio.* This is most obvious and uniform in *Saimiri* (we have

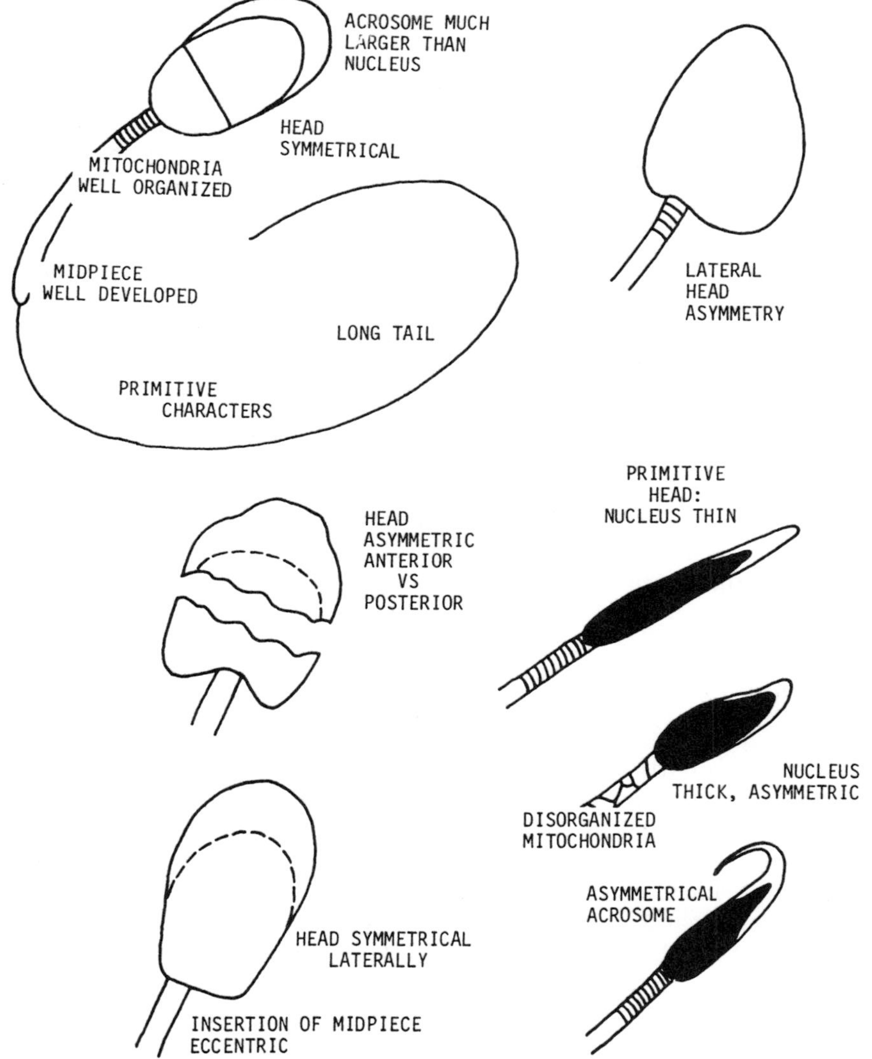

Fig. 7. Diagrammatic representation of some characters used in analysis of comparative sperm structure.

to date only a very restricted sampling from other subfamilies within the Cebidae), but has been noted in only a low percentage of sperm observed from the Callitrichidae.

The data presently available on sperm morphology is rather difficult to interpret. There is a definite indication of a monophyletic origin of most apes and man, as demonstrated by the sharing of nine derived characters (Fig. 6). The status of *Pongo* in this regard is enigmatic, with the repeated disparity of characters effectively divorcing it from the other apes.

At first inspection the cladogram offers strong support for a monophyletic origin of the Old and New World monkeys. There is evidence of sharing of five derived or semiderived characters. The problems of interpretation are, however, demonstrated by the discontinuity of characters 2–5 between genera. This implies that although derived characters, as opposed to primitive characters, are present, the homology of those characters is still open to question; i.e., it is not possible to discount parallelism or convergence as a means of origin.

It is equally evident that no definite distinction can be drawn between the catarrhine and platyrrhine groups. If there were distinct evidence of a polyphyletic origin it would be expected that such a general division would be detectable. At this time it is not practical to join characters 2 and 8 between *Papio* and *Alouatta.* The observed characters do, however, tend to group *Saimiri* with *Cebus,* an observation supporting other reports.

The evidence of the cladogram (Fig. 6), on balance, supports the hypothesis of monophyletic origin of catarrhine and platyrrhine species. If this is interpreted to imply derivation from common lower primate ancestors, then sperm morphology supports a common point of origin which existed well after the initiation of continental drift, i.e., after geographical separation of the continents had occurred. This would favor an east–west migration of ancestral platyrrhines across the South Atlantic Ocean, with subsequent colonization of the South American continent.

Conclusion

The current work does not represent the first attempt to use sperm morphology in assessing evolutionary interrelationships, for it was used many years ago for similar purposes in a study of rodent genera (Friend, 1936). It is expected that sperm morphology will continue to be useful in the future for delineating primate phylogenetic relationships; as additional specimen material becomes available, these will be investigated further, especially with regard to other cebids and lower primates.

Our evidence from sperm morphology could imply an early derivation of the ancestral New World primates from African forms, prior to the isolation of lemuriform primates in Madagascar, with subsequent division of the derived higher primate groups on both continents. To accomplish this, it appears that the most appropriate dispersal route is that of east–west migration of primates via rafting with subsequent colonization of South America. Arguments for and against this possibility based on other evidence have been presented elsewhere (Hoffstetter, 1972; Simpson, 1965).

ACKNOWLEDGMENTS

The authors wish to acknowledge the help of the following persons in providing access to some of the animals used in this work: Dr. Jan Bergeron, Duke University Primate Facility (*Varecia variegatus, Lemur macaco, Galago crassicaudatus*); Kurt Benirschke, M.D., San Diego Zoo (*Alouatta villosa, Leontopithecus rosalia*); David Abbott, MRC Edinburgh (*Callithrix jacchus*); and Dr. L. J. D. Zaneveld, University of Chicago (*Papio cynocephalus*).

References

Allison, A. C., and Hartree, E. F., 1970, Lysosomal enzymes in the acrosome and their possible role in fertilization, *J. Reprod. Fertil.* **21:**501–515.

Bedford, J. M., 1974, Biology of primate spermatozoa, *Contrib. Primatol.* **3:**97–139.

Cohen, A. L., 1977, A critical look at critical point drying—theory, practice and artefacts, *Scanning Electron Microscopy/1977* **1:**525–536.

Flechon, J. E., Kraemer, D. C., and Hafez, E. S. E., 1976, Scanning electron microscopy of baboon spermatozoa, *Folia Primatol.* **26:**24–35.

Friend, G. F., 1936, The sperms of the British Muridae, *Q. J. Microsc. Soc.* **78:**419–443.

Gould, K. G., 1973, Preparation of mammalian gametes and reproductive tract tissues for scanning electron microscopy, *Fertil. Steril.* **24:**448–456.

Gould, K. G., and Martin, D. E., 1978, Comparative morphology of primate spermatozoa using scanning electron microscopy. II. Families Cercopithecidae, Lorisidae, Lemuridae, *J. Hum. Evol.* **7:**637–642.

Gould, K. G., Martin, D. E., and Hafez, E. S. E., 1975, Mammalian spermatozoa: in *SEM Atlas of Mammalian Reproduction* (E. S. E. Hafez, ed.), pp. 42–57, Igaku Shoin, Tokyo.

Gould, K. G., Warner, H., and Martin, D. E., 1978, Rectal probe electroejaculation of primates, *J. Med. Primatol.* **7:**213–222.

Hafez, E. S. E., and Kanagawa, H., 1973, Scanning electron microscopy of human, monkey, and rabbit spermatozoa, *Fertil. Steril.* **24:**776–787.

Hallam, A., 1973, *A Revolution in the Earch Sciences: From Continental Drift to Plate Tectonics,* Oxford University Press, Oxford.

Hoffstetter, R., 1972, Relationships, origins, and history of the ceboid monkeys and caviomorph rodents: A modern reinterpretation, in: *Evolutionary Biology,* Vol. 6 (Th. Dobzhansky, M. K. Hecht, and W. C. Steere, eds.), pp. 323–347, Appleton-Century-Crofts, New York.

Koehler, J. K., and Kinsey, W. H., 1977, Changes in sperm membrane structure during capacitation: A brief review, *Scanning Electron Microscopy/1977* **2:**325–332.

Martin, D. E., and Davidson, M. W., 1976, Differential live-dead stains for bovine and primate spermatozoa, *Proc. Eighth Internatl. Cong. Animal Reprod. Artif. Insem., Krakow* **4:**919–922.

Martin, D. E., and Gould, K. G., 1975, Normal and abnormal hominoid spermatozoa, *J. Reprod. Med.* **14:**204–209.

Martin, D. E., Gould, K. G., and Warner, H., 1975, Comparative morphology of primate spermatozoa using scanning electron microscopy. I. Families Hominidae, Pongidae, Cercopithecidae, and Cebidae, *J. Hum. Evol.* **4**:287–292.

Petter, J. J., 1962, Recherches sur l'ecologie et l'ethologie des Lémuriens Malagaches, *Mém. Mus. Natl. Hist. Nat. Paris, Ser. A* **27**:1–146.

Seligsohn, D., and Szalay, F. S., 1974, Dental occlusion and the masticatory apparatus in *Lemur* and *Varecia:* Their bearing on the systematics of living and fossil primates, in: *Prosimian Biology* (R. D. Martin, G. A. Doyle, and A. C. Walker, eds.), pp. 543–562, Duckworth, London.

Seuanez, H., Carothers, A. D., Martin, D. E., and Short, R. V., 1977, Morphological abnormalities in spermatozoa of man and great apes, *Nature* **270**:345–347.

Simpson, G. G., 1965, *The Geography of Evolution,* Chilton Books, Philadelphia.

Warner, H., Martin, D. E., and Keeling, M. E., 1974, Electroejaculation of the great apes, *Ann. Biomed. Eng.* **2**:419–432.

Evidence from Karyological and Biochemical Studies

VII

The Karyology of South American Primates and Their Relationship to African and Asian Species

19

A. B. CHIARELLI

Introduction

Karyology, the study of chromosomes, is now reaching a level of great detail. Recent studies have yielded a nearly complete knowledge of chromosome ultrastructure and now offer the possibility of interpreting in a reasonable way how the genetic information is organized on them, how it is duplicated, and how it is transferred to individuals of subsequent generations (Du Praw, 1970).

The recently acquired technique of chromosome banding constitutes a projection of this new ultrastructural knowledge into the more traditional study of the chromosomes at the light microscope level (Paris Conference, 1972). The banding technique involves the treatment of chromosome preparations with denaturing agents that, in fact, produce alterations in the chromosome proteins which are the major *structural* components of the chromosomes but have little or nothing to do with the actual base sequence of the DNA molecules. The artifacts on the chromosomes (the bands) most likely reveal the point of attachment of the chromosome filament to the nuclear

A. B. CHIARELLI • Istituto di Antropologia, Università di Firenze, Via del Proconsolo 12, Firenze, Italy. Financial support for this research was provided by the Italian Consiglio Nazionale delle Ricerche (CNR Contract No. 78.02050.04).

membrane (Chiarelli and Brøgger, 1978). This information is of great importance for identifying homologous chromosomes in the same plate, in plates of the same species, and in plates of closely related species, but it is of little use at higher taxonomic levels. The banding *does* appear to show the consistency in the way the chromosome filament is organized but *does not* provide information on the identification of the actual DNA genetic message in the chromosomes such as the more traditional techniques employed in the current study.

The transfer of the chromosome banding homology concept from individuals of the same species or closely related species to phylogenetically more distant species now appears too speculative and unreliable, at least until a proof is provided of a consistent relationship between the banding and the DNA protein complex. It is for this reason that the present discussion of platyrrhine karyology does not utilize recently published data on chromosome banded karyotypes (Dutrillaux, 1979), even if current research in this direction is in progress in my laboratory and collaborative efforts are underway with biological laboratories in Brazil.

Materials and Methods

According to the traditional concept, the importance and interest in the use of chromosomes for taxonomic and phylogenetic purposes relies on three main assumptions. First, chromosomes are linear structures as they appear under the light microscope. Any apparent variation among them can be evaluated, such as a difference in length or in centromere position if evident. Second, chromosomes contain the genetic code (DNA), linearly organized on them. Any variation in chromosome length corresponds to a similar variation in its DNA content. Third, chromosomes are relatively stable structures within the individuals of the same species, due to the constant screening they have to undergo during meiosis. This screening represents a sort of sieve that eliminates anomalous chromosome structures. This stability—which allows relatively few, but easily determinable, possibilities of transgression—represents perhaps the most important element of karyology in determining the systematics and phylogeny of a zoological group. On the basis of the above three assumptions, studies have been undertaken in many different taxonomic groups producing results usually of significant interest for phylogenetic reconstruction (Chiarelli and Capanna, 1973; White, 1973).

Recent research has resulted in a good deal of reliable karyological data for the New World monkeys. These data have been collected by a number of researchers (Bender and Chu, 1963; Bender and Eide, 1963; Bender and Mettler, 1960; Benirschke *et al.*, 1962; Brumback *et al.*, 1971; Chiarelli and Barberis, 1966; Egozcue, 1971; Egozcue and Hagemenas, 1967; Egozcue and Perkins, 1970; Egozcue *et al.*, 1967, 1968*a,b,* 1969; Egozcue and Vilarasau de Egozcue, 1966*a,b,* 1967*a,b*; Hsu and Hampton, 1970; Jones *et al.*, 1973; Kun-

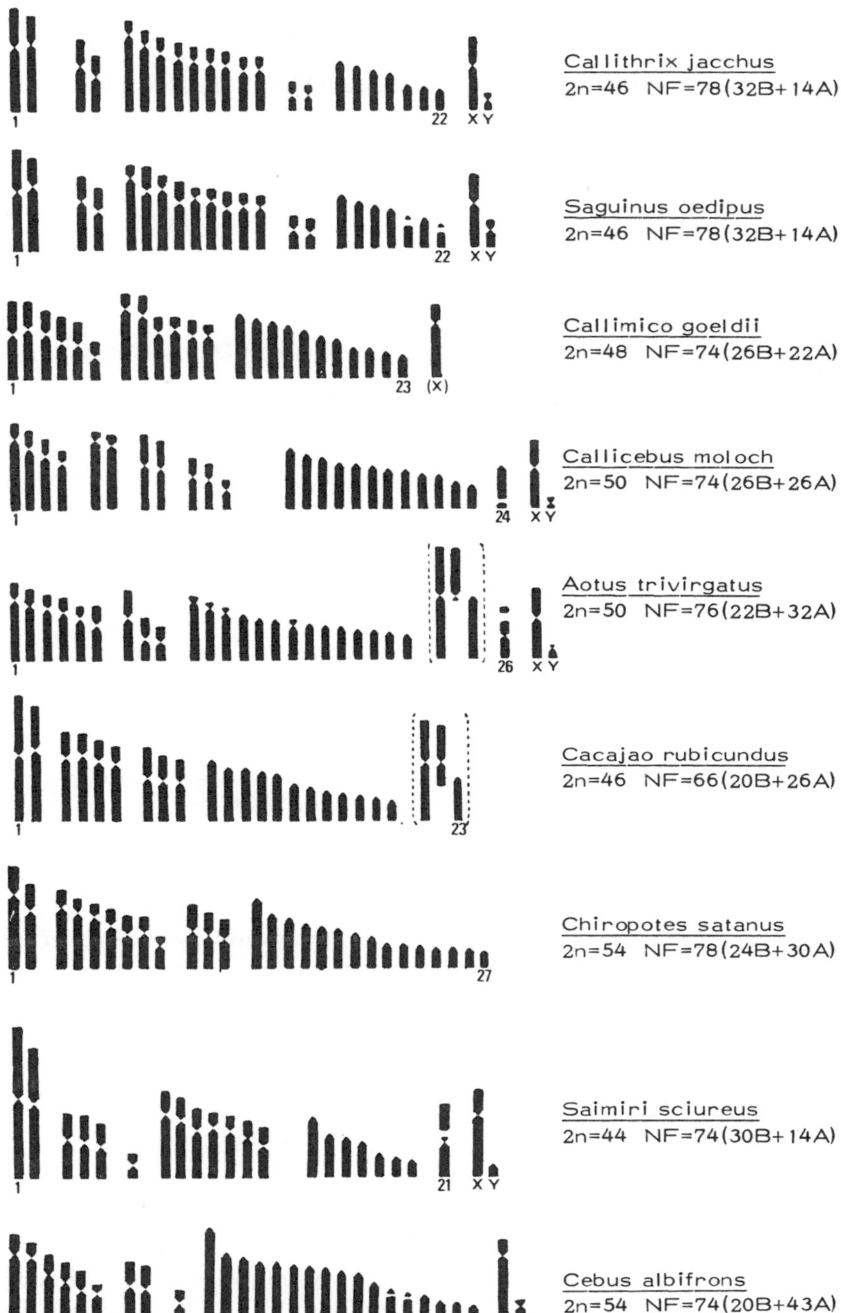

Fig. 1. Idiograms of different species of Platyrrhini and their basic karyotype formulae.

Table 1. Chromosome Number and Karyotype Formulae of the Different Genera of Platyrrhini[a]

Callitrichidae		
Callithrix	46	78 (32B + 14A)
Cebuella	44	78 (32B + 14A)
Leontopithecus	46	78 (32B + 14A)
Saguinus	46	78 (32B + 14A)
Callimiconidae		
Callimico	48	74 (26B + 22A)
Cebidae		
Cebinae		
Callicebus	50	74 (24B + 26A)
Aotinae		
Aotus	50–54	76 (22B + 32A)
Pitheciinae		
Chiropotes	54	78 (25B + 30A)
Cacajao	46	66 (20B + 26A)
Pithecia	48	64 (18B + 28A)
Saimiriinae		
Saimiri	44	74 (30B + 14A)
Cebinae		
Cebus	54	74 (20B + 34A)
Alouattinae		
Alouatta	54	74 (22B + 30A)
Atelinae		
Lagothrix	62	93 (31B + 31A)
Brachyteles	62	88 (13B + 18A)
Ateles	34	66 (32B + 2A)

[a] Karyological data is derived from sources listed in the text.

kel *et al.,* 1973; Low and Benirschke, 1968; Wurster and Benirschke, 1970), but the majority comes from my colleague and friend Bert de Boer, from Utrecht, who has devoted a great part of his scientific activity to the karyology of the New World primates (de Boer, 1974, 1975). In Fig. 1 some idiograms for each genus or group of similar genera within the Platyrrhini are presented. In Table 1 data on the chromosome number and karyotype formulae of New World monkeys is given. A description of each group follows below.

Results

The Callitrichidae have a rather stable karyotype with the $2n$ ranging from 44 to 46 and the fundamental number 78 (32B + 14A). The variation in the chromosome number (from 46 to 44) is due to an evident mechanism of centric fusion between acrocentric chromosomes. The closely related

Callimiconidae consists of a single genus, *Callimico.* Its chromosome number is 48 and its fundamental number is 74 (26B + 22A).

A more complex situation is found in the Cebidae. The different genera belonging to this group, in fact, have a high karyological heterogeneity. For example, two subspecies of *Callicebus moloch* have different chromosome numbers 50 and 46. On the other hand *Callicebus torquatus* has only 20 chromosomes. The fundamental number in *Callicebus,* therefore, does not permit explanation of the karyological variation through possible Robertsonian mechanisms. In this group, as in the well-known case of the Indian muntjac, apparently very complex mechanisms of drastic reduction in the chromosome number are likely to have occurred (Wurster and Benirschke, 1970). In the muntjac the chromosome number varies from 46 in the species *Muntiacus reeversi* to only 7 in *M. muntjak.* The original chromosome formula of *Callicebus* is now judged to have been $2n = 50$ and the fundamental number 74 (24B + 26A). In the Aotinae the genus *Aotus* has an original diploid number of 54 chromosomes, with wide chromosomal polymorphism, which can be attributed mainly to Robertsonian translocation. The fundamental number is 76 with a great number of acrocentric chromosomes (22B + 32A).

In the Pitheciinae the genera *Chiropotes, Cacajao,* and *Pithecia* have chromosome numbers respectively of 54, 46, and 48. Their fundamental numbers are respectively 78, 66, and 64 with many acrocentric chromosomes. The morphological formula for the karyotype of *Chiropotes* is 24B + 30A, for *Cacajao* it is 20B + 26A, and for *Pithecia* it is 18B + 28A. In the Saimiriinae the genus *Saimiri* has a diploid number of 44 and a fundamental number of 74 (30B + 14A) whereas in the Cebinae the chromosome number of *Cebus* is 54; its fundamental number is 74 (20B +34A).

The different species and subspecies of the genus *Alouatta* have chromsome numbers varying from 54 to 44 and a NF varying from 74 to 56. For the species with 54 chromosomes the fundamental number and formula is 74 (22B + 30A). For the genus *Alouatta,* because the number of acrocentric chromosomes is high and constant in the karyotypes, mechanisms of translocation (involving also meta- and submetacentric chromosomes) may have occurred rather than mechanisms of centric fusion. Finally, within the Atelinae there is a surprising reduction of the chromosome number (from 62 to 34), with the fundamental number varying from 93 to 66 (see Table 1). This reduction is evidence that non-Robertsonian mechanisms of variation like those found in *Alouatta* and *Callicebus,* probably took place.

Discussion

If the reductional changes in chromosome number which occurred in the different species of *Callicebus, Alouatta, Ateles, Brachyteles,* and *Lagothrix* are now left aside, the basic chromosome number for the various genera in the whole group of platyrrhine primates ranges from 44 to 62. Furthermore, their fundamental numbers are in the 74 and 78 range and there is never less than 14 pairs of acrocentric chromosomes. These three parameters ($2n$, NF, and minimal number of acrocentric chromosomes) make it possible to characterize a karyotype and permit an easy comparison with karyotypes of other taxonomic groups. By comparing the above-reported generalized formula of the South American monkey karyotype ($2n$ between 44 and 62, NF ranging from 74 to 78, and not less than 14 acrocentric chromosomes) with the karyotype of various lower primate families and Old World catarrhine primates characterized in a similar fashion (Chiarelli *et al.*, 1979), we can identify affinities and foresee what phylogenetic ties exist between these groups (see Tables 1 and 2).

Little similarity exists between the generalized karyotype of platyrrhine monkeys and those of the lower primates here represented by the Lorisoidea, Lemuroidea, and Tarsioidea (Table 2). Turning to the Catarrhini the only group which bears a close resemblence to the platyrrhine generalized formula is the Cercopithecoidea (see Table 2 and Fig. 2). As a consequence, if a phylogenetic relationship had to be determined on karyological grounds, between the New World monkeys and other extant primate groups, the best fit would be with the Cercopithecoidea, In fact, the platyrrhine generalized karotype could be derived directly from the generalized karyotype of the Cercopithecoidea (Fig. 2).

Now, it is well known that the Cercopithecoidea are endemic to the African continent based on the current distribution of Old World monkeys and their rather extensive African fossil record. The ancestors of the Cer-

Table 2. The Karyological Formula of the Platyrrhini Compared with Those of the Lower Primates and the Catarrhini[a,b]

	$2n$	NF	MA
Platyrrhini			
Ceboidea	44–62	74–78	14
Lower primates			
Lorisoidea	38–62	87–102	14
Lemuroidea	44–66	62–70	26
Tarsioidea	80	94	66
Catarrhini			
Cercopithecoidea	42–72	68–112	14
Hylobatidae	44–52	80	0
Hominoidea	46–48	72–80	10

[a] Karyological data for the lower primates and the Catarrhini derived from various sources listed in Chiarelli *et al.* (1979).
[b] The symbols stand for the following: ($2n$) diploid number; (NF) fundamental number; and (MA) minimum acrocentric chromosome number.

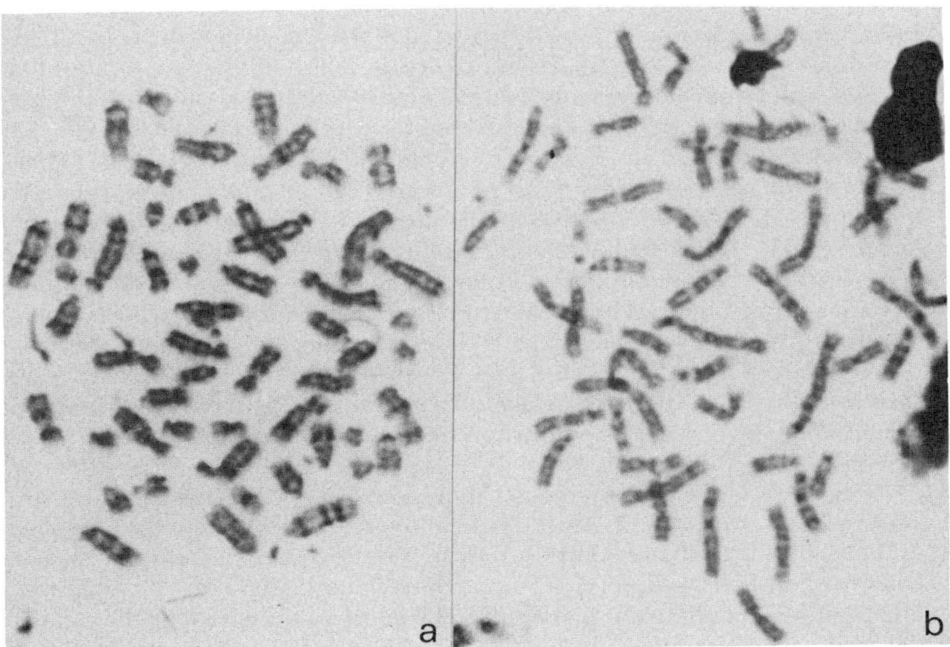

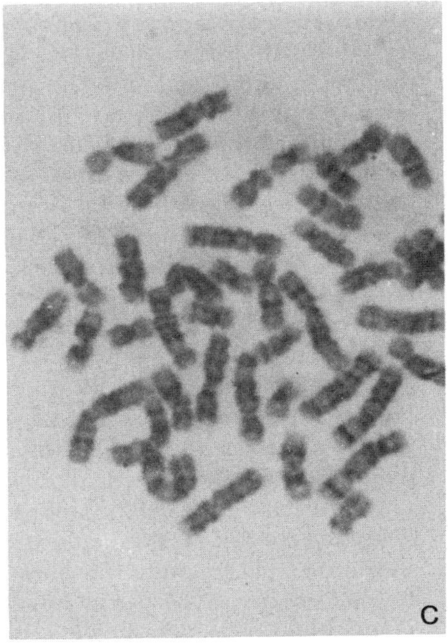

Fig. 2. Banded karyotype of one ceboid primate, *Cebus albifrons* (a) and two cercopithecoid primates, *Cercopithecus talapoin* (b) and *Macaca fascicularis* (c). Note the general similarities in banding patterns among these primates.

copithecoidea were already present in Africa by the early Miocene (Delson, 1975) and even earlier if *Parapithecus* from the Oligocene deposits of the Fayum is considered (Simons, 1969, 1974). It is also interesting to note that the appearance of the earliest primate in South America, *Branisella* (Hoffstetter, 1969; see also Kay, this volume) is roughly contemporaneous with the first record of *Parapithecus* in Africa. With further regard to *Parapithecus,* Genet-Varcin (1963, p. 159) writes "Par ses caractères morphologiques osseux, la mandibule de *Parapithecus* ressemble beaucoup à celle du Platyrrhinien *Callithrix*. . . . " Hence from this perspective there is a strong concordance between the karyological data and morphological observations regarding the phylogenetic affinities of the New World monkeys. Considering the karyological data presented here, perhaps it is best to conclude that the platyrrhine primates are descended from a late Eocene/early Oligocene African early catarrhine stock. In this case an island-hopping route across the South Atlantic Ocean is favored as the appropriate dispersal model for the ancestral Platyrrhini.

There also may be evidence for more than one introduction of ancestral platyrrhines into the South American continent. From studies of the karyological data within the living South American monkeys there appears to be some basis for considering a polyphyletic origin of the Platyrrhini. In spite of an apparent basic similarity in platyrrhine chromosomes revealed by the number of arms, the karyotypes of the South American monkeys present such heterogeneity in the morphology of the chromosomes that it becomes difficult to derive them from one original colonizing form if the time of introduction was about 40 million years B.P. (see Fig. 3). Even if complex mechanisms of dispersion and isolation in the tropical environment of South America were envisaged, it would still be difficult to explain the heterogeneity.

The karyotype of *Callithrix,* for instance, differs from that of *Cebus* to such an extent that an estimate of at least 30 major chromosomal mutations would have to occur in order to explain the change from one karyotype to the other. The magnitude of this change is best illustrated by an out-group comparison. In the study of human karyotypes it was found that 15 major mutations have occurred since the separation of man and the chimpanzee (Chiarelli, 1968) now best estimated at about 10 million years B.P. (Washburn and Moore, 1974). Given an average generation time of 15 years for man/chimp, 10 million years would equal 650,000 generations since their separation with an estimated mutation rate of about one for each 45,000 generations. With regard to *Callithrix* and *Cebus,* if we consider a generation time of five years and 40 million years since separation, the rate of mutation would be one for each 250,000 generations which is more than five times less than the estimated chromosome mutation rate calculated for Chimpanzee and Man. Furthermore, this example still does not take into account the even greater difference in karyotype of the Atelinae whose connection with the rest of the platyrrhines in a monophyletic manner now seems doubtful (as depicted in Fig. 3).

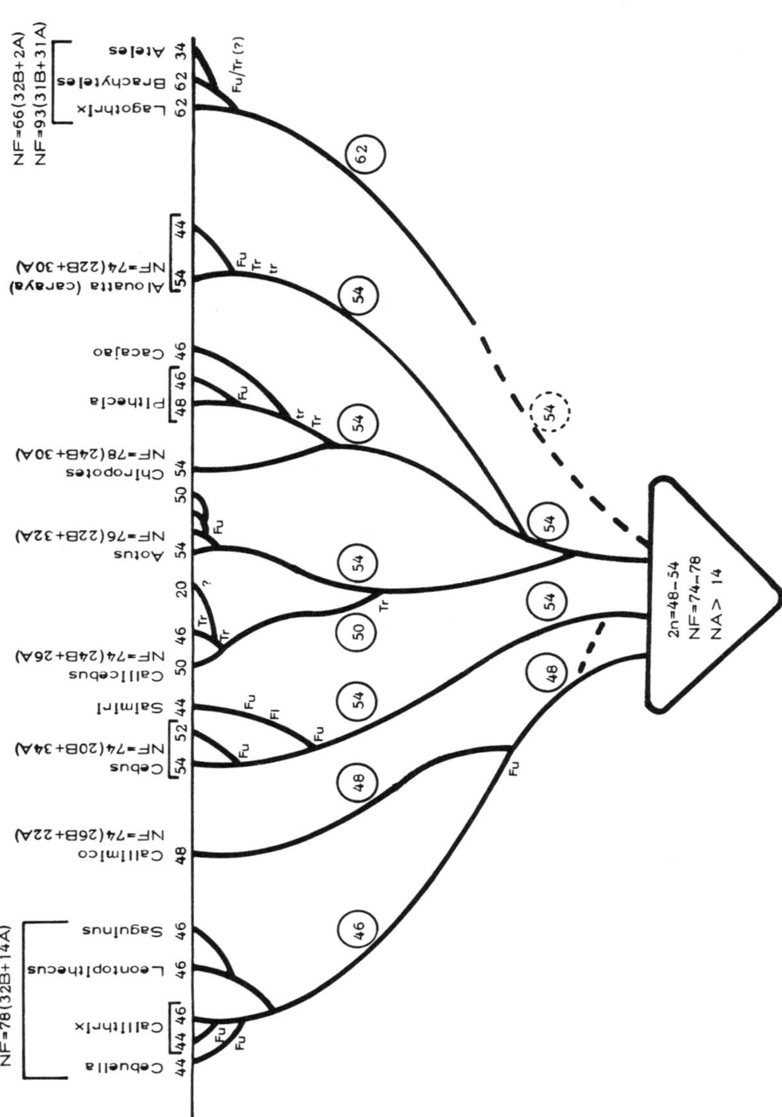

Fig. 3. Tentative reconstruction of the phylogeny of the different genera of the Platyrrhini based on karyological data presented in this study. The top branching can be explained through mechanisms of centric fission (Fi), centric fusion (Fu), pericentric inversion (Pi), translocation of chromosome parts (tr), translocation leading to reduction in chromosome number (Tr) or by more complex mechanisms (?), all with the potential possibility of control, however. The other two levels of interconnection, visualized by the number of chromosomes inside the circle, are more speculative and for the Atelinae appear even impossible on the basis of present knowledge.

Conclusion

The data presented in this study resulting from a gross morphological examination of the New World monkey chromosomes seem to support a polyphyletic origin of the South American primates from the ancestors of a cercopithecoid early catarrhine stock. Since the Cercopithecoidea appear endemic to Africa the karyological data supports an African origin of the Platyrrhini. If the proposal of a polyphyletic origin of the South American primates proves correct then the possibility of more than one rafting from either Africa or North America or even multiple dispersals from both potential source regions would have to be considered (the possibility of deriving the South American primates from *both* Africa and North America seems extremely remote except if some unknown hybridization mechanism has subsequently occurred).

Obviously, new and more refined karyological data are needed to more accurately determine platyrrhine relationships within the Order Primates and to test the possibility of polyphyly within the Platyrrhini. Perhaps a more refined chromosome banding technique could provide more precise information on mutation events at high taxonomic levels. It is to this aim that I and my colleagues, Drs. Saldanha and Koiffman at the cytogenetics laboratories of the University of São Paulo in Brazil, continue our search for new, more powerful, techniques in the study of karyology.

ACKNOWLEDGMENTS

Ms. Chiara Bullo is also acknowledged for providing invaluable assistance during the course of this research.

References

Bender, M. A., and Chu, E. H. Y., 1963, The chromosomes of primates, in: *Evolutionary and Genetic Biology of Primates* (J. Buettner-Janusch, ed.), pp. 261–310, Academic Press, New York.

Bender, M. A., and Eide, P. E., 1963, The karyotype of the woolly monkey (*Lagothrix*), *Mamm. Chrom. Newsl.* **9:**40.

Bender, M. A., and Mettler, L. E., 1960, Chromosome studies of primates. II. *Callithrix, Leontocebus* and *Callimico, Cytologia* **25:**400–404.

Benirschke, K., Anderson, J. M., and Brownhill, L. E., 1962, Marrow chimerism in marmosets, *Science* **138:**513–515.

Brumback, R. A., Staton, R. D., Benjamin, S. A., and Lang, C. M., 1971, The chromosomes of *Aotus trivirgatus* Humboldt, 1812, *Folia Primatol.* **15:**264–273.

Chiarelli, B., 1968, Caryological and hybridological data for the taxonomy and phylogeny of the Old World Primates, in: *Taxonomy and Phylogeny of the Old World Primates with Reference to the Origin of Man* (B. Chiarelli, ed.), pp. 151–186, Rosenberg and Sellier, Torino, Italy.

Chiarelli, B., and Barberis, L., 1966, Some data on the chromosomes of Prosimiae and of New World monkeys, *Mamm. Chrom. Newsl.* **22:**216.

Chiarelli, B., and Brøgger, A., 1978, Superchromosomal organization and its cytogenetic consequences in the eukaryota, *Genetica* **49:**109–126.

Chiarelli, B., and Capanna, E., (eds.), 1973, *Cytotaxonomy and Vertebrate Evolution,* Academic Press, London.

Chiarelli, B., Koen, A. L., and Ardito, G., 1979, *Comparative Karyology of Primates,* Mouton, The Hague.

De Boer, L. E. M., 1974, Cytotaxonomy of the Platyrrhini (Primates), *Genen Phaenen* **17:**1–115.

De Boer, L. E. M., 1975, The somatic chromosome complement and the idiogram of *Pithecia pithecia* pithecia (Linnaeus 1766), *Folia Primatol.* **23:**149–157.

Delson, E., 1975, Evolutionary history of the Cercopithecidae, *Contrib. Primatol.* **5:**167–217.

Du Praw, E. J., 1970, *DNA and Chromosomes,* Holt, Rinehart, and Winston, New York.

Dutrillaux, B., 1979, Chromosomal evolution in primates: Tentative phylogeny from *Microcebus murinus* (Prosimian) to Man, *Hum. Genet.* **48:**251–314.

Egozcue, J., 1971, A note on the chromosomes of *Aotus trivirgatus* Humboldt, 1812, *Folia Primatol.* **15:**274–276.

Egozcue, J., and Hagemenas, F., 1967, The chromosomes of the hooded spider monkey (*Ateles geoffroyi cucullatus*), *Chrom. Newsl.* **8:**12–13.

Egozcue, J., and Perkins, E. M., 1970, The chromosomes of Humboldt's woolly monkey (*Lagothrix lagothricha* Humboldt, 1812), *Folia Primatol.* **12:**77–80.

Egozcue, J., Vilarasau de Egozcue, M., 1966*a,* The chromosomes of *Cebus capucinus, Mamm. Chrom. Newsl.* **20:**71–72.

Egozcue, J., and Vilarasau de Egozcue, M., 1966*b,* The chromosome complement of the howler monkey (*Alouatta caraya* Humboldt, 1812), *Cytogenetics* **5:**20–27.

Egozcue, J., and Vilarasau de Egozcue, M., 1967*a,* The chromosome complement of *Cebus albifrons* (Erxleben, 1777), *Folia Primatol.* **5:**285–294.

Egozcue, J., and Vilarasau de Egozcue, M., 1967*b,* Chromosome evolution in the Cebidae, in: *Neue Ergebnisse der Primatologie* (D. Starck, R. Schneider, and H. J. Kuhn, eds.), pp. 164–166, Fischer, Stuttgart.

Egozcue, J., Vilarasau de Egozcue, M., and Hagemenas, F., 1967, The chromosomes of two species of *Saimiri: S. madeirae juruanus* and *S. boliviensis nigriceps, Mamm. Chrom. Newsl.* **8:**14.

Egozcue, J., Chiarelli, B., and Sarti Chiarelli, M., 1968*a,* The somatic and meiotic chromosomes of *Cebuella pygmaea* (Spix, 1823) with special reference to the behavior of the sex chromosomes during spermatogenesis, *Folia Primatol.* **8:**50–57.

Egozcue, J., Perkins, E. M., and Hagemenas, F., 1968*b,* Chromosomal evolution in marmosets, tamarins and pinches, *Folia Primatol.* **9:**81–94.

Egozcue, J., Perkins, E. M., and Hagemenas, F., 1969, The chromosomes of *Saguinus fuscicollis* illigeri (Pucherian, 1845) and *Aotus trivirgatus* (Humboldt 1811), *Folia Primatol.* **10:**154–159.

Genet-Varcin, E., 1963, *Les Singes Actuels et Fossiles,* University Press, Paris.

Hoffstetter, R., 1969, Un Primate de l'Oligocène inférior sud-américain: *Branisella boliviana* gen. et sp. nov., *C.R. Acad. Sci. Paris, Sér. D* **269:**434–437.

Hsu, T. C., and Hampton, S. H., 1970, Chromosomes of Callithricidae with special reference to an XX/"XO" sex chromosome system in Goeldi's marmoset (*Callimico goeldii* Thomas, 1904), *Folia Primatol.* **13:**183–195.

Jones, T. C., Thorington, R. W., Hu, M. M., Adams, E., and Cooper, R. W., 1973, Karyotypes of squirrel monkeys (*Saimiri sciureus*) from different geographic regions, *Am. J. Phys. Anthropol.* **38:**269–278.

Kunkel, L. M., Heltne, P. G., Borgaonkar, D. S., 1973, Karyotype analyses in the genus *Ateles, Mamm. Chrom. Newsl.* **14:**56–57.

Low, R. J. and Benirschke, K., 1968, Chromosome studies of a marmoset hybrid, *Folia Primatol.* **8:**180–191.

Paris Conference, 1972, *Birth Defects: Original Article Series. The National Foundation* **8:** (7), New York.

Simons, E. L., 1969, The origin and radiation of the Primates, *Ann. N.Y. Acad. Sci.* **167:**319–331.

Simons, E. L., 1974, *Parapithecus grangeri* (Parapithecidae, Old World Higher Primates): New species from the Oligocene of Egypt and the initial differentiation of Cercopithecoidea, *Postilla* **64:**1–12.

Washburn, S. L., and Moore, R., 1974, *Ape Into Man,* Little Brown, Boston.

Wurster, D., and Benirschke, K., 1970, Indian muntjac, *Muntiacus muntjak:* A deer with a low diploid chromosome number, *Science* **168:**1364–1366.

White, M. J. D., 1973, *Animal Cytology and Evolution,* Cambridge University Press, Cambridge.

South American Mammal Molecular Systematics, Evolutionary Clocks, and Continental Drift

20

V. M. SARICH and J. E. CRONIN

Introduction

South America has been an island continent through much of the Tertiary, and the fossil record of many of its surviving mammalian lineages is poor. This state of affairs has made it difficult to develop a coherent picture of the origin, evolution, and systematics of the South American mammalian fauna. In such situations, comparative protein and nucleic studies are of particular utility. The available macromolecular data, though still rather meager, can tell us a great deal about the evolutionary origin and relationships of much of that fauna, and thus yield valuable insights into paleobiogeographical problems, along with testing existing hypotheses derived from paleontological and neontological studies.

We have already written extensively on the origin and intragroup relationships of the New World monkeys (Sarich, 1970; Cronin and Sarich, 1975, 1978; Sarich and Cronin, 1976). In this paper, we will briefly review that material and deal with such molecular information as is available on those

V. M. SARICH • Departments of Anthropology and Biochemistry, University of California, Berkeley, California 94720. J. E. CRONIN • Departments of Anthropology and Organismal and Evolutionary Biology, Peabody Museum, Harvard University, Cambridge, Massachusetts 02138. This research was supported in part by a grant to V.S. from the National Science Foundation (NSF No. 20850).

other groups of placental mammals of relevance to the theme of this volume—Edentata, Hystricognatha, and Cricetidae. We will, in addition, consider the origin of the Anthropoidea in light of the protein sequence and immunological data. Finally, we will apply molecular clock considerations to the relative possible timings of the introductions into South America of the various elements of its fauna, and then attempt a synthesis which should shed appreciable light on questions of geographic origin and relationship to Old World taxa for each of the New World groups.

Utility of Proteins and Nucleic Acids in Evolutionary Biology

Macromolecular data have already proven extremely useful in assessing phylogenetic relationships among numerous taxa. In general, these assessments have been congruent with those developed through traditional comparative anatomical studies. Others have allowed an objective choice to be made among conflicting conclusions drawn from such studies. Examples of the latter include the association of the New and Old World anthropoids (Goodman, 1962), of the pinnipeds with the canoid carnivores (Sarich, 1969), of the giant panda with the bears (Sarich, 1973), of the flying lemur (*Cynocephalus*) with the primate clade (Cronin and Sarich, 1975), and the association of all ratites into a monophyletic clade relative to other birds (Prager *et al.*, 1976). Occasionally the molecular data will provide a result outside the bounds of previous hypotheses, as, for example, the demonstrations that *Pan* and *Gorilla* are no more closely related to one another than either is to *Homo* (Goodman, 1962; Sarich and Cronin, 1977), and that *Rattus* and *Mus* are as distinct from one another in time as the platyrrhines and catarrhines (Sarich, 1972). These latter results point up the fact that controversies have generally developed not so much in the realm of cladistics but about divergence times. The matter of hominid origins is especially noteworthy (Simons, 1976; Walker, 1976; Sarich and Cronin, 1976, 1977).

Primate Origins and Interrelationships

Our conclusions concerning the various extant taxa that might be assigned to the primate clade are given in Figs. 1 and 2. The methodologies and data involved in those constructions have been extensively discussed (Cronin and Sarich, 1975; Sarich and Cronin, 1976; also see the caption for Fig. 1) and need only brief comment here. Exhaustive immunological comparisons involving the albumins of all major groups of mammals clearly associate tree shrews and the flying lemur with forms universally accepted as primates (Lemuroidea, Lorisoidea, *Tarsius,* Anthropoidea), and just as clearly exclude everything else (for example, the various insectivores). In other words, if the universally accepted primates form a clade relative to other mammals, then the tree shrews and flying lemur also belong to that clade. We state our conclusions in this form because the immunological data are of limited utility in developing independent assessments of interordinal relationships among mammals, and appropriate nonmammalian reference species do not exist. Marsupials, for example, are not suitable for assessing intraplacental cladistics because the antisera are made in a placental (the rabbit). Thus, one cannot start from scratch here, but can only hope to place unknowns in relation to some assumed knowns.

On *Cynocephalus,* we have only the immunological data on its albumin and transferrin, and these are concordant in their placement of the flying lemur among the primates. For the tree shrews, on the other hand, the availability of the globin α- and β-chain sequences makes up for the equivocal nature of some of the albumin immunological data. Depending on which series of outside reference albumins is used to assess the amount of change along the *Tupaia* line, one can defend an origin for it at the *Cynocephalus*–other primates node or halfway between that node and the one at which primates associate with other orders such as bats and carnivores. The transferrin data do not improve the situation because this protein has evolved so rapidly as to generally preclude interordinal cross-reactions, and so the globin data, which quite clearly indicate separation of *Tupaia* preceding those among *Tarsius, Lemur, Nycticebus,* and the anthropoids, should be regarded as decisive.

The combination of the available albumin immunological and globin sequence data then suggests that in going back in time from the basic *Cynocephalus*-primate radiation, we encounter the separation of the tree shrews, then various of the nonungulate placentals (bats, carnivores, insectivores), then rodents, ungulates, and, finally, the marsupials. That so many of these can be seen today at the molecular level suggests a substantial temporal existence for each internodal segment with a significant number of amino acid substitutions occurring along it. Ultimately, then, the basal placental radiation can no longer be seen as a Late Cretaceous or Early Paleocene event, but must date back beyond 90 million years ago (M.Y.A.) or so.

Within the primate clade, we cannot at present reliably resolve the details of the separations among the lineages ultimately leading to *Tarsius, Cynocephalus,* Lemuroidea-Lorisoidea, and the Anthropoidea. It is, of course, possible that a divergence situation exists similar to that proposed for *Homo, Pan,* and *Gorilla,* but set so far back in time as to preclude any survival of recognizable derived features to the present at the molecular level. We won't know until a great deal more research, especially with DNA sequences, is carried out with each of the primate and other closely related taxa of concern. We hasten to point out that there is no necessary conflict between this phylogeny, especially as it concerns the flying lemur, and current morphological studies which have not yet produced any derived features consistent with the same pattern. We are all well aware that significant

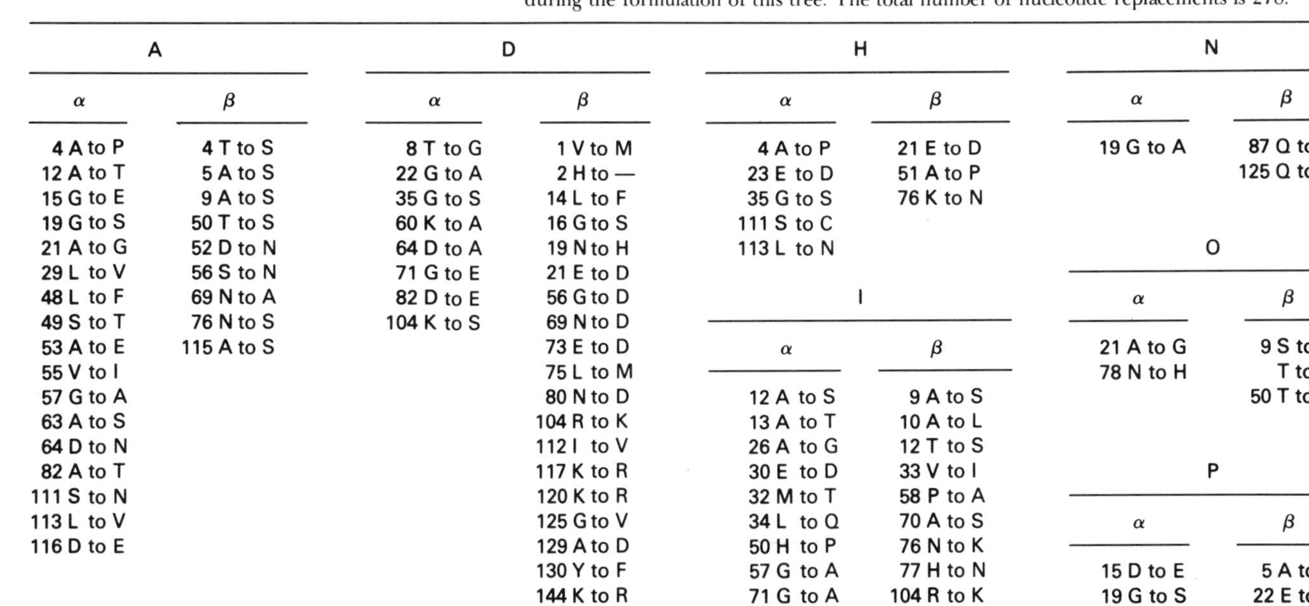

Fig. 1. Placental hemoglobin phylogeny. This tree was developed by ancestral node reconstruction at the codon level using the parsimony criterion. The goal was to see if a cladistic arrangement which lacked certain of the discordant features of the Beard and Goodman (1976) and Beard *et al.* (1976) trees, while remaining equally parsimonious, was possible. Discordance here is measured relative to our immunological data concerning interordinal relationships among the mammals. These discordant features include: (1) the indication of a primate-ungulate clade subsequent to the separations of the rodents and carnivores; (2) the strong association—a lineage with 7 substitutions along it—of *Tarsius* and the Anthropoidea; and (3) the placement of the carnivores outside a rodent-primate clade. The analysis is based upon the following published sequences: *Homo, Gorilla, Macaca mulatta, Macaca fuscata, Cercopithecus, Presbytis, Cebus, Ateles, Lemur, Nycticebus, Tarsius, Tupaia, Erinaceus, Oryctolagus, Bos, Canis,* and *Equus.* The numbers in parentheses along the lineages are the number of nucleotide replacements determined for each lineage whereas the numbers in parentheses after each species name represent the number of substitutions along that lineage from basal node Z.′ The specific nucleotide replacements are detailed below. For maximum clarity we have deleted many of the primate lineages for which sequence data are available. However, all sequences were included during the formulation of this tree. The total number of nucleotide replacements is 278.

A		D		H		N	
α	β	α	β	α	β	α	β
4 A to P	4 T to S	8 T to G	1 V to M	4 A to P	21 E to D	19 G to A	87 Q to T
12 A to T	5 A to S	22 G to A	2 H to —	23 E to D	51 A to P		125 Q to P
15 G to E	9 A to S	35 G to S	14 L to F	35 G to S	76 K to N		
19 G to S	50 T to S	60 K to A	16 G to S	111 S to C			
21 A to G	52 D to N	64 D to A	19 N to H	113 L to N			
29 L to V	56 S to N	71 G to E	21 E to D				
48 L to F	69 N to A	82 D to E	56 G to D				
49 S to T	76 N to S	104 K to S	69 N to D				
53 A to E	115 A to S		73 E to D				
55 V to I			75 L to M				
57 G to A			80 N to D				
63 A to S			104 R to K				
64 D to N			112 I to V				
82 A to T			117 K to R				
111 S to N			120 K to R				
113 L to V			125 G to V				
116 D to E			129 A to D				
			130 Y to F				
			144 K to R				

I	
α	β
12 A to S	9 A to S
13 A to T	10 A to L
26 A to G	12 T to S
30 E to D	33 V to I
32 M to T	58 P to A
34 L to Q	70 A to S
50 H to P	76 N to K
57 G to A	77 H to N
71 G to A	104 R to K

O	
α	β
21 A to G	9 S to T
78 N to H	T to A
	50 T to S

P	
α	β
15 D to E	5 A to G
19 G to S	22 E to D

B

α	β
10 I to V	56 S to G
15 D to E	70 A to S
17 I to V	87 K to A
82 A to D	112 I to V
	116 H to R
	125 Q to E
	126 V to L

C

α	β
15 G to S	2 H to Q
57 G to A	4 T to S
65 A to G	5 A to G
68 K to L	12 T to L
85 D to N	16 G to D
107 V to S	20 V to G
111 S to V	43 E to D
115 S to N	50 T to N
130 A to S	51 A to P
131 N to S	52 D to G
	69 N to H
	72 S to G
	75 L to V
	76 N to H
	111 V to A
	121 E to D
	129 A to S

E

α	β
68 T to K	13 G to A
115 A to S	43 D to E
	69 N to D
	73 D to E

F

α	β
	112 I to C

G

α	β
5 A to T	10 A to L
8 T to A	19 N to K
12 A to T	23 V to F
13 A to F	50 T to S
17 I to L	56 S to G
21 A to G	65 K to A
26 A to G	69 N to Q
30 E to D	71 F to M
34 L to Q	72 S to G
35 S to A	75 L to I
36 F to H	77 L to N
49 S to N	86 A to S
50 H to P	116 H to R
71 G to N	121 E to D
72 H to N	125 Q to A
76 L to V	126 V to A
111 S to L	130 Y to F
131 S to T	143 H to A
133 S to A	

α	β
89 H to Y	
115 A to T	
116 D to E	
129 L to F	
131 S to V	

J

α	β
78 G to T	5 A to G

K

α	β
10 I to V	13 G to A
17 I to V	56 S to G
73 L to V	76 N to D
76 L to M	
78 T to N	
97 N to D	

L

α	β
68 T to N	9 A to S
71 G to A	43 D to E
	69 N to S

M

α	β
15 D to G	5 G to A
23 D to E	A to P
111 C to A	69 S to G
113 H to L	76 D to A
116 D to E	87 K to N

α	β
57 G to A	50 T to S
71 A to S	52 D to S
78 N to S	76 D to N
	112 C to V
	121 E to D
	128 A to S

Q

α	β
53 A to S	5 G to A
73 V to I	6 E to D
129 L to V	19 N to T
	21 D to E
	22 E to D
	52 D to A
	58 P to A
	73 D to E
	75 L to M
	76 D to A
	139 N to T

R

α	β
5 A to G	4 T to S
8 T to S	19 N to D
15 D to G	20 V to L
23 D to Q	21 D to E
48 L to M	22 E to K
57 G to A	33 V to I
67 T to S	43 E to D
115 A to G	50 T to S
	69 N to T

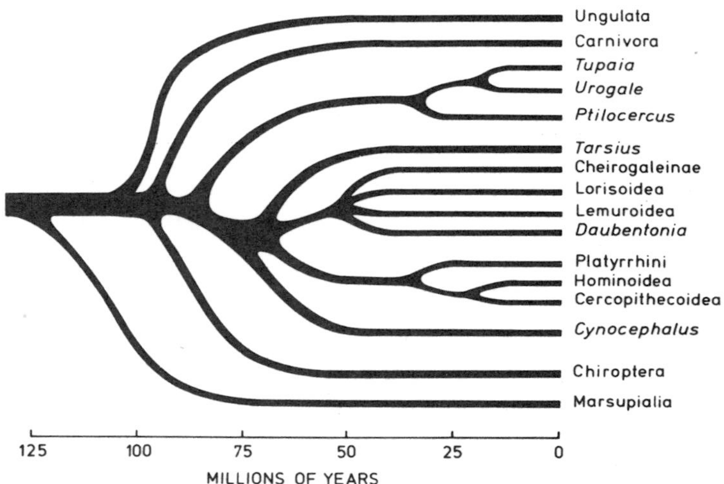

Fig. 2. A mammalian molecular phylogeny. Using the framework developed in Fig. 1, we have added various of the relevant mammalian groups whose placement is relatively secure on the basis of albumin and transferrin immunological comparisons. The Insectivora are a monophyletic clade sharing a lineage with the nonungulate group and diverging at the Chiroptera–Carnivora node.

morphological adaptive change can occur very rapidly, so that a million years may be a long time by those standards and yet a very short one when it comes to documenting the existence of that million years through the molecular changes which will have occurred during them. If indeed the petrosal bulla and carotid circulation of the "true" primates (see Bugge, this volume) are each homologous, derived features unique to them, the molecular picture may simply mean that these developed rather rapidly over a common lineage that existed for only a brief time (brief in geological terms, that is).

Similar problems are associated with efforts to resolve the remaining trifurcation into a series of biologically necessary, though potentially unresolvable, binary divergences. The intellectually current problem is the position of *Tarsius*. The albumin–transferrin picture we have developed is clear enough on this point in not allowing a statistically significant (say more than 5–10 albumin plus transferrin units to change or 2–4 M.Y.) association of any pair of the *Tarsius*–Strepsirhini–Anthropoidea trio to the exclusion of the third. Goodman and his co-workers have argued over the years, again based on their serum protein immunodiffusion comparisons, for a significant *Tarsius*–Anthropoidea association into a haplorhine clade. The problem with that argument is that their data, in contrast to ours, have not been rate-tested; that is, they have not been analyzed cladistically. Thus the marginally greater affinity that Goodman's antisera to whole *Tarsius* serum show for the anthropoid species tested (mean distance to 18 anthropoids = 88.6 ± 6.7; to 7 lower primates = 102.1 ± 2.6) could easily be due to slightly less change in either the *Tarsius* or anthropoids relative to the lower primate lineages tested. Indeed, using Goodman's *Tarsius* arguments on his own data, we could show that it is also more closely related to the lower primates, as eight antisera to whole lower primate sera show stronger cross-reactions with *Tarsius* than with the anthropoids. Eventually this may all be seen as the trivial problem in antiserum nonreciprocity that it probably is, but we emphasize the broader point that one cannot draw cladistic conclusions from data that are not analyzed cladistically.

The available amino acid sequence data make only a marginally positive contribution. The problems are twofold. First, the majority of the data relevant to the problem are derived from proteins entirely unsuitable to the questions involved. They evolve much too slowly to provide precise resolving power. The hemoglobin rate of substitution is about one per lineage per 2–3 M.Y. It is therefore to be expected that lineages could exist for substantial periods of time (2–9 M.Y.) without substitutions occurring. These lineages would of course be invisible. Even more serious are the unanswered questions of the statistical analysis of sequence data. Parsimony is certainly a necessary criterion, but the question of how many total substitution differences between two solutions are necessary before they can be considered statistically different remains open. It is obvious that the amount of demonstrated homoplasy in the globin data should give one pause before attempting to choose among solutions that may differ from one another by only a few substitutions among hundreds.

In Fig. 1, we present the most parsimonious solution that we have been able to achieve from the available hemoglobin sequence data—with the constraint that the solution also be congruent with other cladistic considerations. Thus, we can recognize a basic dichotomy between ungulates and nonungulates, with the rabbit falling in the former clade. The dog and the shrew fall in the latter clade. Once the rabbit and dog are placed, the positions of *Tupaia* and *Tarsuis* become much clearer. The *Tupaia* line (Cronin and Sarich, 1980) shares only 2 nucleotide replacements with the primates after the divergence of the carnivores. The primates share 10 substitutions before *Tarsuis* separates. This is a somewhat earlier separation than we have previously argued (Sarich and Cronin, 1976). Our more recent measurements of the amount of change along the *Tupaia* albumin lineage has also indicated an appreciable common ancestry for the primates subsequent to the separation of the tree shrew line (Cronin and Sarich, 1980).

It is interesting to note that the various lineages in Fig. 1 have accumulated quite similar numbers of nucleotide replacements since the origin of the eutherians. The range is from lows of 33 and 34 for the rabbit and dog to 50 for the horse and hedgehog. The mean is 41.3 and the standard deviation is 6.29 for all 10 species listed. No particularly large rate differentials are noticed. A figure of 2–2.5 M.Y. per NR averaged across the α and β genes gives us a very reasonable 80–100 M.Y. for the basal node of the modern placentals of Fig. 1, much as discussed previously (Cronin and Sarich, 1978, 1980).

The essential point, then, is that an association between any pair of those four primate lines for brief periods of time (say less than 3 M.Y.) is entirely consistent with the molecular data currently available. These data do, however, seriously constrain the possible ranges in time during which any shared morphological features associating such pairs are to be derived. They will thus force a very close working relationship among paleontologists and neontologists to document appropriate derived features existing at appropriate times. For example, Szalay (1975) and Luckett (1975) suggest that there is a number of derived morphological features uniting *Tarsius* with the Anthropoidea. In contrast, Gingerich (1978*b*) and Schwartz *et al.*, (1978) argue for a derivation of the Anthropoidea from an adapid stock, feeling that the various characters listed by Szalay and Luckett are parallelisms or convergences. Thus their sister-groups are Anthropoidea and Lemuriformes with *Tarsius* being an earlier offshoot of a plesiadapiform clade. We do not presume to judge the relative merits of these morphological arguments. It may well be that unequivocally derived morphological characters resolving the basal primate 3- or 4-way adaptive radiation will be as hard to come by as molecular features equal to the same task. We would consider it only intriguing, and in no sense definitive, that one can draw a parsimonious globin tree supportive of the Gingerich–Schwartz–Tattersall–Eldredge position.

Molecular Systematics of the Anthropoidea

The immunological data show that the Anthropoidea share, subsequent to the divergence of the various lower primate lineages, a common ancestry of some 70 units of albumin and transferrin change. This is just about half the average 135–145 units of change seen along the various major primate lineages, including Anthropoidea, since their basal radiation (Cronin and Sarich, 1975; Sarich and Cronin, 1976). We must, therefore, conclude that the catarrhine–platyrrhine separation must have occurred about half as long ago as the basal nontupaiid primate radiation; that is, about 35–40 M.Y.A. Other molecular data (DNA annealing, globin sequence, lysozyme, immunological) are quantitatively consistent with this picture. No protein or nucleic acid datum as yet fails to associate the catarrhines and platyrrhines as a highly derived unit relative to any lower primate.

The length of this common ancestral lineage makes it highly probable that the latest common ancestor of the catarrhines and platyrrhines had already reached a monkey grade of evolution prior to the separation of the two groups. In other words, the New and Old World monkeys had a common ancestor that was a monkey. While no fossil North American primate can be considered of anthropoid grade, there are several genera of undoubted anthropoids in the Oligocene of Africa (e.g., *Apidium, Parapithcus*) that show dental, cranial, and postcranial similarities to modern platyrrhines (Conroy,

1976; Gingerich and Schoeninger, 1976; Gingerich 1978*a*). We suggest that the Fayum contains some forms retaining primitive anthropoid postcranial and dental traits, while others show derived early catarrhine features such as the specialized molar cusp patterning seen in *Propliopithecus* and *Aegyptopithecus*. It is not, then, a very large logical jump to envision, some 5–10 M.Y. earlier, a separation of a somewhat more primitive form leading to the modern platyrrhines—a group which retains substantial portions of the ancestral anthropoid morphology.

As already noted, Gingerich and Schoeninger (1976) see dental and cranial similarities suggesting anthropoid-European adapid affinities. Recent additional anthropoid grade material from the late Eocene of Burma also shows certain similarities to both the Fayum anthropoids and other adapids (Ba Maw, *et al.*, 1979). The origin of the anthropoids could well be Asian or African, and, as Gingerich and Schoeninger (1976) note, the late Eocene-early Oligocene was a time of significant Holarctic mammalian evolution and migration. This could place an appropriate protoplatyrrhine anthropoid stock in North America at a time required by the molecular data, though clearly no such form is present in the known fossil record. Thus a northern origin for the platyrrhines remains a possible, though appreciably less likely alternative to the African ancestry argued for recently by Sarich (1970), Hoffstetter (1972, 1974), and Cronin and Sarich (1975).

Other South American Mammal Groups

The variety of mammalian orders in South America, along with the substantial uncertainties concerning certain of their origins, suggests that the next step in this study should be an inquiry into interordinal relationships among placental mammals in general. Unfortunately, it is just at this level of inquiry that the currently available molecular data have the least to contribute. The distances involved are far too large for DNA annealing and electrophoretic comparisons to be usefully made, and there is a distinct paucity of amino acid sequence data (the relevant mammal data are given in Fig. 1). We do have a large body of albumin immunological data, but, as already mentioned, there are serious interpretative problems involved with their use.

Rodentia: Rates of Change

The rodents, because of their enormous diversity, poor fossil record, and certain consequent biases as to their antiquity, present us with the largest problems. First, where do the rodents fit among the placental mammals? Here the available globin data (including, it should be noted, only one rodent, *Mus*) indicate (Fig. 1) a very early separation from the line later to give rise to tupaiids, carnivores, and primates (no bat, insectivore, or edentate data are available). This placement allows some measure of intrarodent analysis to proceed. Here all we have are the albumin immunological and some few DNA annealing comparisons, and what strikes one first in those data are the very large distances relative to the expectations of most rodent systematists. Thus, for example, the *Rattus–Mus* albumin and DNA distances are comparable to, or larger than, those between catarrhines and platyrrhines, even though a divergence time between the two rodent genera of less than 10 M.Y. has been generally accepted, leading to some unfortunate consequences in the interpretation of recent molecular data (see, for example, Kohne, 1975). We find even larger distances among the New World cricetids, even though their radiation has been seen as occurring within the last 5 M.Y. or so (see, for example, Patterson and Pascual, 1972; Mares, 1975). Such observations have naturally led to the suggestion of accelerated genomic evolution among the rodents, attributed to their short generation lengths. It is, therefore, important to attempt independent estimates of the rates of genomic evolution among the rodents relative to other mammalian lines.

Some time ago, one of us addressed this question in a *Homo–Pan, Rattus–Mus* context (Sarich, 1972), and showed that the albumins of these four, along with those of several other primates and rodents, gave very similar cross-reactions with antisera to three carnivore albumins (*Ursus, Genetta, Hyaena*), thus indicating similar amounts of albumin change along these various lineages from their origin to the present. Indeed, *Mus* albumin gave the strongest cross-reactions, and this indicated conservatism has been confirmed recently using antisera to a number of rodent albumins, where it reacts best among all the murids tested. In this respect, *Mus* albumin is like that of *Aotus* among the simians or *Ursus* among the carnivores. Rodent albumins are, then, on the whole, no more changed from their ancestral sequences than those of the higher primates, and variation in amounts of change along the various rodent albumin lineages sampled seems to be no greater than that found in other vertebrate taxa. The hemoglobin sequence data clearly require an origin for the protorodent lineage significantly predating the protoprimate–protocarnivore separation, giving the rodents a potential existence of at least 80 M.Y., and the incredible diversity of modern forms suggest that we may be able to adequately date divergence events within the order. Given that broad outline, we can begin a consideration of hystricognath and cricetid systematics.

Hystricognatha

The hystricognath data set is appreciably the more extensive and complete one, and those data are presented in Table 1, with the derived phylogeny in Fig. 3. First, note that all the hystricognaths tested, South American and African, clearly form a monophyletic unit relative to other

Table 1. Quantitative Precipitin Cross-Reactions among the Various Hystricognath Albumins[a]

	Antisera									
	0.75	0.75	1.26	0.92	0.98	1.07	1.03	1.07	1.07	Rate
Antigens	*Ca*	*Hy*	*Ba*	*Er*	*Da*	*Ho*	*Cav*	*Ch*	*Cap*	tests
Capybara	100	33	28	36	38	26	32	45	32	+5
Hystrix	29	100	57	42	36	25	37	56	31	−6
Bathyergus	24	44	100	30	31	39	35	32	24	+2
Erethizon	34	39	29	100	48	32	42	52	33	−10
Dasyprocta	37	38	32	45	100	28	43	55	34	−1
Hoplomys	27	25	32	30	28	100	25	34	38	+7
Cavia	37	40	25	29	36	23	100	39	32	0
Chinchilla	38	52	40	54	47	34	49	100	38	−7
Capromys	32	29	41	40	30	29	36	35	100	−3
Myocastor	32	37	38	44	36	62	37	43	45	−2
Petromus	16	27	47	16	25	30	15	21	20	+7
Octodon	40	43	39	48	40	51	39	41	39	−1

Anti-*Capromys*		Anti-*Hoplomys*		Anti-*Dasyprocta*	
Geocapromys	70	*Proechimys*	86	*Cuniculus*	51
Plagiodontia	60	*Diplomys*	54	Anti-*Erethizon*	
		Cercomys	49	*Coendu*	95

[a] All cross-reactions are reported as the percentage of the amount of precipitate given by the homologous antigen. The number above each of the column headings is the factor by which each of the measured cross-reactions given by that antiserum was multiplied to give the reported values. This is the correction for nonreciprocity. The numbers in the last column are the amounts by which the albumins reacted more or less well with a series of nonhystricognath albumin antisera (carnivores and primates) relative to the reaction given by *Cavia*. Nonreciprocity is 3.6% and the F value of the tree calculated from these data is 4.5%. These are the lowest values we have ever obtained for so large a data set.

rodents. Over that period of monophyly we observe about one-third as much albumin immunological change as along an average hystricognath lineage since that separation. The period of common ancestry for the two rodent suborders, on the other hand, cannot be seen as very extensive, as the albumin immunological distances between them tend to approach those between rodents and primates or bats. These considerations bear on the dating of the hystricognath adaptive radiation and consequent continental drift issues. Average immunological distances among the albumins of the four major rodent lineages we have sampled (Hystricognatha, Myomorpha, Geomyoidea, Sciuroidea) are about 180 units (precipitin cross-reaction values of 10–15%), and tests with outside reference species among the carnivores show similar cross-reactions for all four. Though this body of data could certainly be more extensive, there is no indication at present to preclude the formulation of a rodent albumin clock—where the average rodent albumin has accumulated about 90 units of change from the ancestral protorodent condition to the present. As already noted, rodent origins date, by molecular criteria, to at least 80 M.Y.A., and so a good approximation for the rodent albumin rate of

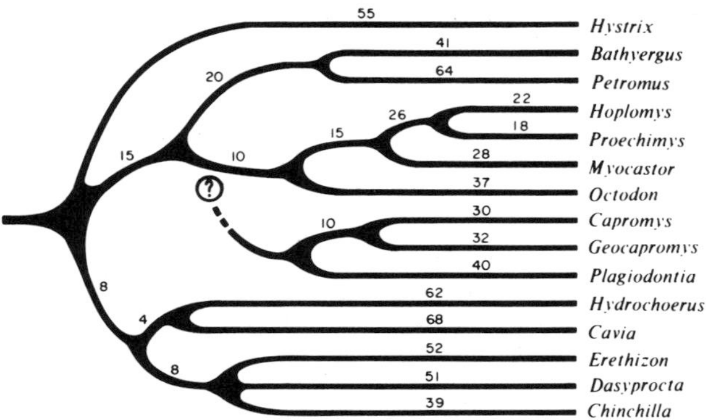

Fig. 3. A molecular phylogeny of the Hystricognatha. The numbers on the lineages represent the amount of albumin change (measured in immunological distance units) which we see as having occurred along them. Each percent of difference in quantitative precipitin cross-reactions is equal to approximately two immunological distance units in microcomplement fixation work. The use of the latter technique was of course impossible because the usual source of complement is guinea pig (*Cavia*) serum. The placement of the "hutias" is ambiguous given our present data, and one could place the origin of the group anywhere along the lineage linking the Old and New World octodontoids. Out best time estimate for the ancestral hystricognath node is 60–65 M.Y.A.

change would be in the area of 1.1 units of change per lineage per million years. The various hystricognath lines average 69 ± 10 units of change since the beginning of their adaptive radiation, thus placing that beginning around 60–65 M.Y.A. This is also consistent with the 20 M.Y. or so which would be calculated for the length in time of the common hystricognath lineage, given the lack of any extensive period of common ancestry indicated for the rodents as a whole.

This extreme antiquity makes at least some of the implications of Fig. 3 easier to accept. Note that neither the African nor the South American hystricognaths form a monophyletic unit relative to the other. Thus Simpson's superfamily Octodontoidea contains both Old and New World forms, an association consistent with our grouping of *Petromus, Thryonomys,* Echimyidae, and Capromyidae in a single clade—except that we also see *Bathyergus* as a member. This pattern is not consistent with the idea of a single migration, in either direction, between Africa and South America, and raises the possibility that the hystricognath adaptive radiation was developing at a time when the two continents were still close enough (along the Brazil–West Africa axis) to make trans-Atlantic contact a reasonable proposition. The simplest model would have an African basic adaptive radiation of the group with at least two lineages providing emigrants to South America. We exclude a South American origin because of a total lack of hystricognaths in Australia and North America (except, of course, for the recent migrant *Erethizon*). The first of these is now extinct in the Old World with its New World descendants seen in a single clade containing the superfamilies Erethizontoidea, Cavioidea, and

Chinchilloidea. The other is represented today in the Old World by *Thryonomys, Petromus,* and *Bathyergus* and in the New World by two clades which we cannot confirm at present as having a period of common ancestry subsequent to their separation from the African forms. The two are Echimyidae plus *Octodon* and *Myocastor,* and the "hutias" (other capromyids). The alternative model, involving a more or less simultaneous invasion of both South America and Africa by several protohystricognath lineages with the subsequent extinction of all of these in the north, would have to be seen as the less probable on simplicity considerations, to which would have to be added the lack of such forms in the fossil record. Of the African forms that have been suggested as possible members of a broader Hystricomorpha, *Pedetes* clearly is not, and we have not yet obtained any ctenodactylid or anomalurid samples.

Edentata

The other clearly ancient group of South American placentals includes the sloths, anteaters, and armadillos. The data of Table 2 (see Fig. 4) argue, though not unambiguously, for a common South American ancestry of the

Table 2. Quantitative Precipitin Cross-Reactions Involving Edentate Albumins

Genus	Anti-*Bradypus*	Anti-*Tamandua*	Anti-*Cabassous*
Bradypus	100	18	28
Choloepus	70	20	27
Tamandua	20	100	22
Myrmecophaga	22	74	22
Cyclopes	26	50	24
Cabassous	25	28	100
Dasypus	29	19	60
Felis	10	36	35
Loxodonta	20	20	17
Orycteropus	14	20	32
Aotus	21	22	28
Homo	10	3	19
Paramanis	10	16	22
Cavia	18	9	14
Procyon	9	14	27
Equus	10	13	25
Syconycteris	8	19	23
Bos	10	9	11
Sus	14	10	9
Trichechus	6	9	2
Balaenoptera	10	7	15

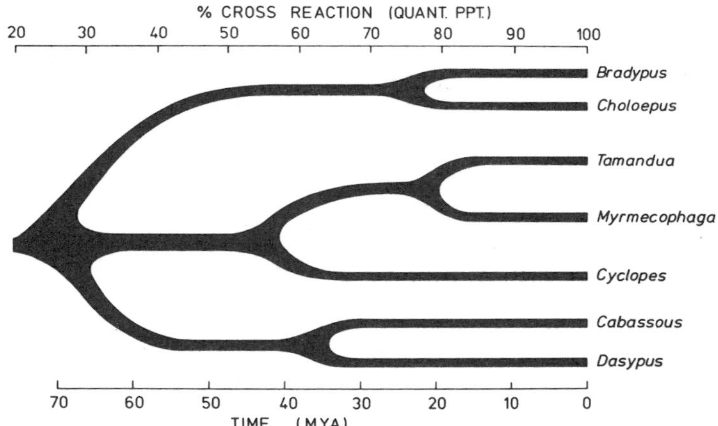

Fig. 4. An Edentata albumin phylogeny. This is a straightforward rendition of the data in Table 2. The tree is constructed according to methods detailed in Sarich and Cronin (1976).

three and their monophyly relative to other placentals. Note that the cross-reactions among the albumins of the three lines average 23%, which is significantly more than that given by any other albumin save some of those we know to be less changed on the basis of other data (*Aotus, Felis, Elephas, Orycteropus*). This is by no means definitive evidence, but it is the best we can do at this time in the absence of any reliable way of comparing the amounts of albumin change among the edentates relative to other placental orders. There is no indication that the pangolins (*Manis*) or aardvark (*Orycteropus*) have any special relationship to the edentates. Again the basic single adaptive radiation was a very early event in the 65–70 M.Y.A. range. The upper bound here, even more than with the hystricognaths, is set by the absence of edentates in Australia—where the marsupials and hylid frogs arrived from South America some 70–75 M.Y.A. (Maxson *et al.,* 1975).

Among the edentates, cladistic analysis does not support the reality of a clade Xenartha (sloths and anteaters), as the three lines are seen as diverging from one another at about the same time. This synchrony derives from the fact that the greater similarity of armadillo albumins to those of sloths and anteaters is paralleled by their greater similarity to placental albumins in general (about 5–6% stronger cross-reactions in both cases). In other words, armadillo albumins have changed somewhat less from their ancestral condition, which naturally increases their similarity to all other albumins relative to those shown by the albumins of sloths and anteaters.

Cricetidae

The final nonprimate group of interest contains the New World cricetids ("hesperomyines" and microtines). Here the molecular data are adequate to

Table 3. Microcomplement Fixation Immunological Distances among Some Cricetid Albumins[a]

Genus	1.09 *C*	0.69 *S*	0.9 *P*	1.1 *On*	1.4 *Or*	0.67 *M*	1.26 *T*
Calomys	0	61	58	69	62	87	132
Sigmodon	63	0	47	55	67	67	99
Peromyscus	62	49	0	17	75	80	98
Onychomys	73	54	29	0	87	85	103
Oryzomys	63	68	68	96	0	91	99
Microtus	93	73	64	80	106	0	97
Tylomys	108	90	87	97	110	90	0
Phyllotis	45	37	29	43	62	75	
Nyctomys	117	87	60	88		63	
Akodon	52	69	46	61	83	86	
Holochilus	101	64	80	112	103	81	
Rhipidomys	62	64	66	91	74	85	
Ichthyomys	63	47	68	74	94	69	
Nectomys	80	71	63	101	123	89	
Zygodontomys	98	84	101	128	101	103	

[a] The distances given by the respective antisera are in the columns. The number above each column is the correction factor for nonreciprocity applied to each raw datum given by that antiserum. The disturbingly large values of some of those factors are a cautioning element relative to any attempts at overly precise phyletic placements of these taxa. The overall picture is clear enough, though—one is dealing with some very ancient separations.

answer the basic question of the antiquity of the group in the New World, though their resolving power with regard to cricetid internal cladistics should not be pushed unduly. The relevant data are presented in Tables 3–7. As noted there, the main problems at present are the lack of appropriate rate test data and some substantial immunological inconsistencies. Counterbalancing these to some extent are the DNA annealing data of Rice (1974), which provide an independent, though limited, source of phylogenetic information. The single intracricetid, nonmicrotine comparison at present available in both systems is that between *Peromyscus* and *Oryzomys*. Here the albumin distance is

Table 4. Some DNA Annealing Comparisons among Rodents[a]

	ΔT_m		ΔT_m		ΔT_m
Peromyscus californicus	0.0	*M. pennsylvanicus*	0.0	*Mus caroli*	0.0
Peromyscus leucopus	4.4	*P. maniculatus*	15.6	*Mus musculus*	4.6
Microtus pennsylvanicus	15.6	*O. palustris*	17.3	*Rattus*	14.0
Oryzomys palustris	17.1	*Mesocricetus*	17.9		

	ΔT_m
Mesocricetus auratus	0.0
Cricetulus griseus	11.5

[a] From Rice (1974).

Table 5. Quantitative Precipitin Data Bearing on Murid and Cricetid Albumin Evolution

Genus	Anti-*Sigmodon*	Anti-*Rattus*	Anti-hystricomorph[a]
Sigmodon	100	—	—
Mesocricetus	44	36	—
Rattus	28	100	—
Mus	—	65	20
Cavia	—	15	—
Peromyscus	—	—	21

[a] Mean for 7 antisera.

72 units and the DNA ΔT_m is 17°C. Even allowing for the indicated relative lack of change along the *Peromyscus* albumin lineage, the albumin to DNA distance ratio of about 5 to 5.5 is still consistent with those seen for other, primarily primate, comparisons between our data and those of Rice (1974). As we have already shown that albumin and DNA evolution has proceeded at similar rates among anthropoid primates and rodents, we believe that this finding can be extended to cover the New World cricetids—in all cases the albumin immunological distance (when corrected for amount of change differences) being about 5 times the DNA annealing ΔT_m measure between the same pairs of taxa.

A combined reading of the albumin and DNA data then sees several lineages among the New World cricetids that have been separated from one another for *at least* the length of time sufficient to produce a ΔT_m of 15°C or an albumin immunological distance of 75 units. These would be *Peromyscus–Onychomys–Phyllotis–Akodon, Calomys–Oryzomys, Sigmodon, Tylomys, Ichthyomys,* and the microtines. According to the molecular data, these lineages must have had separate existences for at least 30 M.Y. and probably closer to 40 M.Y.

Table 6. Other Microcomplement Fixation Comparisons of Cricetid Albumins

Genus	Anti-*Oryzomys albigularis*	Anti-*Peromyscus polionotus*
O. albigularis	0	—
O. subflavius	18	—
O. caligularis	35	—
O. fulvescens	35	—
O. capito	45	—
O. concolor	50	—
P. polionotus	—	0
P. leucopus	—	10
P. maniculatus	—	10
P. gossypinus	—	13
P. californicus	—	32
P. floridanus	—	33

Table 7. Cladistic Events in the Cricetid Rodents—Combined Albumin and DNA

Clades diverging	Approximate time of divergence (M. Y.)[a]
Muridae-Cricetidae	60
Rattus-*Mus*	35
Cricetines-Microtines and Hesperomyines	50
Mesocricetus-*Cricetulus*	30
Peromyscus-*Oryzomys*	45
Peromyscus-*Microtus*	40–45
Peromyscus-*Onychomys*	20

[a] Relative to a beginning of the rodent adaptive radiation 80 M.Y.A.

This follows from our earlier discussions of rodent origins, the lack of any evidence suggesting more rapid molecular evolution in the rodents relative to other groups of mammals or in different rodent groups relative to one another, and the fact that 75–80 units of albumin immunological distance is about 40% of the maximum distances observed within the order. It is not yet at all clear, however, that all the above lineages belong to the same monophyletic unit within the cricetids, and the lack of a clear molecular perspective on this point precludes any realistic discussion of the geographic origins of the family and dispersal within it. We would not be at all surprised to eventually find the cricetid phyletic pattern developing into a later, and perhaps even more complex version of that already detailed for the hystricognaths.

Intra-New-World-Monkey Relationships

Fig. 5 presents a phylogeny of the New World monkeys based on extensive immunological comparisons of their albumins and transferrins. The tree is constructed according to the procedures detailed in Cronin and Sarich (1975) and Sarich and Cronin (1976).

The immunological data, whether analyzed independently for each protein or for the summed distances, provide, with two exceptions, unambiguous placements of the various platyrrhine lineages. All of the lineages leading to the modern New World monkeys are clearly part of a single adaptive radiation that occurred sometime after the catarrhine–platyrrhine divergence. Three secondary radiations leading to modern groups (*Cacajao*-*Pithecia*, *Ateles*-*Alouatta*-*Lagothrix*, the callitrichids including *Callimico*) follow. We can see no evidence associating the callitrichids with any one ceboid lineage to the exclusion of the others. Among the marmosets, *Cebuella* is probably congeneric with *Callithrix*, and *Callimico* is an earlier offshoot of that clade, which then becomes, along with *Leontopithecus* and *Saguinus*, part of the basic

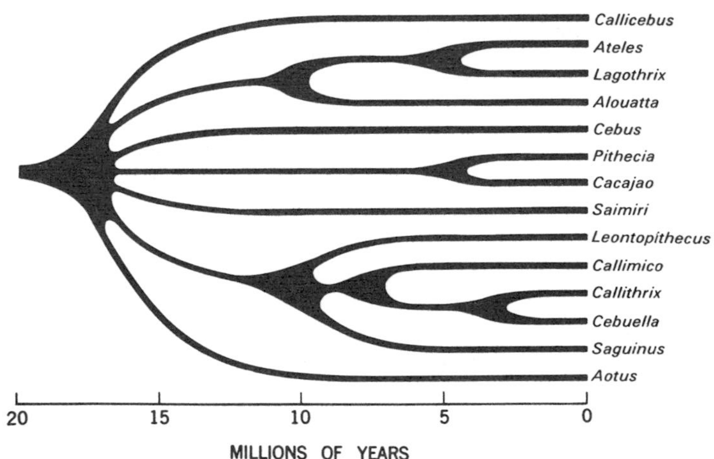

Fig. 5. A molecular phylogeny of the New World monkeys based on immunological comparisons (Cronin and Sarich, 1978).

callitrichid adaptive radiation. We also cannot see in our data any indication of an *Aotus–Callicebus* (Aotinae) or *Cebus–Saimiri* (Cebinae) association.

The albumin and transferrin data sets are discordant in two respects. The first concerns *Aotus*, where the albumins show its lineage as separating prior to an adaptive radiation leading to all other New World monkeys, while the transferrin data are more consistent with a single adaptive radiation including *Aotus*. We tend to favor the transferrin picture because of the extreme and unique indicated conservatism in the evolution of *Aotus* albumin. We see in our data no support for the suggestion of Baba *et al.* (1975) of an *Aotus*–callitrichid association. It is clear, though, that electrophoretic and DNA annealing data would be most welcome here, and we would not be at all surprised to find these affecting our placement of the *Aotus* lineage.

The second disagreement is not cladistic, but quantitative. We have found that transferrins typically evolve about 50% more rapidly than the corresponding albumins. Among the marmosets, however, this situation would appear to be reversed. Mean immunological distances among the three major callitrichid clades are 7–10 units for the transferrins, and about 20 for the albumins. Even so, cladistic analysis of each data set shows the marmosets to be of monophyletic origin relative to the various cebid lineages. In such cases, we tend to favor a placement indicated by averaging those given by the individual molecules, and this is also consistent with the electrophoretic data of Bruce (1977). Within callitrichid distances are, then, about 40% of the maximum intraplatyrrhine distances.

We have also carried out lower taxon comparisons among the callitrichids using the increased resolving power made possible by plasma protein electrophoresis (Cronin and Sarich, 1978). These results are shown in Fig. 6. One might note that intra-*Saguinus* distances range from 0.4 to 1.1. Such values are comparable to those found between species of *Macaca* (Cronin *et al.*, 1980). For the *Callithrix* clade, we see a close association between *C. jacchus* and *Cebuella*, with a *D* of about 1.0. This distance is similar to other closely related pairs of species such as *M. mulatta* and *M. fascicularis* or *C. aethiops* and *C. mitis*. It is entirely possible, indeed likely, that *Cebuella* will turn out to be more closely related to *jacchus* than are other *Callithrix* species.

The immunological and electrophoretic data then suggest that the callitrichids are a rather compact group with all lineages still sharing a common ancestor until the latter part of the Miocene. This implies that the marmoset grade of evolution cannot reasonably be seen as retention of features primitive for the New World as a whole and must be a highly derived state developing long after the basic New World monkey radiation. To continue to view the marmosets as primitive within the cladistic context required by the molecular evidence would require that their features be those of the most recent common ancestor of all extant New World monkeys. This would then imply that the cebid grade was reached independently by at least six different lineages—the cebids clearly not being monophyletic relative to the callitrichids. It would appear much more tenable to view the most recent common ancestor of all the extant ceboids as cebid in its grade of organization, with the marmoset grade then being derived and not primitive.

The fossil record of New World monkeys suggests to us that only one form, *Neosaimiri*, may be ancestral to any extant lineage (Hershkovitz, 1972). *Branisella* appears to be too specialized morphologically to be ancestral to any living New World monkey, but it is clearly an anthropoid primate. We thus have primates present in South America in the earliest Oligocene, while the molecular data would indicate that all the living forms share common ancestry up to the latest Oligocene or earliest Miocene. It may then be that the earliest primates arrived in South America in the latest Eocene, underwent an adaptive radiation, and subsequently all but one of the lineages of this original

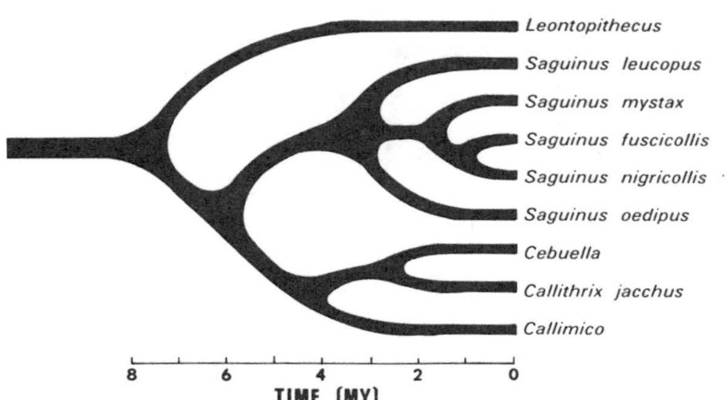

Fig. 6. A molecular phylogeny of the clade Callitrichidae based upon immunological and electrophoretic studies of plasma proteins (Cronin and Sarich, 1978).

radiation became extinct. This one survivor then gave rise to all living lineages beginning in the early Miocene.

Conclusion: Mammalian History in South America

Mammalian evolution on the South American continent is, then, a most complex affair intimately tied in with the vicissitudes of continental drift considerations and chance transoceanic colonizations (Fig. 7). It begins with the entry into South America from the north of marsupials sometime earlier than 75 M.Y.A. while a South America-Antarctica-Australia land connection still existed, and a subsequent dispersal into Australia before the separation between South America and Antarctica-Australia some 70–75 M.Y.A. The lack of early placentals in Australia leads us to conclude that they arrived in South America, again from the north, subsequent to this separation—though obviously the ancient separations among some of the South American forms documented in this paper would suggest not very much later. The edentates are then the only survivors of this first placental colonization of South America.

At least two hystricognaths then arrive at different, somewhat later times,

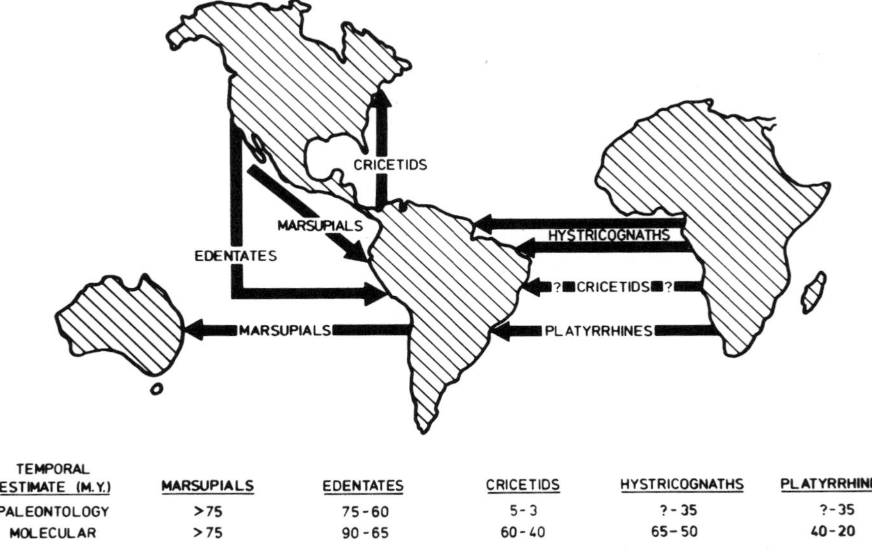

TEMPORAL ESTIMATE (M.Y.)	MARSUPIALS	EDENTATES	CRICETIDS	HYSTRICOGNATHS	PLATYRRHINES
PALEONTOLOGY	>75	75-60	5-3	?-35	?-35
MOLECULAR	>75	90-65	60-40	65-50	40-20

Fig. 7. A summary of the temporal and geographic origins of marsupials, edentates, hystricognath and cricetid rodents and New World monkeys as discussed in the text. The molecularly derived dates are estimates for the entry of each group into South America based on the calculated dates for the relevant cladistic events. The paleontological estimates are dates of the first appearance of the groups in the fossil record in South America.

and undergo major, very successful radiations in South America. They have clearly made an efficient accommodation to the most recent lengthy presence of the more modern cricetids that contrasts with the almost relict status of their Old World relatives. The cricetids, given their total absence in the appropriate North American fossil record, are, then, almost certainly of African origin, though, as discussed above, we cannot as yet begin to molecularly unravel the details of their early history. However, given colonization of the Galapagos, a very small target 1000 km from the coast of South America by two cricetids in the last few million years, we see no great problem in having several cross the Atlantic in the late Eocene and early Oligocene from Africa to the very large target of South America.

The primates are then the latest arrivals from Africa somewhere between a minimum of 20 M.Y.A. and a maximum of 35 M.Y.A. The latter figure is that which we calculate for the catarrhine–platyrrhine separation; the former is our figure for the beginning of the adaptive radiation leading to all modern platyrrhines. Although the presence of *Branisella* certainly leads one to favor the earlier date, we do find the relative lack of derived features common to the platyrrhines disturbing if we are to envision a relatively lengthy stay in South America prior to their most recent adaptive radiation. If the ancestral platyrrhine arrived soon after the catarrhine–platyrrhine separation, then our cladistic picture requires a single surviving line of the first adaptive radiation undergoing a rather late second radiation—and there should be some derived features documenting these events. The problem would be alleviated by postulating an African protoplatyrrhine line surviving until an early Miocene trans-Atlantic passage, though others, already mentioned, would then take its place. This aspect of the primate situation thus remains quite open.

Thus, as few as four trans-Atlantic colonizations over as much as 45 M.Y. might explain much of the South American placental fauna, and we believe that this need not strain credulity or chance beyond their breaking points. With iguanids on Fiji and Madagascar and ancient murids in Australia, a few placentals of African origin in South America would not appear to pose any insurmountable problems.

Acknowledgments

The authors would like to thank A. Wilson of the Department of Biochemistry and R. Palmour of the Department of Genetics, of the University of California, Berkeley, for access to laboratory space. L. Brunker and A. Wilson provided helpful comments in revising the original manuscript. Our thanks go to G. Paine for typing the manuscript and to C. Simmons of the University of California, Berkeley, and W. Powell of the Peabody Museum of Harvard University for their excellent illustrations.

References

Baba, M., Goodman, M., Dene, H., and Moore, G. W., 1975, Origins of the Ceboidea viewed from an immunological perspective, *J. Hum. Evol.* **4:**89–102.

Ba Maw, Ciochon, R. L., and Savage, D. E., 1979, Late Eocene of Burma yields earliest anthropoid primate, *Pondaungia cotteri, Nature (London)* **282:**65–67.

Beard, J. M., and Goodman, M., 1976, The haemoglobins of *Tarsius bancanus,* in: *Molecular Anthropology* (M. Goodman and R. Tashian, eds.), pp. 239–255, Plenum Press, New York.

Beard, J. M., Barnicot, N. A., and Hewett-Emmett, D., 1976, Alpha and beta chains of the major haemoglobin and a note on the minor component of *Tarsius, Nature (London)* **259:**338–339.

Bruce, E. J., 1977, A study of the molecular evolution of primates using the techniques of amino acid sequencing and electrophoresis, Ph.D. Thesis, University of California, Davis.

Conroy, G., 1976, Primate postcranial remains from the Oligocene of Egypt, *Contrib. Primatol.* **8:**1–134, Karger, New York.

Cronin, J. E., and Sarich, V. M., 1975, Molecular systematics of the New World monkeys, *J. Hum. Evol.* **4:**357–375.

Cronin, J. E., and Sarich, V. M., 1978, Marmoset evolution: The molecular evidence in marmosets, in: *Experimental Medicine* (N. Gengozian, and F. Deinhardt, eds.), *Primates in Medicine,* Vol. 10, pp. 12–19, Karger, New York.

Cronin, J. E., and Sarich, V. M., 1980, Tupaiid and Archonta phylogeny: The macromolecular evidence, in: *Comparative Biology and Evolutionary Relationships of Tree Shrews* (W. P. Luckett, ed.), pp. 303–322, Plenum Press, New York.

Cronin, J. E., Cann, R., and Sarich, V. M., 1980, Molecular evolution and systematics of the genus *Macaca,* in: *The Macaques: Studies in Ecology, Behavior and Evolution* (D. Lindberg, ed.), pp. 31–51, Van Nostrand Reinhold, New York.

Gingerich, P. D., 1978*a*, The Stuttgart collection of Oligocene primates from the Fayum Province of Egypt, *Paläontol. Z.* **52:**82–92.

Gingerich, P. D., 1978*b,* Phylogeny reconstruction and the phylogenetic position of *Tarsius,* in: *Recent Advances in Primatology Evolution,* Vol. 3 (D. J. Chivers and K. A. Joysey, eds.), pp. 249–256, Academic Press, New York.

Gingerich, P. D. and Schoeninger, M., 1976, The fossil record and primate phylogeny, *J. Hum. Evol.* **6:**483–505.

Goodman, M., 1962, Evolution of the immunologic species specificity of human serum proteins, *Hum. Biol.* **34:**104–150.

Hoffstetter, R., 1972, Relationships, origins, and history of the ceboid monkeys and caviomorph rodents: A modern reinterpretation, in: *Evolutionary Biology,* Vol. 6 (Th. Dobzhansky, M. K. Hecht, and W. C. Steere, eds.), pp. 323–347, Appleton-Century-Crofts, New York.

Hoffstetter, R., 1974, Phylogeny and geographic deployment of the primates, *J. Hum. Evol.* **3:**327–350.

Hershkovitz, P., 1972, The recent mammals of the Neotropical region: A zoogeographic and ecological review, in: *Evolution, Mammals, and Southern Continents* (A. Keast, F. C. Erk, and B. Glass, eds), pp. 311–431, State University of New York Press, Albany.

Kohne, D., 1975, DNA evolution data and its relevance to mammalian phylogeny, in: *Phylogeny of the Primates* (W. P. Luckett, and F. Szalay, eds.), pp. 249–261, Plenum Press, New York.

Luckett, W. P., 1975, Ontogeny of the fetal membranes and placenta: Their bearing on primate phylogeny, in: *Phylogeny of the Primates* (W. P. Luckett, and F. S. Szalay, eds.), pp. 157–182, Plenum Press, New York.

Mares, M. A., 1975, South American mammal zoogeography: Evidence from convergent evolution in desert rodents, *Proc. Natl. Head. Sci. USA* **72:**1700–1702.

Maxson, L. R., Sarich, V. M., and Wilson, A. C., 1975, Continental drift and the use of albumin as an evolutionary clock, *Nature (London)* **225:**397–398.

Patterson, B., and Pascual, R., 1972, The fossil mammal fauna of South America, in: *Evolution, Mammals, and Southern Continents* (A. Keast, F. C. Erk, and B. Glass, eds.), pp. 247–309, State University of New York Press, Albany.

Prager, E. M., and Wilson, A. C., Osuga, D. T., and Feeney, R. E., 1976, Evolution of flightless land birds on southern continents: Transferrin comparison shows monophyletic origin of ratites, *J. Mol. Evol.* **8:**283–294.

Rice, N. R., 1974, Single-copy relatedness among several species of the Cricetidae (Rodentia), *Carnegie Inst. Wash. Yearb.* **73:**1098–1102.

Sarich, V. M., 1969, Pinniped origins and the rate of evolution of carnivore albumins, *Syst. Zool.* **18:**286–295.

Sarich, V. M., 1970, Primate systematics with special reference to Old World monkeys: A protein perspective, in: *Old World Monkeys* (J. R. and P. H. Napier, eds.), pp. 175–226, Academic Press, New York.

Sarich, V. M., 1972, Generation time and albumin evolution, *Biochem. Gene.* **7:**205–212.

Sarich, V. M., 1973, The giant panda is a bear, *Nature* (*London*) **245:**218–220.

Sarich, V. M., and Cronin, J. E., 1976, Molecular systematics of the primates, in: *Molecular Anthropology* (M. Goodman, and R. Tashian, eds.), pp. 141–170, Plenum Press, New York.

Sarich, V. M., and Cronin, J. E., 1977, Generation length and rates of hominoid molecular evolution, *Nature (London)* **269:**354–355.

Schwartz, J. H., Tattersall, I., and Eldredge, N., 1978, Phylogeny and classification of the primates revisited, *Yearb. Phys. Anthropol.* **21:**95–133.

Simons, E. L., 1976, The fossil record of primate phylogeny, in: *Molecular Anthropology* (M. Goodman, and R. Tashian, eds.), pp. 35–62, Plenum Press, New York.

Szalay, F. S., 1975, Phylogeny of primate higher taxa: The basicranial evidence, in: *Phylogeny of the Primates* (W. P. Luckett, and F. S. Szalay, eds.), pp. 91–125, Plenum Press, New York.

Walker, A., 1976, Splitting times among hominoids deduced from the fossil record, in: *Molecular Anthropology* (M. Goodman, and R. Tashian, eds.), pp. 63–77, Plenum Press, New York.

Biochemical Evidence on the Phylogeny of Anthropoidea

21

M. BABA, L. DARGA, and M. GOODMAN

Introduction

Anthropoidea is the suborder of Primates which includes New and Old World monkeys, apes, and humans. The term "Anthropoidea" was introduced by Mivart and has received general acceptance since Simpson (1945) included it in his classification of mammals (Simons, 1972). However, questions have arisen regarding the monophyletic nature of Anthropoidea. Since the fossil record relating to higher primate origins remains incomplete, there is reasonable doubt that the three superfamilies of Anthropoidea (Ceboidea, Cercopithecoidea, and Hominoidea) descended from a common "stem stock" (Schwartz *et al.*, 1978). Although Simons (1976) views Anthropoidea as a monophyletic assemblage, he notes (1972) that the earliest putative ancestors of Ceboidea, Cercopithecoidea, and Hominoidea do not seem to resemble one another as much as one would expect if all had emerged from a single segment of Paleocene-Eocene lower primates.

In this chapter we explore the phylogenetic relationships of members of the Anthropoidea from the perspective of immunological and biochemical data gathered on primate proteins. Degrees of genetic divergence among primates are assessed by the method of immunodiffusion and by maximum parsimony analysis of amino acid sequence data for homologous proteins

M. BABA • Department of Science and Technology, University Studies/Weekend College Program, College of Lifelong Learning, Wayne State University, Detroit, Michigan 48201. L. DARGA • Department of Anthropology, Oakland University, Rochester, Michigan 48063. M. GOODMAN • Department of Anatomy, School of Medicine, Wayne State University, Detroit, Michigan 48201.

from various extant primate genera. The results argue for the monophyletic origin of the Anthropoidea.

Immunological and Biochemical Analyses

Immunodiffusion analysis assesses genetic divergence between lineages by determining the degree of antigenic distance separating various species. This approach provides a simple, rapid way to determine genetic distances among a large number of species, while still yielding results which correspond closely with those gained by more precise measurements of amino acid substitutions (Goodman, 1974). Procedures involved in antisera production, preparation of trefoil Ouchterlony plates in which the antigen–antibody reactions are carried out, and recording and computer processing of the immunodiffusion data are thoroughly described elsewhere (Goodman and Moore, 1971). Computer processing of the immunodiffusion data yields a dendrogram depicting the probable order of ancestral branching among the species compared. The relationships portrayed in the divergence tree approximate the true cladogeny of the species compared, provided that these species evolved antigenically at roughly comparable rates (Moore, 1971). The possibility of constructing a correct or nearly correct cladogram is increased when the antigenic distance data are based on a spectrum of proteins from the various homologous species rather than just one or two proteins.

The most exact measurements of genetic divergence are based on amino acid sequence data. This is due to the fact that the sequence of coding units (nucleotide triplets or codons) in a gene specifies the sequence of amino acids in a polypeptide chain of a protein. Sequence data for several primate proteins have been analyzed and an evolutionary tree was constructed depicting the phylogenetic relationships among lineages (Goodman *et al.,* 1974, 1975; Beard and Goodman, 1976; Moore *et al.,* 1973).

Ancestral descendant configurations yielding the fewest mutations over the entire tree were calculated, resulting in the most parsimonious tree for each protein. The reconstructed genealogy conforms to the principles of Hennig (1966) since plesiomorphic similarities (primitive traits inherited unchanged from an ancient common ancestor), determine the cladistic relationships of the species in the parsimony tree. Evolutionary rates can vary considerably in different lineages and not adversely affect the correctness of the reconstruction.

Evidence from Immunodiffusion Comparisons

Over 10,000 trefoil Ouchterlony plate comparisons have been carried out using rabbit antisera to whole serum or purified serum proteins of primate, tree shrew, and elephant shrew species. Figure 1 is a divergence tree generated from computer processing of the Ouchterlony data which depicts the major phylogenetic relationships between primate lineages. Within the divergence tree, primates appear as a monophyletic assemblage, more closely related to the tree shrews than to nonprimate eutherian mammals. However, the exact degrees of antigenic distance among various nonprimate eutherian lineages and between these lineages and the tree shrew branch is uncertain at the present time, since, except for the elephant shrew, nonprimate eutherians were not used as donor species in the production of antisera.

The primate assemblage forms two major lineages: the Strepsirhini, including lorisoid and lemuroid branches, and the Haplorhini. The Haplorhini includes two lineages, the Tarsioidea and the Anthropoidea. The latter is composed of two sister groups, the Catarrhini and the Platyrrhini.

The Catarrhini subdivides into two monophyletic superfamilies, the Hominoidea and Cercopithecoidea. The Hominoidea separates into families Hylobatidae and Hominidae. Certain morphological and karyotypic traits have raised doubts concerning the hominoid classification of the hylobatids (von Koenigswald, 1968; Hamerton, 1963; Klinger, 1963; Klinger *et al.*, 1963; Chiarelli, 1966). However, the immunodiffusion data depict a close relationship between the two living hylobatid genera, *Hylobates* and *Symphalangus*, supporting their monophyletic grouping in the Hominoidea. In turn, the immunodiffusion results support dividing the Hominidae into two subfamilies, Ponginae, made up of *Pongo*, and Homininae, for *Homo, Pan*, and *Gorilla*.

The cercopithecoids split into cercopithecine and colobine branches. Within the Colobinae, *Presbytis* and *Nasalis* unite first, followed by *Colobus* and then *Pygathrix*. Within the cercopithecine branch, *Papio* joins *Theropithecus*, followed by the macaques. *Mandrillus sphinx* and *Papio* (= *Mandrillus*?) *leucophaeus* unite early, suggesting a close relationship between these species; these forest baboons then join the branch comprised of *Cercopithecus, Erythrocebus*, and *Cercocebus*. This group then joins the *Papio–Theropithecus–Macaca* assemblage.

Within the platyrrhine region of the tree, degrees of divergence are depicted among ten ceboid lineages. The two species separated by the smallest antigenic distance are *Ateles* and *Lagothrix*. This ateline complex is then joined by *Alouatta*, confirming the notion of common ancestry of Atelinae species and *Alouatta*. The monophyly of an ateline assemblage comprised of *Ateles*-plus-*Lagothrix* lineage which is then joined by *Alouatta* is also supported by Sarich and Cronin's (1976) microcomplement fixation data gathered on primate albumins and transferrins.

The lineage joining the tree after the *Alouatta*–Atelinae complex is composed of the primitive or conservative genera which includes *Aotus, Callicebus*, and the callitrichid genera *Callimico* and *Saguinus*. Their grouping together may simply reflect symplesiomorphic resemblances; on the other hand, if the relationships depicted within this region of the tree prove to be an accurate reflection of the cladogeny of the group, they tend to support the notion that

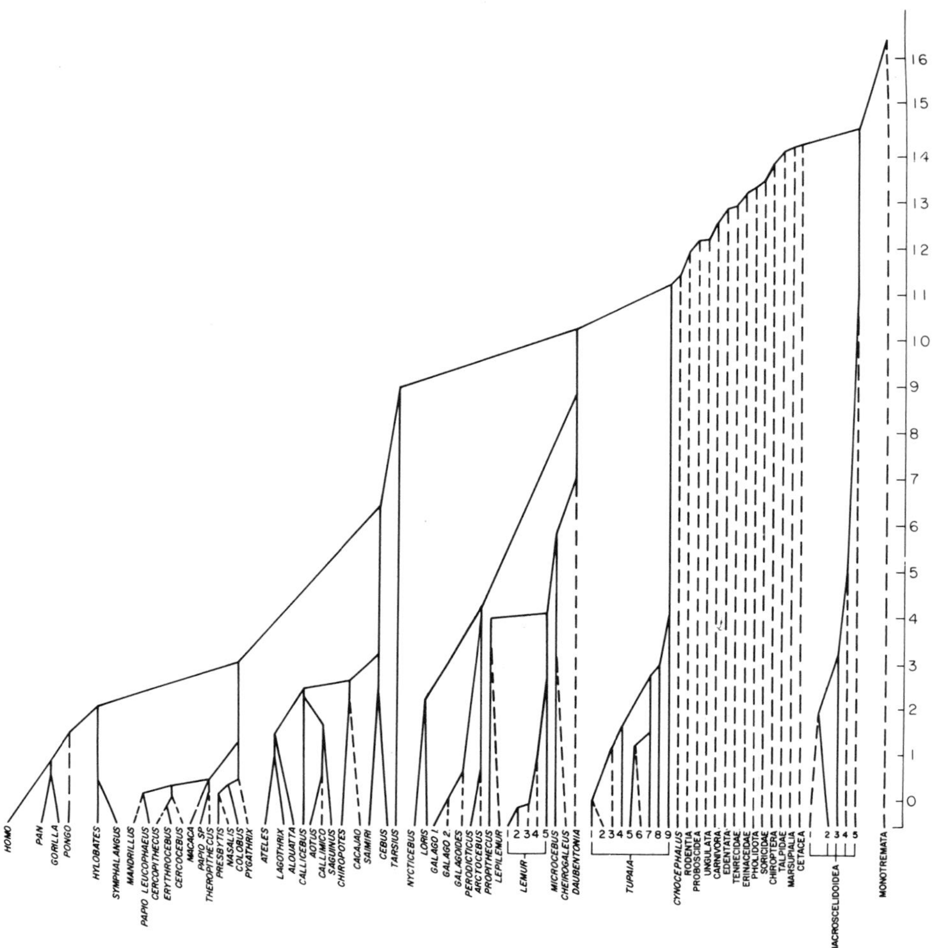

Fig. 1. This divergence tree produced from computer processing of Ouchterlony data using rabbit antisera depicts the degrees of divergence of major primate and mammalian lineages and the branching sequence of the following groups of species: *Homo* (*Homo sapiens*); *Pan* (*Pan troglodytes, P. paniscus*); *Gorilla* (*Gorilla gorilla*); *Pongo* (*Pongo pygmaeus*); *Hylobates* (*Hylobates lar, H. agilis, H. concolor*); *Symphalangus* (*Symphalangus syndactylus*); *Mandrillus* (*Mandrillus sphinx*); *Papio leucophaeus; Cercopithecus* (*Cercopithecus aethiops, C. albogularis, C. diana*); *Erythrocebus* (*Erythrocebus patas*); *Cercocebus* (*Cercocebus galeritus, C. torquatus*); *Macaca* (*Macaca mulatta, M. silenus, M. fascicularis, M. nemestrina, M. speciosa, M. maura, M. fuscata, M. radiata, M. cyclopis, M. sylvana*); *Papio* (*Papio anubis, P. papio, P. hamadryas, P. cynocephalus, P. papio* hybrid); *Theropithecus* (*Theropithecus gelada*); *Presbytis* (*Presbytis entellus, P. melalophos, P. cristatus, P. senex*); *Nasalis* (*Nasalis larvatus*); *Colobus* (*Colobus polykomos, C. badius*); *Pygathrix* (*Pygathrix nemaeus*); *Ateles* (*Ateles geoffroyi, A. fusciceps*); *Lagothrix* (*Lagothrix lagothricha*); *Alouatta* (*Alouatta palliatta*); *Callicebus* (*Callicebus*); *Aotus* (*Aotus trivirgatus*); *Callimico* (*Callimico goeldi*); *Saguinus* (*Saguinus oedipus, S. fuscicollis, S. illigeri, S. nigricollis*); *Chiropotes* (*Chiropotes satanus*); *Cacajao* (*Cacajao rubicundus*); *Saimiri* (*Saimiri sciureus*); *Cebus* (*Cebus albifrons, C. apella*); *Tarsius* (*Tarsius syrichta*); *Nycticebus* (*Nycticebus coucang*); *Loris* (*Loris tardigradus*); *Galago* 1. (*Galago crassicaudatus*); *Galago* 2. (*Galago senegalensis*); *Galagoides*

Cebidae is a paraphyletic or polyphyletic taxon rather than a monophyletic assemblage, since the cebid genera *Aotus* and *Callimico* appear to be more closely related to the tamarin species than to any other cebids. Sarich and Cronin's (1976) microcomplement fixation data on primate albumins and transferrins do not suggest a conservative platyrrhine lineage of a polyphyletic Cebidae. Instead, their divergence tree clearly groups all marmoset and tamarin species (including *Callimico* and *Saguinus*) into one platyrrhine assemblage, while widely separating *Callicebus* and *Aotus*.

The joining of *Cacajao* and *Chiropotes* before either joins other ceboids supports their grouping in the subfamily Pitheciinae. Likewise, the grouping of *Cebus* and *Saimiri* within the tree supports their placement in the subfamily Cebinae, although Hill (1960) noted that "further studies will doubtless necessitate the final severance of *Saimiri* from the present subfamily (Cebinae) . . . ." on the basis of several morphological characters. Sarich and Cronin (1976) separate *Cebus* and *Saimiri* on the basis of microcomplement fixation data for primate albumins and transferrins.

Evidence From Protein Sequence Data

Maximum parsimony analysis of amino acid sequence data is currently underway for several primate protein chains (Goodman, 1976; Goodman *et al.*, 1979). An expanded maximum parsimony analysis of protein sequence data has recently been derived from an analysis of species for which several proteins have been sequenced (α and β hemoglobin chains, myoglobin, α lens crystallin, fibrinopeptide A and B, and cytochrome *c*) (Table 1), and the resultant tree is depicted in Fig. 2. Several branch lines represent two or three closely related species since a single species was not available for which all proteins had been sequenced (see caption).

These combined data strengthen the conclusions derived from analysis of

(*Galagoides demidovii*); *Periodicticus* (*Periodicticus potto*); *Arctocebus* (*Arctocebus calabarensis*); *Propithecus* (*Propithecus verreauxi*); *Lepilemur* (*Lepilemur mustelinus*); *Lemur* (1. *Lemur mongoz*, 2. *Lemur macaco*, 3. *Lemur fulvus*, 4. *Lemur variegatus*, 5. *Lemur catta*); *Microcebus* (*Microcebus murinus*); *Cheirogaleus* (*Cheirogaleus major*); *Daubentonia* (*Daubentonia madagascariensis*); *Tupaia* (1. *Tupaia chinensis*, 2. *T. belangeri*, 3. *T. glis*, 4. *T. longipes*, 5. *T. montana*, 6. *T. tana*, 7. *T. minor*, 8. *T. palawanensis*, 9. *Urogale everetti*); *Cynocephalus*; Nonprimate eutherian mammals: Rodentia (*Rattus, Cavia, Marmota monas, Citellus mexacanus, Cynomys, Eutamias dorsalis*); *Proboscidea* (*Elephas maximus, Loxodonta africana*); Ungulata (*Bos taurus, Rangifer caribou, Tapirus terrestris, Equus*); Carnivora (*Canis familiaris, Potos flavus, Procyon lotor, Ursus arctos, Eumetopias jubatus, Zalophus californias*); Edentata (*Dasypus novemcinctus*); Tenrecidae (*Tenrec ecaudatus, Echinops*); Erinaceidae (*Hemiechinus auritus, H. megalotis, Atelerix, Erinaceus europaeus*); Pholidota (*Manis pentadactyla*); Soricidae (*Suncus murinus, Sorex cinereus*); Chiroptera (*Eptesicus, Leptonycteris nivalis*); Talpidae (*Scapanas aquaticus*); Marsupialia (*Macropus rufus*); Cetacea (*Dalphinapterus leucas*); Macroscelididae (1. *Elephantulus intufi*, 2. *E. myurus*, 3. *Nasilio brachyrhynchus*, 4. *Petrodromus sultan*, 5. *Phynchocyon*); Monotremata (*Tachyglossus setosus*). Heavy lines descend to taxa used as homologous species; dashed lines descend to taxa used only at heterologous species.

Table 1. Maximum Parsimony Analysis of Protein Sequence Data

OTUs	Amino acid sequences employed in a combined alignment					
Homo sapiens	αHb	βHb	Myo	αLen	Fib A-B	Cyt *c*
Pan troglodytes	αHb	βHb	Myo		Fib A-B	Cyt *c*
Gorilla gorilla	αHb	βHb	Myo		Fib A-B	
Pongo pygmaeus	αHb	βHb	Myo		Fib A-B	
Hylobates lar	αHb	βHb			Fib A-B	
Hylobates agilis			Myo			
Macaca mulatta	αHb	βHb		αLen	Fib A-B	Cyt *c*
Macaca fascicularis			Myo			
Macaca fuscata	αHb	βHb				
Cercocebus atys	αHb	βHb				
Papio anubis			Myo			
Mandrillus leucophaeus					Fib A-B	
Cercopithecus aethiops	αHb	βHb			Fib A-B	
Erythrocebus patas			Myo			
Ateles geoffroyi	αHb	βHb			Fib A-B	Cyt *c*
Lagothrix lagothricha			Myo			
Saimiri sciureus	αHb		Myo			
Cebus apella	αHb	βHb			Fib A-B	
Saguinus fuscicollis	αHb	βHb				
Callithrix jacchus			Myo			
Tarsius bancanus	αHb	βHb				
Nycticebus coucang	αHb	βHb	Myo	αLen	Fib A-B	
Loris loris	αHb	βHb				
Lemur fulvus	αHb	βHb		αLen		
Lepilemur mustelinus			Myo			
Oryctolagus cuniculus	αHb	βHb	Myo	αLen	Fib A-B	Cyt *c*
Tupaia glis	αHb	βHb	Myo	αLen		
Mus musculus	αHb	βHb				Cyt *c*
Rattus norvegicus	αHb	βHb		αLen	Fib A-B	Cyt *c*
Elephas maximus	αHb		Myo		Fib A-B	
Loxodonta africana				αLen		
Ovis aries	αHb	βHb	Myo		Fib A-B	Cyt *c*
Bos taurus	αHb	βHb	Myo	αLen	Fib A-B	Cyt *c*
Lama lama and *Camelus dromedarius*	αHb	βHb			Fib A-B	Cyt *c*
Sus scrofa	αHb	βHb	Myo	αLen	Fib A-B	Cyt *c*
Eschrichtius gibbosus			Myo			
Balaenoptera acutorostrata				αLen		
Eschrichtius glaucus						Cyt *c*
Equus caballus	αHb	βHb	Myo	αLen	Fib A-B	Cyt *c*
Canis familiaris	αHb	βHb	Myo	αLen	Fib A-B	Cyt *c*
Meles meles	αHb	βHb	Myo		Fib A-B	
Erinaceus europaeus	αHb	βHb	Myo	αLen		

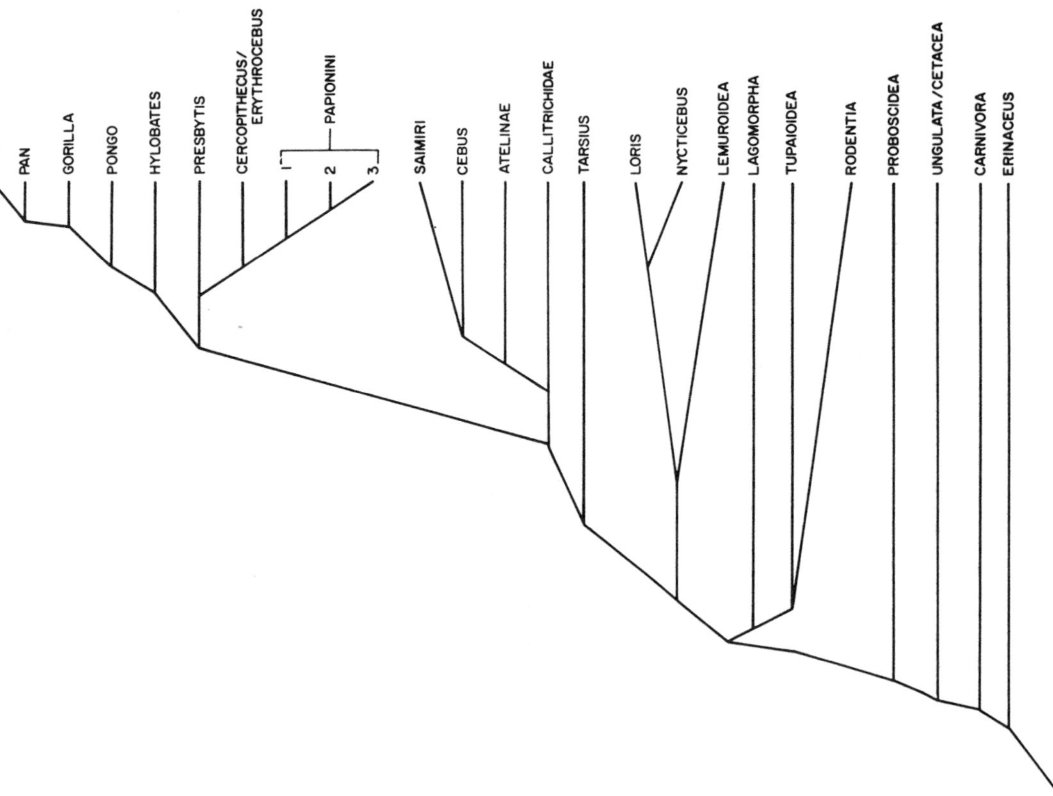

Fig. 2. Maximum parasimony analysis from combined amino acid sequence data. Table lists combinations of protein sequence data that were used for each OTU in the combined alignment.

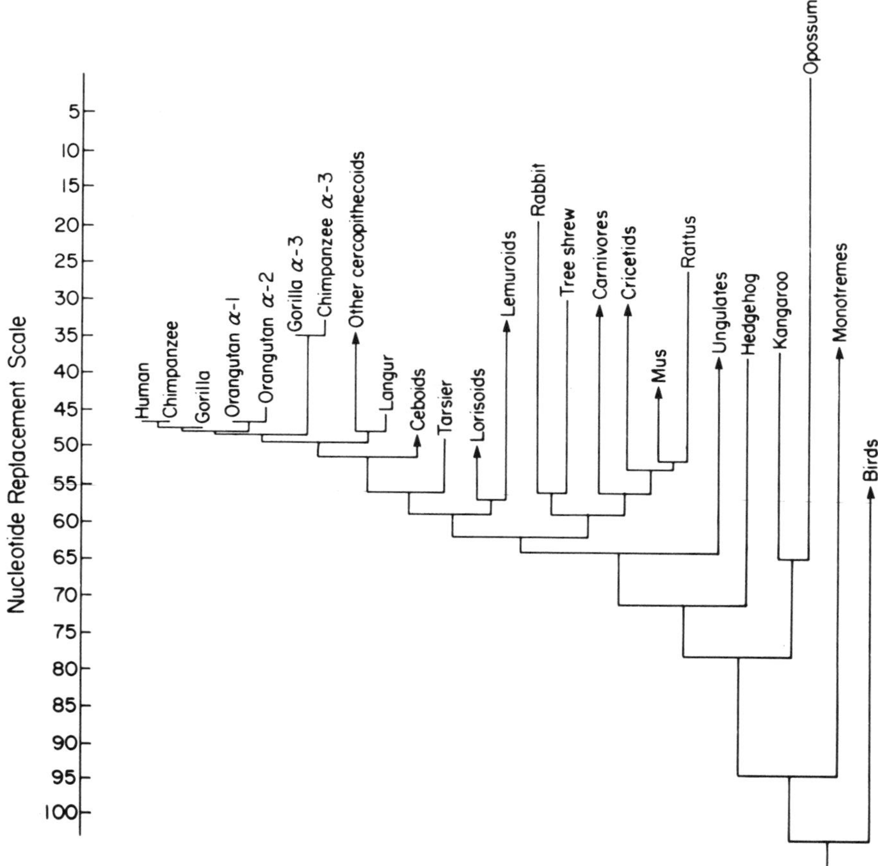

Fig. 3. α Hemoglobin amino acid squences.

the individual chains, with the results depicting the primates as a monophyletic assemblage which branches into Strepsirhini and Haplorhini. In agreement with immunodiffusion data, the Haplorhini appear as a monophyletic lineage splitting into a tarsioid and anthropoid branch. Anthropoidea subdivides into the platyrrhine and catarrhine branches. Within Platyrrhini, *Cebus* joins *Saimiri* and this branch joins Atelinae. *Saguinus* then join this assemblage as the most distantly related species.

Within the Catarrhini, Hominoidea separates from Cercopithecoidea with equally parsimonious solutions occurring when *Hylobates* joins the other hominoids before *Pongo* or, alternatively, if *Hylobates* and *Pongo* unite first and then join the hominoid group. The latter arrangement is in disagreement with the immunodiffusion data and seems to be due to evolution of myoglobin. The data support a close genetic relationship among the members of the Hominidae (*Homo, Pan,* and *Gorilla*).

Within the Old World monkeys, the two macaque species join followed by

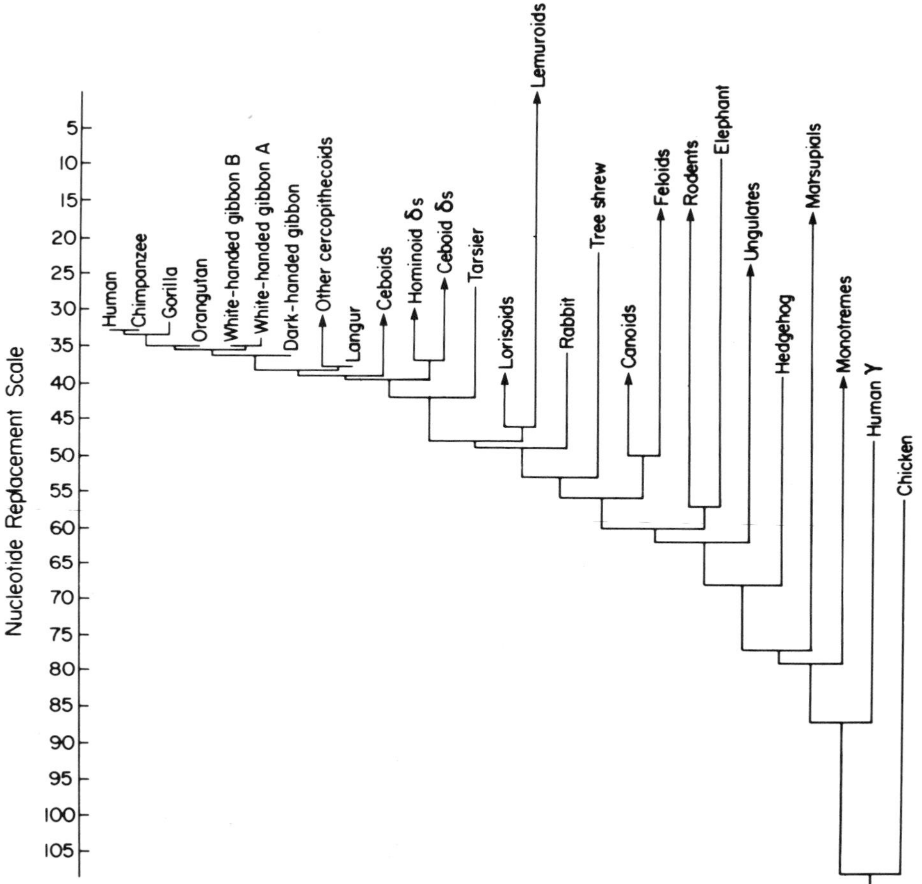

Fig. 4. β hemoglobin amino acid sequences.

a lineage representing several other members of the Papionini, i.e., *Cercocebus, Papio,* and *Mandrillus.* This close association of the members of the subtribe Papionini is also supported by maximum parsimony analysis of hemoglobin α and β chains. The analysis is based on an enlarged collection of 70 contemporary α hemoglobin chains and 87 β chains with sequences inferred at a majority of their residue positions by comparison of the amino acid compositions of peptide fragments to known homologous sequences (see Fig. 3–5).

Monophyly of Anthropoidea

Anthropoidea has generally been accepted as a monophyletic assemblage composed of two coequal sister groups, Catarrhini and Platyrrhini (Simpson, 1945). Although the Platyrrhini diverge markedly from the higher primates of the Old World, all share a number of derived morphological characters

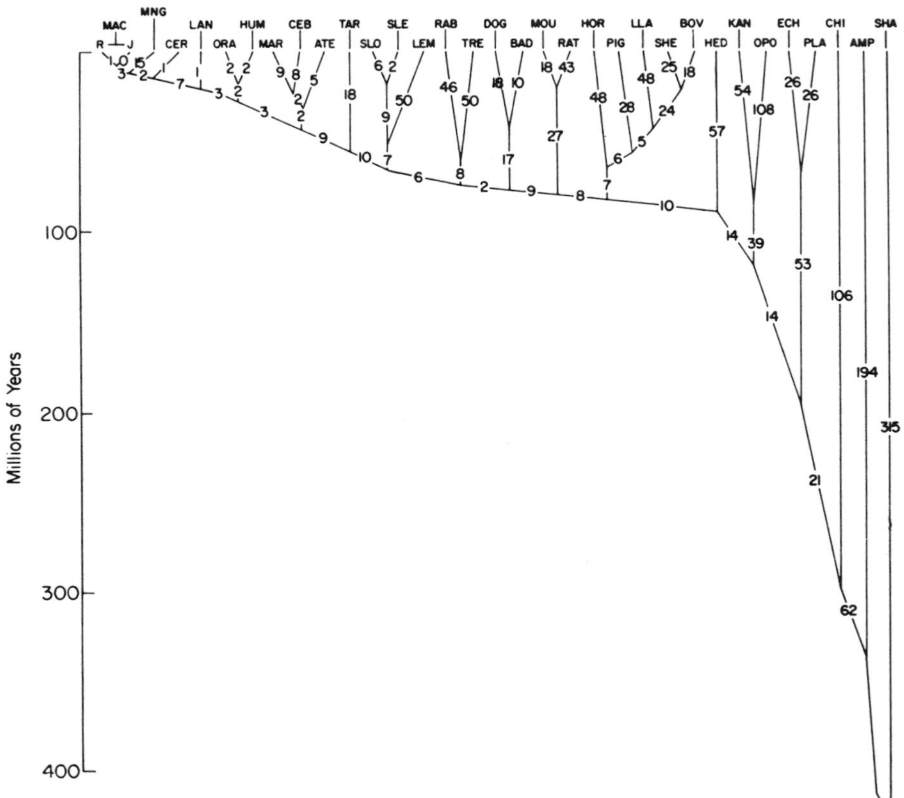

Fig. 5. Most parsimonious tree combined α and β hemoglobin amino acid sequences. Mac R (*Macaca mulatta*); Mac J (*Macaca fuscata*); Mng (*Cercocebus*); Cer (*Cercopithecus*); Lan (*Langur*); Ora (*Orangutan*); Hum (*Homo*); Mar (*Saguinus*); Ceb (*Cebus*); Ate (*Ateles*); Tar (*Tarsius*); Slo (*Nycticebus*); Sle (*Loris*); Lem (*Lemur*); Rab (*Oryctolagus*); Tre (*Tupaia*); Dog (*Canis*); Bad (*Meles*); Mou (*Mus*); Rat (*Rattus*); Hor (*Equus*); Pig (*Sus*); Lla (*Lama*); She (*Ovis*); Bov (*Bos*); Hed (*Hemiechinus*); Kan (*Macropus*); Opo (*Didelphis*); Ech (*Echidna*); Pla (*Ornithorhynchus*); Chi (*Gallus*); Amp (*Rana* and *Taricha*); Sha (*Heterodontus*).

which seem to set them apart as a distant taxonomic unit. These common features include a larger, more convoluted brain, a shortened rostrum and olfactory reduction, nearly complete separation by bony partition between orbits and temporal fossa, flattened nails on digits instead of claws, and forward-directed orbits (Hill, 1957; Simons, 1972). Interpreted phylogenetically however, a monophyletic Anthropoidea requires not only a series of common structural properties but, most importantly, a common ancestor for monkeys, apes, and humans. Furthermore, monophyly requires that all primates descendant from this hypothetical common ancestor be included within the taxon Anthropoidea.

In spite of the general acceptance of Anthropoidea as a valid clade, uncertainty regarding the origins and history of the anthropoid primates have thrown doubt upon the monophyly of the taxon. Romer (1966) suggested that the traditional subdivision of Primates into Prosimii and Anthropoidea was "arbitrary and unnatural," and that given the unclear evolutionary picture depicted by the fossil record, the Primates would be better grouped into five separate suborders (Plesiadapoidea, Lemuroidea, Tarsioidea, Platyrrhini, and Catarrhini). Simons (1972), while generally accepting the existence of Anthropoidea, raised the possibility that the taxon may be polyphyletic. According to Simons, the earliest members of the three groupings comprising Anthropoidea (Ceboidea, Cercopithecoidea, and Hominoidea) do not resemble one another as closely as one would expect if all had been derived from the same small segment of Paleocene-Eocene lower primates. Even more recently, Schwartz *et al.* (1978) noted that the origins of the Platyrrhini and their relationship to other primate lineages are "among the most glaring lacunae in our knowledge of primate phylogeny."

To invalidate Anthropoidea as a monophyletic assemblage, it would be necessary to demonstrate that Platyrrhini and Catarrhini do not exclusively share a common ancestor, i.e., that the common ancestor of these groups did give rise to another lineage, either fossil or extant, that is not now included within Anthropoidea. For example, if it could be demonstrated that any lower primate group, living or extinct, were more closely related to any of the present Anthropoidea than the latter was to other members of the Anthropoidea, then there would be evidence for a polyphyletic Anthropoidea. Currently, the fossil record is inconclusive with regard to this question. Some authors postulate the common descent of basal platyrrhines and catarrhines from an omomyid-like ancestor of the North American Eocene (Simons, 1961; Patterson and Pascual, 1968), possibly the genus *Teilhardina* of Belgium and Wyoming deposits (Simons, 1972, 1976). Gingerich (personal communication) on the other hand, argues that the ancestors of Ceboidea originated in the Old World from an adapid stock and migrated into the New World via a land bridge connecting Asia and North America during the late Eocene or early Oligocene. Other authors comment that the hypothetical ancestor of platyrrhines and catarrhines existed too early to suggest common descent from one stock (Orlosky and Swindler, 1975), or that the relationship between these premonkey ancestors and the New and Old World primates is still open to question (Schwartz *et al.,* 1978).

The question of the monophyletic nature of Anthropoidea has also been raised due to recent evidence on the paleogeographic dynamics of ceboid origins. With the discovery that North America and Europe were probably separated by a vast expanse of ocean after the middle Eocene (Dietz and Holden, 1970; Ramsay, 1971), Sarich (1970) questioned the maintenance of genetic continuity within the hypothetical ancestral stock common to both platyrrhines and catarrhines. Dating the platyrrhine–catarrhine split at 36 million years ago on the basis of his immunological distance values and a molecular clock model of albumin evolution, Sarich (1970) noted that if the traditional hypothesis of ceboid origins were correct, the hypothetical anthropoid stem stock would have been separated geographically for 30 million years before diverging into catarrhine and platyrrhine lineages. Sarich believed that such lengthy separation lessened the likelihood that Ceboidea emerged from a North American lower-primate-like ancestor, thereby questioning the dynamics of the traditional hypothesis of ceboid origins.

Although the fossil evidence is inconclusive with regard to these questions, the biochemical data gives strong support to the validity of monophyletic Anthropoidea, at least as far as extant primates are concerned. Results of immunodiffusion plate comparisons and evidence from protein sequence data show platyrrhine and catarrhine species to be more closely related to one another than either are to any living lower primate. The monophyly of Anthropoidea is also supported by evidence from other immunological techniques. Sarich and Cronin's (1976) albumin plus transferrin dendrogram, based on data gathered using microcomplement fixation, portrays Anthropoidea as a distinct assemblage which branches to form catarrhine and platyrrhine lineages. Furthermore, results from measuring homologies in both repeated (Hoyer and Roberts, 1967) and nonrepeated sequences of DNA (Kohne, 1970; Kohne *et al.*, 1972) from various primates are in agreement with one another and with the protein sequence and immunological data in placing catarrhine and platyrrhine branches closest to one another.

Since biochemical data, especially that derived from maximum parsimony analysis of protein sequence data, are particularly sensitive to differences between *synapomorphic* and *symplesiomorphic* commonalities, it is very likely that the Anthropoidea is a truly monophyletic assemblage.

The Place of Tarsius in Primate Phylogeny

Beyond the question of sister group status for Platyrrhini and Catarrhini, there is the additional problem of the validity of Haplorhini as a true clade. The existence of a haplorhine clade suggests the common origin of Tarsiiformes and Anthropoidea. The phylogenetic relationship of *Tarsius* to other primate lineages has long been a matter of controversy. *Tarsius* has alternately been placed in a separate suborder (Gadow, 1898), grouped with lemurs and lorises in the suborder Prosimii (Simpson, 1945) and assigned to the Haplorhini, a suborder including *Tarsius,* monkeys, apes, and humans (Pocock, 1918; Hill, 1955). Although the tarsier bears a superficial morphological resemblance to certain lemurs, other characteristics (absence of a rhinarium, a localized retinal depression similar to the fovea, and a discoideal, hemochorial placenta) unite this genus to the anthropoids (Beard and Goodman, 1976; Goodman *et al.,* 1979). Furthermore, according to Hershkovitz (1974), the ectotympanic bone in all adults of the higher primates has characteristics that could not have arisen from the ectotympanic condition existing in lemuroids, but probably originated from a primitive tarsier.

The results of biochemical investigations have tended to group *Tarsius* closer to anthropoids than to strepsirhines. With antisera to anthropoid species, *Tarsius* reacts better in immunodiffusion comparisons than strepsirhine species; with antisera to strepsirhine species, *Tarsius* reacts approximately to the same degree as the anthropoids (Dene *et al.,* 1976). The divergence tree of primates from immunodiffusion plate comparisons (Fig. 1) shows that *Tarsius* is about as distant from Anthropoidea as the lorises and lemurs are from one another. These findings are not in agreement with those of Sarich and Cronin (1976) based on microcomplement fixation data using rabbit antisera to primate albumins and transferrins. Their divergence tree places *Tarsius* closer to the lower primate lineages than to the Anthropoidea. However, their results are contradicted by immunodiffusion plate comparisons using chicken antisera to human albumin. The network of comparisons completed for these antisera groups *Tarsius* with the anthropoid lineages.

Maximum parsimony analysis carried out with combined α and β globin alignments agree with the immunodiffusion data in always placing *Tarsius* closer to the anthropoid lineages than to the slow loris (Beard and Goodman, 1976; Goodman *et al.,* 1979). In addition, Hoyer and Roberts (1967) in interspecies comparisons using families of repetitious, polynucleotide sequences of DNA from the human standpoint obtained the following degrees of similarity of heterologous DNAs to human DNA: chimpanzee 100%; gibbon 94%; rhesus 88%; capuchin 83%; tarsier 65%; galago and slow loris, each 58%; tree shrew 28%. As with immunodiffusion data, these results indicate that the sister-group of Anthropoidea is Tarsioidea. All of this evidence would seem to suggest the validity of Pocock's original grouping of *Tarsius* with the anthropoids and supports Hill's (1957) Haplorhini as a monophyletic assemblage, or true clade.

Origins of the Ceboidea

The traditionally accepted view of ceboid origins, founded upon the assumption of continental stability, postulated the independent emergence of New World primates from nonlemuroid lower primate populations (possibly the Omomyidae), widespread in North America and Europe during the Eocene (Le Gros Clark, 1959; Hill, 1962). Ceboid-like descendants of the North American ancestral lower primate stock supposedly either rafted from Central America to South America, since the Bolivar geosyncline precluded overland migration from the Central American peninsula until the Pliocene (Woodring, 1954; Olson, 1964; Harrington, 1962; Whitmore and Stewart, 1965), or "island-hopped" at the end of the Middle Eocene via a series of temporary, discontinuous land connections between Central America and the emergent Columbian Andes (Haffer, 1970). *Rooneyia,* an omomyid of the West Texas Lower Oligocene, displays morphological characters which approach the simian grade of development (Wilson, 1966) and provides evidence of advanced omomyids in the vicinity of the hypothetical route of southern migration. While Rooneyia may not be considered ancestral to Platyrrhini, the omomyid *Washakius* from the Wyoming middle Eocene, possesses a dental type that may represent an ancestral model for *Branisella* (Hoffstetter, 1974). If the traditional view that Omomyidae are ancestral to both catarrhines and platyrrhines (Patterson and Pascual, 1968) is correct, a strictly cladistic classification would require placement of certain of the present Omomyidae within the Anthropoidea (see Fig. 6A). The truly phyletic (clade) Anthropoidea would therefore contain all known species, both extant and fossil, within a monophyletic assemblage. However, if Gingerich (1976*a*) is sustained in his view of anthropoid origins, placement of Omomyidae within Anthropoidea would be incorrect (see Fig. 6C).

With the discovery that South America and Africa were joined as Gondwanaland until the opening of the South Atlantic rift (Upper Jurassic to Middle Cretaceous) (Dietz and Holden, 1970; Ramsay, 1971) came the possibility of an alternative explanation of ceboid origins and provided the impetus for renewed attacks upon the traditional hypothesis. One recent criticism of the traditional view, based upon immunological distance data for primate albumins, has been developed by Sarich (1970). Given a model of protein evolution which assumes a constant rate of nucleotide replacement, Sarich infers that the 59 "immunological distance units" separating platyrrhine and catarrhine albumins suggest that these primate lineages shared a common ancestor for 30 million years subsequent to their divergence from the most

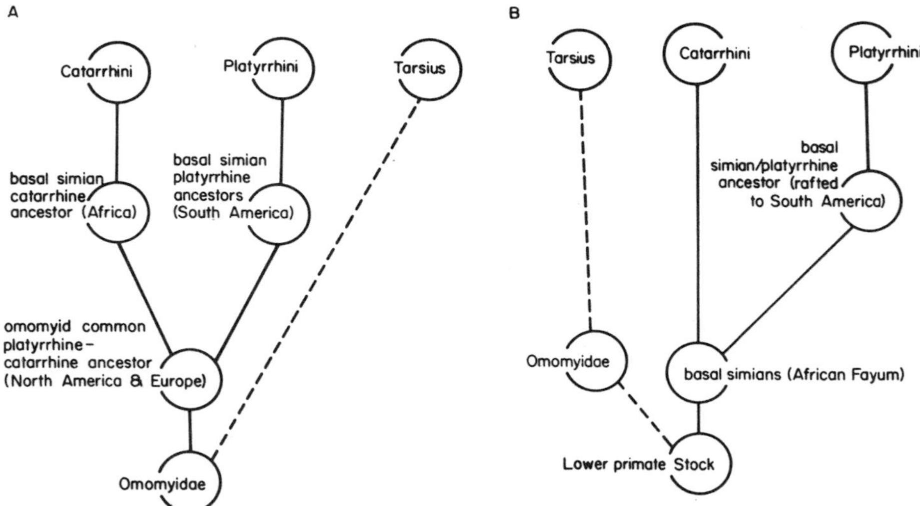

Fig. 6. Alternative hypotheses of ceboid origins. (A) Traditional continental stability hypothesis. (B) Continental mobility hypothesis. (C) Gingerich's hypothesis of ceboid origins. (D) Hershkovitz's alternative continental mobility hypothesis.

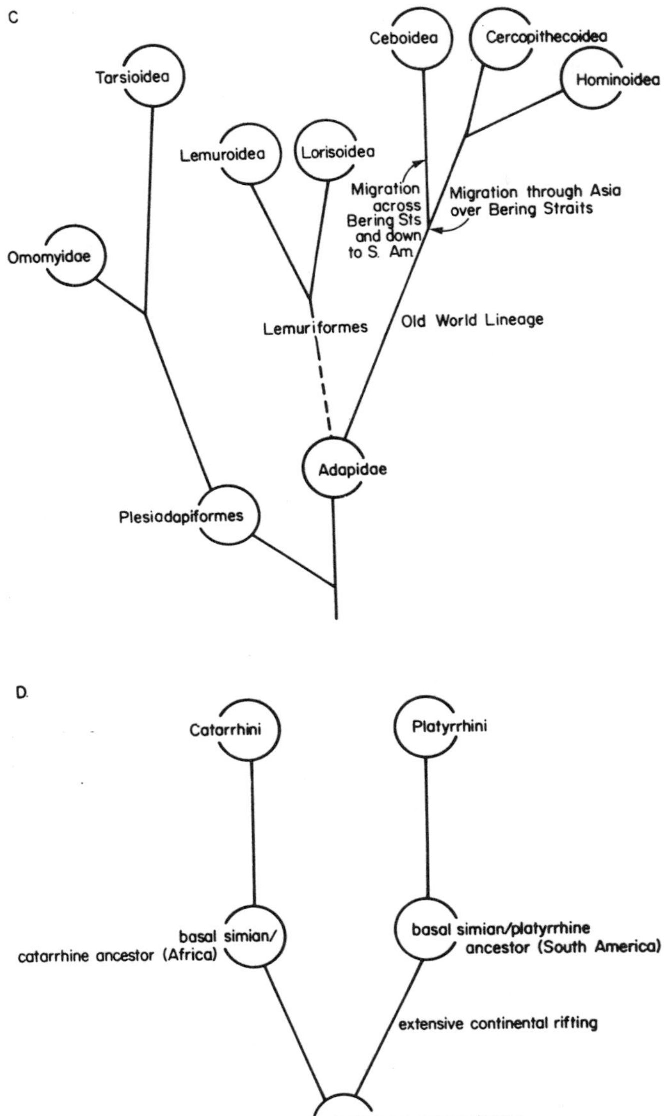

Fig. (*Continued*)

recent common lower primate ancestor. The maintenance of genetic continuity for 30 million years within a common ancestral stock ranging over the northern continents of North America and Europe is considered improbable, particularly since the Atlantic Ocean may have presented a major geographic barrier to gene flow between these continents after the middle Eocene. Rather, an alternative hypothesis of ceboid origins which assumes a single Old World emergence of simian grade primates and subsequent rafting of this basal simian stock from Africa to South America near the Eocene–Oligocene boundary, is more in accordance with the notion of a monophyletic Anthropoidea. According to this second view of ceboid origins, placement of omomyid species within the Anthropoidea is unnecessary, since supporters of this view tend to favor the notion that the common African stem-stock closely resembled extinct monkeys from the Oligocene Fayum (Hoffstetter, 1972), particularly the cercopithecoid subfamily Parapithecinae (see Fig. 6B). If this view were correct, no omomyid species would have to be placed in the Anthropoidea for a cladistic classification, but the Omomyidae along with the tarsier might still be grouped with the Anthropoidea in the taxon Haplorhini. This assumes that Gingerich (1973) will not be sustained in his claim that ancestral lemuroids (Notharctinae) rather than tarsioids were closer to ancestral Anthropoidea.

Sarich's (1970) estimation of a 30-million-year common ancestry for the platyrrhine–catarrhine lineage is not in close agreement with the opinions of paleontologists. Gingerich (1976*b*) in his overall evaluation of the fossil evidence suggests 40 million years ago as the time of the platyrrhine–catarrhine split. According to Simon's (1976) view of the Euronearctica Wasatchian/Sparnacian as the time and place of platyrrhine–catarrhine divergence, Sarich may be 20 million years too late in his molecular clock timing of the split.

The existence of an evolutionary clock is controversial for several reasons which are reviewed by Wilson *et al.* (1977). Although Wilson *et al.* (1977) support the usefulness of a clock model in phylogenetic investigations, they caution against assumption of a metronomic clock, discussing instead the probabilistic nature of the clock based on stochastic variation in evolutionary events (see also Fitch, 1976; Holmquist *et al.*, 1976; Moore *et al.*, 1976).

In addition, if other evidence suggesting that amino acid substitutions are fixed at variable rates for different times is substantiated, the usefulness of the clock in timing evolutionary events could be seriously challenged. One line of evidence pointing to variability in rates of protein evolution is drawn from maximum parsimony analyses of amino acid sequence data from primate α, β,

Table 2. Deceleration in the Rate of Evolution during Descent to the Human α Hemoglobin Chain

Evolutionary period	Age (million years before present)	Nucleotide replacements per 100 codons per 100 million years
Eutherian to primate ancestor	90–65	34
Primate to anthropoid ancestor	65–40	23
Anthropoid to human ancestor	40–0	7

Table 3. Deceleration in the Rate of Evolution during Descent to the Human β Hemoglobin Chain

Evolutionary period	Age (million years before present)	Nucleotide replacements per 100 codons per 100 million years
Eutherian to primate ancestor	90–65	52
Primate to anthropoid ancestor	65–40	25
Anthropoid to human ancestor	40–0	10

and δ hemoglobin chains, and from carbonic anhydrase I and II (Goodman, 1973; Goodman *et al.,* 1974; Tashian *et al.,* 1976; Romero-Herrera *et al.,* 1979). Figures 3–5 are maximum parsimony trees depicting the branching order of major primate lineages for α and β hemoglobin chains. Tables 2–4 compare the rates of molecular change in various regions of these dendrograms, and rates of molecular change in carbonic anhydrase I and II as well. Table 2 shows a pattern of deceleration in the rate of alpha hemoglobin evolution among primate lineages. The rate of nucleotide replacements (NR) falls from 34 nucleotide replacements per 100 codons per 100 million years in the region of the tree separating the eutherian ancestor from the primate ancestor (approximately 25 million years) to 23 nucleotide replacements per 100 codons per 100 million years between the primate ancestor and the anthropoid ancestor (approximately 25 million years) to a low of 7 nucletoide replacements per 100 codons per 100 million years in the line from the anthropoid ancestor to modern humans (approximately 40 million years). The pattern of deceleration is even more dramatic in the tree depicting β and δ hemoglobin evolution. Table 3 shows the rate of nucleotide replacements decreasing from 52 nucleotide replacements per 100 codons per 100 million years (eutherian ancestor to primate ancestor), to 25 nucleotide replacements

Table 4. Nucleotide Replacements in the Evolution of Carbonic Anhydrase I and Carbonic Anhydrase II

Descent to human carbonic anhydrases	Evolutionary time (from fossil record) (million years before present)	Nucleotide replacements (%)
Carbonic anhydrase I		
Eutherian ancestor to catarrhine ancestor	90–35 M.Y.B.P.	19.3
Catarrhine ancestor to chimp-human ancestor	35–15 M.Y.B.P.	12.5
Chimp-human ancestor to human	15–0 M.Y.B.P.	0.0
Catarrhine ancestor to human	35–0 M.Y.B.P.	7.1
Carbonic anhydrase II		
Eutherian ancestor to catarrhine ancestor	90–35 M.Y.B.P.	30.4
Eutherian to anthropoid ancestor	90–50 M.Y.B.P.	37.5
Anthropoid ancestor to catarrhine ancestor	50–35 M.Y.B.P.	11.3
Catarrhine ancestor to human	35–0 M.Y.B.P.	4.9

per 100 codons per 100 million years (primate ancestor to anthropoid ancestor), to 10 nucleotide replacements per 100 codons per 100 million years (anthropoid ancestor to modern human). A similar slowdown in the rate of change during primate evolution can be seen in the phylogenetic tree for carbonic anhydrase I and II (Table 4). As can be seen in Table 4, the deceleration in rate of change in carbonic anhydrase II is particularly noticeable, as is evidenced by the amount of change found between the times of the eutherian and anthropoid ancestors (37.5% of all nucleotide replacements) compared to the change occurring between the time of the anthropoid ancestor and modern humans (approximately 16% of all nucleotide replacements).

Variability in rates of molecular change within primate lineages has been noted by Cronin and Sarich (1975) who found an accelerated rate of albumin evolution in the catarrhine–platyrrhine ancestor. Maximum parsimony analyses of other primate proteins clearly show a deceleration in the rate of change for the anthropoid primates, particularly those within the Hominoidea. These variations in rates of protein evolution suggest that the clock model is not reliable for dating evolutionary events. Considering the variability in rates of change for one protein over time, together with the evidence that the clock model is not metronomic due to stochastic variation in evolutionary processes, one questions the accuracy of Sarich's 36 million year data for the platyrrhine–catarrhine split. Sarich's hypothesis of ceboid rafting from South America became necessary only after his use of the clock model yielded a divergence time far later than that suggested by the traditional hypothesis. Since precise dates based on the clock model of change are open to serious question, one of the most compelling reasons for considering ceboid oceanic rafting is weakened considerably.

In addition to the aforementioned problems with Sarich's timing of the platyrrhine–catarrhine dichotomy, we can add recent criticisms by Simons (1976) pertaining to the improbability of an African origin for platyrrhine primates. After reviewing zoogeographic data relevant to oceanic rafting and colonization by primates, Simons concluded that "no explanation involving transport across wide reaches of ocean is tenable in accounting for the distribution of any primate. . . . it is more probable that the 36 million year ID divergence date of platyrrhines and catarrhines is wrong than that these animals could have rafted the South Atlantic subsequent to it."

Recently, Gingerich (personal communication) proposed another hypothesis of ceboid origins which assumes an Old World emergence of ceboid ancestors, yet avoids the pitfalls of oceanic rafting. According to Gingerich, the earliest ancestors of New World primates could have originated in Eurasia and migrated to North America via a land bridge across the Bering Straits (see Fig. 6C). This hypothesis permits a more recent divergence of Old and New World monkeys, since it is not constrained by the timing of paleogeographic events, i.e., the separation of Europe and North America.

A variation of the hypothesis of African origin for Ceboidea has been advanced by Hershkovitz (1968). This explanation postulates that during the

Cretaceous, lower primates arose in the prerifted southern continents. During the early stages of continental drift, when the South Atlantic did not constitute an impassable barrier, these lower primates radiated throughout South America and Africa. During the Tertiary, congeneric stocks independently evolved to simian grade and established parallel platyrrhine and catarrhine lineages in South America and Africa respectively (see Fig. 6D). Paleontological support for this hypothesis is nonexistent, since fossil evidence for the existence of primates during the Cretaceous is tenuous, and a record of lower primates in either South America or Africa during the Paleocene or Eocene has not been documented (except possibly for the early Eocene of Algeria).

In conclusion, the genealogical relationships of extant primate species suggested by immunological and biochemical data can accomodate any model of ceboid origins which depicts the lineage ancestral to modern Anthropoidea as first separating from Tertiary strepsirhines and tarsioids before diverging to form the Platyrrhini and Catarrhini. Since the major hypotheses of ceboid origins were designed to explain the emergence of just such a monophyletic Anthropoidea, selection of the hypothesis most nearly describing the true phylogeny is not possible on the basis of present biochemical evidence. Moreover, present biochemical evidence does not reveal if the most recent common ancestor of platyrrhines and catarrhines was morphologically still a lower primate or if it existed late enough in the Tertiary to have reached the simian grade and be called a true anthropoid.

References

Beard, J. M. and Goodman, M., 1976, The hemoglobins of *Tarsius bancanus,* in: *Molecular Anthropology* (M. Goodman and R. E. Tashian, eds.), pp. 239–253, Plenum Press, New York.

Chiarelli, B., 1966, Karyology and taxonomy of the catarrhine monkeys, *Am. J. Phys. Anthropol.* **24**:155–170.

Cronin, J. E., and Sarich, V. M., 1975, Molecular systematics of the New World monkeys, *J. Hum. Evol.* **4**:357–375.

Dene, H. T., Goodman, M., and Prychodko, W., 1976, Immunodiffusion evidence on the phylogeny of the Primates, in: *Molecular Anthropology* (M. Goodman and R. E. Tashian, eds.), pp. 171–195, Plenum Press, New York.

Dietz, R. S., and Holden, J. C., 1970, Reconstruction of Pangea: Breakup and dispersion of continents, Permian to Present, *J. Geophys. Res.* **75**:4939–4955.

Fitch, W. M., 1976, Molecular evolutionary clocks, in: *Molecular Evolution* (F. J. Avala, ed.), pp. 162–178, Sunderland, Sinaeur, Massachusetts.

Gadow, H., 1898, *A Classification of Vertebrates, Recent and Extinct,* Black, London.

Gingerich, P. D., 1973, Anatomy of the temporal bone in the Oligocene anthropoid *Apidium* and the origin of the Anthropoidea, *Folia Primatol.* **19**:239–337.

Gingerich, P. D., 1976*a*, Phylogeny reconstruction and the phylogenetic position of *Tarsius,* in: *Recent Advances in Primatology,* Vol. 3, *Evolution* (D. J. Chivers and K. A. Joysy, eds.), pp. 249–255, Academic Press, New York.

Gingerich, P. D., 1976*b*, Cranial anatomy and evolution of early Tertiary Plesiadapidae (Mammalia, Primates), *Mus. Paleontol. Univ. Mich. Papers Paleontol.* **15**:1–141.

Goodman, M., 1973, The chronicle of primate phylogeny contained in proteins, *Symp. Zool. Soc. Lond.* **133**:339–375.

Goodman, M., 1974, Biochemical evidence on hominid phylogeny, *Ann. Rev. Anthropol.* **3**:203–228.

Goodman, M., 1976, Toward a genealogical description of the primates, in: *Molecular Anthropology* (M. Goodman and R. E. Tashian, eds.), pp. 321–353, Plenum Press, New York.

Goodman, M., and Moore, G. W., 1971, Immunodiffusion systematics of the primates. I. The Catarrhini, *Syst. Zool.* **20**:19–62.

Goodman, M., Fariis, W., Moore, G. W., Prychodko, W., and Sorenson, M. W., 1974, Immunodiffusion systematics of the primates. II. Findings on *Tarsius,* Lorisidae, and Tupaiidae, in: *Prosimian Biology* (R. D. Martin, G. A. Doyle, and A. C. Walker, eds.), Duckworth, London.

Goodman, M., Moore, G. W., and Matsuda, G., 1975, Darwinian evolution in the genealogy of hemoglobin, *Nature* **253**:603–608.

Goodman, M., Czelusniak, J., Moore, G. W., Romero-Herrera, A. E., and Matsuda, G., 1979, Fitting the gene lineage into its species lineage: A parsimony strategy illustrated by cladograms constructed from globin sequences, *Syst. Zool.* **28**:132–163.

Haffer, J., 1970, Geologic climatic history and zoogeographic significance of the Uraba region in northwestern Colombia, *Caldasia* **10**:603–636.

Hamerton, J. L., 1963, Primate chromosomes, *Symp. Zool. Soc. London* **10**:221–219.

Harrington, H. J., 1962, Paleogeographical development of South America, *Bull. Am. Assoc. Petrol. Geol.* **46**:1173–1814.

Hennig, W., 1966, *Phylogenic Systematics,* University of Illinois Press, Urbana.

Hershkovitz, P., 1968, The recent mammals of the Neotropical region: A zoogeographic and ecological review, in: *Evolution, Mammals, and Southern Continents* (A. Keast, R. C. Erk, and B. Glass, eds.), pp. 311–431, State University of New York Press, Albany.

Hershkovitz, P., 1974, A new genus of Late Oligocene monkey (Ceboidea, Platyrrhini) with notes on postorbital closure and platyrrhine evolution, *Folia Primatol.* **21**:1–35.

Hill, W. C. O., 1955, *Primates: Comparative Anatomy and Taxonomy.* Vol. II. *Haplorhini: Tarsioidea,* University of Edinburgh Press, Edinburgh.

Hill, W. C. O., 1957, *Primates: Comparative Anatomy and Taxonomy.* Vol. III. *Pithecoidea,* University of Edinburgh Press, Edinburgh.

Hill, W. C. O., 1960, *Primates: Comparative Anatomy and Taxonomy.* Vol. IV, *Platyrrhini, Cebidae,* Part A, University of Edinburgh Press, Edinburgh.

Hill, W. C. O., 1962, *Primates: Comparative Anatomy and Taxonomy.* Vol. V, *Platyrrhini, Cebidae,* Part B, University of Edinburgh Press, Edinburgh.

Hoffstetter, R., 1972, Relationships, origins, and history of the ceboid monkeys and caviomorph rodents: A modern reinterpretation, in: *Evolutionary Biology,* Vol. 6 (Th. Dobzhansky, M. K. Hecht, and W. C. Steere, eds.), pp. 323–347, Appleton-Century-Crofts, New York.

Hoffstetter, R., 1974, Phylogeny and geographical deployment of the primates, *J. Hum. Evol.* **3**:327–350.

Holmquist, R., Jukes, T. H., Moise, H., Goodman, M., and Moore, G. W., 1976, Evolution of globin family genes, convergence of stochastic and augmented maximum parsimony genetic distance for alpha hemoglobin, beta hemoglobin and myoglobin phylogenies, *J. Mol. Biol.* **105**:39–74.

Hoyer, B. H., and Roberts, R. B., 1967, Studies of nucleic acid interactions using DNA-agar, in: *Molecular Genetics, Part II.* (J. H. Taylor, ed.), pp. 425–479, Academic Press, New York.

Klinger, H., 1963, The somatic chromosomes of some primates: *Tupaia glis, Nycticebus coucang, Tarsius bancanus, Cercocebus aterrimus, Symphalangus syndactylus, Cytogenetics* **2**:140–151.

Klinger, H., Hamerton, J. L., Mutton, D., and Lasig, E. M., 1963, The chromosomes of the Hominoidea, in: *Classification and Human Evolution* (S. L. Washburn, ed.), pp. 235–242, Aldine, Chicago.

Kohne, D. E., 1970, Evolution of higher organism DNA, *Quart. Rev. Biophys.* **3**:327–375.

Kohne, D. E., Chiscon, J. A., and Hoyer, B. H., 1972, Evolution of primate DNA sequences, *J. Hum. Evol.* **1**:627–644.

Le Gros Clark, W. E., 1959, *The Antecedents of Man,* University of Edinburgh Press, Edinburgh.

Moore, G. W., 1971, A Mathematical Model for the Construction of Cladograms, *Inst. Stat. Mimeograph Ser.* 731, North Carolina State University, Raleigh.

Moore, G. W., Barnabas, J., and Goodman, M., 1973, A method for constructing maximum parsimony ancestral amino acid sequences on a given network, *J. Theor. Biol.* **38**:459–485.

Moore, G. W., Goodman, M., Callahan, C., Holmquist, R., and Herbert, M., 1976, Estimation of superimposed mutations in the divergent evolution of protein sequences: Stochastic vs. augmented maximum parsimony method-cytochrome C, *J. Mol. Biol.* **26**:111.

Olson, E. C., 1964, The geology and mammalian faunas of the Tertiary and Pleistocene of South America, *Am. J. Phys. Anthropol.* **22**:217–226.

Orlosky, F. J., and Swindler, D. R., 1975, Origins of New World monkeys, *J. Hum. Evol.* **4**:77–83.

Patterson, B., and Pascual, R., 1968, The fossil mammal fauna of South America, in: *Evolution, Mammals and Southern Continents* (A. Keast, R. C. Erk, and B. Glas, eds.), pp. 247–310, State University of New York Press, Albany.

Pocock, R. L., 1918, On the external characters of lemurs and *Tarsius, Proc. Zool. Soc. London,* **1918**:19–53.

Ramsay, A. T. S., 1971, A history of the formation of the Atlantic Ocean, *Adv. Scientist* **27**:239–249.

Romer, A. S., 1966, *Vertebrate Paleontology,* University of Chicago Press, Chicago.

Romero-Herrera, A. E., Lieska, N., Goodman, M., and Simons, E. L., 1979, The use of amino acid sequence analysis in assessing evolution: A critique, *Biochimie* **61**:767–779.

Sarich, V. M., 1970, Primate systematics with special reference to Old World monkeys, in: *Old World Monkeys: Evolution, Systematics and Behavior* (J. R. Napier and P. H. Napier, eds.), pp. 175–266, Academic Press, New York.

Sarich, V. M., and Cronin, J. E., 1976, Molecular systematics of the primates, in: *Molecular Anthroplogy* (M. Goodman and R. E. Tashian, eds.), pp. 141–170, Plenum Press, New York.

Schwartz, J. H., Tattersall, I., and Eldredge, N., 1978, Phylogeny and classification of the primates revisited, *Yearb. Phys. Anthropol.* **21**:95–133.

Simons, E. L., 1961, The dentition of *Ourayia:* Its bearing on relationships of omomyid prosimians, *Postilla* **54**:1–20.

Simons, E. L., 1972. *Primate Evolution,* Macmillan, New York.

Simons, E. L., 1976. The fossil record of primate phylogeny, in: *Molecular Anthropology* (M. Goodman and R. E. Tashian, eds.), pp. 35–60, Plenum Press, New York.

Simpson, G. G., 1945, The principles of classification and a classification of mammals, *Bull. Am. Mus. Nat. Hist.* **85**:1–350.

Tashian, R. E., Goodman, M., Ferrell, R. E., and Tanis, R. J., 1976, Evolution of carbonic anhydrase in primates and other mammals, in: *Molecular Anthropology* (M. Goodman and R. E. Tashian, eds.), pp. 301–319, Plenum Press, New York.

Von Koenigswald, G. H. R., 1968, The phylogenetic position of the Hylobatinae, in: *Taxonomy and Phylogeny of Old World Primates with References to the Origin of Man* (B. Chiarelli, ed.), pp. 271–276, Rosenberg and Sellier, Torino.

Whitmore, F. C., and Stewart, R. A., 1965, Miocene mammals and Central American seaways, *Science* **148**:180–185.

Wilson, J. A., 1966, A new primate from the earliest Oligocene, West Texas: Preliminary report, *Folia Primatol.* **4**:227–248.

Wilson, A. C., Carlson, S. S., and White, T. J., 1977, Biochemical evolution, *Annu. Rev. Biochem.* **46**:573–639.

Woodring, W. P., 1954, Caribbean land and sea through the ages, *Bull. Geol. Soc. Am.* **65**:719–732.

Synthesis, Perspectives, and Conclusions VIII

Phyletic Perspectives on Platyrrhine Origins and Anthropoid Relationships

22

E. DELSON and A. L. ROSENBERGER

Introduction

As the editors of this volume describe in their preface (Ciochon and Chiarelli, 1980), the preceding papers were solicited from researchers in various disciplines so that we could collectively examine a set of interrelated questions: (1) What are the paleontological origins of the New World monkeys?, (2) what is the nature of the phylogenetic affinity between the catarrhine and platyrrhine primates?, and (3) what is the significance of these questions, and their resolution, for understanding the influence of continental drift upon the modern distributional patterns of the anthropoid primates? We have been asked to evaluate the status of Questions 1 and 2, which are essentially phylogenetic problems, on the basis of the foregoing contributions as well as our own respective researches. We have attempted to do so by reiterating some of the more salient arguments in capsule form and pointing out what we feel are their strengths and weaknesses (see summary in Tables I–III). Our conclusion—in brief—is that a substantial set of first steps has been taken, largely due to the multi-disciplinary persuasion of the contributors to this volume, but many important problems remain: the data on living platyrrhine comparative morphology is still meager; the fossil record of platyrrhines is sparse but tantalizing; comparisons of early catarrhines and platyrrhines have

E. DELSON • Department of Vertebrate Paleontology, American Museum of Natural History, New York, New York 10024, and Department of Anthropology, Lehman College, CUNY, Bronx, New York 10468. A. L. ROSENBERGER • Department of Anthropology, University of Illinois at Chicago Circle, Box 4348, Chicago, Illinois 60680.

hardly begun; too little is still known of omomyid (and adapid) crania and postcrania; the somewhat better-known adapids (not to mention the rarer omomyids) are still poorly understood phyletically; and, especially, without a clear genealogical picture of platyrrhine, catarrhine, and interanthropoid relationships, no scientific model of their deployment can be synthesized.

The papers in this volume reflect a diversity of methods that is both healthy and indicative of the breadth of the attack on these problems, and we doubt that procedural and philosophical differences are significantly responsible for the lack of a consensus on a number of fundamental issues. However, much of the data that has been generated comes in the wake of the featured debates of the last decade, contrasting the strepsirhine–haplorhine, simio–lemuriform and prosimian–anthropoid dichotomous models of primate evolution. It seems timely now to recast our questions, and perhaps our search for fossils, if we are to make more rapid progress toward solving the problems of platyrrhine origins and platyrrhine–catarrhine relationships.

Consideration of these two questions began in the late 19th Century. Anatomists early recognized some major distinctions between New and Old World monkeys, reconciling them as examples of convergent evolution. *Tarsius* was also seen to have closer ties to the anthropoids than to Lemuriformes on the basis of placentation and cerebral arteries. Meanwhile, some paleontologists proposed that the "lemur-like" *Notharctus* was ancestral to platyrrhines. Later, this view was extended to view tarsiiforms as catarrhine ancestors, implying anthropoid polyphyly. Thus, both the approaches and the hypotheses of this volume are rooted in the earliest interpretive works on primate evolution.

Platyrrhine Relationships

The primary orientation of this volume, which focuses on issues of anthropoid origins from the perspective of the New World monkeys, is appropriate for a number of reasons. Not only are the platyrrhines more conservative than catarrhines in many aspects of their morphology, but they have also been shown to represent actual, rather than purely hypothetical, analogs of early catarrhine behaviors and adaptations (e.g., Fleagle, 1980). Nevertheless, fundamental to their heuristic utilization as models of the extinct early catarrhines is the development of a coherent picture of platyrrhine genealogy, which seems far from achieving a uniformity of opinion. For example, the prolonged debate over the ancestral or derived nature of marmoset morphology has important implications for understanding the evolutionary transition marking the rise of the anthropoids. Were primitive platyrrhines, and protoanthropoids, small-bodied, scansorial, claw-bearing frugivore-insectivores (i. e., marmoset-like) or not? If not, what taxon or phyletic group does most closely approximate our expectations of the kind of animal that was an early anthropoid? Perhaps even more important is an appreciation of the morphological pattern thought to have characterized the earliest New World monkeys, for that suite of features is prerequisite to the establishment of the phylogenetic relationships of the catarrhines and platyrrhines.

Whereas a number of contributors to this volume have concluded that the claw-bearing marmosets, Callitrichinae, are quite derived in aspect (e. g., Luckett, 1980; Bugge, 1980; Maier, 1980; Kay, 1980; Hoffstetter, 1980; Gantt, 1980; Martin and Gould, 1980; see also Rosenberger, 1977, 1979), marking somewhat of a transition from the prevailing opinion of preceeding decades (e.g., Le Gros Clark, 1959; Napier and Napier, 1967; Hershkovitz, 1977 and before), the details of marmoset and nonmarmoset interrelationships are not agreed upon or even well established in certain cases. To some extent, this is due to a genuine lack of information and the still underdeveloped interest in platyrrhine biology. On the other hand, it seems true also that most current students continue to employ the conventional marmoset vs. nonmarmoset perspective for framing their questions and interpreting their data. Rosenberger (1977, 1979) and some others (e. g., Egozcue and Perkins, 1971; Romero-Herrera *et al.,* 1976, 1978; Dene *et al.,* 1976) have contested the phylogenetic accuracy of that distinction, and we have attempted to document (e. g., Szalay and Delson, 1979) an alternative dichotomy based upon a cladistic split between atelids (*Aotus, Callicebus,* saki–uakaris, atelines) and cebids (cebines and callitrichines). Thus far, this notion has received little support from immunological efforts, although the DNA sequencing data of Romero-Herrera and colleagues uphold the major outlines of this interpretation as a parsimonius possibility. The biomolecular-based contributions of this volume (Sarich and Cronin, 1980; Baba *et al.,* 1980) are not mutually consistent and present a number of significant problems. It seems especially important, for example, to determine why the albumin and transferrin data seem to have low resolving power beyond a few almost universally accepted phyletic groupings (*Pithecia+Cacajao*; *Ateles+Lagothrix+Alouatta*; Callitrichinae), and why the Cronin–Sarich estimates of divergence times predict that no relatives of the living forms would exist prior to 15–20 million years ago. The fossil record establishes almost uneqivocally that platyrrhines were present as early as 35 million years ago and that species exceedingly like, if not ancestral to, the living squirrel monkey (*Dolichocebus*) and the owl monkey (*Tremacebus*) existed 20–25 million years ago. Given our limited knowledge in this area, we note only that the validity of the molecular clock must continue to be seriously questioned, especially since internal analyses have shown that many of the macromolecules used in clock construction do not evolve at mutually consistent rates (Corruccini *et al.,* 1979).

Platyrrhine Origins

As several authors state or imply, the question of platyrrhine origins may be evaluated within the framework of either of two alternative phyletic approaches: (1) ancestor–descendant, lineage hypotheses or (2) sister–taxon, cladistic hypotheses. The latter, of course, is an indirect approach to the issue of *origins,* but represents a less complex first step which may remain the only course when the nature of the data so dictates (e. g., in neontological work) or when ancestor–descendant hypotheses are nullified. It is worth pointing out in this regard that except for amino acid sequencing, which assesses the transformation of specific, unit character states from one condition to another, none of the molecular evidence is truly comparable to the essence of cladistic analysis, the inference of shared, homologous derived features. Thus, although couched in cladistic terminology in that clusters of taxa are recognized as "clades" (implying a unique common ancestry) the foundation of such analyses (e. g., Sarich and Cronin, 1980) is essentially phenetic. We do not wish to minimize the significance of phenetic studies, but merely point out that *we* prefer them to play an auxiliary role in the establishment of genealogical relationships.

While a number of authors have suggested definite scenarios of platyrrhine origins, we consider all of these as highly speculative or lacking in robusticity. Proponents of a polyphyletic origination model (e.g., Chiarelli, 1980; Perkins and Meyer, 1980) bear the burden of refuting the contradictory morphological evidence which implies that platyrrhines are in fact monophyletic (see Table I). This objection stands irrespective of the ancestral stock(s) from which these workers would derive the New World monkeys. In advocating a dual origin involving both adapids *and* omomyids, Perkins and Meyer have essentially resurrected the early 20th century hypotheses noted above. In this form, however, it is based on neontological rather than paleontological evidence, thus having little resolution as far as descent is concerned.

Hoffstetter's (1980 and before) argument for the descent of platyrrhines from catarrhines via the Parapithecidae has been specifically considered by a number of workers (Rosenberger, 1979; Szalay and Delson, 1979; Kay, 1980). All of these are firmly in opposition, citing the uniquely derived attributes of parapithecids (relative to eucatarrhines) or platyrrhines (relative to catarrhines) which militate against Hoffstetter's hypothesis (see Table I). Although *Parapithecus* and *Apidium* may resemble *some* platyrrhines in *some* features, these appear to be conservative retentions from the last common ancestor of anthropoids and thus do not signify a special relationship between parapithecids and New (or Old) World monkeys. Furthermore, it is becoming increasingly well established (Szalay and Delson, 1979; Fleagle and Simons, 1979) that parapithecids are dentally derived by comparison to other Fayum catarrhines but are more conservative than cercopithecids and pongines in lacking ischial callosities and in retaining P2. In sum, the evidence suggests that parapithecids are a collateral branch of the catarrhines which did not give rise to any of the living anthropoids.

A similar set of anatomical features and phyletic arguments are applicable to any hypothesis which postulates the descent of platyrrhines from a bona fide catarrhine stock (e.g., Falk, 1980). Even the most basic of catarrhine

Table I. Some Characters of Selected Higher Primate Morphotypes[a]

	Character	Source
Platyrrhines		
D	Hypoconulid absent on M_3	Kay (1980)
D	Metaconules highly reduced with paraconules probably absent	Rosenberger (1979)
D	Zygomatico-parietal pterion with lateral orbital fissure	Rosenberger (1977)
D	Intraplacental maternal vessels present; placental hematopoiesis present	Luckett (1980)
D	Reduction of nasal wing cartilages; enlarged embryonic nasal capsule	Maier (1980)
Catarrhines		
D	Presence of facet "X" on lower molars	Kay (1980)
D	Presence of hypoconulid on $M_{1,2}$	Szalay and Delson (1979)
D	Loss of lateral orbital fissure	Cartmill (1980)
D	Reduction of presphenopalatine lamina of palatine	Cartmill (1980)
D	Placental disk villous; cytotrophoblastic shell well developed	Luckett (1980)
D	Narrow internarial septum with reduction of wing cartilages and olfactory scrolls; loss of vomeronasal organ of Jacobson	Maier (1980)
Anthropoids		
D	I^2 conical and robust	Rosenberger and Szalay (1980)
?D	Thickened enamel on lower anterior premolar	Kay (1980)
P	Loss of P1; mesiodistally "crowded" premolars	Kay (1980)
D	Symphyseal fusion	Kay (1980) (?D; and adapids)
P	Type IIB enamel prism pattern	Gantt (1980)
D	Postorbital septum complete or nearly so	Cartmill (1980) (D; and tarsiers)
D	Trabeculate hypotympanic sinus	Rosenberger and Szalay (1980)
D	Loss of stapedial artery	Bugge (1980); Rosenberger and Szalay (1980)
?A	Ophthalmic artery arises from internal carotid	Bugge (1980) (D)
D	Presence of transverse central cerebral sulcus	Falk (1980)
D	Expanded visual cortex and associated sulci	Falk (1980)
D	Reduced lesser trochanter of femur	Ford (1980) (?D; ?P)
D	Loss of femoral third trochanter	Ford (1980) (?D; ?P)
D	Distal femoral epiphysis anteroposteriorly compressed	Ford (1980) (?D; ?P)
P	Karyotypic similarity	Chiarelli (1980)
D	Primordial amniotic cavity present: amniogenesis by cavitation; bidiscoidal hemochorial placenta; blastocyst attachment by embryonic pole; primary and secondary yolk sac present; trabecular disk uterus simplex; rudimentary villous anchoring; no head-to-head sperm agglutination; sublingua absent	Luckett (1980)

[a] Our interpretation of the polarity or status [see Rosenberger (1979) for methods] of each feature is indicated in the left column according to the following conventions: A, ancestral, shared with a sister taxon; D, uniquely derived by comparison to sister taxon; C, convergent, nonhomologous similarity; P, phenetic similarity whose phyletic significance we cannot infer. The right column lists the sources for each character and their interpretation when different from our own. Although we have not attempted to assess character correlation in this tabulation, several sets of character states have been grouped for convenience.

molar patterns (excluding the two poorly-known forms *Oligopithecus*—which we consider probably nonanthropoid—and *Pondaungia*) is too derived to have been ancestral to that of the platyrrhines unless a number of reversals can be documented (see Table I). This implies that any presumptive platyrrhine ancestor inhabiting the Old World would not be regarded as a catarrhine (even on the basis of the Atlantic Ocean as a major diagnostic feature), but rather as a protoanthropoid.

Anthropoid Origins

Given that none of the known anthropoids is ancestral to platyrrhines (or to catarrhines), the next questions for consideration relate to the monophyly of anthropoids and their relationships to other primates. The majority of authors in this volume have accepted the concept that anthropoids are monophyletic (Table I), thus implying the prior existence of an ancestral species which displayed at least some of the characteristic anthropoid morphology. On the other hand, no authoritative response has yet been counterposed to widespread doubts as to anthropoid monophyly (e.g., Simpson, 1945; Gazin, 1958; Simons, 1972; Cachel, 1979). Other than brief reviews such as this one, there is still no published, detailed objective analysis of the anthropoid morphotype which goes beyond conventional wisdom and the *scala naturae,* such as that provided by Le Gros Clark (1959). The concern over monophyly has largely been based upon the supposition that the postorbital septum evolved convergently among platyrrhines and catarrhines, coupled with a healthy mistrust of the zoogeographic requirements engendered by the monophyly hypothesis. The ontogenetic and distributional patterning of the bony mosaic at the pterion among all primates is a topic worthy of detailed analysis. Major distinctions do contrast platyrrhines and catarrhines (see also Rosenberger, 1977; Cartmill, 1980), and these probably do bear on the evolution of postorbital closure.

Nonetheless, following the consensus of this volume, we may turn to an assessment of the ancestry of the earliest anthropoid, a problem much debated of late as a result of the prominent controversy among primate systematists during the past decade. Most of the morphological and biochemical evidence seems to support the view that the haplorhine primates (anthropoids plus tarsiiforms) are also monophyletic (Table II and below; see Rosenberger and Szalay, 1980; Kay, 1980; Hoffstetter, 1980). Gingerich (1980), however, argues that this interpretation is incorrect. He suggests, alternatively, that the living lemuriforms are more closely related to anthropoids than is *Tarsius,* and that Eurasian adapids were ancestral to both the living strepsirhines and the anthropoids. Cartmill and Kay (1978) have provided a shred of indirect support for Gingerich's thesis by questioning the traditional acceptance of a close relationship between lemuriforms and adapids and hinting that the latter

Table II. Some Characters Common to Tarsiiforms and Anthropoids[a]

Tarsiiforms and anthropoids			
1.	P	Semispatulate incisors variably present	Orlosky (1980); Rosenberger and Szalay (1980)
2.	P	Mesiodistally "crowded" premolars	Kay (1980) (D)
3.	P	Nannopithex-fold replaced by postprotocrista (variably)	Kay (1980) (D)
4.	P	Premetacristid well developed on $M_{2,3}$ (variably)	Kay (1980) (D)
5.	P	Trigonid low, talonid basin expanded (variably)	Kay (1980) (D)
6.	P	Reduced lower third molars (variably)	Kay (1980) (D)
7.	D	Short, deep, low-hafted facial skull	Rosenberger and Szalay (1980)
8.	D	Apical interorbital septum	Luckett (1980)
9.	D	Diminished nasal fossa; probable lack of olfactory recess	Rosenberger and Szalay (1980)
10.	D	Reduced stapedial artery; enlarged promontory artery	Rosenberger and Szalay (1980)
11.	D	Medially positioned carotid foramen	Rosenberger and Szalay (1980)
12.	D	Anteromedially enlarged hypotympanic sinus	Rosenberger and Szalay (1980)
13.	?D	Downturned humeral trochlea	Rosenberger and Szalay (1980)
14.	D	Enlarged occipital lobes; reduced olfactory lobes	Rosenberger and Szalay (1980)
15.	?D	Loss of coronolateral sulcus	Rosenberger and Szalay (1980)
Tarsius and Anthropoids			
1.	?D*	Haired rhinarium; fused nasal processes	Luckett (1980)
2.	D*	No choriovitelline placenta; rudimentary allantois; well-developed body stalk; ovarian bursa reduced or absent; primordial amniotic cavity transitory; invasive attachment; monodiscoidal hemochorial placenta	Luckett (1980)
3.	C	Postorbital septum	Cartmill (1980) (D)
4.	P*	Presence of fovea centralis	Cartmill (1980) (D)
5.	C	Anterior position of carotid foramen	Cartmill (1980) (D)
6.	A	Incipient enlargement of internal carotid artery	Bugge (1980) (D)

[a] "Variable" features are not present in all taxonomic groups; asterisked features not observable in fossils. For key to symbols see also notes to Table I.

group may be closer to the haplorhine clade, with the lemuriforms and plesiadapiforms being somewhat further removed. We regard both of these as less likely hypotheses (see also Rosenberger and Szalay, 1980), further suggesting that other resemblances between adapids and lemuriforms (e.g., the freely suspended ectotympanic and the lack of an ossified annulus membrane [=? reduced linea semicircularis]) may well turn out to be synapomorphies.

Gingerich's (1980 and before) hypothesis of adapid-anthropoid ties is predicated upon (1) the presence of more than a dozen itemized points of resemblance shared between them (Table III); (2) recognition of presumed morphologically intermediate forms that are difficult to allocate [e.g., *Pro-*

Table III. Some Characters Common to Adapids and Anthropoids

1.	C	Body size greater than 500 g	Gingerich (1980)[a]
2.	C	Tendency to fuse the mandibular symphysis	Gingerich (1980); Kay (1980) (?D)
3.	C	Vertical, spatulate incisors	Gingerich (1980)
4.	C	I_1 smaller than I_2	Gingerich (1980)
5.	C	Interlocking canine occlusion	Gingerich (1980)
6.	C	Canines moderately large and projecting	Orlosky (1980) (P)
7.	C	Canines sexually dimorphic	Gingerich (1980)
8.	A	Canine-premolar "honing"	Gingerich (1980); Kay (1980)
9.	P	Molarized P4	Gingerich (1980)
10.	P	Tendency toward quadrate lower molars	Gingerich (1980)
11.	A	Nontubular [partially free] ectotympanic[b]	Gingerich (1980)
12.	P	Relatively short calcaneum	Gingerich (1980)
13.	A	Unfused tibia-fibula	Gingerich (1980)

[a] None of the characters enumerated by Gingerich were stipulated as shared, derived conditions—merely as similarities indicative of close relationship. See also notes to Table I.
[b] Gingerich has claimed that the ectotympanic is partially free in early anthropoids. See text, p. 453 for our refutation of this claim.

toadapis ("*Cercamonius*") *brachyrynchus, Amphipithecus, Hoanghonius, Oligopithecus* and *Pondaungia*]; (3) the intermediate stratigraphic position of these dubious taxa and the continuous nature of the Paleogene primate record; and (4) the geographic distribution of adapids and early anthropoids. As examples of "extrinsic" nongenetic evidence, we regard the last three points as having only a secondary relevance to the issue. A phyletic hypothesis should be based upon testable statements about homologous similarities. Other forms of information may sharpen the argument but cannot supersede morphology and genealogical reasoning, either positive or negative. Moreover, the fossil records of adapids and omomyids are in fact replete with stratigraphic and morphologic gaps; uncertainty about the evolutionary significance of incomplete fossils should militate against their being used in grand hypotheses; and the temporal sequence of *taxa* has far less significance than the temporal sequence of *characters,* which tells us little in this case.

The morphological evidence for adapid-anthropoid links also suffers upon close scrutiny. Many of the characters involved are probably correlated, a point often glossed over by most workers, including ourselves (e.g., features 2–4, 5–8, and 12–13 of Table III), so their sheer number is not as impressive as it might seem. Some of these resemblances are likely to represent convergences on the anthropoid condition (characters 2, 3–4, 5–8, and 9–10; see Cartmill and Kay, 1978; Kay, 1980; Rosenberger and Szalay, 1980) or are primitive for the euprimates (features 10, 12, and 13) or otherwise are of limited genealogical value (condition 1). Some authors have employed terms such as *spatulate* incisors, *molarized* premolars, and *quadrate* lower molars in describing shared character states among these primates. Such biologically imprecise terms do not permit clear understanding of the details of any potential similarity, so that determination of homology vs. convergence is not possible in these cases. Moreover, one feature (number 11 of Table III) is based on a specimen which we suggest may be misidentified. The only evidence that any anthropoids ever had a free, intrabullar ectotympanic comes from a broken bone allocated by Gingerich (1973) to *Apidium* (in part on the basis of its recovery alongside a molar of that genus). Such features of this presumed squamous temporal fragment as (1) the orientation of the "zygomatic process"; (2) the morphology of the "postglenoid process" and its surrounding anatomy; (3) the very large size of the bone by comparison to other fragments of *Apidium*; and (4) the extreme lateral position of the "ectotympanic" inferred by Gingerich lead us to doubt that this bone derives from a primate, much less represents the otherwise well-known *Apidium phiomense*. Finally, Gingerich and other proponents of the adapid ancestry hypothesis have not adequately dealt with much of the positive evidence supporting the tarsiiform-anthropoid theory [although Gingerich (1980) has made several important points in this vein]. If we are to believe that Adapidae is the sister-taxon of anthropoids, we must be persuaded by morphological and systematic argument that the characters identified as haplorhine synapomorphies (Table II) are either conservative retentions or nonhomologous (convergent) shared traits. To ignore counterarguments is not to refute them.

As noted above, we believe that the tarsiiform hypothesis of anthropoid origins, which presumes that the protoanthropoid was omomyid-derived, is the best available interpretive scheme for explaining the bulk of the evidence. The strength of this hypothesis lies in the complementary nature of the results from character analyses of a variety of data sets obtained from both extant and extinct taxa (Table II) combined with the phenetic support from biomolecular studies (for example, see Baba *et al.*, 1980). Moreover, the incorporation of nullifying counterarguments against the adapid-anthropoid alternative scheme allows us to reject opposing interpretations based on the same anatomical systems (see above). Clearly, additional work can further sharpen this hypothesis by excluding many of the known genera or lineages from potential ancestral status (see Kay, 1980; Rosenberger and Szalay, 1980) and by the recovery of more informative cranial and postcranial remains.

Cartmill (Cartmill, 1980; Cartmill and Kay, 1978) has attempted to go beyond this conservatively vague statement of tarsiiform-anthropoid affinities in offering the intriguing hypothesis that *Tarsius* itself, rather than some unknown or unrecognized tarsiiform or omomyid, is most closely related to anthropoids. Some of the evidence against this view has been presented by Rosenberger and Szalay (1980), but Cartmill (1980) has marshalled additional points in support. It appears to us that the key to this question lies in comparisons between *Tarsius* and microchoerine omomyids, some of which share with *Tarsius* such derived features (by comparison to strepsirhines and/or *Rooneyia*) as tibio-fibular fusion and major calcaneal elongation (see Gingerich, 1980; Szalay and Delson, 1979), a narrow interorbital region and somewhat enlarged orbits (Cartmill and Kay, 1978), and a secondarily narrowed external auditory tube and reduced subtympanic recess of the bulla (Rosenberger and

Szalay, 1980). If these homologies and polarities are correct, the hypothesis most compatible with the many autapomorphies of *Tarsius* would recognize microchoerines, rather than anthropoids, as the closest relatives of tarsiers. This concept, supported by Simons (1961) but rejected by Szalay (1976), requires further analysis before it will be widely accepted.

Furthermore, we remain unconvinced that Cartmill's (1980) admittedly fragile reconstruction of orbit and eyeball evolution among the haplorhines is correct (see also Rosenberger and Szalay, 1980). The enormous bony ring and flanges which make up the tarsier eye socket resemble those of *Aotus,* whose ocular and orbital morphology is derived among platyrrhines (Rosenberger, 1979). Whatever advantage a postorbital enclosure might provide when a retinal fovea is present, as it is in *Tarsius* (apparently) and in anthropoids other than *Aotus,* the anatomical association of these two structures need not be causally linked. Although Cartmill (1980) implied that all strepsirhines have a tapetum lucidum while all haplorhines (save *Aotus*) possess a fovea, the literature is replete with queries to this simple picture. Pariente (1979) has reported foveae in *Lemur catta* and *Hapalemur griseus,* both of which lack an anthropoid-like postorbital septum, while Wolin and Massopust (1970) indicate doubts about the presence of a true fovea in *Tarsius* and the distribution of tapeta in strepsirhines. Cartmill (1980) has suggested that *Tetonius,* an early omomyid, may have possessed a tapetum on the basis of its relatively large orbital size, but we offer an alternative interpretation. Cartmill and Kay (1978) indicated that smaller species have relatively larger orbits than do larger relatives, and most mammals do not have either a tapetum or a fovea, suggesting this lack to be the ancestral condition. If *Tetonius* (and by implication other omomyids) were diurnal animals lacking either derived feature, the eyes would have been large to gather the unconcentrated light, especially in a small animal which was vision-oriented. Such a conservative omomyid might give rise to diurnal foveate anthropoids, while the microchoerines and *Tarsius* might have evolved parallel, canalized specializations independently, involving both the fovea and the postorbital septum.

Conclusion

In summation, we agree with the majority of authors in this volume in supporting strict monophyly of both catarrhines and platyrrhines (although we offer a different internal arrangement of the ceboids). Anthropoids, too, are most likely monophyletic, with the earliest representatives presenting at least some of the synapomorphies listed in Table I. Such an early anthropoid would not have been greatly similar in dental details to any of the known Oligocene to modern platyrrhines or catarrhines. Comparing the several most widely accepted hypotheses of origin for ancestral anthropoids, we think that the tarsiiform genealogical tie is the most firmly established (Table II). Not only does morphology (Table III and refutation above) not support a set of homologous synapomorphies between adapids and anthropoids, but there are important stratigraphic lacunae in the supposed continuum as well. All suggested refinements of the tarsiiform–anthropoid concept suffer from significant difficulties and appear to be based mainly on negative evidence, essentially related to our limited knowledge of omomyid morphology and interrelationships. Microchoerines may be the sister-taxon of modern *Tarsius,* but it is doubtful that this clade is especially close to the protoanthropoids. On the other hand, we suggest that features of the anterior dentition in forms such as *Arapahovius* (perhaps *Tetonius*) and *Ourayia,* which are little if at all known cranially or postcranially, probably include derived homologies shared with anthropoids. Thus we take the conservative stand that the ancestral higher primate originated somewhere in or near the Omomyidae.

Finally, in the spirit of speculation (and of paleogeography, to which this book is dedicated), we offer our current deployment scenario, already put forward in essence by Szalay and Delson (1979). It appears that the east Asian ?adapid *Lushius* and omomyid *Altanius* have their most significant morphological resemblances to western North American anaptomorphine omomyids, lending primate support to the Bering connection as a mammalian migration route during the Eocene. Similarly, a primate connection between eastern Asia and Africa is suggested by (1) the disjunct presence of *Hoanghonius* and *Oligopithecus,* both probably nonanthropoid; and (2) a possible phyletic link between the still poorly-known *Pondaungia* of Burma and the Fayum catarrhines (see Szalay and Delson, 1979; Gingerich, 1980; Kay, 1980). Recent studies of Mediterranean rodents (Adrover *et al.,* 1978), Turkish embrithopods (Sen and Heintz, 1979) and Pakistani proboscideans and cetaceans (West, 1980) suggest further links of these regions and taxa to Fayum relatives. Thus, as Gingerich (1980) delineates in his Fig. 4 (but with different taxa involved), some early euprimates could have occupied a single biotic community spanning the circum-Pacific region and differentiated there into the protoanthropoid stock. With the apparent world-wide oceanic regression during the late Eocene, the formative catarrhine branch (of which *Pondaungia* may represent an offshoot) might have crossed the narrowing western Tethys and entered Africa, while the protoplatyrrhines managed to cross into South America from the north (see also Wood, 1980).* As with all paleogeographic hypotheses, this one is not easily amenable to testing in the precise manner applicable to morphological theories, but must stand or fall on consensus

*Many authors in this volume have preferred a trans-Atlantic rafting dispersal of protoanthropoids from Africa to South America. We reject such dispersal not only because of the problems of dehydration, salt poisoning, and exposure facing any rafted primate unable to estivate, but also on phyletic grounds. No known Old World anthropoid is conservative enough to be ancestral to platyrrhines, even the earliest of which lack several of the catarrhine derived characters found in Fayum and Pondaung fossils. Thus, rafting requires postulation of an unknown source group, as well as serendipitous paleocontinental relationships and paleooceanographic conditions.

analyses of a variety of data. We await the next incarnation of this volume (or at least of the questions it has posed) for such a consensual evaluation.

ACKNOWLEDGMENTS

We thank Drs. Ciochon and Chiarelli for inviting us to prepare this summary, which represents solely our own views, and our colleagues who wrote stimulating papers and permitted us to analyze them in this prepublication manner. This study was supported (in part) by research grants to E. D. from the PSC-BHE Faculty Research Award Program of City University of New York (Nos. 12188 and 12985) and the National Science Foundation (BNS79-15091) and by a postdoctoral fellowship to A.L.R. from the Department of Anatomical Sciences, State University of New York, Stony Brook.

References

Adrover, R., Hugueney, M., Moya, S., and Pons, J., 1978, Paguera II, nouveau gisement de petits mammifères (Mammalia) dans l'Oligocène de Majorque (Baléares, Espagne), *Nouv. Arch. Mus. Hist. Nat. Lyon,* fasc. 16, suppl., pp. 13–15.

Baba, M., Darga, L., and Goodman, M., 1980, Biochemical evidence on the phylogeny of Anthropoidea, in: *Evolutionary Biology of the New World Monkeys and Continental Drift* (R. L. Ciochon and A. B. Chiarelli, eds.), p. 423–443, Plenum Press, New York.

Bugge, J., 1980, Comparative anatomical study of the carotid circulation in New and Old World primates. Implications for their evolutionary history, in: *Evolutionary Biology of the New World Monkeys and Continental Drift* (R. L. Ciochon and A. B. Chiarelli, eds.), pp. 293–316, Plenum Press, New York.

Cachel, S., 1979, A functional analysis of the primate masticatory system and the origin of the anthropoid post-orbital septum, *Am. J. Phys. Anthropol.* **50**:1–18.

Cartmill, M., 1980, Morphology, function, and evolution of the anthropoid postorbital septum, in: *Evolutionary Biology of the New World Monkeys and Continental Drift* (R. L. Ciochon and A. B. Chiarelli, eds.), pp. 243–274, Plenum Press, New York.

Cartmill, M., and Kay, R., 1978, Cranio-dental morphology, tarsier affinities and primate suborders, in: *Recent Advances in Primatology,* Vol. 3 (D. J. Chivers and K. A. Joysey, eds.), pp. 205–214, Academic Press, London.

Chiarelli, A. B., 1980, The karyology of South American primates and their relationship to African and Asian species, in: *Evolutionary Biology of the New World Monkeys and Continental Drift* (R. L. Ciochon and A. B. Chiarelli, eds.), pp. 387–398, Plenum Press, New York.

Ciochon, R. L., and Chiarelli, A. B., 1980, Preface, in: *Evolutionary Biology of the New World Monkeys and Continental Drift* (R. L. Ciochon and A. B. Chiarelli, eds.), pp. vii–ix, Plenum Press, New York.

Corruccini, R. S., Cronin, J., and Ciochon, R. L., 1979, Scaling analysis and congruence among anthropoid primate macromolecules, *Hum. Biol.* **51**:167–185.

Dene, H. T., Goodman, M., and Prychodko, W., 1976, Immunodiffusion evidence on the phylogeny of the primates, in: *Molecular Anthropology* (M. Goodman and R. Tashian, eds.), pp. 171–195, Plenum Press, New York.

Egozcue, J., and Perkins, E. M., 1971, Chromosomal evolution in the Platyrrhini, *Proc. Third Internat. Cong. Primatol.*, Vol. 2, pp. 131–134, S. Karger, Basel.

Falk, D., 1980, Comparative study of the endocranial casts of New and Old World monkeys, in: *Evolutionary Biology of the New World Monkeys and Continental Drift* (R. L. Ciochon and A. B. Chiarelli, eds.), pp. 275–292, Plenum Press, New York.

Fleagle, J. G., 1980, Locomotor behavior of the earliest anthropoids, *Z. Morphol. Anthropol.* **71**:149–156.

Fleagle, J. G., and Simons, E. L., 1979, Anatomy of the bony pelvis in parapithecid primates, *Folia Primatol.* **31**:176–186.

Ford, S. M., 1980, Phylogenetic relationships of the Platyrrhini: The evidence of the femur, in: *Evolutionary Biology of the New World Monkeys and Continental Drift* (R. L. Ciochon and A. B. Chiarelli, eds.), pp. 317–329, Plenum Press, New York.

Gantt, D. G., 1980, Implications of enamel prism patterns for the origin of the New World monkeys, in: *Evolutionary Biology of the New World Monkeys and Continental Drift* (R. L. Ciochon and A. B. Chiarelli, eds.), pp. 201–217, Plenum Press, New York.

Gazin, C. L., 1958, A review of the Middle and Upper Eocene primates of North America, *Smithsonian Misc. Coll.* **136**:1–112.

Gingerich, P. D., 1973, Anatomy of the temporal bone in the Oligocene anthropoid *Apidium* and the origin of the Anthropoidea, *Folia Primatol.* **19**:329–337.

Gingerich, P. D., 1980, Eocene Adapidae, paleobiogeography and the origin of South American Platyrrhini, in: *Evolutionary Biology of the New World Monkeys and Continental Drift* (R. L. Ciochon and A. B. Chiarelli, eds.), pp. 123–138, Plenum Press, New York.

Hershkovitz, P., 1977, *Living New World Monkeys (Platyrrhini)*, Vol. 1, University of Chicago Press, Chicago.

Hoffstetter, R., 1980, Origin and deployment of New World monkeys emphasizing the southern continents route, in: *Evolutionary Biology of the New World Monkeys and Continental Drift* (R. L. Ciochon and A. B. Chiarelli, eds.), pp. 103–122, Plenum Press, New York.

Kay, R. F., 1980, Platyrrhine origins: A reappraisal of the dental evidence, in: *Evolutionary Biology of the New World Monkeys and Continental Drift* (R. L. Ciochon and A. B. Chiarelli, eds.), pp. 159–187, Plenum Press, New York.

Le Gros Clark, W. E., 1959, *The Antecedents of Man*, University of Chicago Press, Chicago.

Luckett, W. P., 1980, Monophyletic or diphyletic origins of Anthropoidea and Hystricognathi: Evidence of the fetal membranes, in: *Evolutionary Biology of the New World Monkeys and Continental Drift* (R. L. Ciochon and A. B. Chiarelli, eds.), pp. 347–368, Plenum Press, New York.

Maier, W., 1980, Nasal structures in Old World and New World Primates, in: *Evolutionary Biology of the New World Monkeys and Continental Drift* (R. L. Ciochon and A. B. Chiarelli, eds.), pp. 219–241, Plenum Press, New York.

Martin, D. E., and Gould, K. C., 1980, Comparative study of the sperm morphology of South American primates and those of the Old World, in: *Evolutionary Biology of the New World Monkeys and Continental Drift* (R. L. Ciochon and A. B. Chiarelli, eds.), pp. 369–386, Plenum Press, New York.

Napier, J. R., and Napier, P. R., 1967, *A Handbook of Living Primates*, Academic Press, New York.

Orlosky, F. J., 1980, Dental evolutionary trends of relevance to the origin and dispersion of the platyrrhine monkeys, in: *Evolutionary Biology of the New World Monkeys and Continental Drift* (R. L. Ciochon and A. B. Chiarelli, eds.), pp. 189–200, Plenum Press, New York.

Pariente, G., 1979, The role of vision in prosimian behavior, in: *The Study of Prosimian Behavior* (G. A. Doyle and R. D. Martin, eds.), pp. 411–459, Academic Press, New York.

Perkins, E. M., and Meyer, W. C., 1980, The phylogenetic significance of the skin of primates: Implications for the origin of New World monkeys, in: *Evolutionary Biology of the New World Monkeys and Continental Drift* (R. L. Ciochon and A. B. Chiarelli, eds.), pp. 331–346, Plenum Press, New York.

Romero-Herrera, A. E., Lehmann, H., Joysey, K. A., and Friday, A. E., 1976, Evolution of myoglobin amino acid sequences in primates and other vertebrates. in: *Molecular Anthropology* (M. Goodman and R. Tashian, eds.), pp. 289–300, Plenum Press, New York.

Romero-Herrera, A. E., Lehmann, H., Joysey, K. A., and Friday, A. E., 1978, On the evolution of myoglobin, *Philos. Trans. R. Soc. London, Ser. B* **283**:61–183.

Rosenberger, A. L., 1977, *Xenothrix* and ceboid phylogeny, J. *Hum. Evol.* **6**:541–561.

Rosenberger, A. L., 1979, Phylogeny, evolution and classification of New World monkeys (Platyrrhini, Primates), Ph.D. thesis, C.U.N.Y., New York.

Rosenberger, A. L., and Szalay, F. S., 1980, On the tarsiiform origins of Anthropoidea, in: *Evolutionary Biology of the New World Monkeys and Continental Drift* (R. L. Ciochon and A. B. Chiarelli, eds.), pp. 139–157, Plenum Press, New York.

Sarich, V. M., and Cronin, J. E., 1980, South American mammal molecular systematics, evolutionary clocks, and continental drift, in: *Evolutionary Biology of the New World Monkeys and Continental Drift* (R. L. Ciochon and A. B. Chiarelli, eds.), pp. 399–421, Plenum Press, New York.

Sen, S., and Heintz, E., 1979, *Palaeoamasia kansui* Ozansoy, 1966, embrithopode (Mammalia) de l'Eocène d'Anatolie, *Ann. Paléontol. (Vertébrés)* **65**:73–91.

Simons, E. L., 1961, Notes on Eocene tarsioids and a revision of some Necrolemurinae, *Bull. Br. Mus. (Nat. Hist.), Geology* **5**:43–49.

Simons, E. L., 1972, *Primate Evolution: An Introduction to Man's Place in Nature,* Macmillan, New York.

Simpson, G. G., 1945, The principles of classification and a classification of mammals, *Bull. Am. Mus. Nat. Hist.* **85**:1–350.

Szalay, F. S., 1976, Systematics of the Omomyidae (Tarsiiformes, Primates): Taxonomy, phylogeny and adaptations, *Bull. Am. Mus. Nat. Hist.* **156**:157–450.

Szalay, F. S., and Delson, E., 1979, *Evolutionary History of the Primates,* Academic Press, New York.

West, R. M., 1980, Middle Eocene large mammal assemblage with Tethyan affinities, Ganda Kas region, Pakistan, *J. Paleontol.* **54**:508–533.

Wolin, L. R., and Massopust, L. C., 1970, Morphology of the primate retina, in: *The Primate Brain* (C. R. Noback and W. Montagna, eds.), pp. 1–28, Appleton-Century-Crofts, New York.

Wood, A. E., 1980, The origin of the caviomorph rodents from a source in Middle America: A clue to the area of origin of the platyrrhine primates, in: *Evolutionary Biology of the New World Monkeys and Continental Drift* (R. L. Ciochon and A. B. Chiarelli, eds.), pp. 79–91, Plenum Press, New York.

Paleobiogeographic Perspectives on the Origin of the Platyrrhini

23

R. L. CIOCHON and A. B. CHIARELLI

Introduction

This chapter will explore the various hypotheses and scenarios that have been advanced concerning the paleobiogeographic source of origin of the Platyrrhini. The specific questions to be addressed include: (1) From what geographical region or regions were the immediate ancestors of the New World monkeys derived? (2) When and by what means did these first platyrrhines reach the island continent of South America? (3) What influence did continental drift have on the geographic source of origin and mode of dispersal of the Platyrrhini? Nearly all of the preceding chapters in this volume have attempted to specifically answer one or more of these questions. It is our purpose herein to provide a background for the discussion of all these paleobiogeographic proposals and then to present a consensus-oriented maximum-parsimony model of platyrrhine origins and dispersal.

Continental Drift

The changing positions of the continents throughout the Mesozoic and Cenozoic Eras have had a major impact on the past and present biogeographic

---

R. L. CIOCHON • Department of Anthropology and Sociology, University of North Carolina at Charlotte, Charlotte, North Carolina 28223. A. B. CHIARELLI • Istituto di Antropologia, Università di Firenze, Via del Proconsolo, 12, 50122 Firenze, Italy. This research was supported (in part) by a grant to R.L.C. from the L.S.B. Leakey Foundation.

distribution of life on this planet. The true significance of this impact has only recently come to light, resulting in a whole series of reinterpretations of the evolutionary history and biogeography of many of the earth's living and extinct organisms. The concept of continental drift and plate tectonics has now been successfully applied to the evolution and distribution of angiosperm plants (Axelrod, 1978), laroniinine spiders (Platnick, 1976), marine invertebrates (Valentine, 1973; Schram, 1977), ostariophysan fishes (Novacek and Marshall, 1976), early tetrapods (Milner and Panchen, 1973), fossil reptiles (Colbert, 1973; Galton, 1977), fossil birds (Cracraft, 1973; Rich, 1978), vertebrate distribution (Cracraft, 1974), Mesozoic mammals (Lillegraven *et al.*, 1979), marsupials (Tedford, 1974), neotropical floras and faunas (Keast, 1972, 1977; Hershkovitz, 1972), hystricomorph rodents (Lavocat, 1974a, 1977), and fossil primates (Walker, 1972; Hoffstetter, 1972, 1977; Szalay, 1975), to name only a few. In almost every instance, the application of this "new paleogeography," to borrow the phrase of Tedford (1974), has resulted in a much more understandable and justifiable synthesis of previously existing data.

It is the synthetic and unifying nature of the "new palegeography," which has truly revolutionized, and perhaps forever changed, our conceptualizations of the evolution and distribution of the earth's organisms. It is therefore not at all surprising that most of the contributors to this volume have used the concept of mobile continents to better decipher the evolutionary origins of the New World monkeys.

Historical Biogeography

The application of continental drift, plate tectonics, and sea-floor spreading to paleobiogeographic problems has also brought about the introduction of several new principles of historical biogeography. Terms such as "Noah's arks" and "beached Viking funeral ships" (McKenna, 1972a, 1973) will be added to the list of such Simpsonian principles as sweepstakes routes, waif dispersal, filter bridges, and land corridors (Simpson, 1940, 1953, 1978). Indeed, some of these principles of historical biogeography bear directly on how the first primates reached the South American continent.

The nearly universal acceptance of the continental-drift paradigm in earth history and the recent desire by many researchers to develop more testable or falsifiable [in the sense of Popper (1959, 1963)] models of past biogeographic events has brought about a sort of minirevolution in the field of historical biogeography. No longer are many workers content with the analysis of their data following the traditional or narrative approach to biogeography as conceptualized by Darlington (1957, 1959, 1965), supported by Mayr (1965) and Briggs (1966) among many others, and vigorously defended by Darlington (1970). In place of the traditional approach to biogeography, two more rigorous methods have been developed. One can be referred to as "phylogenetic or cladistic biogeography" and the other as "vicariance biogeography."

Cladistic biogeography as we wish to term it here is a new, more formalized theory of biogeography (see Brundin, 1972; Ross, 1974; Cracraft, 1974, 1975; Ashlock, 1974; Morse and White, 1979) which has as its primary tenet the construction of the most parsimonious hypothesis regarding the location of an ancestral group based on a definable, testable concept of phylogenetic relationship. Thus, reconstruction of the center of origin and pathways for dispersal of a group is deduced (a deductive inference) from prior phylogenetic analysis. Historical biogeography, in this sense, "... endeavors primarily to reconstruct the phylogeny of organisms, especially of higher taxa, and to place this phylogeny in geographic perspective" (Cracraft, 1974, p. 215). Not all biogeographers feel that phylogeny should play such a central role in biogeographic reconstruction (for example, see Briggs, 1966; Darlington, 1970). In the past, many have reconstructed the center of origin of a particular group based primarily on patterns of species or generic diversity and on present-day diversity gradients, paying little if any attention to the group's phylogenetic history. Since the biogeographic event concerning the origin of the Platyrrhini, which we are attempting to reconstruct, occurred many millions of years in the past (certainly prior to the early Oligocene), we feel that it is absolutely essential to use all potential sources of biogeographic data, especially those derived from phylogenetic analysis. In this regard, we feel that the cladistic biogeographic approach has considerable merit.

Vicariance biogeography, according to Nelson and Platnick (1978), is as different from traditional biogeography in approach as cladistic analysis is from phenetic analysis. In other words, vicariance biogeography represents a nearly complete reformulation of the basic principles of historical biogeography. The vicariance paradigm, as currently formulated, owes its existence to the energetic writings of Leon Croizat (for example, see Croizat, 1958, 1962; Croizat *et al.,* 1974). For more than 30 years, he has expressed the view that the distribution of animal and plant life on the earth is not necessarily the result of innumerable dispersal events from various centers of origins but rather the result of geologic or geographic changes in the environment (vicariant events) on a worldwide basis which in themselves bring about episodes of allopatric speciation. Thus, tectonic change, not dispersal, is the basis of vicariance biogeography. When one considers that Croizat [a complete bibliography of his works appears in Nelson (1973)] formulated the vicariance paradigm during a period when most biologists considered the concept of mobile continents a near impossibility, it is a credit to his perseverance as a scientist that vicariance biogeography has become an acceptable (and in some cases all-inclusive) alternative in historical biogeographic analysis today.

During the past five years, a small group of workers (Nelson, 1974, 1975, 1978; Rosen, 1975, 1978; Platnick, 1976; Platnick and Nelson, 1978) have discussed, refined, and applied the vicariance paradigm in a number of useful ways. For instance, Rosen (1978, p. 187) has pointed out the importance of recognizing "... that geology and biogeography are both parts of natural history and, if they represent the independent and dependent variables respectively in a cause and effect relationship, that they can be reciprocally illuminating." Nelson (1974, p. 557) has suggested that "estimating the time of splitting of lineages through a study of vicariance in relation to dated barriers is an alternative to the traditional approach through paleontology, which tends to underestimate the absolute age, or age of origin, of lineages." Finally, Platnick and Nelson (1978, p. 1) contrast dispersal with vicariance by stating that "dispersal models explain disjunctions by dispersal across preexisting barriers, vicariance models by the appearance of barriers fragmenting the ranges of ancestral species." The concept of vicariance insofar as it can be applied to the origin of the Platyrrhini will be discussed further in an upcoming section.

Phylogenetic Background

Before any potential biogeographic models of platyrrhine origins are presented, the evidence derived from the various morphological and phylogenetic studies of this volume must be considered. Delson and Rosenberger (1980), in the preceding chapter, have summarized and interpreted much of this information in a well-thought-out and reasonable fashion. We will briefly summarize their interpretations here.

Nearly all of the morphological and biochemical evidence presented in this volume supports the concept of a monophyletic origin of the Anthropoidea. There is also a considerable body of evidence favoring both catarrhine and platyrrhine monophyly. This relationship is depicted in the cladogram presented in Fig. 1. Note how platyrrhines and catarrhines are each uniquely derived in their own right from a precatarrhine–preplatyrrhine anthropoid stock. The existence of this basal anthropoid group can be reconstructed from the large number of synapomorphies uniting Platyrrhini and Catarrhini in the Anthropoidea. Figure 1 illustrates the existence of at least 12 derived features that can be used to characterize the Anthropoidea as a whole, whereas platyrrhines and catarrhines are each distinguished by 5 and 6 derived features, respectively. All of these derived characters are presented in Table I of Delson and Rosenberger (1980) (see Chapter 22). They have been assembled from a careful analysis of the contributions* in this volume to-

*One minor criticism which some may wish to level at the analysis of Delson and Rosenberger (1980) concerns the methods they use for selecting individual characters from all those presented by the contributors. Little effort has been made by the authors to present exactly how the primitive vs. derived characters were sorted out. Our insistence on brevity is perhaps the major contributing factor to this situation. In any case, the reader is still left with the choice of accepting their summary conclusions based on the degree of confidence that the reader chooses to have in their authority. Fortunately, it was our confidence in their authority which prompted us in our editorial capacity to solicit this contribution.

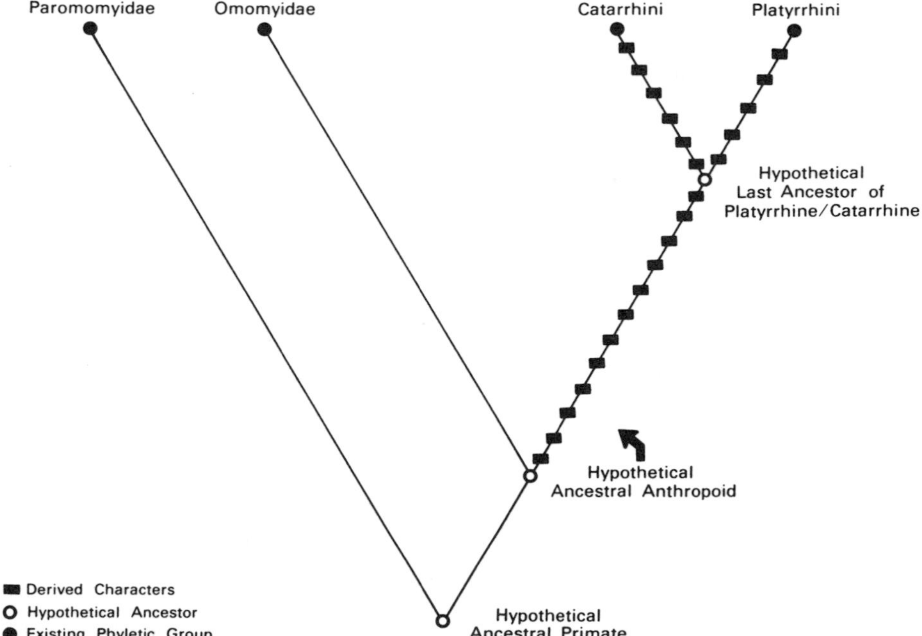

Fig. 1. Abbreviated cladogram of the primates showing the relationship of platyrrhine and catarrhine primates and the existence of a preplatyrrhine–precatarrhine ancestral anthropoid stock. Derived characters depicted in this cladogram have been compiled by Delson and Rosenberger (1980) from the various analyses presented in this volume and their own respective research efforts (see Table I in Chapter 22). The establishment of the derived anthropoid, platyrrhine, and catarrhine characters and their graphic depiction in this cladogram are put forth as a preliminary first attempt at understanding the relationships of the various higher-primate groups. No derived characters have been graphically presented for the sister group of the Anthropoidea, the Omomyidae, or for the Paromomyidae, though their relationship in this cladogram is confirmed by many chapters in this volume and by additional sources (for example, see Szalay and Delson, 1979). It should be pointed out that Gingerich (1980 and elsewhere) has cogently argued for considering the Eocene Adapidae the sister group of the Anthropoidea. This viewpoint is not represented in this cladogram for reasons discussed in the text.

gether with the addition of several outside sources. They are also supported by the biochemical studies of Baba *et al.* (1980) and Sarich and Cronin (1980). The list is by no means complete; as Delson and Rosenberger (1980) state, it is only a first effort. Yet, we feel that their analysis, even in its preliminary form, points out some very interesting trends.

The anthropoid synapomorphies depicted in Fig. 1 are twice as numerous as and are more significant in terms of level of organization (structual–functional transformations) than the synapomorphies characterizing either the Platyrrhini or the Catarrhini. We feel that these basic relationships will remain the same even as new characters are discovered and added to the list. Furthermore, we interpret this arrangement to mean that the transition from lower primates to higher primates was a fundamentally more significant evolutionary event than the origin of either of the two major divisions of the Anthropoidea. Following this rationale and its related graphic depiction in Fig. 1, we are compelled to view the evolutionary origin and early ancestry of the platyrrhines and catarrhines as very closely linked. This relationship has interesting implications for the biogeographic origin of the New World monkeys which will be discussed later.

From arguments presented above and throughout this volume, we may now infer that the paleontological origins of the New World monkeys will most likely be traced to a preplatyrrhine–precatarrhine early anthropoid stock. One of the more prominent issues in primate evolutionary biology today concerns the origin of this early anthropoid stock from the lower primates. A number of contributors to this volume support the view that the Anthropoidea are most closely related to the tarsiiform primates and can be grouped together in the monophyletic taxon Haplorhini [see Delson and Rosenberger (1980) for arguments and summary]. This would imply an origin of the anthropoids from an early tarsiiform group such as the Omomyidae (Rosenberger and Szalay, 1980). A contrary view is expressed by Gingerich (1980), who suggests derivation of the anthropoids from early lemuriform primates such as the Adapidae (see also Gingerich, 1973, 1975, 1977). Kay (1980), in his review of the dental evidence for platyrrhine origins, suggests that the arguments favoring an omomyid derivation vs. an adapid derivation are not compelling in either case.

We conclude that the balance of evidence (especially the soft anatomy) presented by the contributors to this volume and reviewed by Delson and Rosenberger (1980) does favor the omomyid or tarsiiform hypothesis of anthropoid origins. Figure 1 depicts this relationship as we view it. We do not choose to argue as strongly for this position as Delson and Rosenberger (1980) have done. Rather, we suggest that the apparent discrepancies between the two opposing viewpoints (adapid origins vs. omomyid origins) may be due as much to the differing methods of analysis employed, such as the stratophenetic approach (Gingerich, 1979) vs. the cladistic approach (Szalay and Delson, 1979), as to the actually somewhat fragmentary fossilized data base.

Resolving the issue of platyrrhine, catarrhine, and anthropoid origins would naturally be greatly aided by the recovery of a more complete fossil primate record. Unfortunately, this fossil record, as it now stands, does little more than sketch the paleohistory of these groups in the broadest possible terms. The first record of a platyrrhine primate in South America occurs in the early Oligocene deposits of Salla-Luribay, Bolivia (Hoffstetter, 1969, 1974). Although rather extensive sediments of Paleocene and Eocene age are known from South America, no earlier record of primates has yet been documented, which has been interpreted by some to mean that none was in fact present. In Africa, the first record of a catarrhine primate likewise occurs in the early Oligocene deposits of the Fayum region of Egypt (Simons, 1962, 1971). Unlike South America, the African Paleocene and Eocene fossil record is very poorly known. At present, the only documented recovery of a somewhat questionable primate (certainly a nonanthropoid) comes from the Eocene deposits of Algeria (Sudre, 1975). In North America and Eurasia, primates abound in the Paleocene and Eocene, yet all are thought to represent lower (nonanthropoid) primates with the probable exception of two late Eocene taxa from Southeastern Asia. These two primates from Burma, *Amphipithecus* and *Pondaungia,* have for a long time been considered potential early anthropoids (Colbert, 1937; Pilgrim, 1927). Recent discoveries of new, more complete specimens have reinforced that opinion (Ba Maw *et al.,* 1979; Thaw Tint *et al.,* 1981). Neither *Amphipithecus* nor *Pondaungia* can be considered a potential direct ancestor of any known platyrrhine or catarrhine (in fact, *Pondaungia* may itself *be* an aberrant catarrhine). Yet, both of these taxa can be regarded as the only known occurrence of Eocene anthropoids. This places the earliest record of the higher primates in Asia. It may also be an indication that even earlier basal precatarrhine–preplatyrrhine anthropoids were evolving on the continent of Asia in the latter part of the Eocene.

Deployment Scenarios of the Platyrrhini

There are but two basic geographic sources* from which to derive the immediate ancestors of the Platyrrhini: (1) the North American continent (see Fig. 2) and (2) the African continent (see Fig. 4). There are also a number of peripheral or intermediate geographical regions which have been suggested as possible centers of evolution for the ancestral New World monkeys, but in each case, the pathway of eventual dispersal would certainly have passed through either North America (including what little was present of Central or Middle America) or Africa. Based on the presence of the first platyrrhine in South America in the earliest Oligocene, we surmise that this dispersal event very probably occurred some time during the Eocene and therefore certainly involved the crossing of some sort of an oceanic barrier (since South America remained an island continent until the Pliocene). Apart from these basic points of agreement, divided opinions exist over exactly how the first anthropoids might have reached the South American continent. Contributors to this volume have supported source areas of origin in North or Middle America (Orlosky, 1980; Perkins and Meyer, 1980; Rosenberger and Szalay, 1980; Gantt, 1980; Wood, 1980; McKenna, 1980), Asia via the Bering Bridge to North America (Gingerich, 1980; Delson and Rosenberger, 1980), and di-

*To our knowledge, no one has yet proposed a derivation of the Platyrrhini from the continent of Antarctica. As Tarling (1980) has shown, some sort of land connection existed between Antarctica and South America throughout the Cretaceous and possibly into the early Tertiary. The absence of any living or extinct record of primates (except *Homo sapiens*) on either Antarctica or Australia coupled with the generalized patterns of biotic distribution which characterize these southern continents (see Brundin, 1966; Keast *et al.,* 1972; Craw, 1979) would, in our opinion, constitute falsification of this proposal.

rectly from Africa (Tarling, 1980; Lavocat, 1980; Hoffstetter, 1980; Maier, 1980; Bugge, 1980; Luckett, 1980; Martin and Gould, 1980; Chiarelli, 1980; Sarich and Cronin, 1980).

When considering the various platyrrhine sources of origin and dispersal scenarios outlined in the preceding chapters, it is important to distinguish exactly how each author has defined what constitutes an ancestral platyrrhine. The chapter by Delson and Rosenberger (1980) has done much to clarify this in a summary fashion. With consensus on the demonstration of anthropoid monophyly, we feel that platyrrhine origins and catarrhine origins should no longer be considered as two isolated independent events. Any potential source area of one group should probably be tied to the source area of the other. With these points in mind, we will now review the various paleobiogeographic models of platyrrhine origins.

North-American-Origin Model

The traditional geographic source of origin for the ancestors of the New World monkeys has been the North American continent. The dispersal scenario usually associated with this model involves island-hopping by waif dispersal* across the Caribbean Sea to South America (Fig. 2). Simpson (1945) was the first to formally advocate this model, though in concept it existed a good deal earlier (see Matthew, 1915; Gregory, 1920).

Most advocates of the North-American-origin model suggest that the

*Waif dispersal as defined by Simpson (1978, p. 321) ". . . refers to the occasional occurrence of one or a few members of a species outside its usual or previous range." Waif dispersal is commonly visualized as occurring along a sweepstakes route such as a series of widely separated oceanic islands (Simpson, 1953). Sweepstakes dispersal is very nearly the equivalent of waif dispersal; Simpson (1978, p. 321) suggests that it should refer to ". . . geographic spread of a group of organisms across a barrier, such as an ocean or strait, for terrestrial organisms where the probability of such spread is very small but not zero, analogous to the probability of holding the winning ticket in a sweepstakes." McKenna (1973, p. 297) considers that "sweepstakes dispersal is inherently improbable although not impossible at any one time, but with sufficient time sweepstakes dispersal becomes probable, although random and unpredictable." Since South America remained an island continent until the Pliocene, the ancestral Platyrrhini had to reach this continent via an oceanic barrier. For terrestrial mammals in the size range of the ancestral New World monkeys (probably squirrel-sized), the crossing of this barrier was almost certainly made by waif dispersal across an oceanic sweepstakes route of tectonically active island arcs (island-hopping). Near plate margins, these volcanic island chains appear to have formed intermittently and occasionally coalesced into temporary land bridges, only to later drop back below sea level. Initial transport from the mainland to an island and any subsequent over-water inter-island migrations probably occurred via a series of successive rafting episodes. It is possible that other dispersal mechanisms such as the innovative "Noah's ark" concept of McKenna (1972a, 1973) or a rather fanciful mechanism we propose of occasional passive transport via large predatory birds could also have been responsible. Whatever the actual dispersal mechanism, we envision the process of spread along the sweepstakes route to have taken many generations, with the probable result of a single colonizing group reaching the South American continent some time in the late Eocene.

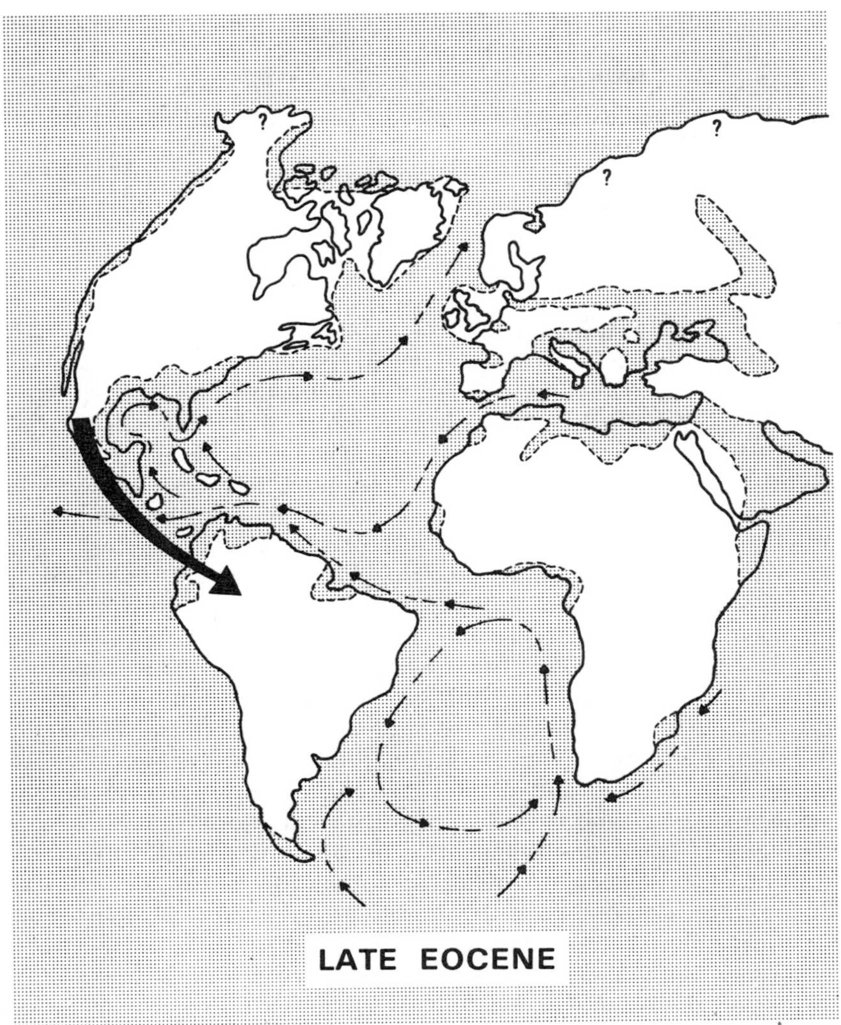

Fig. 2. Model illustrating derivation of the Platyrrhini from the North American continent in the late Eocene. Stippled areas represent oceans or land surfaces transgressed by epicontinental seas. The heavy arrow indicates the direction of potential dispersal of the ancestral platyrrhines, not necessarily the precise pathway for this dispersal, which could have occurred via any of the reconstructed island arc chains in the Caribbean. Note the distance between the southernmost extent of land on the North American continent and the northernmost extent of land on the South American continent. Also note the direction of oceanic circulation patterns (small arrows) in and around the Caribbean. Configuration of continents in this reconstruction is after Sclater *et al.*, (1977); position of epicontinental seas based on Bond (1978) and references therein; surface oceanic circulation patterns derived from Berggren and Hollister (1974); island arcs in the Caribbean reconstructed from Tomblin (1975) and Malfait and Dinkelman (1972). To focus primary emphasis on the North-American-derivation model, note that none of the potential island chains in the South Atlantic Ocean which subsequently appear in Fig. 4 has been included in this reconstruction.

Eocene Omomyidae, a tarsiiform primate family, Holarctic in distribution, comprise the basal stock of both the Platyrrhini and the Catarrhini (McKenna, 1967; Szalay, 1975; Orlosky and Swindler, 1975; Simons, 1976; Szalay and Delson, 1979). In the early to middle Eocene, continuity between the North American and European populations of omomyids, which would eventually become ancestral platyrrhines and catarrhines, was maintained by gene flow across the North Atlantic Bridge. Some time before the middle Eocene, this bridge was submerged, probably due to a combination of crustal rifting and marine transgressions (see Pitman and Talwani, 1972; Fitch *et al.,* 1974; McKenna, 1975). The dissolution of this transcontinental land bridge would have at first severely impaired and later brought a cessation of gene flow between the North American and European Omomyidae. The possibility that gene flow was also maintained via the Bering Bridge both during and after the closure of the North Atlantic Bridge exists, but is complicated by the imposing barrier of the Turgai Strait (also termed the Ouralian Sea), which bisected the Eurasian land mass throughout much of the Eocene.

As Simons (1976) has argued, if one accepts on morphological grounds that the Omomyidae (perhaps in the form of a *Teilhardina* species) represent the ancestral populations from which the platyrrhines and catarrhines were independently derived, then on geological grounds, the submergence of the North Atlantic Bridge marks the point of the platyrrhine–catarrhine dichotomy. McKenna (1972b, 1975) has pinpointed this interruption in the North Atlantic Bridge as occurring late in the early Eocene or about 50 million years B.P. All subsequent evolution of the ancestral Platyrrhini in North America and their eventual dispersal into South America would therefore have occurred as an independent parallel event to the evolution of the Catarrhini in Europe and their separate dispersal into Africa.

The North-American-origin model of the Platyrrhini is grounded in a long tradition of North American paleontology dating back at least to Matthew (1915). Proponents of this traditional view attempt to derive all the major South American faunal elements from North American ancestors and disperse them southward in a series of successive "invasions." In concept, this fact alone constitutes no reasonable basis for rejection; yet, the fact that this model was conceived prior to the acceptance of continental drift does present some interesting conceptual difficulties.

It is now considered virtually certain that during the Eocene, the continent of South America occupied a position considerably more distant from North and Middle America than it does today. In fact, the global position of South America in the Eocene has been demonstrated to lie equidistant between the continents of North America and Africa (Sclater *et al.,* 1977; Tarling, 1980; Lavocat, 1977) (see also Figs. 2 and 4). Furthermore, Ladd (1976) presents evidence that during the Eocene, South America may have temporarily reversed its direction of drift and actually moved away from North America for a period of time. Therefore, any colonization of South America from North or Middle America during the Eocene would have been over a considerable oceanic distance. The existence of volcanic island arcs in the Eocene Caribbean is documented (see Tomblin, 1975; Malfait and Dinkelman, 1972), which makes a variety of island stepping-stone routes possible. Yet, the exact location of these routes and the distance between islands in the reconstructed chains remain quite conjectural (for example, compare Lloyd, 1963; Malfait and Dinkelman, 1972; Weyl, 1974; Tomblin, 1975). The presence of oceanic circulation patterns in the Eocene which flowed both east to west *across* these potential Caribbean dispersal routes and south to north *against* the routes also seems well substantiated (see Figs. 2 and 4—based on Berggren and Hollister, 1974) (also see Frakes and Kemp, 1973; Holcombe and Moore, 1977; discussion by Lavocat, 1980). These patterns of paleocurrent flow in the Caribbean would argue strongly against rafting as a means of dispersal between preexisting volcanic island arcs. Since no direct land connection between Middle America and South America is known to have existed until the Pliocene, we are left with a rather improbable situation. Nevertheless, we feel that McKenna (1980) has very expertly argued that in spite of these paleogeographic factors, some interchange between North America and South America did occur prior to the development of the Pliocene Panamanian Bridge. Whether or not the ancestral platyrrhines were among this faunal interchange remains to be demonstrated.

The North-American-origin model as we see it attempts to derive the Platyrrhini from middle Eocene North American omomyids (perhaps *Teilhardina*) which do not share, or are not known to share, the derived dental and skeletal features characteristic of the ancestral anthropoid stock as defined by contributors to this volume (see Fig. 1). The absence of a potential platyrrhine ancestor in North America coupled with the considerable oceanic distance separating Middle America and South America and the unfavorable prevailing oceanic circulation patterns, in our opinion, very nearly falsify this model of platyrrhine origins.

Asian-Origin Model

A variant of the North-American-origin model depicts the ancestral stock of the Platyrrhini developing in Asia and dispersing to North America via the Bering Bridge in the late Eocene and ultimately reaching South America by waif dispersal across the island arcs of the Caribbean (see Fig. 3). The ancestral catarrhines also disperse from a similar source area in Asia via a land corridor in the south and cross the reduced western extent of the Tethys Sea into Africa. The conceptual framework for this model was originally suggested by Gingerich (1977) and has recently been further refined and substantiated by him (see Gingerich, 1980). As a deployment scenario, it has also been proposed by Delson and Rosenberger (1980) and Szalay and Delson (1979, p. 519), though aspects of their version differ somewhat from that presented here. It appears to us that the unifying conceptual basis for all these

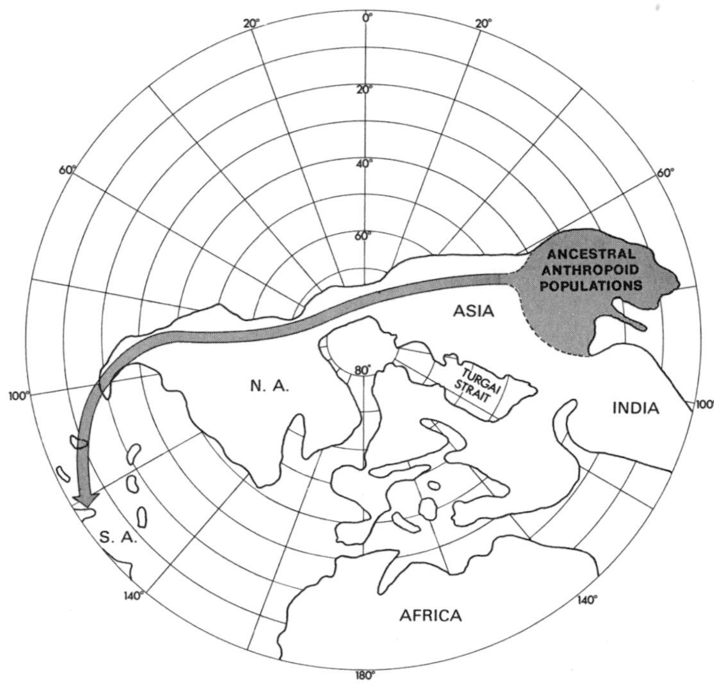

Fig. 3. Model illustrating the derivation of the Platyrrhini from the Asian continent in the late Eocene. Areas below sea level in this polar projection are indicated by the presence of longitude and latitude demarcations. The stippled area represents the approximate range of ancestral anthropoid populations in Asia, and the arrow indicates the potential pathway of dispersal across the Bering Bridge through western North America to South America via an island-hopping route through the Caribbean. Note that the over-water distance from the southernmost extent of North America to the South American continent is still quite large despite the presence of inferred island arcs in the Caribbean. Configuration of continents in this reconstruction is derived from a North polar stereographic projection provided by A. G. Smith and Briden (1976); position of the Indian subcontinent is after Johnson *et al.* (1976) and Bingham and Klootwijk (1980); location and extent of the Bering Bridge based in part on McKenna (1972b); position of epicontinental seas derived from Bond (1978), Sahni and Kumar (1974), and Vinogradov (1967); island arcs in the Caribbean reconstructed from Tomblin (1975) and Malfait and Dinkelman (1972).

authors' proposals stems from the viewpoint that the ancestries of the platyrrhines and catarrhines are very closely linked. Since it no longer seems tenable to derive these two groups from widely separate and isolated geographical areas, hypotheses such as the Asian-origin model provide a logical and testable alternative.

As previously indicated, Gingerich (1980) favors derivation of the ancestral anthropoid stock from the Eocene Adapidae. In his view of the Asian-origin model, the presence of transitional adapid–anthropoid primates in the late Eocene of Burma (*Pondaungia* and *Amphipithecus*) and in the early

Oligocene of Africa (*Oligopithecus*) provides evidence that the origin of the higher primates from an advanced "protosimian" (preplatyrrhine–precatarrhine) stock occurred in either southern Asia or Africa or perhaps simultaneously in both areas. Deriving the ancestral platyrrhines from a source area in southern Asia and dispersing them to North America is then based on (1) the more primitive and generalized appearance of the transitional adapid–anthropoids, *Amphipithecus* and *Pondaungia*; (2) the apparent open nature of the high-latitude Bering Bridge during a warm interval in the late Eocene; (3) the existence of a marked mammalian faunal similarity between Eurasia and North America in the late Eocene; and (4) the unexpected occurrence of *Mahgarita,* an adapine primate of Eurasian aspect, in the late Eocene of Texas.

Gingerich (1980) considers the Asia-based model of platyrrhine origins (see his Fig. 4) the most parsimonious hypothesis based on current data. However, as a second only slightly less likely alternative, he suggests that ". . . it is also possible that the adapid–protosimian stock ancestral to Ceboidea crossed the South Atlantic directly from Africa to South America" (p. 131). Finally, as a third considerably less likely alternative, Gingerich (1980, p. 135) cautiously states that ". . . it is possible that higher primates originated in southern North America and subsequently migrated in the late Eocene from North America into South America and also from North America across the Bering route into south Asia and ultimately Africa." We feel that presentation of alternative models in this manner sets the stage for future hypothesis-testing which will eventually result in the development of a single falsifiable model.

Delson and Rosenberger (1980) take a different tack from Gingerich in their apparent support of the Asian origin model of the Platyrrhini. They favor derivation of the ancestral Anthropoidea from the Omomyidae, a view which we also support. Instead of specifically stating that the ancestral anthropoids differentiated in southern Asia as we present in Fig. 3, they offer the scenario that these ancestral higher primates actually occupied a single biotic community which spanned the circum-Pacific region across the Bering Bridge all the way from Asia to North America. In the late Eocene, the formative catarrhine branch of this community then entered Africa from Asia, while the protoplatyrrhine branch dispersed to South America from the north via a Caribbean island-hopping route (Delson and Rosenberger, 1980). In support of their model, Delson and Rosenberger emphasize (1) the significant morphological resemblance between Asian and western North American anaptomorphine omomyids in the Eocene, (2) the potentiality of the Bering Bridge as a mammalian migration route throughout the Eocene, (3) the possibility of a phyletic link between *Pondaungia* and the Fayum catarrhines, and (4) the strength of faunal links between African Fayum mammalian taxa, including the nonanthropoid *Oligopithecus,* with Mediterranean and south Asian mammalian taxa, including *Hoanghonius*.

We suggest that both of these versions of the Asian-origin model as proposed by Delson and Rosenberger (1980) and Gingerich (1980) have considerable merit, since each supports the existence of a preplatyrrhine–precatarrhine stem anthropoid stock (as depicted in Fig. 1) and each version either directly or indirectly supports a post-middle Eocene platyrrhine–catarrhine divergence date. The phylogenetic arguments reviewed in an earlier section, in our opinion, make these two assumptions a requisite basis for any viable paleobiogeogrpahic model of platyrrhine origins. However, both Gingerich and Delson and Rosenberger still must contend with the arguments presented under the North-American-origin model indicating the potential difficulties to be encountered in dispersing the ancestral platyrrhines across the Caribbean via an island sweepstakes route.

African-Origin Model

The African-origin model of the Platyrrhini attempts to derive the ancestral New World monkeys directly from the African continent via a sweepstakes route across a much less expansive Eocene South Atlantic Ocean (see Fig. 4). Given the long-acknowledged morphological similarity of platyrrhines and catarrhines, it would seem that such a hypothesis, at least in principle, would have been proposed many years ago. However, it was not until the nearly universal acceptance of the continental-drift paradigm during the last decade that this model of platyrrhine origins began to receive serious consideration. Its list of adherents now includes Sarich (1970), Hoffstetter (1972, 1974, 1977), Cracraft (1974), Cronin and Sarich (1975), Thorington (1976), Lavocat (1974b, 1977), Hershkovitz (1972, 1977), and Washburn and Moore (1980), among others.

Many supporters of the African-origin model favor derivation of the ancestral anthropoid stock from Eurasian omomyid tarsiiforms in the middle to late Eocene. Soon after this stem anthropoid stock had populated the African continent either as a precatarrhine–preplatyrrhine early anthropoid or possibly as a protoanthropoid, it dispersed to South America through rafting and island-hopping across the South Atlantic in the late Eocene or earliest Oligocene. Since no Paleocene or Eocene fossil record of any consequence exists over the entire expanse of the African continent, there is no way to substantiate this scenario. The early to middle Oligocene deposits of the Fayum region of Egypt have produced a rich primate fauna, yet these all appear to be bona fide catarrhines. Nevertheless, Hoffstetter (1980 and elsewhere) has attempted to derive the ancestral platyrrhines from one group of the Fayum primates, the Parapithecidae. As Hoffstetter, among others, has argued, the parapithecids may indeed represent the most primitive (least derived) members of the Catarrhini, but this alone does not make them appropriate ancestors for the earliest Platyrrhini (see arguments by Delson and Rosenberger, 1980; Kay, 1980; Szalay and Delson, 1979). If the existence of a precatarrhine–preplatyrrhine anthropoid as depicted in Fig. 1 is proven correct, then an African-based branch of this stock would be a more appropriate

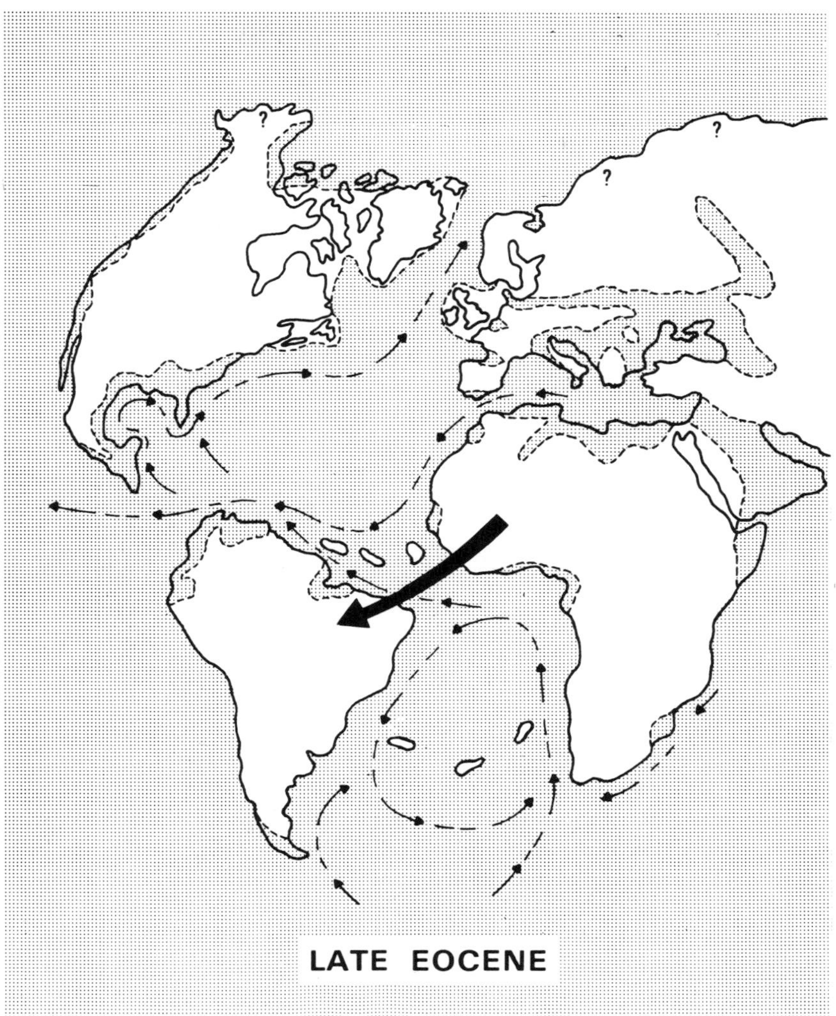

Fig. 4. Model illustrating derivation of the Platyrrhini from the African continent in the late Eocene. Stippled areas represent oceans or land surfaces transgressed by epicontinental seas. As in Fig. 2, the heavy arrow indicates the direction of potential dispersal of the ancestral platyrrhines, not necessarily the exact pathway for this dispersal, which was evidently via island-hopping and most probably across the more northern portion of the South Atlantic. Compare the distance between the closest adjacent points of Africa and South America with the oceanic distance separating South America and North America. Also compare the direction of oceanic circulation patterns (small arrows) in the northern South Atlantic with the patterns of circulation in the Caribbean. Configuration of the continents in this reconstruction is after Sclater *et al.* (1977); position of epicontinental seas based on Bond (1978) and references therein; surface oceanic circulation patterns derived from Berggren and Hollister (1974); position of oceanic islands in the South Atlantic based on Tarling (1980), Sibuet and Mascle (1978), and Van Andel *et al.* (1977). As was the case with Fig. 2, to focus primary emphasis on the African-derivation model, note that none of the potential island arcs in the Caribbean which appeared in the former figure has been included in this reconstruction.

structural ancestor for the Platyrrhini. Without the aid of a Paleocene or Eocene primate fossil record in Africa, there is, of course, no way to test this hypothesis. It could just as easily be argued (and indeed has been) that anthropoids originated in Africa from unknown Eocene tarsiiforms or lemuriforms and dispersed to South America via the South Atlantic route to become ancestors of the New World monkeys and to south Asia via a crossing of the Tethys to become ancestors of the presumed anthropoids of Burma's Pondaung Hills.

Before being completely carried away with proposals for hypothetical platyrrhine-deployment scenarios and for equally hypothetical platyrrhine ancestors, we feel that it is necessary to mention one potential source of information that is *not* highly disputed and which *does* appear to lend credence to the African-origin model. Current evidence now indicates that the Platyrrhini and Catarrhini represent two equal subdivisions of the Anthropoidea that arose from a common early anthropoid ancestral stock (see Fig. 1). Therefore, it is proper, and indeed essential, to consider the Platyrrhini as the sister group of the Catarrhini and vice versa. The current predominant distribution of these two groups in the continents of South America and Africa, respectively, targets these regions as their potential areas of differentiation. This disjunct distribution and sister-group relationship takes on added significance when it is recalled that another group of mammals, the South American caviomorph rodents and African phiomorph rodents, exhibit a nearly identical distribution and phyletic relationship (Lavocat, 1980). When these similarities are further explored utilizing the method of Hennig (1966) for determining the biogeographic origin of a group called the "search for the sister group" and the concept of Croizat *et al.* (1974) of "generalized patterns of biotic distribution" or "generalized tracks," a pattern emerges which supports a wholly southern-continents origin of platyrrhines, catarrhines, caviomorphs, and phiomorphs irrespective of hypothetical deployment scenarios or of hypothetical ancestors.

Historically, the main obstacle that has been used to argue against derivation of the platyrrhine primates and caviomorph rodents from a source in Africa is the reported great size of the marine barrier separating that continent from South America in the early Tertiary. Actually, the exact size of the equatorial South Atlantic during the later Eocene–earliest Oligocene (the most probable time a crossing would have taken place) is still a matter of some debate. There does appear to be a consensus that the final separation of the contiguous subaerial land masses of South America and Africa occurred in the early late Cretaceous (Larson and Ladd, 1973; Sclater *et al.*, 1977; Sibuet and Mascle, 1978; Tarling, 1980). Initially, this separation appears to have been in the form of a series of continuous rift valleys which were subjected to periodic marine incursions. Similar marine incursions are known to have flooded the subsiding continental margins of South America and Africa as early as Albian times (ca. 105 million years B.P.), though these episodes did not represent the final rifted separation (Förster, 1978). The final contact of land between the two continents, probably in the form of an uplifted land bridge between Brazil and West Africa, occurred in the Turonian (ca. 92 million years B.P.) (Kennedy and Cooper, 1975; Reyment *et al.*, 1976). After that point, progressive development of the intracontinental rift coupled with a worldwide rise in sea level firmly established a pattern of continuous circulation between the Central and South Atlantic.

Because the rate of sea-floor spreading in the South Atlantic has not been constant since the final separation of South America and Africa (Herz, 1977; Sibuet and Mascle, 1978), precise estimates of the size of the marine barrier separating the closest approach of the two continents in the late Eocene–earliest Oligocene vary somewhat. Based on calculations from scaled paleocontinental maps, Funnell and Smith (1968) estimate the oceanic barrier at 1000 km, A. G. Smith and Briden (1976) at 1450 km, Sclater *et al.* (1977) at 1600 km, Tarling (1980) at 1650 km, and Ladd *et al.* (1973) at 1750 km. For comparison, the current size of the marine barrier separating South America and Africa is approximately 3200 km. The sheer size of this marine barrier even in the late Eocene would have acted as a nearly impenetrable filter if it had not been for the presence of a series of oceanic islands which very possibly acted as stepping stones.

The existence of volcanic islands in the early Tertiary South Atlantic is based on the cyclic occurrence of tectonically active processes associated with sea-floor spreading in the region. McKenna (1980, p. 53) has suggested that island chains in the equatorial and southern South Atlantic were very possibly brought about by (1) excessive production of Mid-Atlantic Ridge lavas, (2) temporarily elevated sections of fracture zones, and (3) off-ridge volcanic activity. The geographical distribution and petrographic–geochemical nature of existing islands in the South Atlantic Ocean (see Baker, 1973) bear out McKenna's proposals. The exact location of these potentially subaerial island chains would have been along the Ceará and Sierra Leone Rises in the equatorial South Atlantic and along the Rio Grande–Walvis Ridge further to the south (see Fig. 4). The structure and geologic history of the Ceará and Sierra Leone Rises is detailed by Kumar and Embley (1977) and Sibuet and Mascle (1978). Evidence is provided which indicates that both these features of the newly formed equatorial South Atlantic were volcanic in origin and potentially above sea level sporadically since their formation about 80 million years B.P. Tarling (1980, p. 34) suggests that both the Ceará Rise and the Sierra Leone Rise were "... strong positive features in the early Tertiary." He further states that "it seems probable, therefore, that oceanic islands, some hundreds of square kilometers in size, existed offshore from Brazil and West Africa, and that the mid-oceanic rise was also subaerial for much of this period" (p. 34). Distances between islands in this oceanic chain may have been as little as 200 km (Tarling, 1980).

In the southern portion of the South Atlantic, oceanic island chains potentially existed along the Rio Grande Rise and Walvis Ridge spanning the distance between present-day southern Brazil and South-West Africa. Studies of the depositional history and geochemistry of this region (Van Andel *et al.*, 1977; P. J. Smith, 1977) firmly establish the barrier nature and volcanic origin of these features and suggest the occurrence of subaerial volcanic islands or seamounts. The formation of the Walvis–Rio Grande Rises was in the early Cretaceous (Sclater *et al.*, 1977), which suggests that they predate the equatorial Ceará–Sierra Leone Rises by at least 40 million years. This factor, among others, led Tarling (1980, p. 33) to state that "the Walvis–Rio Grande Rise . . . is likely to have been a much more rugged, intermittent migration route with the gradual decrease in its elevation leading to greater difficulty in 'island-hopping,' but the occasional volcanic eruption creating temporary new land and providing improved connections for short periods." McKenna (1980) offers the analogy that a portion of these rises might be considered a southern version of Iceland. The subsidence of the Rio Grande Rise–Walvis Ridge barrier in the earliest Oligocene established deepwater circulation in the southern South Atlantic for the first time (Van Andel *et al.*, 1977) and brought about the end of any potential island-hopping route in this region.

Even with the probable existence of volcanic island chains in both the Eocene South Atlantic and Caribbean, the colonizing ancestors of the Platyrrhini would have still been forced to make some sort of oceanic crossing. Arguments for and against the concept of an oceanic crossing have been particularly heated with regard to the African-origin model (compare Hoffstetter, 1974; Simons, 1976). The obvious mechanism for the crossing of a substantial marine barrier would be by means of rafting. Darlington (1957, p. 15) quite simply characterizes rafting as the ". . . dispersal of land animals across water on floating objects." Without question, rafting as a mode of dispersal has occurred in the past. The existence of land mammals on islands separated by large marine barriers from all other land masses can often be explained by no other means. Several examples from the literature also document the existence of long-distance rafting as a mode of dispersal. Ridley (1930) cites the occurrence of a raft consisting of driftwood and living clumps of lalang grass and bamboos which was apparently ridden by a crocodile from the coast of Java to the Cocos Islands, some 1100 km in distance. Powers (1911, pp. 304–305) provides an extraordinary example of a floating island seen in 1892 off the coast of North America hundreds of miles out to sea in the Atlantic:

> When first noticed in July in latitude 39.5° N., longitude 65° W. (about 400 km west of Cape Cod) the island was about 9,000 square feet in extent, with trees thirty feet in height upon it, which made it visible for seven miles. It had apparently become detached from the coast of this country and been carried out to sea by the Gulf Stream. It was again seen in September in latitude 45° 29 N., longitude 42° 39 W., (about 600 km west of Newfoundland) after it had passed through a severe storm. By this time it must have traveled over 1,000 nautical miles (= 1850 km approx.), and it may have eventually arrived at the coast of Europe.

Such examples of floating islands are rare, but given the probability of chance dispersal over long periods of geologic time (see Simpson, 1952), such unlikely events as the rafting of small primates across marine gaps separating oceanic islands become possible at least in theory. It should be noted here that physiological and behavioral objections have been presented to primate rafting over long distances,* yet these would appear to apply equally to Caribbean dispersal routes from North America and to the equatorial South Atlantic dispersal route from Africa.

Rafts like those described by Ridley (1930) and Powers (1911) do not necessarily require the action of great flooded rivers undercutting their banks during storm conditions to start them on their way. Rather, they can be formed and floated into the ocean anywhere that marshes or shallow lakes drain seaward (Darlington, 1957). Paleogeographic data indicate that such conditions almost certainly prevailed along the coast of West Africa in the late Eocene–earliest Oligocene. Potential rafts therefore may very possibly have existed. Once the seaward drift of a raft had begun, the paleocurrents which flowed from along the coast of West Africa in a westerly direction across the South Atlantic (Berggren and Hollister, 1974) (see also Fig. 4) could have carried the raft or floating island to one of the equatorial volcanic islands or possibly all the way to the coast of South America. If this raft or floating island had vertically standing trees on it like the one described by Powers (1911), these could have acted as a sail, since prevailing surface winds in the Eocene and Oligocene blew in a westerly direction off the African continent (Frakes and Kemp, 1973). Even today, at a distance more than twice the size of the Eocene South Atlantic, floating drift from the Niger and Congo Rivers carried by the South Equatorial Current across the South Atlantic has been reported thrown up on the coast of Brazil (Darlington, 1957). Scheltema (1971) has shown from studies of pelagic marine larvae that such a passive crossing of the present-day South Atlantic can take as little as 60 days. Perhaps some time in the late Eocene–earliest Oligocene, such a waif-dispersal scenario via a rafting and island-hopping route across the South Atlantic was responsible for the introduction of the ancestral New World monkeys into South America.

In summary, we feel that the African-origin model of the Platyrrhini as presented here has much to offer, since it not only satisfies preexisting

*Simons (1976, pp. 51–52) presents the following excellent critique:

> The physiological and behavioral objections to primate rafting over long distances are great. The most crucial objection would be dehydration due to lack of water, accelerated by heat, sun, and lack of shelter. Exposed to salt spray, vegetation would wither and cease to provide a food or water source, thus upsetting the water/salt balance. Small monkeys do not normally utilize tree holes and thus would be exposed during the day to heat and wind stresses. Platyrrhines are known to be sensitive to heat stress and are easily susceptible to sun stroke. Lack of food and water, salt imbalance, and the aforementioned environmental stresses would render most small primates comatose or unconscious in 4–6 days. Primates would likely jump off a raft when it was forming, or, if not, the probable isolated rafting group would be only a few individuals. Mother–son and sister–brother incest avoidance occurs in marmosets and cebids and might further impede colonization. Also, because they are social animals, small group size and the characteristic slow birth rate might lead to abnormal behavior. Finally, the sudden adjustment to a new environment long-distance rafting necessarily requires (new predators, new foods) argues against a successful colonization.

phylogenetic requirements of deriving both the Platyrrhini and the Catarrhini from a common geographical source but also provides a deployment scenario for dispersal of the Platyrrhini to South America over a potentially viable and ultimately verifiable oceanic sweepstakes route.

Vicariance-Origin Model

An alternate form of the African-origin model has been proposed by Hershkovitz (1972, 1977) as a possible alternative scenario for the origin of the Platyrrhini and Catarrhini. This model, depicted in Fig. 5, is described here by Hershkovitz (1972, p. 323):

> . . . prosimians arose in the rifted South American–African continents, possibly during the Cretaceous, and spread across both continents during early drift stages. Platyrrhines and catarrhines then evolved independently in isolated South America and Africa, respectively, perhaps during the early Tertiary from closely related, possibly congeneric prosimian stocks. This hypothesis accounts for basic similarities without the need for initial convergence.

Hershkovitz (1977, p. 67) later described an extension of the aforedescribed model:

> . . . this hypothesis assumes that the unknown haplorhine forerunner of platyrrhines and catarrhines had already evolved to simian grade in the rifted but not yet widely drifted South America–Africa. The timing of the geological events may be questioned, but nothing in the fossil record denies the sequence of events outlined here.

The essence of the Hershkovitz model is his belief that platyrrhines and catarrhines appear so fundamentally similar that they must have had a common center of origin and were most probably derived from a pre-simian (tarsioid, haplorhine) stock (see Hershkovitz, 1977). Given the present-day disjunct distribution of platyrrhines and catarrhines in the continents of South America and Africa, respectively, Hershkovitz reasoned that their obvious area of origin would be in the combined Afro–South American continents. Unfortunately, as Hershkovitz (1977) himself admits, this Afro–South American-based derivation model of the Platyrrhini–Catarrhini does not appear to fit the current paleocontinental configurations as proposed by Ladd *et al.* (1973) or Sclater *et al.* (1977), nor is there any direct fossil evidence to support it.

An inherent implication of the Hershkovitz model is that the tectonically active process of intracontinental rifting brought about the isolation of ancestral platyrrhines in South America and ancestral catarrhines in Africa. We therefore feel that his proposals can be interpreted as support for a vicariance model of platyrrhine origins. Though Hershkovitz has made no statement to this effect, we have nonetheless taken the liberty of including his model under the present subheading. In support of this, we would like to quote one final

LATE CRETACEOUS

Fig. 5. Model adapted from Hershkovitz (1977) illustrating the origin of the Platyrrhini and Catarrhini in the rifted but not yet widely drifted southern continents from a currently unknown lower primate stock in the late Cretaceous. This model stresses the independent evolution of platyrrhines and catarrhines in South America and Africa, respectively, from a "pre-simian" (tarsioid, haplorhine) primate that had a pan-Afro–South American distribution toward the end of the Cretaceous. Since no actual dispersal of the New World monkeys into South America is implied by this model and since tectonic change in the form of an active Mid-Atlantic Rift can be viewed as the independent variable determining the origin of the Platyrrhini, this model appears to fit the vicariance paradigm. Indeed, proponents of vicariance biogeography, such as Nelson (1974) and Croizat (1971, 1979), have used arguments very similar to these proposed by Hershkovitz (1972, 1977) to explain the disjunct distribution of certain South American and African faunal elements. For further discussion, see the text. Configuration of the continents in this reconstruction is based on Sclater *et al.* (1977) and Tarling (1980); note that no oceans or epicontinental seas are indicated.

statement by Hershkovitz which, in our view, bridges the gap between his innovative proposals of the early 1970's and those currently made by Rosen (1978), a proponent of the vicariance paradigm. Referring to the hypotheses of platyrrhine–catarrhine origins presented above, Hershkovitz (1972, p. 324) states that they are ". . . based on an idealized construction of evolutionary and geological sequences, (which) may be inconsonant with the supposed chronology of either primate evolution [= cladistic sequences] or of continental drift [= vicariant events], but not both." Rosen (1978) would certainly argue that they should be consonant with both! (Phrases appearing in brackets in the quote were added by us for emphasis.)

Utilizing arguments presented by Croizat (1971, 1979) and Nelson (1974), a vicariance model of platyrrhine origins can be constructed. Following the depiction by Nelson (1974, p. 556, Fig. 1) of a vicariant event fostering the development of the current disjunct distribution of a hypothetical South American–African taxon, the following scenario can be visualized: The ancestral platyrrhine–catarrhine species, either an early anthropoid or a lower primate (Species A), exhibited a pan-South American–African distribution at some point in the past before the breakup of these continents. The descendant species of Platyrrhini (Species Al) and Catarrhini (Species A2) appeared when a marine barrier developed between South America and Africa (the result of sea-floor spreading and continental drift), causing the splitting (vicariance) and allopatric speciation of the ancestral species. This vicariance scenario would fit the distributional and limited paleontological evidence of platyrrhines and catarrhines. Unfortunately, as Nelson (1974), Croizat *et al.* (1974), and Rosen (1978) have argued, the timing of the development of the barrier is the causal geographic factor in any vicariance model of speciation. Geological, geophysical, and invertebrate paleontological data have already been presented (for example, see Sclater *et al.,* 1977; Tarling, 1980; Sibuet and Mascle, 1978; Kennedy and Cooper, 1975) indicating that the final separation (development of the marine barrier) between South America and Africa occurred in the Turonian, at least 90 million years B.P. Since the oldest known primate, *Purgatorius*, thought to be very near the basal radiation of that order, does not appear in the fossil record until the latest Cretaceous of North America, at most 70 million years B.P. (Van Valen and Sloan, 1965; Clemens, 1974), it is difficult to argue for the vicariance origin of platyrrhines and catarrhines in the southern continents *much* earlier in time. It is indeed possible that lower primates older than *Purgatorius* may one day be found in the late Cretaceous of South America or Africa or both, which would then lend support to the vicariance model. However, as was presented in Fig. 1, results of a cladistic analysis performed by Delson and Rosenberger (1980) indicate that the Platyrrhini and Catarrhini shared some period of common ancestry as early anthropoids before their divergence. Use of this phylogenetic evidence as support for the vicariance model would mean that anthropoids (= higher primates) had developed in the prerifted South American–African continent some 20–25 million years prior to the earliest known occurrence of a lower primate anywhere else in the world. At present, this does not seem a likely possibility.

We conclude that evidence derived from the primate paleontological record, the cladistic relationships of platyrrhines and catarrhines, and the timing of intracontinental rifting of South America from Africa tends to falsify the vicariance model of platyrrhine origins. As we present in the following section, we do support the view that the origin of the platyrrhines and catarrhines was very likely a southern-continents event. However, waif dispersal from Africa in the late Eocene, not vicariance at a much earlier date, is the most probable mechanism for explaining the arrival of the first platyrrhines in South America. As Ferris (1980, p. 72) suggests, "... vicariance biogeographers do not deny the possibility of dispersal over barriers, but they consider it a random event." In essence, then, we are discussing the reality of oceanic dispersal vs. the reality of vicariance of the pan-Afro–South American biota. In a similar discussion concerning the congruence between biological and geological relationships, Rosen (1978, p. 186) concludes that "a decision concerning the nature of this correspondence, whether by dispersal or vicariance or some combination of the two, must be a parsimony decision concerned with minimizing the number of separate assumptions entailed by the different types of explanations." Therefore, we suggest that a vicariance origin of the Platyrrhini based on the megavicariant event of an intracontinental rifting is not the most parsimonious decision at present. We would nevertheless support the view that microvicariant events, such as the transgression–regression of the Tethys Sea and Turgai Strait and the opening–closing of the Bering and North Atlantic Bridges, rather than dispersal *per se,* had a major impact on other phases of Paleogene primate evolution.

Maximum-Parsimony Model of Platyrrhine Origins

The principle of parsimony dictates that a particular model is preferable to all others provided that the known data do not call for its rejection. Furthermore, following the hypotheticodeductive method as conceived by Popper (1959, 1963) and recently analyzed by Kitts (1977), it is now possible to formalize the process of verification of a parsimony model. To this end, we have selected a recently proposed technique for historical biogeographic analysis which meets these and other criteria, and we will apply it here to the development of a maximum-parsimony model of platyrrhine origins and dispersal.

Morse and White (1979, p. 357) have presented the following methodological outline for historical biogeographic analysis in comparative biology:

(1) Infer sequential hypothetical ancestors of the taxa (ultimately species) under study according to the "ex-group comparison" method of Ross (1974) or the "sister group" method of Hennig (1965, 1966) and Brundin (1966).
(2) Note the geographical distribution of each non-hypothetical taxon studied.
(3) Infer the distribution of each hypothetical ancestor, beginning with the most recent ancestors and working backward in time, according to the principles of phylogenetic compatibility. . . .
(4) Having completed this hypothesis forming phase through the oldest ancestor of the group, trace the implied vicariances (instances of allopatric speciation) and dispersals in a narrative summary. . . .
(5) Test the hypotheses as appropriate. If any are falsified infer alternate ones by repeating the necessary steps.

In essence, this technique of Morse and White has already been applied in various ways throughout the preceding reviews of the paleobiogeographic models of platyrrhine origins. What we plan to present here is a shortened but more formalized version of the model which our studies indicate is the most parsimonious one.

Cladistic Biogeography

Following the conclusions reached by Delson and Rosenberger (1980) and supported by many contributors to this volume, the Anthropoidea can be viewed most parsimoniously as a monophyletic group with the Platyrrhini and Catarrhini as two equal subdivisions each derived with respect to one another and to their hypothetical early anthropoid ancestor (see Fig. 1). Platyrrhines and catarrhines can therefore be considered sister groups of one another. Hennig (1966) has adopted a cladistic-based method for deciphering the biogeographic origin of a group which he has termed the "search for the sister group" (see also Ashlock, 1974). He suggests that the first critical question to be posed concerning the biogeographic origin of any group is: what is the group's closest sister group and where does that sister group occur? If this question were asked concerning the Platyrrhini, the answer would be the Catarrhini and Africa. Naturally, the answers would be the Platyrrhini and South America if a similar question were posed concerning the Catarrhini. As we see it, the central point to be grasped from this discussion concerns the application of a strictly cladistic approach to a problem in historical biogeography. None of Hennig's other biogeographic methods such as the Progression Rule, the Phylogenetic Intermediate Rule, or the Multiple Sister-Group Rule (for an excellent summary, see Ashlock, 1974) can be applied to the wholly disjunct distribution of platyrrhines and catarrhines. Furthermore, application of the new proposals of Platnick and Nelson (1978) for historical biogeographic analysis does little to resolve the issue of the biogeographic origin of the Platyrrhini.* In our opinion, the fact that the platyrrhines inhabit South America distinguishes them in principle from the Catarrhini of Africa in a fashion similar to that of a series of derived morphological

*Platnick and Nelson (1978) argue that a two-taxon, two-area pattern like the one exhibited by platyrrhines and catarrhines is not immediately resolvable by their method of analysis for historical biogeography. Instead, they suggest that the three-taxon statement should be regarded as the most basic unit of analysis in biogeography. Since the Omomyidae are the sister group of the Platyrrhini–Catarrhini (see Fig. 1), it is therefore the third taxon available for analysis. Unfortunately, it has a documented Holarctic (pan-European–Asian–North American) distribution which does little to resolve the paleobiogeographic problem at hand. The Omomyidae could be the ancestral stock from which the potential early anthropoids *Amphipithecus* (see Simons, 1971) and *Pondaungia* (see Szalay and Delson, 1979) were derived. The absence of omomyids from the Paleocene–Eocene fossil record of Africa is a moot point, since virtually no fossiliferous deposits are known; in contrast, their absence from the Paleogene of South America appears to be well documented. Thus, the occurrence of Omomyidae in all the northern continents and possibly in Africa cannot be used to falsify the paleobiogeographic model outlined above. Insofar as the omomyids can be judged ancestral to the Burmese early anthropoids, this evidence can add support to the maximum-parsimony model depicted in Fig. 7.

features. Thus, the biogeographic distribution of taxa in itself can be an important source of information.

In our construction of a maximum-parsimony model of platyrrhine origins, we will employ the cladistic biogeographic approach (see Morse and White, 1979). As previously discussed, this method is based on the prior development and utilization of a testable model of phylogenetic relationship. As Brundin (1972, p. 74) has stated, "... a careful establishment of strict monophyly and sister-group relationship is a necessary prerequisite for a realistic interpretation of a distribution pattern." The cladogram we presented in Fig. 1 fits the phylogenetic requirements proposed by Brundin (1972) and Morse and White (1979). To place this cladogram in a geographic perspective, we have applied it to a paleogeographic model of the world in the late Eocene (see Fig. 6). Ross (1974) has discussed this method for inferring distribution and direction of dispersal from phylogenetic analysis, though in concept it is based on the work of Kinsey (1936). It is also remarkably similar to procedures outlined by Hennig (1965, 1966) and Ashlock (1974). It is, of course, possible to argue that there are other ways to apply this cladogram to the late Eocene paleogeographic map, yet we conclude that this is the most parsimonious application based on its congruence with Hennig's (1966) "search for the sister group" and the occurrence of the earliest documented record of anthropoids* in southeastern Asia.

Following the procedures outlined by Morse and White (1979) coupled with the geographic interpretations derived from Fig. 6, we have been able to construct a maximum-parsimony model of platyrrhine origins and dispersal (Fig. 7). This model combines phylogenetic relationships, areal relationships, and probable dispersal patterns. The apparent congruence (see Fig. 7) between the biological cladogram of taxa and the geological cladogram of areas corroborates this maximum-parsimony model. Rosen (1978) has recently argued that there should be nothing fundamentally different in cladograms expressing biological relationships vs. geological relationships of a particular group. We agree completely with his assessment.

*The aspect of this model which may be most easily falsified concerns the origin of the Anthropoidea in Asia. As previously stated in the text, many authors do regard *Amphipithecus* and *Pondaungia* as probable anthropoids, and these taxa have been recovered *only* from the South-East Asian late Eocene deposits of Burma. Unfortunately, the fossil record of Africa during this same period of time is almost entirely unknown. It is possible that the Eocene anthropoids of Burma enjoyed a pan-Asian–African distribution. However, paleogeographic evidence in the form of the epicontinental transgressions of the Tethys Sea and Turgai Strait effectively isolated Africa from faunal interchange with either Europe or Asia throughout most of the Eocene. Therefore, if anthropoid origins were in Africa, the Tethys Sea and Turgai Strait would have restricted any range expansion into Asia until the late Eocene. Since the lack of any Eocene African fossil evidence can neither support nor deny this possibility, it is not falsifiable at present. With regard to the Pondaung primates of Burma, their presence in the late Eocene of Asia does provide falsifiable evidence in support of an Asian origin of the Anthropoidea as depicted in Figs. 6 and 7. We feel that this is the most parsimonious hypothesis until evidence is produced in the future to deny it.

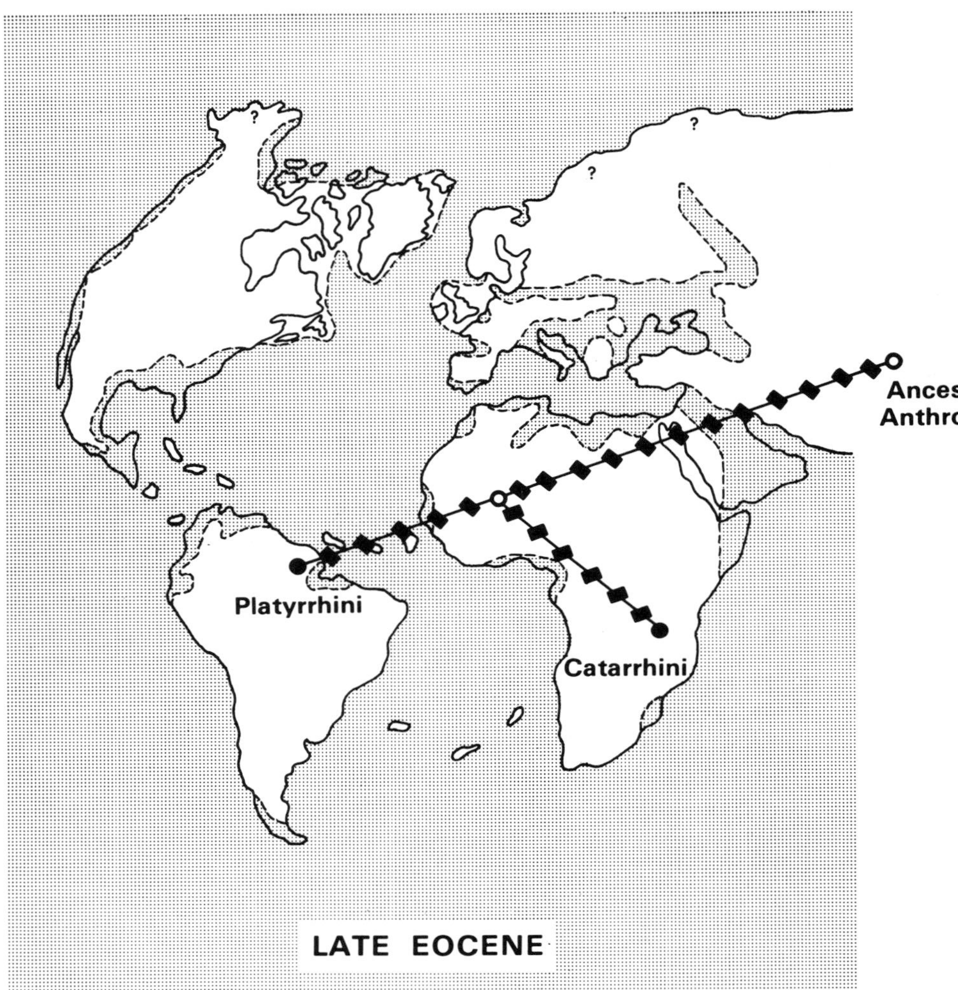

Fig. 6. Cladogram of the higher primates originally presented in Fig. 1 overlaid on a paleogeographic model of the world in the late Eocene. Applying the cladogram to the model in this particular manner represents the most reasonable and falsifiable combination of phylogenetic and paleogeographic information attainable at present. This wedding of phylogenetic and geographic data results in a paleobiogeographic scenario in concert with an origin of the Anthropoidea in Asia, the existence of a precatarrhine–preplatyrrhine early anthropoid stock in Africa, and the derivation of the Platyrrhini directly from the African continent via a sweepstakes dispersal route through island-hopping across the South Atlantic Ocean. Thus, the African-derivation model of the Platyrrhini presented in Fig. 4 is supported by this scenario. For background information concerning the paleogeographic reconstruction depicted here, see the references in the Figs. 2 and 4 captions.

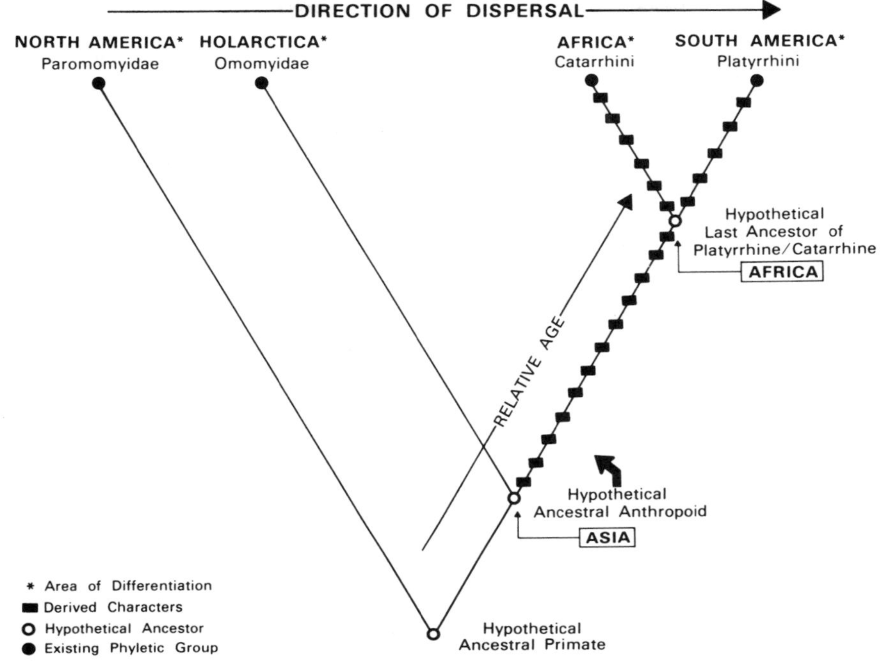

Fig. 7. Maximum-parsimony model in graphic form of the phylogenetic and biogeographic relationships of the primates as outlined in this chapter. The basic phylogenetic relationships expressed in this cladogram were originally presented and discussed in Fig. 1, adapted in part to a late Eocene paleogeographic model of the world in Fig. 6, and here combine the results of systematics, phylogeny, and geography in a synthetic and testable model. The horizontal arrow at top of figure points out inferred overall direction of dispersal for the existing phyletic groups, and the oblique arrow indicates decreasing relative age. Note that the ancestral platyrrhines appear most parsimoniously derived from a precatarrhine anthropoid from the African continent. A study by Brundin (1972) provides the conceptual basis for this maximum-parsimony model.

Dispersal Scenario

It is easily discernible from information presented in Figs. 6 and 7 that this maximum-parsimony model supports an African origin of the Platyrrhini from a preplatyrrhine–precatarrhine early anthropoid stock via waif dispersal across an equatorial South Atlantic inter-island sweepstakes route. Arguments in support of this proposal have already been presented in the section on the African-origin model. Since any major opposition to this maximum-parsimony model will almost certainly center around the likelihood of the proposed dispersal scenario, we feel it is necessary to present once again the comments of the geophysicist D. H. Tarling (1979):

> The implication of these models for migration routes in the Eocene–earliest Oligocene is that uplifted oceanic islands probably existed between Africa and South

> America, with marine gaps of probably 200 km or less between them. There is much less likelihood of island arc connections between North and South America at this time, as the Antilles, formed at 80 m.y., only lead toward the marine conditions of the Gulf of Mexico, and the southward growth of Central America had only just commenced. Migration from Africa would also be aided by westward flowing ocean currents, but these would inhibit migrations from North to South America. The palaeogeography, based on the palaeomagnetic record, therefore strongly supports trans-South Atlantic connections between islands in both the equatorial Atlantic and possibly across the Walvis–Rio Grande Rises in Eocene–earliest Oligocene times and suggests that North to South American crossings are much less probable, although not impossible.

Earlier, J. Tuzo Wilson, developer of the transform fault concept of plate tectonics, had also argued for the existence of island chains in the formative equatorial Atlantic. Wilson (cited in Chace and Manning, 1972, p. 6) states:

> . . . Ascension Island is only the latest in a series of islands whose remains form scattered seamounts and ridges from Ascension Island to the Cameroons in one direction (The Guinea Rise) and in the other direction to the north-east corner of Brazil.

Due to the nature of these arguments, we think it is incumbent upon those researchers who wish to argue against this maximum-parsimony model of platyrrhine origins to attempt initially to falsify the accumulating body of geological data on which it is based.

Falsifiability of the Model

No matter how many data can be assembled in support of a particular hypothesis, it will remain unproven with the criterion of falsifiability being the ultimate test of its adequacy. In its present form, this maximum-parsimony model of platyrrhine origins is not fully falsifiable. We suggest, as others have in the past, that no paleobiogeographic model can be fully falsifiable. However, if a single well-preserved higher primate jaw were recovered from Eocene deposits in West Africa, this discovery alone might lend direct biological support to our model. For example, if the jaw proved to be a preplatyrrhine/precatarrhine anthropoid or possibly an incipient platyrrhine, then this maximum-parsimony model would become testable and falsifiable on the basis of both paleontological *and* geological data. With this example in mind, it is interesting to note that Conroy (1980, p. 450), in a study of auditory structures and primate evolution, suggests that the last common ancestor of platyrrhines and catarrhines has yet to be found because ". . . it still lies buried in the sands of the African Paleogene!"

To conclude, most aspects of this maximum-parsimony model do represent the consensus of contributors to this volume. It is directly supported by Lavocat (1980), Hoffstetter (1980), Maier (1980), Bugge (1980), Luckett (1980), Martin and Gould (1980), Chiarelli (1980), Sarich and Cronin (1980), and Tarling (1980) and discussed as an alternative proposal in Gingerich (1980), Rosenberger and Szalay (1980), and McKenna (1980). We fully believe that its list of supporters will expand just as the continents themselves continue to drift.

ACKNOWLEDGMENTS

We would like to thank Ms. Joy Myers, who greatly assisted one of us (R. L. C.) throughout all stages of the preparation of this chapter. We would also like to acknowledge Ms. Doris Carter, who typed and retyped the manuscript, and Ms. Evelyn Oates, who drew the figures. Finally, we wish to thank Dr. Eric Delson for his many useful comments throughout the course of this study.

References

Ashlock, P. D., 1974, The uses of cladistics, *Annu. Rev. Ecol. Syst.* **5**:81–99.

Axelrod, D. I., 1978, The roles of plate tectonics in angiosperm history, in: *Historical Biogeography, Plate Tectonics, and the Changing Environment* (J. Gray and A. J. Boucot, eds.), pp. 435–447, Oregon State University Press, Corvallis.

Baba, M., Darga, L., and Goodman, M., 1980, Biochemical evidence on the phylogeny of Anthropoidea, in: *Evolutionary Biology of the New World Monkeys and Continental Drift* (R. L. Ciochon and A. B. Chiarelli, eds.), pp. 423–443, Plenum Press, New York.

Baker, P. E., 1973, Islands of the South Atlantic, in: *The Ocean Basins and Margins,* Vol. 1, *The South Atlantic* (A. E. M. Nairn and F. G. Stehli, eds.), pp. 493–553, Plenum Press, New York.

Ba Maw, Ciochon, R. L., and Savage, D. E., 1979, Late Eocene of Burma yields earliest anthropoid primate, *Pondaungia cotteri, Nature* **282**:65–67.

Berggren, W. A., and Hollister, C. D., 1974, Paleogeography, paleobiogeography, and the history of circulation in the Atlantic Ocean, in: *Studies in Paleo-oceanography* (W. W. Hay, ed.), *Soc. Econ. Paleontol. Mineral. Spec. Publ.* **20**:126–186.

Bingham, D. K., and Klootwijk, C. T., 1980, Paleomagnetic constraints on Greater India's underthrusting of the Tibetan Plateau, *Nature* **284**:336–338.

Bond, G., 1978, Evidence for late Tertiary uplift of Africa relative to North America, South America, Australia and Europe, *J. Geol.* **86**:47–65.

Briggs, J. C., 1966, Zoogeography and evolution, *Evolution* **20**:282–289.

Brundin, L., 1966, Transantarctic relationships and their significance, as evidenced by the chironomid midges, with a monograph of the subfamily Podonominae, Aphroteniinae and the austral Heptagyiae, *Kl. Sven. Vetenskaps akad. Handl., Ser. 4* **11**:1–472.

Brundin, L., 1972, Phylogenetics and biogeography, a reply to Darlington's "practical criticism" of Hennig-Brundin, *Syst. Zool.* **21**:69–79.

Bugge, J., 1980, Comparative anatomical study of the carotid circulation in New and Old World primates: Implications for their evolutionary history, in: *Evolutionary Biology of the New World Monkeys and Continental Drift* (R. L. Ciochon and A. B. Chiarelli, eds.), pp. 293–316, Plenum Press, New York.

Chace, F. A., Jr., and Manning, R. B., 1972, Two new caridean shimps, one representing a new family, from marine pools on Acension Island (Crustacea: Decapoda: Natantia), *Smithson. Contrib. Zool.* **131**:1–18.

Chiarelli, A. B., 1980, The karyology of South American primates and their relationship to African and Asian species, in: *Evolutionary Biology of the New World Monkeys and Continental Drift* (R. L. Ciochon and A. B. Chiarelli, eds.), pp. 387–398, Plenum Press, New York.

Clemens, W. A., 1974, *Purgatorius,* an early paromomyid primate (Mammalia), *Science* **184**:903–905.

Colbert, E. H., 1937, A new primate from the upper Eocene Pondaung formation of Burma, *Am. Mus. Novit.* **No. 951**:1–18.

Colbert, E. H., 1973, Continental drift and the distributions of fossil reptiles, in: *Implications of Continental Drift to the Earth Sciences,* Vol. 1 (D. H. Tarling and S. K. Runcorn, eds.), pp. 395–412, Academic Press, London.

Conroy, G. C., 1980, Ontogeny, auditory structures, and primate evolution, *Am. J. Phys. Anthropol.* **52**:443–451.

Cracraft, J., 1973, Continental drift, paleoclimatology, and the evolution and biogeography of birds, *J. Zool.* **169**:455–545.

Cracraft, J., 1974, Continental drift and vertebrate distribution, *Annu. Rev. Ecol. Syst.* **5**:215–261.

Cracraft, J., 1975, Historical biogeography and Earth history: Perspectives for future synthesis, *Ann. Missouri Bot. Gard.* **62**:494–521.

Craw, R. C., 1979, Generalized tracks and dispersal in biogeography: A response to R. M. McDowall, *Syst. Zool.* **28**:99–107.

Croizat, L., 1958, *Panbiogeography,* 3 vols., published by the author, Caracas.

Croizat, L., 1962, *Space, Time, Form: The Biological Synthesis,* published by the author, Caracas.

Croizat, L., 1971, De la "pseudovicariance" et de la "disjonction illusoire," *Anu. Soc. Broteriana* **37**:113–140.

Croizat, L., 1979, Review of *Biogeographie: Fauna und Flora der Erde und ihre geschichtliche Entwicklung* (P. Bănărescu and N. Boşcaiu), *Syst. Zool.* **28**:250–252.

Croizat, L., Nelson, G., and Rosen, D. E., 1974, Centers of origin and related concepts, *Syst. Zool.* **23**:265–287.

Cronin, J. E., and Sarich, V. M., 1975, Molecular systematics of the New World monkeys, *J. Hum. Evol.* **4**:357–375.

Darlington, P. J., Jr., 1957, *Zoogeography: The Geographical Distribution of Animals,* Wiley, New York, 675 pp.

Darlington, P. J., Jr., 1959, Area, climate, and evolution, *Evolution* **13**:488–510.

Darlington, P. J., Jr., 1965, *Biogeography of the Southern End of the World,* Harvard University Press, Cambridge, 236 pp.

Darlington, P. J., Jr., 1970, A practical criticism of Hennig–Brundin "phylogenetic systematics" and Antarctic biogeography, *Syst. Zool.* **19**:1–18.

Delson, E., and Rosenberger, A. L., 1980, Phyletic perspectives on platyrrhine origins and anthropoid relationships, in: *Evolutionary Biology of the New World Monkeys and Continental Drift* (R. L. Ciochon and A. B. Chiarelli, eds.), pp. 445–458, Plenum Press, New York.

Ferris, V. R., 1980. A science in search of a paradigm?—Review of the symposium, "Vicariance Biogeography: A Critique," *Syst. Zool.* **29**:67–76.

Fitch, F. J., Miller, J. A., Warrell, D. M., and Williams, S. C., 1974, Tectonic and radiometric age comparisons, in: *The Ocean Basins and Margins,* Vol. 2, *The North Atlantic* (A. E. M. Nairn and F. G. Stehli, eds.), pp. 485–538, Plenum Press, New York.

Förster, R., 1978, Evidence for an open seaway between northern and southern proto-Atlantic in Albian times, *Nature* **272**:158–159.

Frakes, L. A., and Kemp, E. M., 1973, Paleogene continental positions and evolution of climate, in: *Implications of Continental Drift to the Earth Sciences,* Vol. 1 (D. H. Tarling and S. K. Runcorn, eds.), pp. 539–559, Academic Press, London.

Funnell, B. M., and Smith, A. G., 1968, Opening of the Atlantic Ocean, *Nature* **219**:1328–1333.

Galton, P., 1977, The upper Jurassic ornithopod dinosaur *Dryosaurus* and a Laurasia–Gondwanaland connection, in: *Paleontology and Plate Tectonics* (R. M. West, ed.), *Milwaukee Public Mus. Spec. Publ. Biol. Geol.* **2**:41–54.

Gantt, D. G., 1980, Implications of enamel prism patterns for the origin of New World monkeys, in: *Evolutionary Biology of the New World Monkeys and Continental Drift* (R. L. Ciochon and A. B. Chiarelli, eds.), pp. 201–217, Plenum Press, New York.

Gingerich, P. D., 1973, Anatomy of the temporal bone in the Oligocene anthropoid *Apidium* and the origin of Anthropoidea, *Folia Primatol.* **19**:329–337.

Gingerich, P. D., 1975, A new genus of Adapidae (Mammalia, Primates) from the late Eocene of southern France and its significance for the origin of higher primates, *Contrib. Mus. Paleontol. Univ. Michigan* **24**:163–170.

Gingerich, P. D., 1977, Radiation of Eocene Adapidae in Europe, *Géobios, Mém. Spéc.* **1**:165–185.

Gingerich, P. D., 1979, The stratophenetic approach to phylogeny reconstruction in vertebrate paleontology, in: *Phylogenetic Analysis and Paleontology* (J. Cracraft and N. Eldredge, eds.), pp. 41–77, Columbia University Press, New York.

Gingerich, P. D., 1980, Eocene Adapidae, paleobiogeography, and the origin of South American Platyrrhini, in: *Evolutionary Biology of the New World Monkeys and Continental Drift* (R. L. Ciochon and A. B. Chiarelli, eds.), pp. 123–138, Plenum Press, New York.

Gregory, W. K., 1920, On the structure and relations of *Notharctus*, an American Eocene primate, *Mem. Am. Mus. Nat. Hist., N. S.* **3**:51–243.

Hennig, W., 1965, Phylogenetic systematics, *Annu. Rev. Entomol.* **10**:97–116.

Hennig, W., 1966, The Diptera fauna of New Zealand as a problem in systematics and zoogeography, *Pac. Insects Monogr.* **9**:1–81.

Hershkovitz, P., 1972, The recent mammals of the Neotropical region: A zoogeographic and ecological review, in: *Evolution, Mammals, and Southern Continents* (A. Keast, F. C. Erk, and B. Glass, eds.), pp. 311–431, State University of New York Press, Albany.

Hershkovitz, P., 1977, *Living New World Monkeys (Platyrrhini) with an Introduction to the Primates,* Vol. 1, University of Chicago Press, 1117 pp.

Herz, N., 1977, Timing of spreading in the South Atlantic: Information from Brazilian alkalic rocks, *Geol. Soc. Am. Bull.* **88**:101–112.

Hoffstetter, R., 1969, Un primate de l'Oligocene inférieur Sud-Américain: *Branisella boliviana* gen. et sp. nov., *C. R. Acad. Sci. Paris, Sér. D* **269**:434–437.

Hoffstetter, R., 1972, Relationships, origins, and history of the ceboid monkeys and caviomorph rodents: A modern reinterpretation, in: *Evolutionary Biology,* Vol. 6 (Th. Dobzhansky, M. K. Hecht, and W. C. Steere, eds.), pp. 323–347, Appleton-Century-Crofts, New York.

Hoffstetter, R., 1974, Phylogeny and geographical deployment of the primates, *J. Hum. Evol.* **3**:327–350.

Hoffstetter, R., 1977, Phylogénie de primates: Confrontation des résultats obtenus par les diverses voies d'approche du probleme, *Bull. M. Soc. Anthropol. Paris* **4**(Sér. 13):327–346.

Hoffstetter, R., 1980, Origin and deployment of New World monkeys emphasizing the southern continents route, in: *Evolutionary Biology of the New World Monkeys and Continental Drift* (R. L. Ciochon and A. B. Chiarelli, eds.), pp. 103–122, Plenum Press, New York.

Holcombe, T. L., and Moore, W. S., 1977, Paleocurrents in the eastern Caribbean: Geologic evidence and implications, *Mar. Geol.* **23**:35–56.

Johnson, B. D., Powell, C. McA., and Veevers, J. J., 1976, Spreading history of the eastern Indian Ocean and Greater India's northward flight from Antarctica and Australia, *Geol. Soc. Am. Bull.* **87**:1560–1566.

Kay, R. F., 1980, Platyrrhine origins: A reappraisal of the dental evidence, in: *Evolutionary Biology of the New World Monkeys and Continental Drift* (R. L. Ciochon and A. B. Chiarelli, eds.), pp. 159–187, Plenum Press, New York.

Keast, A., 1972, Continental drift and the evolution of the biota on southern continents, in: *Evolution, Mammals, and Southern Continents* (A. Keast, F. C. Erk, and B. Glass, eds.), pp. 23–87, State University of New York Press, Albany.

Keast, A., 1977, Zoogeography and phylogeny: The theoretical background and methodology to the analysis of mammal and bird fauna, in: *Major Patterns in Vertebrate Evolution* (M. K. Hecht, P. C. Goody, and B. M. Hecht, eds.), pp. 249–312, Plenum Press, New York.

Keast, A., Erk, F. C., and Glass, B., (eds.), 1972, *Evolution, Mammals, and Southern Continents,* State University of New York Press, Albany, 543 pp.

Kennedy, W. J., and Cooper, M., 1975, Cretaceous ammonite distributions and the opening of the South Atlantic, *Geol. Soc. London J.* **131**:283–288.

Kinsey, A. C., 1936, *The Origin of Higher Categories in Cynips, Indiana University Pub., Sci. Ser.,* No. 4.

Kitts, D. B., 1977, Karl Popper, verifiability, and systematic zoology, *Syst. Zool.* **26:**185–194.

Kumar, N., and Embley, R. W., 1977, Evolution and origin of Ceará Rise: An aseismic rise in the western equatorial Atlantic, *Geol. Soc. Am. Bull.* **88**:683–694.

Ladd, J. W., 1976, Relative motion of South America with respect to North America and Caribbean tectonics, *Geol. Soc. Am. Bull.* **87**:969–976.

Ladd, J. W., Dickson, G. O., and Pittman, W. C., III, 1973, The age of the South Atlantic, in: *The Ocean Basins and Margins,* Vol. 1, *The South Atlantic* (A. E. M. Nairn and and F. G. Stehli, eds.), pp. 555–573, Plenum Press, New York.

Larson, R. L., and Ladd, J. W., 1973, Evidence for the opening of the South Atlantic in the early Cretaceous, *Nature* **246**:209–212.

Lavocat, R., 1974a, What is an hystricomorph?, in: *The Biology of Hystricomorph Rodents* (I. W. Rolands and B. J. Weir, eds.), *Symp. Zool. Soc. London,* No. 34, pp. 7–20 [see also discussion, pp. 55–60], Academic Press, London.

Lavocat, R., 1974b, The interrelationships between the African and South American rodents and their bearing on the problem of the origin of South American monkeys, *J. Hum. Evol.* **3**:323–326.

Lavocat, R., 1977, Sur l'origine des faunes sud-américaines de Mammifères du Mésozoïque terminal et du Cénozoïque ancien, *C. R. Acad. Sci. Paris, Sér. D* **285**:1423–1426.

Lavocat, R., 1980, The implications of rodent paleontology and biogeography to the geographical sources and origin of platyrrhine primates, in: *Evolutionary Biology of the New World Monkeys and Continental Drift* (R. L. Ciochon and A. B. Chiarelli, eds.), pp. 93–102, Plenum Press, New York.

Lillegraven, J. A., Kraus, M. J., and Bown, T. M., 1979, Paleogeography of the world of the Mesozoic, in: *Mesozoic Mammals* (J. A. Lillegraven, Z. Kielan-Jaworowska, and W. A. Clemens, eds.), pp. 277–308, University of California Press, Berkeley.

Luckett, W. P., 1980, Monophyletic or diphyletic origins of Anthropoidea and Hystricognathi: Evidence of the fetal membranes, in: *Evolutionary Biology of the New World Monkeys and Continental Drift* (R. L. Ciochon and A. B. Chiarelli, eds.), pp. 347–368, Plenum Press, New York.

Lloyd, J. L., 1963, Tectonic history of the south Central-American orogen, in: *Backbone of the Americas—A Symposium* (Childs and Beebe, eds.), pp. 88–100, American Association of Petroleum Geologists, Tulsa.

Maier, W., 1980, Nasal structures in Old World and New World primates, in: *Evolutionary Biology of the New World Monkeys and Continental Drift* (R. L. Ciochon and A. B. Chiarelli, eds.), pp. 219–241, Plenum Press, New York.

Malfait, B. T., and Dinkelman, M. G., 1972, Circum-Caribbean tectonic and igneous activity and the evolution of the Caribbean plate, *Geol. Soc. Am. Bull.* **83**:251–272.

Martin, D. E., and Gould, K. C., 1980, Comparative study of the sperm morphology of South American primates and those of the Old World, in: *Evolutionary Biology of the New World Monkeys and Continental Drift* (R. L. Ciochon and A. B. Chiarelli, eds.), pp. 369–386, Plenum Press, New York.

Matthew, W. D., 1915, Climate and evolution, *Ann. N. Y. Acad. Sci.* **24**:171–318.

Mayr, E., 1965, What is a fauna?, *Zool. Jahrb. Syst. Geogr.* **92**:473–486.

McKenna, M. C., 1967, Classification, range, deployment of the prosimian primates, *Colloq. Int. C. N. R. S.* **163**:603–613.

McKenna, M. C., 1972a, Possible biological consequences of plate tectonics, *BioScience* **22**:519–525.

McKenna, M. C., 1972b, Was Europe connected directly to North America prior to the middle Eocene?, in: *Evolutionary Biology,* Vol. 6 (Th. Dobzhansky, M. K. Hecht, and W. C. Steere, eds.), pp. 179–189, Appleton-Century-Crofts, New York.

McKenna, M. C., 1973, Sweepstakes, filters, corridors, Noah's arks, and beached Viking funeral ships in palaeogeography, in: *Implications of Continental Drift to the Earth Sciences*, Vol. 1 (D. H. Tarling and S. K. Runcorn, eds.), pp. 295–308, Academic Press, London.

McKenna, M. C., 1975, Fossil mammals and early Eocene North Atlantic land continuity, *Ann. Missouri Bot. Gard.* **62**:335–353.

McKenna, M. C., 1980, Early history and biogeography of South America's extinct land mammals, in: *Evolutionary Biology of the New World Monkeys and Continental Drift* (R. L. Ciochon and A. B. Chiarelli, eds.), pp. 43–77, Plenum Press, New York.

Milner, A. R., and Panchen, A. L., 1973, Geographic variation in the tetrapod faunas of the upper Carboniferous and lower Permian, in: *Implications of Continental Drift to the Earth Sciences,* Vol. 1 (D. H. Tarling and S. K. Runcorn, eds.), pp. 353–368, Academic Press, London.

Morse, J. C., and White, D. F., Jr., 1979, A technique for analysis of historical biogeography and other characters in comparative biology, *Syst. Zool.* **28**:356–365.

Nelson, G., 1973, Comments on Leon Croizat's biogeography, *Syst. Zool.* **22**:312–320.

Nelson, G., 1974, Historical biogeography: An alternative formalization, *Syst. Zool.* **23**:555–558.

Nelson, G., 1975, Biogeography, the vicariance paradigm, and continental drift, *Syst. Zool.* **24**:490–504.

Nelson, G., 1978, From Candolle to Croizat: Comments on the history of biogeography, *J. Hist. Biol.* **11**:269–305.

Nelson, G., and Platnick, N. I., 1978, The perils of plesiomorphy: Widespread taxa, dispersal, and phenetic biogeography, *Syst. Zool.* **27**:474–477.

Novacek, M. J., and Marshall, L. G., 1976, Early biogeographic history of ostariophysan fishes, *Copeia* **1**:1–12.

Orlosky, F. J., 1980, Dental evolutionary trends of relevance to the origin and dispersion of the platyrrhine monkeys, in: *Evolutionary Biology of the New World Monkeys and Continental Drift* (R. L. Ciochon and A. B. Chiarelli, eds.), pp. 189–200, Plenum Press, New York.

Orlosky, F. J., and Swindler, D. R., 1975, Origins of New World monkeys, *J. Hum. Evol.* **4**:77–83.

Perkins, E. M., and Meyer, W. C., 1980, The phylogenetic significance of the skin of primates: Implications for the origin of New World monkeys, in: *Evolutionary Biology of the New World Monkeys and Continental Drift* (R. L. Ciochon and A. B. Chiarelli, eds.), pp. 331–346, Plenum Press, New York.

Pilgrim, G. E., 1927, A *Sivapithecus* palate and other primate fossils from India, *Mem. Geol. Surv. India* **14**:1–24.

Pitman, W. C., III, and Talwani, M., 1972, Sea-floor spreading in the North Atlantic, *Geol. Soc. Am. Bull.* **83**:619–646.

Platnick, N. I., 1976, Drifting spiders or continents?: Vicariance biogeography of the spider family Laroniinae (Araneae: Gnaphosidae), *Syst. Zool.* **25**:101–109.

Platnick, N. I., and Nelson, G., 1978, A method of analysis for historical biogeography, *Syst. Zool.* **27**:1–16.

Popper, K. R., 1959, *The Logic of Scientific Discovery,* Harper and Row, New York, 480 pp.

Popper, K. R., 1963, *Conjectures and Refutations,* Harper and Row, New York, 417 pp.

Powers, S., 1911, Floating islands, *Pop. Sci. Monthly* **79**:303–307.

Reyment, R. A., Bengtson, P., and Tait, E. A., 1976, Cretaceous transgressions in Nigeria and Seripe-Alagoas (Brazil), *An. Acad. Brasil Ciên.* **48**:253–264.

Rich, P. V., 1978, Fossil birds of old Gondwanaland: A comment on drifting continents and their passengers, in: *Historical Biogeography, Plate Tectonics and the Changing Environment* (J. Gray and A. J. Boucot, eds.), pp. 321–332, Oregon State University Press, Corvallis.

Ridley, H. N., 1930, *The Dispersal of Plants Throughout the World,* L. Reeve, Ashford, Kent.

Rosen, D. E., 1975, A vicariance model of Caribbean biogeography, *Syst. Zool.* **24**:431–464.

Rosen, D. E., 1978, Vicariant patterns and historical explanation in biogeography, *Syst. Zool.* **27**:159–188.

Rosenberger, A., and Szalay, F. S., 1980, On the Tarsiiform origins of Anthropoidea, in: *Evolutionary Biology of the New World Monkeys and Continental Drift* (R. L. Ciochon and A. B. Chiarelli, eds.), pp. 139–157, Plenum Press, New York.

Ross, H. H., 1974, *Biological Systematics,* Addison-Wesley, Reading, Massachusetts, 345 pp.

Sahni, A., and Kumar, V., 1974, Paleogene palaeobiogeography of the Indian subcontinent, *Palaeogeogr. Palaeoclimatol. Palaeoecol.* **15**:209–226.

Sarich, V. M., 1970, Primate systematics with special reference to Old World monkeys, in: *Old World Monkeys* (J. R. Napier and P. H. Napier, eds.), pp. 175–226, Academic Press, New York.

Sarich, V. M., and Cronin, J. E., 1980, South American mammal molecular systematics, evolutionary clocks, and continental drift, in: *Evolutionary Biology of the New World Monkeys and Continental Drift* (R. L. Ciochon and A. B. Chiarelli, eds.), pp. 399–421, Plenum Press, New York.

Scheltema, R. S., 1971, Larval dispersal as a means of genetic exchange between geographically separated populations of shallow-water benthic marine gastropods, *Biol. Bull.* **140**:284–322.

Schram, F. R., 1977, Paleozoogeography of late Paleozoic and Triassic Malacostraca, *Syst. Zool.* **26**:367–379.

Sclater, J. G., Hellinger, S., and Tapscott, C., 1977, The paleobathymetry of the Atlantic Ocean from the Jurassic to the Present, *J. Geol.* **85**:509–552.

Sibuet, J. C., and Mascle, J., 1978, Plate kinematic implications of Atlantic equatorial fracture zone trends, *J. Geophys. Res.* **83**:3401–3421.

Simons, E. L., 1962, Two new primate species from the African Oligocene, *Postilla* **No. 64**: 1–12

Simons, E. L., 1971, Relationships of *Amphipithecus* and *Oligopithecus, Nature* **232**:489–491.

Simons, E. L., 1976, The fossil record of primate phylogeny, in: *Molecular Anthropology* (M. Goodman and R. E. Tashian, eds.), pp. 35–62, Plenum Press, New York.

Simpson, G. G., 1940, Mammals and land bridges, *J. Wash. Acad. Sci.* **30**:137–163.

Simpson, G. G., 1945, The principles of classification and a classification of the mammals, *Bull. Am. Mus. Nat. Hist.* **85**:1–350.

Simpson, G. G., 1952, Probabilities of dispersal in geologic time, in: *The Problem of Land Connections Across the South Atlantic, with Special Reference to the Mesozoic* (E. Mayr, ed.), *Bull. Am. Mus. Nat. Hist.* **99**:163–176.

Simpson, G. G., 1953, *Evolution and Geography,* Oregon State System of Higher Education, Eugene, 64 pp.

Simpson, G. G., 1978, Early mammals in South America: Fact, controversy, and mystery, *Proc. Am. Philos. Soc.* **122**:318–328.

Smith, A. G., and Briden, J. C., 1976, *Mesozoic and Cenozoic Paleocontinental Maps,* Cambridge University Press, Cambridge, 63 pp.

Smith, P. J., 1977, Origin of the Rio Grande rise, *Nature* **269**:651–652.

Sudre, J., 1975, Un prosimien du Paléogène ancien du Sahara Nord-Occidental: *Azibius trerki* n. g. n. sp., *C. R. Acad. Sci. Paris, Ser. D* **280**:1539–1542.

Szalay, F. S., 1975, Phylogeny, adaptations, and dispersal of the Tarsiiform primates, in: *Phylogeny of the Primates* (W. P. Luckett and F. S. Szalay, eds.). pp. 357–404, Plenum Press, New York.

Szalay, F. S., and Delson, E., 1979, *Evolutionary History of the Primates,* Academic Press, New York, 580 pp.

Tarling, D. H., 1979, Continental drift and the positioning of the circum-Atlantic continents throughout the last 100 million years of Earth history, in: *Abstracts of the VII Congress of the International Primatological Society* (M. Moudgal, ed.), Bangalore, India.

Tarling, D. H., 1980, The geologic evolution of South America with special reference to the last 200 million years, in: *Evolutionary Biology of the New World Monkeys and Continental Drift* (R. L. Ciochon and A. B. Chiarelli, eds.), pp. 1–41, Plenum Press, New York.

Tedford, R. H., 1974, Marsupials and the new paleogeography, in: *Paleogeographic Provinces and Provinciality* (C. A. Ross, ed.), *Soc. Econ. Paleont. Mineral. Spec. Publ.* **21**:109–126.

Thaw Tint, Ba Maw, Savage, D. E., and Ciochon, R. L., 1981, New discovery of *Amphipithecus* from the late Eocene of Burma: Anthropoid origins in Asia?, *Science* (submitted).

Thorington, R. W., 1976, The systematics of New World monkeys, in: *First Inter-American Conference on Conservation and Utilization of American Nonhuman Primates in Biomedical Research,* Pan African Health Organization, Sci. Publ. No. 317, pp. 8–18.

Tomblin, J. F., 1975, The Lesser Antilles and Aves Ridge, in: *The Ocean Basins and Margins,* Vol. 3, *The Gulf of Mexico and the Caribbean* (A. E. M. Nairn and F. G. Stehli, eds.), pp. 467–500, Plenum Press, New York.

Valentine, J. W., 1973, Plates and provinciality, a theoretical history of environmental discontinuities, in: *Organisms and Continents Through Time* (N. F. Hughes, ed.), *Spec. Pap. Palaeontol.* **12**:79–92.

Van Andel, T. H., Thiede, J., Sclater, J. G., and Hay, W. W., 1977, Depositional history of the South Atlantic Ocean during the last 125 million years, *J. Geol.* **85**:651–698.

Van Valen, L., and Sloan, R. E., 1965, The earliest primates, *Science* **150**:743–745.

Vinogradov, A. P. (ed.), 1967, *Atlas of the Lithological-Paleogeographical Maps of the U.S.S.R.*, Vol. IV (Paleogene, Neogene and Quaternary) I-55, Ministry of Geology of the U.S.S.R., Academy of Sciences of the U.S.S.R.

Walker, A., 1972, The dissemination and segregation of early primates in relation to continental configuration, in: *Calibration of Hominoid Evolution* (W. W. Bishop and J. A. Miller, eds.), pp. 195–218, Scottish Academic Press, Glasgow.

Washburn, S. L., and Moore, R., 1980, *Ape Into Human: A Study of Human Evolution,* Little, Brown, Boston, 194 pp.

Weyl, R., 1974, Die paläogeographische Entwicklung Mittelamerikas, *Zentralbl. Geol. Palaeontol. Teil 1* **5/6**:432–466.

Wood, A. E., 1980, The origin of the caviomorph rodents from a source in Middle America: A clue to the area of origin of the platyrrhine primates, in: *Evolutionary Biology of the New World Monkeys and Continental Drift* (R. L. Ciochon and A. B. Chiarelli, eds.), pp. 79–91, Plenum Press, New York.

Concluding Remarks 24

R. L. CIOCHON and A. B. CHIARELLI

> The source of the South American Primates (Ceboidea or Platyrrhini) and early Rodentia and their direct successors (Caviomorpha) is at present one of the most uncertain and disputed problems in the history of South American mammals . . . (p. 319)
>
> G. G. Simpson (1978)

Introduction

In 1945 George Gaylord Simpson proposed that from data bearing on the present day distribution of the Ceboidea and from aspects of their comparative anatomy there could be little doubt that all the New World monkeys represented a radiation from a single ancestral stock isolated in South America during the early Tertiary. Simpson further suggested that "the most reasonable hypothesis is that ceboids arose from one of the Paleocene or Eocene prosimian stocks of North America and that their early deployment, or indeed almost all of their history, occurred in the more tropical parts of South America, where Tertiary fossils are extremely rare. This is, however, only a hypothesis" (Simpson, 1945, p. 185). Prior to the proposal of this hypothesis, which was certainly a reasonable one at the time, many accounts of platyrrhine origins had been primarily anecdotal in nature. Take, for instance, the proposal of Sera (1938) who concluded that the Platyrrhini had recently evolved from an aquatic ancestor or had passed through an aquatic stage during their more recent evolution based on the form of the platyrrhine nose and laryngeal cartilages, the shape of the external ears, the short auditory

R. L. CIOCHON • Department of Anthropology and Sociology, University of North Carolina at Charlotte, Charlotte, North Carolina 28223. A. B. CHIARELLI • Istituto di Antropologia, Università di Firenze, Via del Proconsolo, 12, 50122 Firenze, Italy.

passage, the lobulations of the kidney and the structure of the female genital passages.

In the decades following the presentation of the Simpsonian viewpoint on platyrrhine origin and dispersal, there has been a nearly universal acceptance of this hypothesis by anthropologists, paleontologists, and anatomists on both sides of the Atlantic [for example, see Hill (1955), Gazin (1958), Le Gros Clark (1962), and Simons (1972)]. Perhaps this wide acceptance of an hypothesis favoring a North American origin of the Platyrrhini was due in large part to the fact that no other plausible alternative existed. That situation changed with the advent of the 1970s and the development of a workable model of plate tectonics and continental drift. Two French paleontologists, Lavocat (1969) and Hoffstetter (1971, 1972), began proposing the innovative alternative of an African origin of the Platyrrhini and other South American faunal elements via a transatlantic dispersal route. This bold proposal immediately polarized arguments on the subject.

Purpose of the Volume

With debate on both sides of the issue of platyrrhine origins growing throughout the 1970s this edited volume was conceived as a possible vehicle to channel, define, and, hopefully, resolve these arguments. Scientists from a variety of fields covering a wide range of topics were asked to contribute papers focusing on the potential phylogenetic affinity and geographical source of the ancestral platyrrhine primate (see Fig. 1). It was our wish that this volume approach the topic of platyrrhine origins from every possible perspective. Unfortunately, we found that not all the papers solicited were totally germane to solving the issue at hand. Furthermore, the basic approach and premises of some contributors, in our opinion, clouded their ability to deal forthrightly with the theme of the volume which lends an uneven appearance to some contributions. This is to be expected in any large edited volume, and we, in our editorial capacity, made every possible attempt to standardize the approach of each contribution, eliminate any overlapping and repetitive information and unify the taxonomic nomenclature throughout the volume.

In the end we found it was not possible to obtain well-conceived and carefully written papers on all the topics presented in Fig. 1. In particular we feel the following omissions should be pointed out since some may provide areas for potentially significant future research.

1. No paper appears comparing the paleobotanical or neobotanical record of South America with Africa and North America. Almost certainly the ancestral platyrrhines lived in tropical or subtropical forests and ate vegetative matter. Therefore, the recent and past distribution of these plants could bear on the issue of New World monkey dispersal routes. Unfortunately, no botanist could be found on relatively short notice who was willing to tackle this

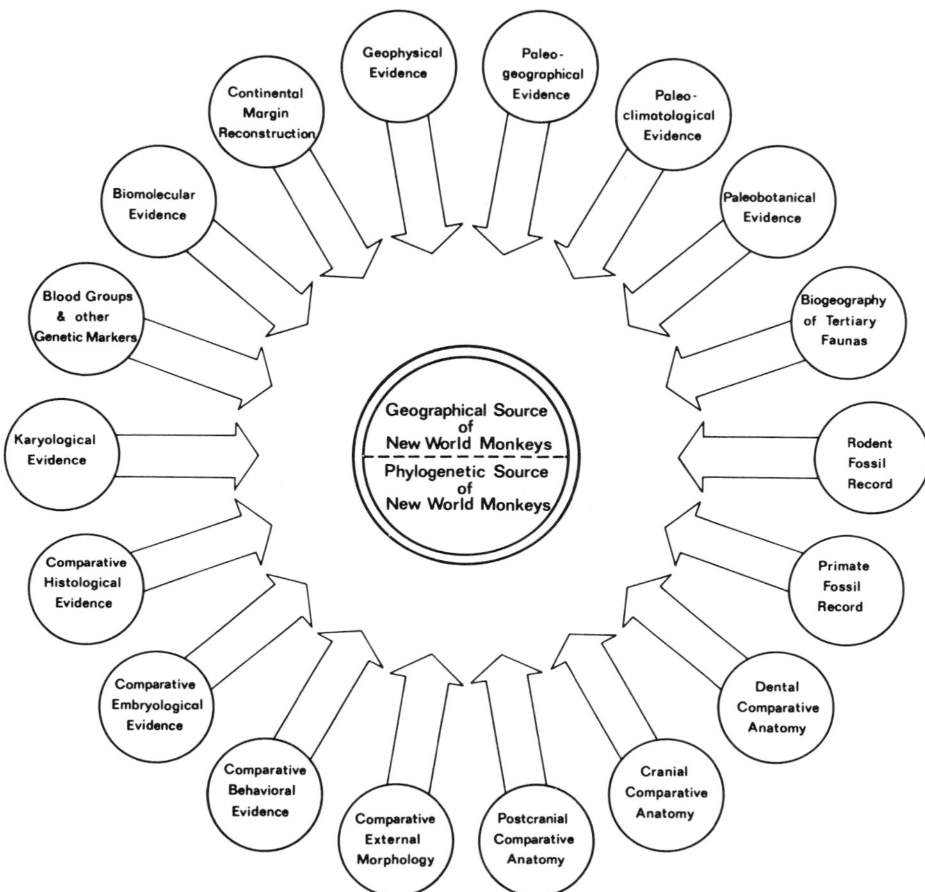

Fig. 1. Graphic representation of evidence from a wide variety of sources originally considered for inclusion in this volume.

problem. P. V. Arrigoni of the Istituto Botanico Università, Firenze, Italy, is now beginning research in this area.

2. No paper presents data on epigenetic characters of platyrrhines such as blood groups, cranial discrete traits and dermatoglyphics comparing this information with that obtained for catarrhines and lower primate groups. Several attempts were made to solicit papers on these topics, but we finally had to conclude that this information is not strictly relevant to intergroup comparisons of high taxonomic categories such as Platyrrhini vs. Catarrhini. For ascertaining primate phylogenetic relationships this data source appears most useful at or below the familial level.

3. Several papers were solicited which would have compared the social behavior and/or locomotor behavior of New World monkeys with selected

catarrhines and various lower primates. Unfortunately, conversations with potential contributors revealed that little information which could impact on the phylogenetic affinity or geographical source of the Platyrrhini would result from such comparisons. It was decided that such behavioral evidence, though obviously quite important in understanding the current adaptations of an organism, does not yet exist in sufficient quantity to be useful in reconstructing events that occurred many millions of years in the past.

4. Finally, no paper appears where the host–parasite relationships of the Platyrrhini are elucidated and compared with those of catarrhines and lower primates. This parasitological data can provide an independent source of verification regarding the host's phylogeny (see Mayr, 1957) and has been used with some success in verifying the relationships of South American and African rodents (Durette-Desset, 1971; Lavocat, 1974). Certain nonpathogenic parasites are known to evolve at slower more conservative rates than their host species (Cameron, 1956) which make them excellent candidates for reconstructing the phylogenetic and biogeographic relationships of their hosts. In this regard several parasitologists were contacted with the aim of soliciting such a study for the Platyrrhini. It was found that, like so many other aspects of New World monkey comparative biology, very little data existed on the subject. Therefore no study could be attempted. We feel this area of platyrrhine biology deserves future attention and may one day yield important supporting evidence.

Results of the Symposium and Volume

As detailed in the volume preface, symposia in Turin, Italy, and Bangalore, India, were held to promote discussion and debate on the issue of platyrrhine origins. Many of the contributors had completed their manuscripts prior to the convening of these symposia. It is noteworthy to mention that after the two symposia more than half of the contributors wished to significantly revise their manuscripts in light of these deliberations.

At the final symposium in Bangalore a consensus was reached that the issue of platyrrhine origins actually consisted of a series of interrelated questions (see Preface). Succinctly stated, these questions can be phrased in the following manner:

1. What precise phylogenetic relationship do the Platyrrhini bear to the Catarrhini and to the generally accepted concept of the Anthropoidea?
2. What is the probable paleontological source of origin of the Platyrrhini?
3. What is the most probable dispersal scenario and geographical source(s) of origin for the Platyrrhini?

The first 21 chapters of this volume attempt to specifically address the preceding questions from a wide variety of perspectives. Chapters 22 and 23 then present critically focused, consensus-oriented summaries of all the preceding chapters, once again specifically addressing the above questions. Here we will combine the summaries of these two concluding chapters together with a certain degree of editorial license to formulate the following scenario for the origin and dispersal of the Platyrrhini.

It is our contention that the Anthropoidea represent a monophyletic group whose ancestry extends back into the Eocene. Platyrrhini and Catarrhini are two separate and equal divisions of this group and are each strictly monophyletic in their own right. However, the derived characters (anthropoid synapomorphies) uniting platyrrhines and catarrhines *vis-à-vis* lower primates are quite numerous and more significant in terms of level of organization than the derived characters which distinguish these two groups (see Table 1 in Delson and Rosenberger, 1980 and Fig. 1 in Ciochon and Chiarelli 1980). Therefore, a preplatyrrhine/precatarrhine early anthropoid stock must have existed for a period of time in the past. The existence of this group is theoretically suggested by the large number of synapomorphies uniting platyrrhines with catarrhines and paleontologically documented by the presence of early anthropoids in the late Eocene of Burma such as *Amphipithecus* and *Pondaungia* (see Ba Maw *et al.,* 1979; Thaw Tint *et al.,* 1981).

The paleontological source of origin of this early anthropoid stock and therefore of both platyrrhines and catarrhines most probably lies in an early tarsiiform group such as the Omomyidae. An adapid–anthropoid tie, however, cannot be fully discounted and we therefore do not wish to argue as strongly against this possibility as do Delson and Rosenberger (1980). We suggest the actual paleontological evidence favoring an omomyid origin of the Anthropoidea vs. an adapid origin is not particularly strong in either case. Rather, it is the neontological evidence favoring a tarsier–anthropoid tie which tips the balance in favor of an omomyid derivation of the Anthropoidea. Future paleontological discoveries of early anthropoid remains could very easily alter this balance.

The final question concerns the establishment of the geographical source or sources of origin of the Platyrrhini and a probable dispersal scenario. Since a growing body of data (including nearly universal agreement by the contributors to this volume) now support strict platyrrhine monophyly, it is reasonable to assume the New World monkeys had only one geographical source of origin and therefore probably arrived in South America through one dispersal event.* Beginning with the premise that platyrrhines and catarrhines

*If the Platyrrhini indeed reached the South American continent by one dispersal event some 40 million years ago, then their entire morphological heterogeneity and genetic diversity—well documented in this volume—would have developed subsequent to that event. A probable mechanism for the evolution of this diversity (which may even have been significantly greater in the past—see Rosenberger, 1979) was by dispersal and differentiation in small isolated groups in

shared a period of common ancestry prior to their differentiation and given their current disjunct distribution in the New and Old World tropics, several different geographical sources and dispersal scenarios can be envisioned (see Ciochon and Chiarelli, 1980). We feel that arguments stemming from direct analysis of the phylogeny must play a central role in selecting the most parsimonious paleobiogeographic hypothesis followed then by evidence derived from geology and other related sciences. Since the evidence presented above closely links the ancestry of the Platyrrhini with that of the Catarrhini, it is possible to argue that they shared a common geographical source of origin in the Southern continents (see Ciochon and Chiarelli, 1980). Nearly all documented Paleogene and most Neogene anthropoid evolution and differentiation has occurred in the Old World tropics, primarily on the African continent. We therefore suggest that the last common ancestor of the Platyrrhini was derived from a precatarrhine African-based early anthropoid stock. We conclude that this precatarrhine (and preplatyrrhine) early anthropoid, possibly related to the *ancestors* of the Pondaung primates, reached Africa from Asia in the early Late Eocene (see Fig. 6 in Ciochon and Chiarelli, 1980). Populations of these early anthropoids spread to all available habitats throughout the African continent; some later became the ancestors of the Catarrhini while others made the transatlantic crossing to South America to become the ancestors of the Platyrrhini. The crossing to South America was accomplished by waif dispersal across a much less expansive Eocene equatorial Atlantic Ocean that was punctuated by uplifted oceanic islands separated by marine gaps of 200 km or less (Ciochon and Chiarelli, 1980; see also Tarling, 1980).

The probable existence of an omomyid-derived preplatyrrhine/ precatarrhine anthropoid stock in Africa during the late Eocene coupled with evidence derived from sea-floor spreading indicating a narrow equatorial Atlantic, the presence of tectonically active island chains and favorable east to west oceanic paleocurrents and winds in our opinion makes this scenario for the origin and dispersal of the Platyrrhini the most parsimonious model. Of course, not all the contributors reached a conclusion similar to the one presented here. Nevertheless, we feel it represents a reasonable working hypothesis that is subject to future testing and scrutiny. In its current form it is not clearly falsifiable by evidence now at hand; yet we are most certain that future paleontological discoveries, more complete paleogeographic data, in-depth comparative anatomical studies, and new strides in the fields of genetic biology will turn this situation about. It is further hoped that the presentation of this consensus-oriented model will not in any way lessen debate on the issue of platyrrhine origins. As mentioned in the preface, this collection of papers

---

the equatorial forests of South America. These microevolutionary events staged throughout millions of years of isolation on the island continent of South America today provide platyrrhine morphologists and genetic biologists with a unique opportunity to study animal speciation, adaptation, and differentiation under nearly ideal conditions.

was assembled to *promote* discussion on this subject. We feel hopeful, at the very least, that the volume has accomplished that task.

References

Ba Maw, Ciochon, R. L., and Savage, D. E., 1979, Late Eocene of Burma yields earliest anthropoid primate, *Pondaungia cotteri, Nature* **282**:65–67.

Cameron, T. W., 1956, *Parasites and Parasitism,* Wiley, New York.

Ciochon, R. L., and Chiarelli, A. B., 1980, Paleobiogeographic perspectives on the origin of the Platyrrhini, in: *Evolutionary Biology of the New World Monkeys and Continental Drift* (R. L. Ciochon and A. B. Chiarelli, eds.), pp. 459–493, Plenum Press, New York.

Delson, E., and Rosenberger, A. L., 1980, Phyletic perspectives on platyrrhine origins and anthropoid relationships, in: *Evolutionary Biology of the New World Monkeys and Continental Drift* (R. L. Ciochon and A. B. Chiarelli, eds.), pp. 445–458, Plenum Press, New York.

Durette-Desset, M. C., 1971, Essai de classification des Nématodes Heligmosomes. Correlations avec la paléobiogéographie des hôtes, *Mem. Mus. Natl. Hist. Nat., Paris* **69**:1-126.

Gazin, L. 1958, A Review of the Middle and Upper Eocene primates of North America, *Smithson. Misc. Publs.* **136**(1):1-112.

Hill, W. C. O., 1955, *Primates: Comparative Anatomy and Taxonomy.* Vol 2, *Haplorhini: Tarsioidea,* Edinburgh University Press, Edinburgh.

Hoffstetter, R., 1971, Le peuplement mammalien de l'Amérique du Sud. Rôle des continents austraux comme centres d'origine, de diversification et de dispersion pour certains groupes mammaliens, *An. Acad. Brasil. Ciên. (Suppl.)* **43**:125-144.

Hoffstetter, R., 1972, Relationships, origins, and history of the ceboid monkeys and caviomorph rodents: A modern reinterpretation, in: *Evolutionary Biology,* Vol. 6 (Th. Dobzhansky, M. K. Hecht, and W. C. Steere, eds.), pp. 323-347, Appleton-Century-Crofts, New York.

Lavocat, R., 1969, Le systématique des Rongeurs Hystricomorphes et la dérive des continents, *C. R. Acad. Sci. Paris, Ser. D.* **269**:1496-1497.

Lavocat, R., 1974, What is an hystricomorph? in: *The Biology of Hystricomorph Rodents* (I. W. Roland and B. J. Weir, eds.), *Symp. Zool. Soc. Lond.,* No. 34, pp. 7-20 (see also discussion, pp. 55–60), Academic Press, London.

Le Gros Clark, W. E., 1962, *The Antecedents of Man,* 2nd edn., Quadrangle Books, Chicago.

Mayr, E., 1957, Evolutionary aspects of host specificity among parasites of vertebrates, in: *First Symposium on Host Specificity among Parasites of Vertebrates,* Inst. Zool. Univ. Neuchatel, Switzerland.

Rosenberger, A. L., 1979, Phylogeny, evolution and classification of New World monkeys (Platyrrhini, Primates), Ph.D. Thesis, C.U.N.Y., New York.

Sera, G. L., 1938, Alcuni caratteri anatomici delle Platirrine ed il recente abbandono da parte di esse delle'abitato acquatico, *Arch. Zool. Ital. Torino* **25**:201-218.

Simons, E. L., 1972, *Primate Evolution, An Introduction to Man's Place in Nature,* Macmillan, New York.

Simpson, G. G., 1945, The principles of classification and a classification of the mammals, *Bull. Am. Mus. Nat. Hist.* **85**:1-350.

Simpson, G. G., 1978, Early mammals in South America: Fact, controversy, and mystery, *Proc. Am. Philos. Soc.* **122**:318-328.

Tarling, D. H., 1980, The geologic evolution of South America with special reference to the last 200 million years, in: *Evolutionary Biology of the New World Monkeys and Continental Drift* (R. L. Ciochon and A. B. Chiarelli, eds.), pp. 1–41, Plenum Press, New York.

Thaw Tint, Ba Maw, Savage, D. E., and Ciochon, R. L., 1981, New discovery of *Amphipithecus* from the Late Eocene of Burma: Anthropoid origins in Asia? *Science* (submitted).

Author Index

Page numbers in italics refer to citations in the references, whereas page numbers in roman type refer to citations in the text.

Subject Index

Page numbers in italics followed by an *f* or *t* indicate information contained in figures or tables, respectively.